The Nuclear Shell Model

Kris L. G. Heyde

The Nuclear Shell Model

Study Edition

Second Corrected and Enlarged Edition
With 201 Figures, Worked Examples,
100 Problems, and a 3½" MS-DOS Diskette

Springer-Verlag
Berlin Heidelberg New York
London Paris Tokyo
Hong Kong Barcelona
Budapest

Professor Dr. *Kris L. G. Heyde*

Laboratorium voor Theoretische Fysica en Laboratorium voor Kernfysica,
Vakgroep Subatomaire en Stralingsfysica, Universiteit Gent,
Proeftuinstraat 86, B-9000 Gent, Belgium

The first edition of this volume appeared in the series
Springer Series in Nuclear and Particle Physics

ISBN 3-540-58072-7 2. Auflage Springer-Verlag Berlin Heidelberg New York
ISBN 0-387-58072-7 2nd Edition Springer-Verlag New York Berlin Heidelberg

ISBN 3-540-51581-X 1. Auflage Springer-Verlag Berlin Heidelberg New York
ISBN 0-387-51581-X 1st Edition Springer-Verlag New York Berlin Heidelberg

CIP data applied for

Typesetting: Springer TEX in-house system
SPIN: 10102040 56/3140 - 5 4 3 2 1 0 – Printed on acid-free paper

To
Daisy, Jan and Mieke

Preface to the Second Edition

In the present edition, a number of new features have been added.

First of all, a number of typographical errors that had crept into the text have been corrected. More importantly, a number of new examples, figures and smaller sections have been added. In evaluating the two-body matrix elements which characterize the residual interaction, attention has been paid to the multipole expansion and insight into the importance of various multipoles is presented. The example of ^{18}O is now worked out for all the different angular momentum states in the section on configuration mixing. Some additional comments on how to determine one- and two-body matrix elements in j^n configurations, on isospin and the application of isospin to the study of light odd-odd nuclei are included. In Chap. 3, a small section on the present use of large-scale shell model calculations and a section on experimental tests of how a nucleon actually moves inside the nucleus (using electromagnetic probing of nucleonic motion) has been added. In Chap. 4, some recent applications of the study of quadrupole motion in j^n particle systems (with reference to the Po, Rn, Ra nuclei) are presented. In the discussion of magnetic dipole moments, the effects and importance of collective admixtures are pointed out and discussed. In Chap. 5, some small additions relating to the particle-hole conjugation and to the basic Hartree-Fock theory have been made. In Chap. 7, which concentrates on pairing and the interactions between nucleons in open shells, a small section on the effects of superfluidity in neutron stars is given. As far as the Interacting Boson Model approximation is concerned a section giving some of the main ideas on how symmetries can be used to study such an interacting boson model has also been included. Here, some of the basic aspects of dynamical symmetries in the nucleus are presented. An extra appendix on group theory and symmetries has been added.

One of the major modifications in this new edition is that the FORTRAN source files of the programs have been put onto a 3 1/2" MS-DOS diskette, while extra programs have been added offering an extended package for particle-core coupling, including a description of particle-core coupling and instructions for using the code in Chap. 9.

A new Chap. 10 "Shell structure: a major theme in the binding of various physical systems" is added. This extends somewhat the scope of the book to other fields as shell-effects have been observed in systems as diversified as positronium, quark-antiquark bound systems (Charmonium), ..., as well as in metallic cluster systems. It also presents the very powerful methods inherent in shell-structure as a first step to understanding a large variety of physical phenomena.

The number of problems is almost doubled, the more difficult ones being marked by *, and the problems are arranged at the end of each chapter, separately. In some of the more difficult problems, hints are given on how they might be solved. Working through these is one of the important ways in which this edition can be of benefit. The reader should feel free to contact the author with respect to hints on solving the problems, solutions, ...

Beginning with Chap. 2 a summary is provided at the end of each chapter, in which the major goals and features of the chapter are recapitulated. These summaries may also help the reader to better appreciate the text and check his own impression with those given.

With all these new features built in, and in particular with the source programs on disc, it is felt that the present edition should accomplish its aim more effectively conveying some of the basic features of the nuclear shell model, not only in an academic and theoretical manner but also with enough possibilities to test ones understanding in a "hands-on" fashion through the examples worked out in the text, by solving the problems and using the codes.

I have benefited greatly from advice given by those people who tried to implement the book as a text in teaching a course on the nuclear shell model. The encouragement I received from Prof. A. Richter, Prof. M. Irvine and Dr. H. J. Kölsch of Springer-Verlag was of paramount importance in the realization of this new edition.

Gent, June 1994 *Kris Heyde*

Preface to the First Edition

This book is aimed at enabling the reader to obtain a working knowledge of the nuclear shell model and to understand nuclear structure within the framework of the shell model. Attention is concentrated on a coherent, self-contained exposition of the main ideas behind the model with ample illustrations to give an idea beyond formal exposition of the concepts.

Since this text grew out of a course taught for advanced undergraduate and first-year graduate students in theoretical nuclear physics, the accent is on a detailed exposition of the material with step-by-step derivations rather than on a superficial description of a large number of topics. In this sense, the book differs from a number of books on theoretical nuclear physics by narrowing the subject to only the nuclear shell model. Most of the expressions used in many of the existing books treating the nuclear shell model are derived here in more detail, in a practitioner's way. Due to frequent student requests I have expanded the level of detail in order to take away the typical phrase "... after some simple and straightforward algebra one finds ...". The material could probably be treated in a one-year course (implying going through the problem sets and setting up a number of numerical studies by using the computer codes provided). The book is essentially self-contained but requires an introductory course on quantum mechanics and nuclear physics on a more general level. Because of this structure, it is not easy to pick out certain chapters for separate reading, although an experienced practitioner of the shell model could do that.

After introductory but necessary chapters on angular momentum, angular momentum coupling, rotations in quantum mechanics, tensor algebra and the calculation of matrix elements of spherical tensor operators within angular momentum coupled states, we start the exposition of the shell model itself. Chapters 3 to 7 discuss the basic ingredients of the shell model exposing the one-particle, two-particle and three-particle aspects of the nuclear interacting shell-model picture. After studying electromagnetic properties (one-body and two-body moments and transition rates), a short chapter is devoted to the second quantization, or occupation number representation, of the shell model. In later chapters, the elementary modes of excitation observed in closed shell nuclei (particle-hole excitations) and open shell nuclei (pairing properties) are discussed with many applications to realistic nuclei and nuclear mass regions. In Chap. 8, a state-of-the-art illustration of present day possibilities within the nuclear shell model, constructing both the residual interaction and the average field properties is given. This chapter has a somewhat less pedagogical orientation than the first seven chapters. In the final chapter, some simple computer codes are included and discussed. The set of appendices constitutes an integral part of the text, as well as a number of exercises.

Several aspects of the nuclear interacting many-body system are not discussed or only briefly mentioned. This is due to the choice of developing the nuclear shell model as an in-depth example of how to approximate the interactions in a complicated many-fermion system. Having studied this text, one should be able, by using the techniques outlined, to study other fields of nuclear theory such as nuclear collective models and Hartree-Fock theory.

This book project grew out of a course taught over the past 8 years at the University of Gent on the nuclear shell model and has grown somewhat beyond the original concept. Thereby, in the initial stages of teaching, a set of unpublished lecture notes from F. Iachello on nuclear structure, taught at the "Kernfysisch Versneller Instituut" (KVI) in Groningen, were a useful guidance and influenced the first chapters in an important way. I am grateful to the many students who, by encouraging more and clearer discussions, have modified the form and content in almost every aspect. The problems given here came out of discussions with them and out of exam sets: the reader is encouraged to go through them as an essential step in mastering the content of the book.

I am most grateful to my colleagues at the Institute of Nuclear Physics in Gent, in particular in the theory group, in alphabetic order, C. De Coster, J. Jolie, J. Moreau, J. Ryckebusch, P. Van Isacker, D. Van Neck, J. Van Maldeghem, H. Vincx, M. Waroquier, and G. Wenes who contributed, maybe unintentionally, to the present text in an important way. More particularly, I am indebted to M. Waroquier for the generous permission to make extensive use of results obtained in his "Hoger Aggregaat" thesis about the feasability of performing shell-model calculations in a self-consistent way using Skyrme forces. I am also grateful to C. De Coster for scrutinizing many of the formulas, reading and critizing the whole manuscript.

Also, discussions with many experimentalists, both in Gent and elsewhere, too many to cite, have kept me from "straying" from the real world of nuclei. I would like, in particular, to thank J.L. Wood, R. F. Casten and R.A. Meyer for insisting on going ahead with the project and Prof. M. Irvine for encouragement to put this manuscript in shape for the Springer Series.

Most of my shell-model roots have been laid down in the Utrecht school; I am most grateful to P.J. Brussaard, L. Dieperink, P. Endt, and P.W.M. Glaudemans for their experience and support during my extended stays in Utrecht.

Gent, March 1990 *K.L.G. Heyde*

Contents

Introduction

Approaching the atomic nucleus at low excitation energy (excitation energy less than the nucleon separation energy) can be done on a non-relativistic level. If we start from an A-nucleon problem interacting via a given two-body potential $V_{i,j}$, the non-relativistic Hamiltonian can be written as

$$H = \sum_{i=1}^{A} t_i + \tfrac{1}{2} \sum_{i,j=1}^{A} V_{i,j} \ ,$$

where t_i is the kinetic energy of the nucleon motion. Much experimental evidence for an average, single-particle independent motion of nucleons exists, a point of view that is not immediately obvious from the above Hamiltonian. This idea acts as a guide making a separation of the Hamiltonian into A one-body Hamiltonians (described by an average one-body potential U_i) and residual interactions. This can be formally done by writing

$$H = H_0 + H_{\text{res}} \ , \quad \text{with}$$

$$H_0 = \sum_{i=1}^{A} \{t_i + U_i\} \ , \quad \text{and}$$

$$H_{\text{res}} = \tfrac{1}{2} \sum_{i,j=1}^{A} V_{i,j} - \sum_{i=1}^{A} U_i \ .$$

It is a task to determine U_i as well as possible such that the residual interaction H_{res} remains as a small perturbation on the independent A-nucleon system. This task can be accomplished by modern Hartree-Fock methods where the residual interaction with which one starts is somewhat more complicated such as the Skyrme-type interactions (two-body plus three-body terms) which have been used with considerable success. This process of going from the two-body interaction $V_{i,j}$ towards a one-body potential is drawn schematically.

Besides specific nucleon excitations in particular mass regions, coherent nucleon motion appears that is not easily handled properly within a shell-model basis. Nuclear collective vibrations, rotations and all possible transitional types of excitations have been observed in rare-earth and actinide nuclei. In these situations, model concepts stemming from analogies with well-known classic systems have been used and developed to a high level of art (Bohr-Mottelson). It is now a major

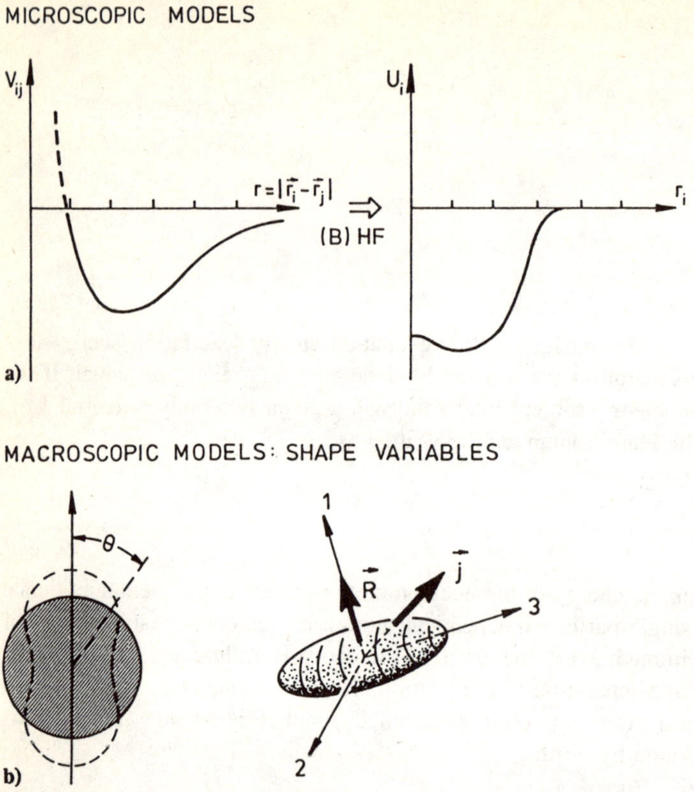

(**a**) In a schematic way, we illustrate the fact that starting from the two-body interactions $V_{i,j}$ in the nuclear A-body problem, one can construct an average one-body field, expressed by U_i. In this illustration, specific radial shapes for both $V_{i,j}$ (short-range repulsion – long-range attractive one-pion exchange tail OPEP) and U_i (Woods-Saxon type) are given. The connection is established through the (Brueckner) Hartree-Fock method
(**b**) Illustration of some specific collective modes of motion when a macroscopic (shape) model is considered to describe the nuclear A-body problem. We illustrate both the case for vibrational excitations (*left-hand part*) and for collective rotational motion (*right-hand part*). Here, the 3-axis is the nuclear symmetry axis and $\boldsymbol{R}$ denotes the collective rotational angular momentum ($\boldsymbol{j}$ denotes any intrinsic angular momentum)

task to try to bridge the space between pure shell-model methods and macroscopic methods of nuclear collective motion. Over the last years, making use of symmetry aspects of the nuclear many-body problem and its microscopic foundations, the interacting boson model has helped to carry out this bridging programme in an important way although many open problems remain to be solved.

In this book, we shall mainly concentrate on the nuclear shell model proper with, at the end, some more advanced topics. The main aim, however, is to bring in the necessary elements of technique in order to understand and also handle the nuclear shell model with some success. One consequence is that we have to leave out collective nuclear models, but many techniques can easily be adopted in that field, too, once the nuclear shell model has been worked through in detail. In order not to interrupt the shell model discussion by technical accounts of angular momentum, tensor operators, matrix elements and the like, we start with introductory chapters on angular momentum in quantum mechanics and on rotation in quantum mechanics. These chapters are self-contained but not exhaustive (proofs are usually left out), since they serve only as an avenue to the shell model. In a long Chap. 3, we discuss the one-particle average field, two-particle identical nucleon systems and their properties (wave function construction, residual interactions, configuration mixing, ...), three-particle identical nucleon systems, non-identical proton-neutron systems and isospin. Chapter 4 concentrates on electromagnetic properties in the nuclear shell model with major emphasis on one- and two-particle moments (μ, Q) and transition rates ($E2$ and $M1$ transitions). In Chap. 5, we discuss second quantization methods to be able to reformulate the nuclear shell model with its study of elementary modes of excitation in a Hartree-Fock framework: particle-hole excitations near doubly-closed shell nuclei (Chap. 6) and particle-particle excitations in open shells (pairing correlations in Chap. 7). In this chapter, we briefly discuss some recent applications of pairing aspects of the nuclear many-body system in a broken-pair and interacting boson model description of low-lying collective excitations. In Chap. 8, advanced topics on state-of-the-art shell model calculations, using a single interaction type in order to fix both the average field and the residual interactions, are discussed. This chapter brings the reader into contact with present day shell model methods. Problem sets as well as appendices that treat a number of more technical problems are included at the end. Also, a set of elementary FORTRAN programs are added that allow the calculation of the most-used angular momentum coefficients ($3j$, $6j$); calculation of radial integrals for harmonic oscillator wave functions and a diagonalization program using the Jacobi method for diagonalizing small matrices. A code that calculates the two-body δ-interaction matrix elements, including spin exchange, is also given.

Note that in these chapters the subject is intentionally restricted to the nuclear shell model to allow an in-depth, technical but also broad treatment. We hope to thus be able to prepare students for their own research in this field.

1. Angular Momentum in Quantum Mechanics

Before starting a detailed discussion of the underlying mechanisms that establish the nuclear shell model, not only for the single-particle degrees of freedom and excitations but also in order to study nuclei where a number of valence nucleons (protons and/or neutrons) are present outside closed shells, the quantum mechanical methods are discussed in some detail. We discuss both angular momentum in the framework of quantum mechanics and the aspect of rotations in quantum mechanics with some side-steps to elements of groups of transformations. Although in these Chaps. 1 and 2, not everything is proved, it should supply all necessary tools to tackle the nuclear shell model with success and with enough background to feel at ease when manipulating the necessary "Racah"-algebra (Racah 1942a, 1942b, 1943, 1949, 1951) needed to gain better insight into how the nuclear shell model actually works. These two chapters are relatively self-contained so that one can work through them without constant referral to the extensive literature on angular momentum algebra. More detailed discussions are in the appendices. We also include a short summary of often used expressions for the later chapters.

1.1 Central Force Problem and Orbital Angular Momentum

A classical particle, moving in a central one-particle field $U(r)$ can be described by the single-particle Hamiltonian (Brussaard, Glaudemans 1977)

$$H = \frac{\boldsymbol{p}^2}{2m} + U(r) . \qquad (1.1)$$

In quantum mechanics, since the linear momentum $\boldsymbol{p}$ has to be replaced by the operator $-\mathrm{i}\,\hbar\boldsymbol{\nabla}$, this Hamiltonian becomes

$$H = -\frac{\hbar^2}{2m}\Delta + U(r) , \qquad (1.2)$$

where Δ is the Laplacian operator. The orbital angular momentum itself is defined as

$$\boldsymbol{l} = \boldsymbol{r} \times \boldsymbol{p} , \qquad \text{or} \qquad (1.3)$$

$$\boldsymbol{l} = -\mathrm{i}\,\hbar\boldsymbol{r} \times \boldsymbol{\nabla} , \qquad (1.4)$$

as the corresponding quantum mechanical angular momentum operator. The components can be easily obtained in an explicit way by using the determinant notation for $\boldsymbol{l}$, i.e.,

$$-\mathrm{i}\,\hbar \begin{vmatrix} \boldsymbol{i} & \boldsymbol{j} & \boldsymbol{k} \\ x & y & z \\ \dfrac{\partial}{\partial x} & \dfrac{\partial}{\partial y} & \dfrac{\partial}{\partial z} \end{vmatrix}, \tag{1.5}$$

where $\boldsymbol{i}, \boldsymbol{j}, \boldsymbol{k}$ denote unit vectors in the x, y and z direction, respectively.

The commutation rules between the different components of the angular momentum operator can be easily calculated using the relations

$$\left[x,\, p_x\right] = x p_x - p_x x = \mathrm{i}\,\hbar\,, \tag{1.6}$$

which leads to the results

$$\left[l_x,\, l_y\right] = \mathrm{i}\,\hbar l_z\,, \tag{1.7}$$

with cyclic permutations.

We can furthermore define the operator that expresses the total length of the angular momentum as

$$\boldsymbol{l}^2 = l_x^2 + l_y^2 + l_z^2\,, \tag{1.8}$$

which has the following commutation relations with the separate components

$$\left[\boldsymbol{l}^2,\, l_i\right] = 0 \qquad (i \equiv x,\, y,\, z)\,. \tag{1.9}$$

If we now try to determine the one-particle Schrödinger equation that corresponds to the central force problem of (1.1), we can use a shorthand method for evaluating the operator $\boldsymbol{l}^2$. Starting from the commutation relations (1.6), one can show that

$$\boldsymbol{l}^2 = (\boldsymbol{r} \times \boldsymbol{p}) \cdot (\boldsymbol{r} \times \boldsymbol{p}) = r^2 p^2 - \boldsymbol{r}(\boldsymbol{r} \cdot \boldsymbol{p}) \cdot \boldsymbol{p} + 2\mathrm{i}\,\hbar \boldsymbol{r} \cdot \boldsymbol{p}\,, \tag{1.10}$$

and using

$$\boldsymbol{r} \cdot \boldsymbol{p} = -\mathrm{i}\,\hbar r \frac{\partial}{\partial r}\,, \tag{1.11}$$

one obtains

$$\boldsymbol{l}^2 = \boldsymbol{r}^2 \boldsymbol{p}^2 + \hbar^2 \frac{\partial}{\partial r}\left(r^2 \frac{\partial}{\partial r}\right). \tag{1.12}$$

The kinetic energy operator of (1.1) then becomes

$$\boldsymbol{T} = \frac{\boldsymbol{p}^2}{2m} = \frac{\boldsymbol{l}^2}{2mr^2} - \frac{\hbar^2}{2mr^2} \frac{\partial}{\partial r}\left(r^2 \frac{\partial}{\partial r}\right). \tag{1.13}$$

We shall now briefly recapitulate the solutions to the central one-body Schrödinger equation, solutions that form a basis of eigenfunctions of the operators $\boldsymbol{H}$, $\boldsymbol{l}^2$ and l_z simultaneously. So, we can write still in a rather general way that

$$H\varphi(\boldsymbol{r}) = E\varphi(\boldsymbol{r}) , \tag{1.14}$$

$$\boldsymbol{l}^2\varphi(\boldsymbol{r}) = \hbar^2\lambda\varphi(\boldsymbol{r}) . \tag{1.15}$$

The Schrödinger equation (1.14) now becomes [using (1.15)]

$$\left[-\frac{\hbar^2}{2mr^2}\frac{\partial}{\partial r}\left(r^2\frac{\partial}{\partial r}\right) + \frac{\lambda\hbar^2}{2mr^2} + U(r)\right]\varphi(\boldsymbol{r}) = E\varphi(\boldsymbol{r}) . \tag{1.16}$$

Using a separable solution of the type

$$\varphi(\boldsymbol{r}) \equiv R(r)Y(\theta,\varphi) = \frac{u(r)}{r}\cdot Y(\theta,\varphi) , \tag{1.17}$$

the radial equation becomes

$$-\frac{\hbar^2}{2m}\frac{d^2u(r)}{dr^2} + \left[\frac{\lambda\hbar^2}{2mr^2} + U(r)\right]u(r) = Eu(r) , \tag{1.18}$$

and its solution, in particular, depends on the choice of the form of the central potential $U(r)$. This particular problem will be discussed in Chap. 3.

The eigenfunctions for the angular part of (1.16) can be obtained most easily by rewriting the angular momentum $\boldsymbol{l}^2$ operator explicitly in a basis of spherical coordinates (Fig. 1.1). One works as follows:

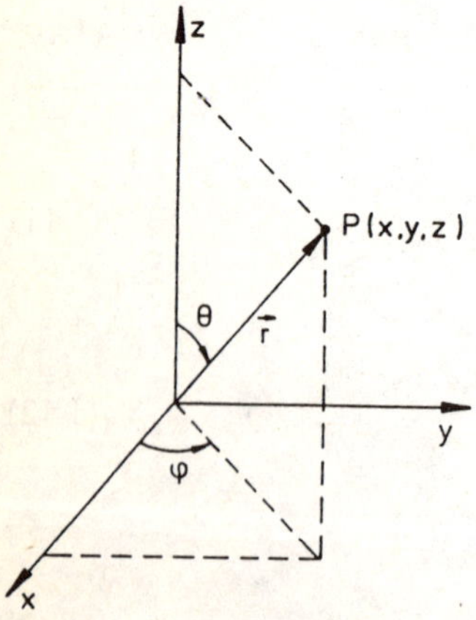

Fig. 1.1. The cartesian and spherical coordinates for the point $P(\boldsymbol{r})$ (x, y, z) and (r, θ, φ) for which the orbital angular momentum is analyzed

i) rewrite the cartesian components l_x, l_y, l_z as a function of the spherical coordinates (r, θ, φ)

$$\begin{aligned} l_x &= \mathrm{i}\,\hbar\left(\sin\varphi\frac{\partial}{\partial\theta} + \cot\theta\cos\varphi\frac{\partial}{\partial\varphi}\right) , \\ l_y &= \mathrm{i}\,\hbar\left(-\cos\varphi\frac{\partial}{\partial\theta} + \cot\theta\sin\varphi\frac{\partial}{\partial\varphi}\right) , \\ l_z &= -\mathrm{i}\,\hbar\frac{\partial}{\partial\varphi} . \end{aligned} \tag{1.19}$$

For more details, see (Sect. 1.3).

ii) rewrite the length of the angular momentum $\boldsymbol{l}^2$ as a function of the spherical coordinates

$$\begin{aligned} \boldsymbol{l}^2 &= l_x^2 + l_y^2 + l_z^2 \\ &= -\hbar^2\left\{\left(\sin\varphi\frac{\partial}{\partial\theta} + \cot\theta\cos\varphi\frac{\partial}{\partial\varphi}\right)\left(\sin\varphi\frac{\partial}{\partial\theta} + \cot\theta\cos\varphi\frac{\partial}{\partial\varphi}\right)\right. \\ &\quad + \left(-\cos\varphi\frac{\partial}{\partial\theta} + \cot\theta\sin\varphi\frac{\partial}{\partial\varphi}\right) \\ &\quad \left.\times\left(-\cos\varphi\frac{\partial}{\partial\theta} + \cot\theta\sin\varphi\frac{\partial}{\partial\varphi}\right) + \frac{\partial^2}{\partial\varphi^2}\right\} \\ &= -\hbar^2\left\{\frac{1}{\sin\theta}\frac{\partial}{\partial\theta}\left(\sin\theta\frac{\partial}{\partial\theta}\right) + \frac{1}{\sin^2\theta}\frac{\partial^2}{\partial\varphi^2}\right\} . \end{aligned} \tag{1.20}$$

We now determine the angular momentum eigenfunctions starting from (1.15, 20). Using a separable form

$$Y(\theta, \varphi) = \Phi(\varphi)\cdot\Theta(\theta) , \tag{1.21}$$

one gets the two differential equations

$$\frac{d^2\Phi}{d\varphi^2} + m^2\Phi = 0 , \tag{1.22}$$

$$\frac{1}{\sin\theta}\frac{d}{d\theta}\left(\sin\theta\frac{d}{d\theta}\Theta\right) - \frac{m^2}{\sin^2\theta}\Theta + \lambda\Theta = 0 . \tag{1.23}$$

In order to obtain these two equations, one uses a separation method for the variables θ and φ as is discussed in introductory courses on quantum mechanics (Flügge 1974). The solutions to (1.22), using the condition of uniqueness of the solutions, become

$$\Phi(\varphi) = \mathrm{e}^{im\varphi} \qquad m = 0, \pm1, \pm2, \ldots . \tag{1.24}$$

Putting now $\lambda = l(l+1)$ and $\xi = \cos\theta$, one recognizes in (1.23) the differential equation for the associated Legendre polynomials P_l^m (Edmonds 1957), i.e., one has $(0 \le |m| \le l)$

$$P_l^m(\xi) = \left(1-\xi^2\right)^{m/2} \frac{d^m}{d\xi^m} P_l(\xi) , \tag{1.25}$$

with

$$P_l(\xi) = \frac{1}{2^l l!} \frac{d^l}{d\xi^l} \left(\xi^2 - 1\right)^l . \tag{1.26}$$

(P_l are the Legendre polynomials).

Finally, the solution to (1.15) becomes

$$Y_l^m(\theta,\varphi) = \sqrt{\frac{(2l+1)}{4\pi}\left(\frac{(l-m)!}{(l+m)!}\right)}(-1)^m \, \mathrm{e}^{im\varphi} \, P_l^m(\cos\theta) , \tag{1.27}$$

with correct angular normalization, and using $m \ge 0$. For negative values of m, one has the relation

$$Y_l^{-m}(\theta,\varphi) = (-1)^m Y_l^m(\theta,\varphi)^* .$$

These functions are well known as the spherical harmonics. Using the above solutions, the angular momentum eigenvalue equations can be written

$$l^2 Y_l^m(\theta,\varphi) = \hbar^2 l(l+1) Y_l^m(\theta,\varphi) , \tag{1.28}$$

$$l_z Y_l^m(\theta,\varphi) = \hbar m Y_l^m(\theta,\varphi) . \tag{1.29}$$

We give here some of the most used spherical harmonics.

$$Y_0^0 = \frac{1}{\sqrt{4\pi}} , \tag{1.30}$$

$$\begin{aligned} Y_1^0 &= \sqrt{\frac{3}{4\pi}} \cos\theta , \\ Y_1^{\pm 1} &= \mp\sqrt{\frac{3}{8\pi}} \mathrm{e}^{\pm i\varphi} \sin\theta , \end{aligned} \tag{1.31}$$

$$\begin{aligned} Y_2^0 &= \sqrt{\frac{5}{16\pi}} \left(3\cos^2\theta - 1\right) , \\ Y_2^{\pm 1} &= \mp\sqrt{\frac{15}{8\pi}} \mathrm{e}^{\pm i\varphi} \cos\theta \sin\theta , \\ Y_2^{\pm 2} &= \sqrt{\frac{15}{32\pi}} \mathrm{e}^{\pm 2i\varphi} \sin^2\theta . \end{aligned} \tag{1.32}$$

In Fig. 1.2, we illustrate some typical linear combinations, which are called the s, p and d functions (Weissbluth 1974). These functions play a major role when describing electronic bonds in molecule formation. These are the combinations

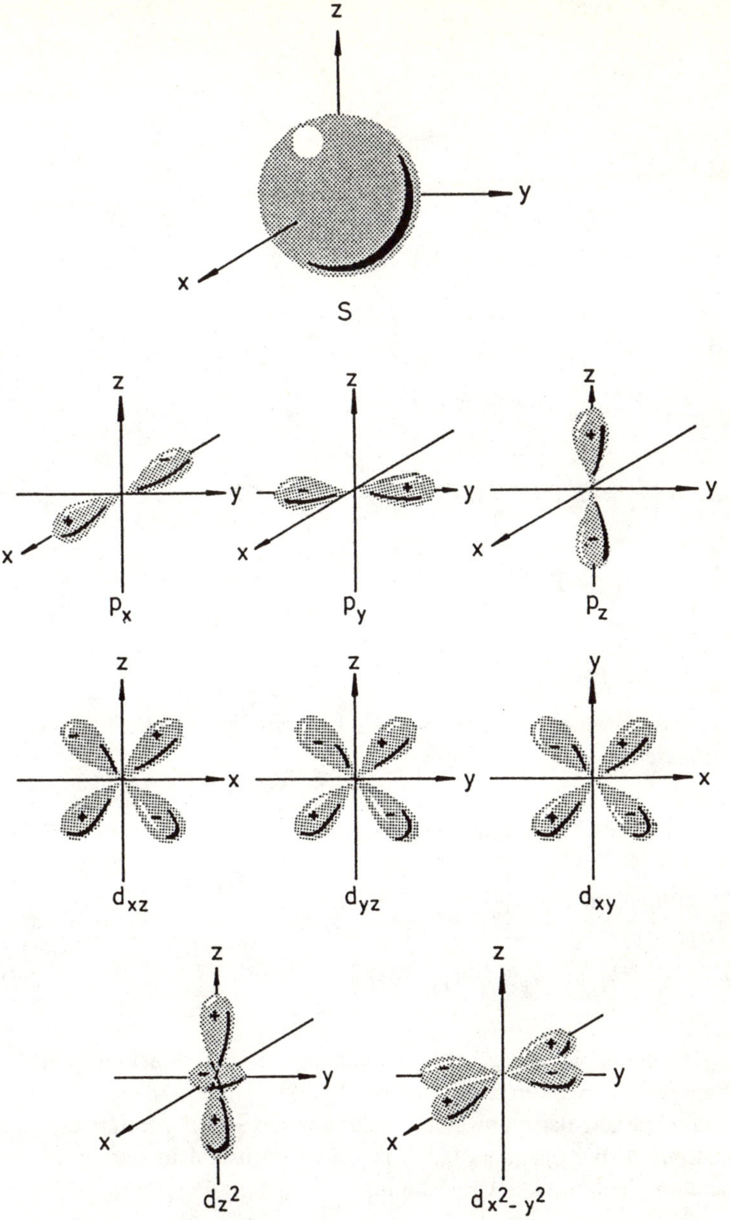

Fig. 1.2. Polar diagrams for s ($l = 0$), p ($l = 1$) and d ($l = 2$) angular wave functions. These represent real combinations of the Y_0^0, Y_1^μ and Y_2^μ spherical harmonics. The figure is taken from C.J. Ballhausen and H.B. Gray "Molecular Orbital Theory", W.A. Benjamin, Inc.

$$s = \sqrt{4\pi} Y_0^0 ,$$

$$p_x = \sqrt{\frac{4\pi}{3}} \sqrt{\frac{1}{2}} \left(-Y_1^1 + Y_1^{-1} \right) r = x ,$$

$$p_y = \sqrt{\frac{4\pi}{3}} \sqrt{\frac{1}{2}} i \left(Y_1^1 + Y_1^{-1} \right) r = y ,$$

$$p_z = \sqrt{\frac{4\pi}{3}} Y_1^0 \cdot r = z ,$$

$$d_{z^2} = \sqrt{\frac{4\pi}{5}} Y_2^0 \cdot r^2 = \frac{1}{2} (3z^2 - r^2) , \tag{1.33}$$

$$d_{x^2 y^2} = \sqrt{\frac{4\pi}{5}} \sqrt{\frac{1}{2}} \left(Y_2^2 + Y_2^{-2} \right) \cdot r^2 = \frac{1}{2} \sqrt{3} (x^2 - y^2) ,$$

$$d_{xy} = \sqrt{\frac{4\pi}{5}} \sqrt{\frac{1}{2}} \mathrm{i} \left(-Y_2^2 + Y_2^{-2} \right) \cdot r^2 = \sqrt{3} xy ,$$

$$d_{yz} = \sqrt{\frac{4\pi}{5}} \sqrt{\frac{1}{2}} \mathrm{i} \left(Y_2^1 + Y_2^{-1} \right) \cdot r^2 = \sqrt{3} yz ,$$

$$d_{zx} = \sqrt{\frac{4\pi}{5}} \sqrt{\frac{1}{2}} \left(-Y_2^1 + Y_2^{-1} \right) \cdot r^2 = \sqrt{3} zx .$$

The angular momentum eigenfunctions are a set of orthogonal functions on the unit sphere expressed by

$$\int_0^{2\pi} \int_0^{\pi} Y_l^{m*}(\theta, \varphi) Y_{l'}^{m'}(\theta, \varphi) \sin\theta \, d\theta \, d\varphi = \delta_{ll'} \delta_{mm'} . \tag{1.34}$$

Also, an addition theorem exists

$$P_l(\cos\theta_{12}) = \frac{4\pi}{2l+1} \sum_{m=-l}^{+l} Y_l^{m*}(\Omega_1) Y_l^m(\Omega_2) , \tag{1.35}$$

where Ω_1, Ω_2 are the angles (θ_1, φ_1), (θ_2, φ_2) defining the two directions and θ_{12} is the angle between the two direction vectors $\boldsymbol{\Omega}_1, \boldsymbol{\Omega}_2$.

We now introduce angular momentum ladder operators $l\pm$, operators that are linear combinations of the operators l_x, l_y but are very useful in setting up the angular momentum algebra relations. Defining

$$l_\pm \equiv l_x \pm \mathrm{i} l_y , \tag{1.36}$$

we shall determine the action of these ladder operators acting on the spherical harmonics. Therefore, we just need to evaluate the commutation relations of the ladder operators among themselves and with l_z. One can easily evaluate that

$$[l_+, l_-] = 2\hbar l_z , \quad [l_z, l_+] = \hbar l_+ , \quad [l_z, l_-] = -\hbar l_- . \tag{1.37}$$

Using the spherical coordinates and the explicit forms of l_x, l_y, l_z (1.17–19), one can rewrite l_+, l_- and l_z as

$$\begin{aligned} l_+ &= \hbar e^{i\varphi}\left(\frac{\partial}{\partial\theta} + i\cot\theta\frac{\partial}{\partial\varphi}\right) \\ l_- &= -\hbar e^{-i\varphi}\left(\frac{\partial}{\partial\theta} - i\cot\theta\frac{\partial}{\partial\varphi}\right) \\ l_z &= -i\hbar\frac{\partial}{\partial\varphi} . \end{aligned} \tag{1.38}$$

Knowing the explicit form of the spherical harmonics $Y_l^m(\theta,\varphi)$, the action of the operators $l\pm$, l_z on these functions can, in principle, be calculated in a straightforward but tedious way. We shall discuss a more elegant method in evaluating the action of $l\pm$ on the spherical harmonics. We start from the eigenvalue equation (1.29) on which we act with the operator l_+ giving

$$l_+l_zY_l^m(\theta,\varphi) = m\hbar l_+Y_l^m(\theta,\varphi) . \tag{1.39}$$

Now, using the commutation relations (1.37), this relation can be rewritten as

$$l_+l_zY_l^m(\theta,\varphi) = \left(l_z - \hbar\right)l_+Y_l^m(\theta,\varphi) = m\hbar l_+Y_l^m(\theta,\varphi) , \tag{1.40}$$

or,

$$l_z\left(l_+Y_l^m(\theta,\varphi)\right) = (m+1)\hbar\left(l_+Y_l^m(\theta,\varphi)\right) . \tag{1.41}$$

This indicates that $l_+Y_l^m(\theta,\varphi)$ is an eigenfunction of the operator l_z with eigenvalue $(m+1)\hbar$ and thus the ladder operator l_+ effectively adds one unit $\hbar$ to the original m-projection. Likewise, l_- subtracts one unit $\hbar$ from the original m-projection and gives

$$l_z\left(l_-Y_l^m(\theta,\varphi)\right) = (m-1)\hbar\left(l_-Y_l^m(\theta,\varphi)\right) . \tag{1.42}$$

This allows for the relations

$$\begin{aligned} l_+Y_l^m(\theta,\varphi) &= \alpha(l,m)Y_l^{m+1}(\theta,\varphi) , \\ l_-Y_l^m(\theta,\varphi) &= \beta(l,m)Y_l^{m-1}(\theta,\varphi) . \end{aligned} \tag{1.43}$$

The factors $\alpha(l,m)$, $\beta(l,m)$ can be determined by calculating the norm of expressions (1.43). Thus,

$$\int Y_l^{m*}(\theta,\varphi)l_-l_+Y_l^m(\theta,\varphi)d\Omega = |\alpha(l,m)|^2 , \tag{1.44}$$

since the spherical harmonics form an orthonormal set of eigenfunctions. We can now evaluate the operator expression l_-l_+ explicitly as follows. We start from

$$\boldsymbol{l}^2 = l_x^2 + l_y^2 + l_z^2 = \tfrac{1}{2}\left(l_+l_- + l_-l_+\right) + l_z^2 , \tag{1.45}$$

and using the commutation relations (1.37), this simplifies into

$$\boldsymbol{l}^2 = l_-l_+ + l_z\left(l_z + \hbar\right) , \tag{1.46}$$

giving rise to the equality

$$l_- l_+ = \boldsymbol{l}^2 - l_z(l_z + \hbar) . \tag{1.47}$$

Here, one also needs to impose the conditions

$$l_+ Y_l^l = l_- Y_l^{-l} = 0 .$$

This relation (1.47) used in (1.44) gives the result

$$\hbar^2[l(l+1) - m(m+1)] = |\alpha(l,m)|^2 , \tag{1.48}$$

or

$$l_+ Y_l^m(\theta,\varphi) = \hbar\{l(l+1) - m(m+1)\}^{1/2} Y_l^{m+1}(\theta,\varphi) . \tag{1.49}$$

Similarly, for the other ladder operator l_- one has

$$l_- Y_l^m(\theta,\varphi) = \hbar\{l(l+1) - m(m-1)\}^{1/2} Y_l^{m-1}(\theta,\varphi) . \tag{1.50}$$

1.2 General Definitions of Angular Momentum

In Sect. 1.1, we have derived the angular momentum operator $\boldsymbol{l}(l_x, l_y, l_z)$ explicitly, starting from the one-body central force problem. This method only allows for entire values of the angular momentum eigenvalue l and m. It is now possible to define angular momentum in a more general but abstract way starting from the commutation rules (1.7,9). If we construct general operators $\boldsymbol{J}^2$, $J_i (i \equiv x, y, z)$ which fulfill the relations

$$\begin{aligned} \left[\boldsymbol{J}^2, J_i\right] &= 0 \qquad (i \equiv x, y, z) , \\ \left[J_x, J_y\right] &= \mathrm{i}\,\hbar J_z , \end{aligned} \tag{1.51}$$

and cyclic permutations, the operator $\boldsymbol{J}^2$ defines a general angular momentum operator. The eigenvectors are now defined as the abstract vectors in a Hilbert space carrying two quantum numbers, i.e., the quantum number defining the length j and the quantum number defining the projection of $j(m)$, since the quantum numbers corresponding to the full set of commuting operators define the state vector uniquely. Thus, we have the eigenvector relations

$$\boldsymbol{J}^2|j,m\rangle = \hbar^2 j(j+1)|j,m\rangle , \tag{1.52}$$

$$J_z|j,m\rangle = \hbar m|j,m\rangle . \tag{1.53}$$

Using the ladder operators, we can also write

$$J_\pm|j,m\rangle = \hbar\{j(j+1) - m(m\pm 1)\}^{1/2}|j,m\pm 1\rangle , \tag{1.54}$$

$$J_z|j,m\rangle = \hbar m|j,m\rangle , \tag{1.55}$$

as the defining expressions for a general angular momentum operator.

1.2.1 Matrix Representations

In the discussion of Sect. 1.1, the angular momentum operators had an explicit expression in terms of the coordinates and derivatives to these coordinates (differential form). In the more general case, as discussed above, we can derive a matrix representation of the operators J_x, J_y, J_z or J_+, J_-, J_z and $\boldsymbol{J}^2$ within the space spanned by the state vectors $|j,m\rangle$. As an example, we use the five state vectors $|j,m\rangle$ for the case of $j = 2(-2 \leq m \leq +2)$. We denote the state vectors in column vector form as

$$\begin{vmatrix} 1 \\ 0 \\ 0 \\ 0 \\ 0 \end{vmatrix} , \quad \begin{vmatrix} 0 \\ 1 \\ 0 \\ 0 \\ 0 \end{vmatrix} , \ldots , \quad \begin{vmatrix} 0 \\ 0 \\ 0 \\ 0 \\ 1 \end{vmatrix} , \tag{1.56}$$

$$|j,2\rangle , \quad |j,1\rangle , \ldots , \quad |j,-2\rangle .$$

The action of the ladder operators now leads to

$$J_+|j,m\rangle = a_{\bar{m},\bar{m}+1}|j,m+1\rangle , \tag{1.57}$$

where $a_{\bar{m},\bar{m}+1}$ defines the following matrix representation of J_+, i.e.,

$$J_+ \Rightarrow \begin{vmatrix} 0 & a_{1,2} & 0 & 0 & 0 \\ 0 & 0 & a_{2,3} & 0 & 0 \\ 0 & 0 & 0 & a_{3,4} & 0 \\ 0 & 0 & 0 & 0 & a_{4,5} \\ 0 & 0 & 0 & 0 & 0 \end{vmatrix} . \tag{1.58}$$

Similarly, one gets a matrix representation for J_- following from

$$J_-|j,m\rangle = b_{\bar{m},\bar{m}-1}|j,m-1\rangle , \tag{1.59}$$

and for J_z since

$$J_z|j,m\rangle = \hbar m \delta_{m,m'}|j,m\rangle . \tag{1.60}$$

1.2.2 Example for Spin $\frac{1}{2}$ Particles

The angular momentum representation for spin $\frac{1}{2}$ particles (electron, proton, ...), using the general method as outlined in Sect. 1.2, now gives in a simple application the construction of the 2×2 spin matrices. We briefly recapitulate the spin $\frac{1}{2}$ angular momentum commutation relations.

$$\left[s_+, s_-\right] = 2\,\hbar s_z , \quad \left[s_z, s_+\right] = \hbar s_+ , \quad \left[s_z, s_-\right] = -\,\hbar s_- . \tag{1.61}$$

Defining $\boldsymbol{s} = \hbar/2\boldsymbol{\sigma}$, these commutation relations become

$$\left[\sigma_+, \sigma_-\right] = 4\sigma_z , \quad \left[\sigma_z, \sigma_+\right] = 2\sigma_+ , \quad \left[\sigma_z, \sigma_-\right] = -2\sigma_- . \tag{1.62}$$

The matrix representations are spanned in the two-dimensional space defined by the state vectors

$$\begin{bmatrix}1\\0\end{bmatrix}, \quad \begin{bmatrix}0\\1\end{bmatrix}, \tag{1.63}$$

which correspond to the states $|\frac{1}{2}, +\frac{1}{2}\rangle$ and $|\frac{1}{2}, -\frac{1}{2}\rangle$, respectively. One often denotes the former by $\chi^{1/2}_{+1/2}$ or $\alpha(s)$ and the latter by $\chi^{1/2}_{-1/2}$ or $\beta(s)$ in literature on angular momentum (Edmonds 1957, de-Shalit, Talmi 1963, Rose, Brink 1967, Brussaard, Glaudemans 1977). The ladder operator relations (1.54) for the specific case of spin 1/2 particles become

$$\begin{aligned} s_+|\tfrac{1}{2}, -\tfrac{1}{2}\rangle &= \hbar|\tfrac{1}{2}, +\tfrac{1}{2}\rangle , \\ s_-|\tfrac{1}{2}, +\tfrac{1}{2}\rangle &= \hbar|\tfrac{1}{2}, -\tfrac{1}{2}\rangle , \end{aligned} \tag{1.64}$$

or

$$\begin{aligned} \sigma_+|\tfrac{1}{2}, -\tfrac{1}{2}\rangle &= 2|\tfrac{1}{2}, +\tfrac{1}{2}\rangle , \\ \sigma_-|\tfrac{1}{2}, +\tfrac{1}{2}\rangle &= 2|\tfrac{1}{2}, -\tfrac{1}{2}\rangle . \end{aligned} \tag{1.65}$$

One immediately gets the σ_x, σ_y and σ_z "operators" as

$$\sigma_x \equiv \begin{pmatrix}0 & 1\\1 & 0\end{pmatrix}, \quad \sigma_y \equiv \begin{pmatrix}0 & -\mathrm{i}\\\mathrm{i} & 0\end{pmatrix}, \quad \sigma_z \equiv \begin{pmatrix}1 & 0\\0 & -1\end{pmatrix}. \tag{1.66}$$

Finally, we give a number of interesting properties for the Pauli spin $\frac{1}{2}$ matrices without proof:

$$\text{i) } \sigma_x^2 = \sigma_y^2 = \sigma_z^2 = \mathbb{1} , \tag{1.67}$$

where $\mathbb{1}$ denotes the 2×2 unit matrix.

$$\text{ii) } \{\sigma_x, \sigma_y\} = 0 , \tag{1.68}$$

and cyclic where $\{A, B\}$ is the anti-commutator defined as $AB + BA$.

$$\text{iii) } (\boldsymbol{\sigma} \cdot \boldsymbol{A})(\boldsymbol{\sigma} \cdot \boldsymbol{B}) = \boldsymbol{A} \cdot \boldsymbol{B} + \mathrm{i}\boldsymbol{\sigma} \cdot (\boldsymbol{A} \times \boldsymbol{B}) , \tag{1.69}$$

if both $\boldsymbol{A}$ and $\boldsymbol{B}$ commute with $\boldsymbol{\sigma}$.

1.3 Total Angular Momentum for a Spin $\frac{1}{2}$ Particle

The total wave function characterizing a particle with intrinsic spin $\frac{1}{2}$ (electron, proton, neutron,...) which, at the same time, carries orbital angular momentum can be written as the product wave function

$$\begin{aligned} \varphi(\boldsymbol{r}, \boldsymbol{\sigma}) &= \psi\left(lm_l, \tfrac{1}{2}m_s\right) \\ &= R_{nl}(r) Y_l^{m_l}(\theta, \varphi) \chi^{1/2}_{m_s}(\boldsymbol{\sigma}) , \end{aligned} \tag{1.70}$$

where $R_{n,l}(r)$ describes the solution of the radial Schrödinger equation (1.18) using a given potential $U(r)$, n describes a radial quantum number counting the number of nodes, and l the orbital angular momentum eigenvalue. Furthermore, $Y_l^{m_l}(\theta, \varphi)$ (with $-l \leq m_l \leq +l$) describes the angular part of the wave function and $\chi_{m_s}^{1/2}(\sigma)$ ($m_s = \pm\frac{1}{2}$) the intrinsic spin wave function. Since $\chi_{m_s}^{1/2}$ can be written as a state vector in a two-dimensional space, it is more correct to speak of (1.70) as a state vector than as a wave function.

The following eigenvalue equations are fulfilled for the state vectors (1.70):

$$\begin{aligned} \boldsymbol{l}^2\psi\left(lm_l, \tfrac{1}{2}m_s\right) &= \hbar^2 l(l+1)\psi\left(lm_l, \tfrac{1}{2}m_s\right) , \\ l_z\psi\left(lm_l, \tfrac{1}{2}m_s\right) &= \hbar m_l\psi\left(lm_l, \tfrac{1}{2}m_s\right) , \\ \boldsymbol{s}^2\psi\left(lm_l, \tfrac{1}{2}m_s\right) &= \hbar^2\tfrac{3}{4}\psi\left(lm_l, \tfrac{1}{2}m_s\right) , \\ s_z\psi\left(lm_l, \tfrac{1}{2}m_s\right) &= \hbar m_s\psi\left(lm_l, \tfrac{1}{2}m_s\right) . \end{aligned} \tag{1.71}$$

We now define the operator

$$\boldsymbol{J} = \boldsymbol{l} + \boldsymbol{s} = \boldsymbol{l} + \hbar/2\boldsymbol{\sigma} . \tag{1.72}$$

Using the definitions for a general angular momentum operator (1.51), one can show that the operator $\boldsymbol{J}(\boldsymbol{J}^2, J_x, J_y, J_z)$ is indeed an angular momentum operator since the commutation relations

$$\begin{aligned} \left[J_x, J_y\right] &= \mathrm{i}\,\hbar J_z , \dots , \\ \left[\boldsymbol{J}^2, J_i\right] &= 0 , \end{aligned} \tag{1.73}$$

hold (verify this explicitly).

By construction $\boldsymbol{l}^2$, l_z, $\boldsymbol{s}^2$ and s_z form a set of commuting operators (the orbital angular momentum operators and the intrinsic angular momentum "spin" operators act in totally different spaces, the former in the space of coordinates (x, y, z), the latter in an abstract space, spanned by the unit vectors $\chi_{m_s}^{1/2}$).

One can now show that the operators $\boldsymbol{J}^2$, J_z, $\boldsymbol{l}^2$ and $\boldsymbol{s}^2$ also form a set of commuting operators indicating that it is also possible to describe the full state of the particle with orbital and intrinsic spin in a basis characterized with quantum numbers relating to $\boldsymbol{J}^2$, J_z, $\boldsymbol{l}^2$ and $\boldsymbol{s}^2$, respectively. Since $J_z = l_z + s_z$, in the above case, a fixed m-eigenvalue will occur but not necessarily a fixed m_l, m_s value. This follows from the fact that $\boldsymbol{J}^2$ does not commute with l_z or with s_z but only with the sum $l_z + s_z$. (The proof of this is left as an exercise.)

We now study the effect of acting with $\boldsymbol{J}^2$ on the state vectors that are eigenvectors of the "uncoupled" ($\boldsymbol{l}^2$, l_z, $\boldsymbol{s}^2$, s_z) basis. Since one can write $\boldsymbol{J}^2$ as

$$\boldsymbol{J}^2 = \boldsymbol{l}^2 + \boldsymbol{s}^2 + 2l_z s_z + l_+ s_- + l_- s_+ , \tag{1.74}$$

acting on the vectors $\psi(lm_l, \frac{1}{2}m_s)$ one obtains

$$J^2\psi\left(lm_l, \tfrac{1}{2}m_s\right) = \hbar^2\left\{l(l+1) + \tfrac{3}{4} + 2m_l m_s\right\}\psi\left(lm_l, \tfrac{1}{2}m_s\right)$$
$$+ \alpha\psi\left(l\,m_l + 1, \tfrac{1}{2}\,m_s - 1\right) + \beta\psi\left(l\,m_l - 1, \tfrac{1}{2}\,m_s + 1\right) . \tag{1.75}$$

From (1.75) it becomes clear that the eigenvectors $\psi(lm_l, \frac{1}{2}m_s)$ are not in general eigenvectors of $\boldsymbol{J}^2$, although they are eigenvectors of $\boldsymbol{l}^2$, $\boldsymbol{s}^2$ and J_z. The right-hand side of (1.75) represents a 2×2 matrix spanned by the configurations $\psi(lm_l, \frac{1}{2}m_s)$ with a fixed value of total magnetic quantum number m $(= m_l + m_s)$. By diagonalizing this matrix one obtains two eigenvalues of j, i.e., $j = l + \frac{1}{2}$ and $j = l - \frac{1}{2}$. Of course, in the two extreme cases (see the problem set) $j = l + \frac{1}{2}$, $m = l + \frac{1}{2}$ and $j = l + \frac{1}{2}$, $m = -l - \frac{1}{2}$, only one component, $\psi(l, m_l = l, \frac{1}{2}, m_s = +\frac{1}{2})$ and $\psi(l, m_l = -l, \frac{1}{2}, m_s = -\frac{1}{2})$, respectively, results.

As an example, we take the case of $l = 4$, $s = \frac{1}{2}$ that can be combined to form both the $j = \frac{9}{2}$ and $j = \frac{7}{2}$ total angular momenta. The states obtained are

$$\begin{aligned}
&\psi\left(l = 4,\ s = \tfrac{1}{2},\ j = \tfrac{9}{2},\ m = +\tfrac{9}{2}\right)\\
&\quad = \psi\left(l = 4,\ m_l = 4,\ s = \tfrac{1}{2},\ m_s = +\tfrac{1}{2}\right) ,\\
&\psi\left(l = 4,\ s = \tfrac{1}{2},\ j = \tfrac{9}{2},\ m = +\tfrac{7}{2}\right)\\
&\quad = \alpha\psi\left(l = 4,\ m_l = 4,\ s = \tfrac{1}{2},\ m_s = -\tfrac{1}{2}\right)\\
&\qquad + \beta\psi\left(l = 4,\ m_l = 3,\ s = \tfrac{1}{2},\ m_s = +\tfrac{1}{2}\right) ,\\
&\psi\left(l = 4,\ s = \tfrac{1}{2},\ j = \tfrac{7}{2},\ m = +\tfrac{7}{2}\right)\\
&\quad = \beta\psi\left(l = 4,\ m_l = 4,\ s = \tfrac{1}{2},\ m_s = -\tfrac{1}{2}\right)\\
&\qquad - \alpha\psi\left(l = 4,\ m_l = +3,\ s = \tfrac{1}{2},\ m_s = +\tfrac{1}{2}\right) ,\\
&\quad \ldots\\
&\psi\left(l = 4,\ s = \tfrac{1}{2},\ j = \tfrac{9}{2},\ m = -\tfrac{9}{2}\right)\\
&\quad = \psi\left(l = 4,\ m_l = -4,\ s = \tfrac{1}{2},\ m_s = -\tfrac{1}{2}\right) .
\end{aligned} \tag{1.76}$$

In general, the eigenvectors of $\boldsymbol{J}^2$, J_z, $\boldsymbol{l}^2$, $\boldsymbol{s}^2$, denoted by $\psi(ls = \frac{1}{2}, jm)$ can be expanded in the eigenvectors of $\boldsymbol{l}^2$, l_z, $\boldsymbol{s}^2$, s_z that are given by $\psi(lm_l, \frac{1}{2}\,m_s)$ as follows:

$$\psi\left(ls = \tfrac{1}{2},\ jm\right) = \sum_{m_l, m_s} \langle lm_l, \tfrac{1}{2}m_s | ls = \tfrac{1}{2}, jm\rangle \psi\left(lm_l, \tfrac{1}{2}m_s\right) . \tag{1.77}$$

The coefficients that establish the transformation of one complete basis to the other complete basis $\langle lm_l, \frac{1}{2}m_s | ls = \frac{1}{2}, jm\rangle$ are denoted as Clebsch-Gordan

Table 1.1 Analytic expressions for the Clebsch-Gordan coefficients appearing in (1.79) for coupling the orbital angular momentum l with the intrinsic spin $s = 1/2$ to a total angular momentum $j = l \pm 1/2$

j	$m_s = +\frac{1}{2}$	$m_s = -\frac{1}{2}$
$l+\frac{1}{2}$	$\left(\frac{l+1/2+m}{2l+1}\right)^{1/2}$	$\left(\frac{l+1/2-m}{2l+1}\right)^{1/2}$
$l-\frac{1}{2}$	$\left(\frac{l+1/2-m}{2l+1}\right)^{1/2}$	$-\left(\frac{l+1/2+m}{2l+1}\right)^{1/2}$

(Clebsch 1872, Gordan 1875) or vector-coupling coefficients. In the new, "coupled" basis of (1.77) one has the following eigenvalue relations

$$\begin{aligned}
\boldsymbol{l}^2\psi\left(ls=\tfrac{1}{2},jm\right) &= \hbar^2 l(l+1)\psi\left(ls=\tfrac{1}{2},jm\right),\\
\boldsymbol{s}^2\psi\left(ls=\tfrac{1}{2},jm\right) &= \hbar^2\tfrac{3}{4}\psi\left(ls=\tfrac{1}{2},jm\right),\\
\boldsymbol{J}^2\psi\left(ls=\tfrac{1}{2},jm\right) &= \hbar^2 j(j+1)\psi\left(ls=\tfrac{1}{2},jm\right),\\
J_z\psi\left(ls=\tfrac{1}{2},jm\right) &= \hbar m\psi\left(ls=\tfrac{1}{2},jm\right).
\end{aligned} \tag{1.78}$$

Written explicitly, the total state vector for a spin $s = \frac{1}{2}$ fermion particle becomes

$$\begin{aligned}
\psi\left(nls=\tfrac{1}{2},jm\right) = R_{nl}(r)\Big\{&\langle lm-\tfrac{1}{2},\tfrac{1}{2}+\tfrac{1}{2}|l\tfrac{1}{2},jm\rangle Y_l^{m-1/2}(\theta,\varphi)\chi^{1/2}_{+1/2}(\boldsymbol{\sigma})\\
&+\langle lm+\tfrac{1}{2},\tfrac{1}{2}-\tfrac{1}{2}|l\tfrac{1}{2},jm\rangle Y_l^{m+1/2}(\theta,\varphi)\chi^{1/2}_{-1/2}(\boldsymbol{\sigma})\Big\},
\end{aligned} \tag{1.79}$$

with the $\langle\ldots|\ldots\rangle$ Clebsch-Gordan coefficients given in Table 1.1.

1.4 Coupling of Two Angular Momenta: Clebsch-Gordan Coefficients

In this section, we concentrate on the coupling of two distinct angular momenta, but now for the more general case of angular momentum operators $\boldsymbol{J}_1$, $\boldsymbol{J}_2$ that may have an m-projection J_{1z}, J_{2z} representing both integer or half-integer values. In that case, the total angular momentum operator is expressed by the sum $\boldsymbol{J} = \boldsymbol{J}_1 + \boldsymbol{J}_2$. The set of four commuting operators

$$\left\{\boldsymbol{J}_1^2, J_{1z}, \boldsymbol{J}_2^2, J_{2z}\right\}, \tag{1.80}$$

are characterized by the common eigenvectors, expressed by the product state vectors

$$|j_1m_1, j_2m_2\rangle \equiv |j_1m_1\rangle|j_2m_2\rangle \ . \tag{1.81}$$

The other set of commuting operators

$$\left\{J^2, J_z, J_1^2, J_2^2\right\} , \tag{1.82}$$

can also be characterized by a set of common eigenvectors, being linear combinations of the eigenvectors (1.81), such that they form eigenvectors of J^2. One also denotes them as the "coupled" and "uncoupled" eigenvectors that are related via the expression

$$|j_1j_2; jm\rangle = \sum_{\substack{m_1,m_2 \\ (m_1+m_2=m)}} \langle j_1m_1, j_2m_2|j_1j_2, jm\rangle|j_1m_1\rangle|j_2m_2\rangle \ . \tag{1.83}$$

The overlap coefficients $\langle\ldots|\ldots\rangle$, going from one basis to the other are called the Clebsch-Gordan coefficients. To simplify the notation one frequently uses an abbreviation in the ket side of the bracket, i.e.,

$$\langle j_1m_1, j_2m_2|j_1j_2, jm\rangle \to \langle j_1m_1, j_2m_2|jm\rangle \ . \tag{1.84}$$

In order to determine the relative phases for the Clebsch-Gordan coefficients, we shall use the Condon-Shortley phase convention (Condon, Shortley 1935, Edmonds 1957), which is basically defined as follows.

i) When acting with the ladder operator $J_\pm$ on the eigenvectors $|j_1j_2; jm\rangle$, we define the phase as

$$J_\pm|j_1j_2; jm\rangle = e^{i\delta}\,\hbar\big((j \mp m)(j \pm m + 1)\big)^{1/2}|j_1j_2; jm \pm 1\rangle \ , \tag{1.85}$$

with $e^{i\delta} = +1$.

ii) By acting with the operator J^2 ($J = J_1 + J_2$) on the states with the extreme projection quantum number $M = j_1 + j_2$ or $M = -j_1 - j_2$, one gets the result that

$$J^2|j_1j_2; jm = \pm(j_1 + j_2)\rangle = (j_1 + j_2)(j_1 + j_2 + 1)\,\hbar^2|j_1j_2, jm = \pm(j_1 + j_2)\rangle \ , \tag{1.86}$$

indicating that the state in (1.86) is an eigenstate of J^2 and J_z with eigenvalues $(j_1+j_2)(j_1+j_2+1)\,\hbar^2$ and $M = \pm\,\hbar(j_1+j_2)$, respectively. Thus we get, by applying (1.83) for this particular case

$$\begin{aligned}|j_1j_2; j = j_1 + j_2,\ m = \pm(j_1 + j_2)\rangle \\ = e^{i\alpha}\,|j_1m_1 = \pm j_1\rangle|j_2, m_2 = \pm j_2\rangle \ ,\end{aligned} \tag{1.87}$$

which, with the choice $e^{i\alpha} = +1$, gives the aligned Clebsch-Gordan coefficients

$$\langle j_1m_1 = \pm j_1, j_2m_2 = \pm j_2|j = j_1 + j_2,\ m = \pm(j_1 + j_2)\rangle = +1 \ . \tag{1.88}$$

iii) By acting now with the ladder (lowering) operator $J_- = J_{1-} + J_{2-}$, on (1.87), and relating that result to the explicit form of the state vector (1.83) with $j = j_1 + j_2$, $m = j_1 + j_2 - 1$; we get

$$J_-|j_1 j_2; j = j_1 + j_2, m = j_1 + j_2\rangle = \left(J_{1-} + J_{2-}\right)|j_1, m_1 = j_1\rangle|j_2, m_2 = j_2\rangle , \tag{1.89}$$

and

$$\begin{aligned}|j_1 j_2; j &= j_1 + j_2, m = j_1 + j_2 - 1\rangle \\ &= \langle j_1 j_1 - 1, j_2 j_2 | j = j_1 + j_2, m = j_1 + j_2 - 1\rangle |j_1, j_1 - 1\rangle |j_2, j_2\rangle \\ &\quad + \langle j_1 j_1, j_2 j_2 - 1 | j = j_1 + j_2, m = j_1 + j_2 - 1\rangle |j_1, j_1\rangle |j_2, j_2 - 1\rangle .\end{aligned} \tag{1.90}$$

This leads to the identification

$$\begin{aligned}\langle j_1 j_1 - 1, j_2 j_2 | j = j_1 + j_2, m = j_1 + j_2 - 1\rangle &= \left(j_1/(j_1 + j_2)\right)^{1/2} \\ \langle j_1 j_1, j_2 j_2 - 1 | j = j_1 + j_2, m = j_1 + j_2 - 1\rangle &= \left(j_2/(j_1 + j_2)\right)^{1/2} .\end{aligned} \tag{1.91}$$

With the value of $m = j_1 + j_2 - 1$, another state can be constructed, i.e., $|j = j_1 + j_2 - 1, m = j_1 + j_2 - 1\rangle$ which should be constructed from the same uncoupled states that appear in (1.90). By imposing the condition of orthogonality, one can deduce both the absolute value and the relative phase of the Clebsch-Gordan coefficients

$$\begin{aligned}&\langle j_1 j_1 - 1, j_2 j_2 | j = j_1 + j_2 - 1, m = j_1 + j_2 - 1\rangle , \quad \text{and} \\ &\langle j_1 j_1, j_2 j_2 - 1 | j = j_1 + j_2 - 1, m = j_1 + j_2 - 1\rangle .\end{aligned} \tag{1.92}$$

The absolute phases are now defined by the condition that for any given total J and projection M one has

$$\langle JM | J_{1z} | J - 1, M\rangle \geq 0 . \tag{1.93}$$

This condition, written out for the Clebsch-Gordan coefficients making up the states $|JM\rangle$ and $|J - 1, M\rangle$ becomes

$$\sum_{m_1, m_2} m_1 \langle j_1 m_1, j_2 m_2 | JM\rangle \langle j_1 m_1, j_2 m_2 | J - 1 M\rangle \geq 0 . \tag{1.94}$$

The above condition (1.94) can be shown to be equivalent to the condition (Brussaard 1967)

$$\langle j_1 j_1, j_2 J - j_1 | J, M = J\rangle \geq 0 \qquad \text{for each } J . \tag{1.95}$$

These are now the phase conditions of (i), (ii) and (iii) that uniquely define the Clebsch-Gordan coefficients and thus also the coupled state vectors of (1.83).

1.5 Properties of Clebsch-Gordan Coefficients

Since the Clebsch-Gordan coefficients serve as expansion coefficients for a given eigenvector in a specified ortho-normal basis, there exist orthogonality relations that are given by

$$\begin{aligned}&\sum_{m_1,m_2} \langle j_1m_1, j_2m_2|jm\rangle\langle j_1m_1, j_2m_2|j'm'\rangle = \delta_{jj'}\delta_{mm'} , \quad \text{and}\\ &\sum_{j,m}\langle j_1m_1, j_2m_2|jm\rangle\langle j_1m_1', j_2m_2'|jm\rangle = \delta_{m_1m_1'}\delta_{m_2m_2'} .\end{aligned} \tag{1.96}$$

Interesting symmetry relations exist when interchanging the two angular momenta that become coupled, e.g., (de-Shalit, Talmi 1963)

$$\langle j_1m_1, j_2m_2|jm\rangle = (-1)^{j_1+j_2-j}\langle j_2m_2, j_1m_1|jm\rangle . \tag{1.97}$$

In interchanging either j_1 and j or j_2 and j, more complex relations result, since angular momenta are coupled in a certain 'direction' e.g. j_1 with j_2 to form the angular momentum j and in that order. More symmetric ways of coupling can be made.

Two angular momentum states $|jm\rangle$ can be coupled to a total angular momentum of zero. The normalized state then becomes

$$|\Phi_0\rangle = \sum_m (2j+1)^{-1/2}(-1)^{j-m}|j,m\rangle|j,-m\rangle . \tag{1.98}$$

Using the Wigner $1j$-symbol (Brussaard 1967, Wigner 1959)

$$\begin{pmatrix} j \\ mm' \end{pmatrix} = (-1)^{j+m}\delta_{m,-m'} = (-1)^{j-m'}\delta_{m,-m'} , \tag{1.99}$$

it follows that the combination

$$(2j+1)^{1/2}|\Phi_0\rangle = \sum_{m_1,m_2} \begin{pmatrix} j \\ m_1m_2 \end{pmatrix} |jm_1\rangle|jm_2\rangle , \tag{1.100}$$

forms an angular momentum invariant.

Using the same method, one can construct out of the three subsystems j_1, j_2 and j_3 a system with total angular momentum zero. The state vector thus constructed becomes

$$|\Psi_0\rangle = \sum_{m_1,m_2,m_3} \begin{pmatrix} j_1 & j_2 & j_3 \\ m_1 & m_2 & m_3 \end{pmatrix} |j_1m_1\rangle|j_2m_2\rangle|j_3m_3\rangle , \tag{1.101}$$

with the coefficients $\begin{pmatrix} \cdots \\ \cdots \end{pmatrix}$, the Wigner $3j$-symbols (Wigner 1959, de-Shalit and Talmi 1963, Brussaard 1967). By constructing the state $|\Psi_0\rangle$ by first coupling the individual angular momenta j_1 and j_2 to an angular momentum j_3 and then subsequent to the third angular momentum j_3 to form a state of total angular momentum zero, a relation between the Wigner $3j$-symbol and the Clebsch-Gordan coefficients is obtained. This relation is given by

$$\begin{pmatrix} j_1 & j_2 & j_3 \\ m_1 & m_2 & m_3 \end{pmatrix} = \frac{(-1)^{j_1-j_2-m_3}}{\sqrt{2j_3+1}} \langle j_1 m_1, j_2 m_2 | j_3 \; -m_3 \rangle \,, \tag{1.102}$$

symmetry properties under the interchange of any two angular momenta of the set (j_1, j_2, j_3) become very simple:

$$\begin{aligned} \begin{pmatrix} j_1 & j_2 & j_3 \\ m_1 & m_2 & m_3 \end{pmatrix} &= \begin{pmatrix} j_2 & j_3 & j_1 \\ m_2 & m_3 & m_1 \end{pmatrix} = \begin{pmatrix} j_3 & j_1 & j_2 \\ m_3 & m_1 & m_2 \end{pmatrix} \\ &= (-1)^{j_1+j_2+j_3} \begin{pmatrix} j_1 & j_3 & j_2 \\ m_1 & m_3 & m_2 \end{pmatrix} \\ &= (-1)^{j_1+j_2+j_3} \begin{pmatrix} j_3 & j_2 & j_1 \\ m_3 & m_2 & m_1 \end{pmatrix} = \ldots \,, \end{aligned} \tag{1.103}$$

or, a phase factor +1 for an *even* permutation and a phase factor $(-1)^{j_1+j_2+j_3}$ for an *odd* permutation. Moreover, one gets the relation

$$\begin{pmatrix} j_1 & j_2 & j_3 \\ -m_1 & -m_2 & -m_3 \end{pmatrix} = (-1)^{j_1+j_2+j_3} \begin{pmatrix} j_1 & j_2 & j_3 \\ m_1 & m_2 & m_3 \end{pmatrix} . \tag{1.104}$$

The former orthogonality relations (1.96) for the Clebsch-Gordan coefficients now rewritten in terms of the Wigner $3j$-symbols become

$$\sum_{m_1,m_2} \begin{pmatrix} j_1 & j_2 & j_3 \\ m_1 & m_2 & m_3 \end{pmatrix} \begin{pmatrix} j_1 & j_2 & j_3' \\ m_1 & m_2 & m_3' \end{pmatrix} = \frac{1}{2j_3+1} \delta_{j_3 j_3'} \delta_{m_3 m_3'} \,, \tag{1.105}$$

$$\sum_{m_1,m_2,m_3} \begin{pmatrix} j_1 & j_2 & j_3 \\ m_1 & m_2 & m_3 \end{pmatrix}^2 = 1 \,, \tag{1.106}$$

$$\sum_{j_3,m_3} (2j_3+1) \begin{pmatrix} j_1 & j_2 & j_3 \\ m_1 & m_2 & m_3 \end{pmatrix} \begin{pmatrix} j_1 & j_2 & j_3 \\ m_1' & m_2' & m_3 \end{pmatrix} = \delta_{m_1 m_1'} \delta_{m_2 m_2'} \,. \tag{1.107}$$

Extensive sets of tables of Wigner $3j$-symbols exist (e.g. Rotenberg et al. 1959).

Explicit calculations of the Wigner $3j$-symbol are easily performed using the expression (de-Shalit, Talmi 1963).

$$\begin{aligned} \begin{pmatrix} j_1 & j_2 & j_3 \\ m_1 & m_2 & m_3 \end{pmatrix} &= \delta_{m_1+m_2,-m_3} \Big(\big((j_1+j_2-j_3)!\,(j_2+j_3-j_1)! \\ &\times (j_3+j_1-j_2)! \big) / (j_1+j_2+j_3+1)! \Big)^{1/2} \\ &\times \big((j_1+m_1)!(j_1-m_1)!(j_2+m_2)!(j_2-m_2)!(j_3+m_3)!(j_3-m_3)! \big)^{1/2} \\ &\times \sum_t (-1)^{j_1-j_2-m_3+t} \big(t!(j_1+j_2-j_3-t)!(j_3-j_2+m_1+t)! \\ &\times (j_3-j_1-m_2+t)!(j_1-m_1-t)!(j_2+m_2-t)! \big)^{-1} . \end{aligned} \tag{1.108}$$

with $(-m)! = 0$ if m is positive, t is entire and $0! = 1$, so the following conditions hold

$$\begin{aligned}
&t \geq 0\\
&j_1 + j_2 - j_3 \geq t\\
&-j_3 + j_2 - m_1 \leq t\\
&-j_3 + j_1 + m_2 \leq t\\
&j_1 - m_1 \geq t\\
&j_2 + m_2 \geq t\,.
\end{aligned} \tag{1.109}$$

In Chap. 9, a FORTRAN program is given that evaluates (1.108) numerically. As an example, we evaluate the $3j$-symbol

$$\begin{pmatrix} j & 1 & j \\ -m & 0 & m \end{pmatrix}.$$

The conditions (1.109) give the restrictions on t

$$\begin{aligned}
&t \geq 0;\ 1 \geq t;\ -j + 1 + m \leq t;\ 0 \leq t;\\
&j + m \geq t;\ 1 \geq t \quad \text{or} \quad t = 0, 1\,.
\end{aligned}$$

Calculating in detail, one gets

$$\begin{aligned}
\begin{pmatrix} j & 1 & j \\ -m & 0 & m \end{pmatrix} &= \big((2j-1)!/(2j+2)!\big)^{1/2}\\
&\quad \times \big((j-m)!(j+m)!(j+m)!(j-m)!\big)^{1/2}\\
&\quad \times (-1)^{j-1-m}\Big[\big((j-1-m)!(j+m)!\big)^{-1} - \big((j-m)!(j+m-1)!\big)^{-1}\Big]\\
&= (-1)^{j-m} m / \big(j(j+1)(2j+1)\big)^{1/2}\,.
\end{aligned}$$

1.6 Racah Recoupling Coefficients: Coupling of Three Angular Momenta

In the case of a system described by three independent angular momentum operators $\boldsymbol{J}_1$, $\boldsymbol{J}_2$, $\boldsymbol{J}_3$; one can again form the total angular momentum operator $\boldsymbol{J}$ defined as

$$\boldsymbol{J} = \boldsymbol{J}_1 + \boldsymbol{J}_2 + \boldsymbol{J}_3\,. \tag{1.110}$$

The six commuting operators

$$\left\{\boldsymbol{J}_1^2, J_{1z}, \boldsymbol{J}_2^2, J_{2z}, \boldsymbol{J}_3^2, J_{3z}\right\}, \tag{1.111}$$

have a set of common eigenvectors, the product vectors

$$|j_1 m_1\rangle |j_2 m_2\rangle |j_3 m_3\rangle\,. \tag{1.112}$$

For the three angular momentum operators, it is now possible to form three sets of commuting operators:

$$\left\{\boldsymbol{J}^2, J_z, \boldsymbol{J}_1^2, \boldsymbol{J}_2^2, \boldsymbol{J}_3^2, \boldsymbol{J}_{12}^2\right\}, \tag{1.113}$$

$$\left\{\boldsymbol{J}^2, J_z, \boldsymbol{J}_1^2, \boldsymbol{J}_2^2, \boldsymbol{J}_3^2, \boldsymbol{J}_{23}^2\right\}, \tag{1.114}$$

$$\left\{\boldsymbol{J}^2, J_z, \boldsymbol{J}_1^2, \boldsymbol{J}_2^2, \boldsymbol{J}_3^2, \boldsymbol{J}_{13}^2\right\}, \tag{1.115}$$

with the eigenvectors

$$|(j_1 j_2) J_{12} j_3; JM\rangle , \tag{1.116}$$

$$|j_1 (j_2 j_3) J_{23}; JM\rangle , \tag{1.117}$$

$$|(j_1 j_3) J_{13} j_2; JM\rangle , \tag{1.118}$$

respectively.

It is possible to use a diagrammatic way of expressing the vector coupled eigenstates (Brussaard, Glaudemans 1977), by using lines and arrows for a given angular momentum and the order in which they are coupled. The intermediate angular momentum is shown by the dashed line (Fig. 1.3).

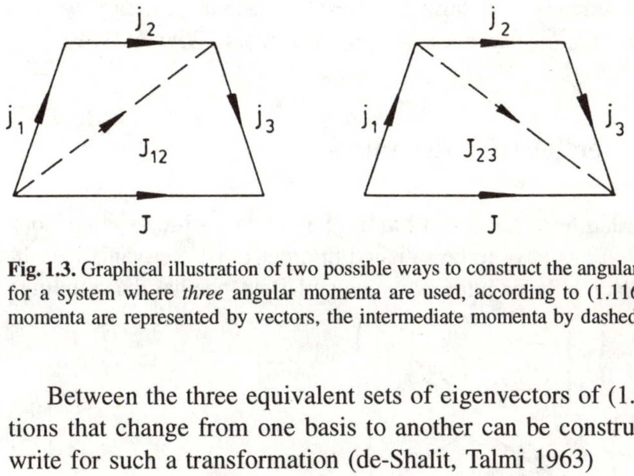

Fig. 1.3. Graphical illustration of two possible ways to construct the angular momentum wave functions for a system where *three* angular momenta are used, according to (1.116) and (1.117). The angular momenta are represented by vectors, the intermediate momenta by dashed-line vectors

Between the three equivalent sets of eigenvectors of (1.116–118), transformations that change from one basis to another can be constructed. We can formally write for such a transformation (de-Shalit, Talmi 1963)

$$|j_1 (j_2 j_3) J_{23}; JM\rangle = \sum_{J_{12}} \langle (j_1 j_2) J_{12} j_3; J | j_1 (j_2 j_3) J_{23}; J\rangle \times |(j_1 j_2) J_{12} j_3; JM\rangle . \tag{1.119}$$

It can easily be shown that the transformation coefficients in (1.119) and in similar relations do not depend on the projection quantum number M. Now by explicitly carrying out the recoupling from the states $|j_1(j_2 j_3)J_{23}; JM\rangle$ to the coupling scheme $|(j_1 j_2)J_{12} j_3; JM\rangle$ (Appendix B), one obtains the detailed form of the recoupling coefficient of (1.119). In this particular situation, a full sum over all magnetic quantum numbers of products of *four* Wigner $3j$-symbols results. The latter, defined as an angular momentum invariant quantity (no longer dependent on the specific orientation of a quantization axis), the Wigner $6j$-symbol, leads to the following result (Wigner 1959, Brussaard 1967)

$$\langle j_1(j_2j_3)J_{23};J|(j_1j_2)J_{12}j_3;J\rangle = (-1)^{j_1+j_2+j_3+J}\hat{J}_{12}\hat{J}_{23}\begin{Bmatrix} j_1 & j_2 & J_{12} \\ j_3 & J & J_{23} \end{Bmatrix}, \qquad (1.120)$$

(using the notation $\hat{J} \equiv (2J+1)^{1/2}$).

The precise definition of the $6j$-symbol in terms of the $3j$-symbols reads (de-Shalit, Talmi 1963)

$$\begin{Bmatrix} j_1 & j_2 & j_3 \\ l_1 & l_2 & l_3 \end{Bmatrix} = \sum_{\text{all } m_i, m_i'} (-1)^{\Sigma j_i + \Sigma l_i + \Sigma m_i + \Sigma m_i'} \begin{pmatrix} j_1 & j_2 & j_3 \\ m_1 & m_2 & m_3 \end{pmatrix}$$
$$\times \begin{pmatrix} j_1 & l_2 & l_3 \\ -m_1 & m_2' & -m_3' \end{pmatrix} \begin{pmatrix} l_1 & j_2 & l_3 \\ -m_1' & -m_2 & m_3' \end{pmatrix} \begin{pmatrix} l_1 & l_2 & j_3 \\ m_1' & -m_2' & -m_3 \end{pmatrix} \qquad (1.121)$$

and very much resembles a "contraction of tensors" (one sums over projection quantum numbers $m_1, m_2, \ldots, m_3'$, both of which always show up in different $3j$-symbols with opposite sign). We show in Chap. 2 that, indeed, the $6j$-symbol is a full contraction not on cartesian but on spherical tensors (Wigner 1959).

1.7 Symmetry Properties of $6j$-Symbols

Because of the very structure of the definition in (1.121), in each $6j$-symbol four angular momentum couplings have to be satisfied in order to be non-vanishing. In shorthand notation, replacing the angular momenta with dots, one has the couplings

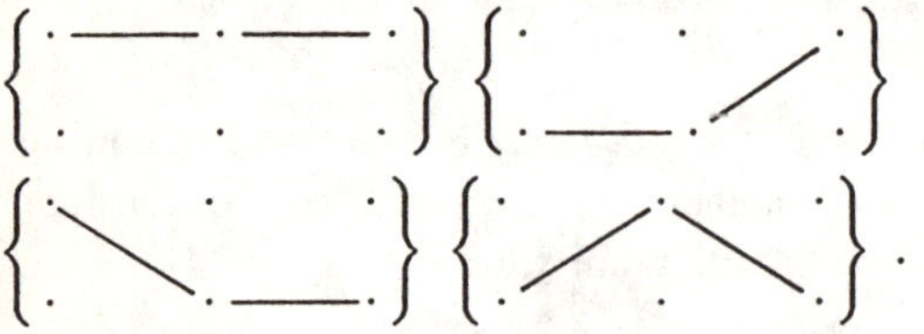

(1.122)

Here we quote some often used symmetry properties. A more detailed account can be found in various texts (de-Shalit, Talmi 1963, Edmonds 1957, Rose, Brink 1967, Brussaard 1967, Brink, Satchler 1962)

$$\text{i)} \quad \begin{Bmatrix} j_1 & j_2 & j_3 \\ l_1 & l_2 & l_3 \end{Bmatrix} = 0, \qquad (1.123)$$

unless the triangular (coupling) conditions $(j_1j_2j_3)$, $(j_1l_2l_3)$, $(l_1l_2j_3)$, $(l_1j_2l_3)$ are fulfilled

ii)
$$\begin{Bmatrix} j_1 & j_2 & j_3 \\ l_1 & l_2 & l_3 \end{Bmatrix} = \begin{Bmatrix} j_2 & j_3 & j_1 \\ l_2 & l_3 & l_1 \end{Bmatrix} = \begin{Bmatrix} j_1 & j_3 & j_2 \\ l_1 & l_3 & l_2 \end{Bmatrix}$$
$$= \begin{Bmatrix} l_1 & l_2 & j_3 \\ j_1 & j_2 & l_3 \end{Bmatrix} = \begin{Bmatrix} j_1 & l_2 & l_3 \\ l_1 & j_2 & j_3 \end{Bmatrix} = \dots , \qquad (1.124)$$

iii) orthogonality relation

$$\sum_j (2j+1) \begin{Bmatrix} j_1 & j_2 & j \\ j_3 & j_4 & j' \end{Bmatrix} \begin{Bmatrix} j_1 & j_2 & j \\ j_3 & j_4 & j'' \end{Bmatrix} = \delta_{j'j''}(2j'+1)^{-1} , \qquad (1.125)$$

iv) special case

$$\begin{Bmatrix} j_1 & j_2' & j_3 \\ j_2 & j_1' & 0 \end{Bmatrix} = (-1)^{j_1+j_2+j_3} \left(\hat{j}_1 \hat{j}_2\right)^{-1} \delta_{j_1 j_1'} \delta_{j_2 j_2'} , \qquad (1.126)$$

v) Explicit form: Racah formula (de-Shalit, Talmi 1963)

$$\begin{Bmatrix} j_1 j_2 j_3 \\ l_1 l_2 l_3 \end{Bmatrix} = \Delta(j_1 j_2 j_3)\Delta(j_1 l_2 l_3)\Delta(l_1 j_2 l_3)\Delta(l_1 l_2 j_3)$$
$$\times \sum_t (-1)^t (t+1)! \big[(t-j_1-j_2-j_3)!(t-j_1-l_2-l_3)!$$
$$\times (t-l_1-j_2-l_3)!(t-l_1-l_2-j_3)!(j_1+j_2+l_1+l_2-t)!$$
$$\times (j_2+j_3+l_2+l_3-t)!(j_3+j_1+l_3+l_1-t)!\big]^{-1} , \qquad (1.127)$$

with

$$\Delta(abc) = \left[(a+b-c)!(b+c-a)!(c+a-b)!/(a+b+c+1)!\right]^{1/2} , \qquad (1.128)$$

and the condition of having non-negative values of the integer in the factorial expression in (1.127).

1.8 Wigner 9j-Symbols: Coupling and Recoupling of Four Angular Momenta

Similarly to the methods used in Sect. 1.6, we can construct the total angular momentum operator corresponding to the sum of the four independent angular momentum operators as

$$\boldsymbol{J} = \boldsymbol{J}_1 + \boldsymbol{J}_2 + \boldsymbol{J}_3 + \boldsymbol{J}_4 . \qquad (1.129)$$

In constructing the total set of commuting operators one has in the uncoupled representation,

$$\left\{\boldsymbol{J}_1^2, J_{1z}, \boldsymbol{J}_2^2, J_{2z}, \boldsymbol{J}_3^2, J_{3z}, \boldsymbol{J}_4^2, J_{4z}\right\} , \qquad (1.130)$$

which have as eigenvectors the product vectors

$$|j_1 m_1\rangle |j_2 m_2\rangle |j_3 m_3\rangle |j_4 m_4\rangle . \qquad (1.131)$$

In the coupled representation, one needs two intermediate angular momentum operators for which a large choice exists. Coupling pairwise, one has three possibilities

$$
\begin{aligned}
&\boldsymbol{J}^2, J_z, \boldsymbol{J}_{12}^2, \boldsymbol{J}_{34}^2, \boldsymbol{J}_1^2, \boldsymbol{J}_2^2, \boldsymbol{J}_3^2, \boldsymbol{J}_4^2\,,\\
&\boldsymbol{J}^2, J_z, \boldsymbol{J}_{13}^2, \boldsymbol{J}_{24}^2, \boldsymbol{J}_1^2, \boldsymbol{J}_2^2, \boldsymbol{J}_3^2, \boldsymbol{J}_4^2\,,\\
&\boldsymbol{J}^2, J_z, \boldsymbol{J}_{14}^2, \boldsymbol{J}_{23}^2, \boldsymbol{J}_1^2, \boldsymbol{J}_2^2, \boldsymbol{J}_3^2, \boldsymbol{J}_4^2\,,
\end{aligned}
\tag{1.132}
$$

(Fig. 1.4), with eigenvectors

$$
\begin{aligned}
&|(j_1j_2)J_{12}(j_3j_4)J_{34}; JM\rangle\,,\\
&|(j_1j_3)J_{13}(j_2j_4)J_{24}; JM\rangle\,,\\
&|(j_1j_4)J_{14}(j_2j_3)J_{23}; JM\rangle\,,
\end{aligned}
\tag{1.133}
$$

respectively.

There exist other possibilities, too, however, such as

$$
\boldsymbol{J}^2, J_z, \boldsymbol{J}_{12}^2, \boldsymbol{J}_{123}^2, \boldsymbol{J}_1^2, \boldsymbol{J}_2^2, \boldsymbol{J}_3^2, \boldsymbol{J}_4^2\,,
\tag{1.134}
$$

shown in Fig. 1.4. The latter method is probably the best adapted to extend coupling to n angular momenta by successive coupling of an extra angular momentum to the former $n-1$ system (Yutsis et al. 1962). Here, too, many possible recoupling schemes and recoupling coefficients can be obtained (Edmonds 1957). Here we only discuss recoupling between the states of (1.133) since they lead to the Wigner $9j$-symbol, e.g., (de-Shalit, Talmi 1963)

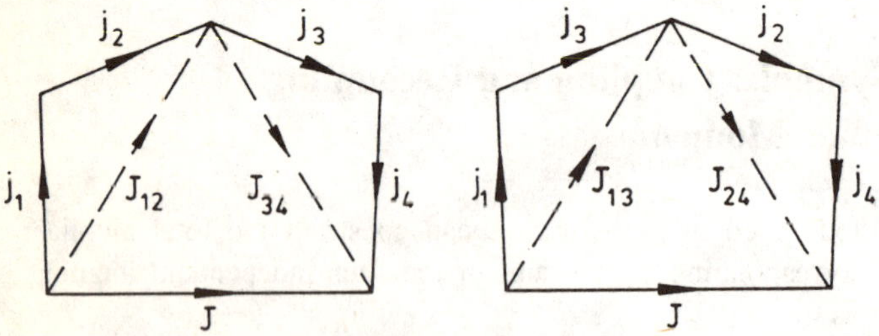

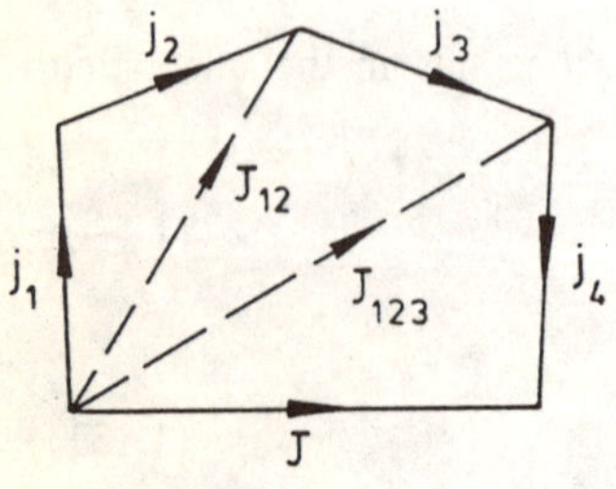

Fig. 1.4. Graphical illustration of possible ways to construct the angular momentum wave functions for a system where *four* angular momenta are used, according to (1.133). In the lower part, we present the more general way of constructing the four-angular momentum system by specifying J_{12}, J_{123} as intermediate angular momenta, respectively

$$|(j_1 j_3) J_{13} (j_2 j_4) J_{24}; JM\rangle = \sum_{J_{12}, J_{34}} \hat{J}_{13} \hat{J}_{24} \hat{J}_{12} \hat{J}_{34}$$

$$\times \begin{Bmatrix} j_1 & j_2 & J_{12} \\ j_3 & j_4 & J_{34} \\ J_{13} & J_{24} & J \end{Bmatrix} |(j_1 j_2) J_{12} (j_3 j_4) J_{34}; JM\rangle \,. \tag{1.135}$$

Precise definitions of the Wigner 9j-symbol as a full contraction over products of 6 3j-symbols can be found (Edmonds 1957, de-Shalit, Talmi 1963). In the present context, where we shall concentrate on the nuclear shell model, we quote a special case that often occurs:

$$\begin{Bmatrix} j_1 & j_2 & J \\ j_4 & j_3 & k \end{Bmatrix} = (-1)^{j_2+J+j_3+k} \hat{J} \hat{k} \begin{Bmatrix} j_1 & j_2 & J \\ j_3 & j_4 & J \\ k & k & 0 \end{Bmatrix} \,. \tag{1.136}$$

Also, we point out that the general expression of (1.135) can be used when recoupling from a $(j\,j)$ coupling basis into an (LS) coupling basis if we consider cases with two fermions. Thus one can relate the states $|(l_1 l_2) L(\frac{1}{2}\,\frac{1}{2}) S; JM\rangle$ and $|(l_1 \frac{1}{2}) j_1 (l_2 \frac{1}{2}) j_2; JM\rangle$ by the transformation

$$|(l_1 l_2) L \left(\tfrac{1}{2}\,\tfrac{1}{2}\right) S; JM\rangle = \sum_{j_1, j_2} \hat{L} \hat{S} \hat{j}_1 \hat{j}_2 \begin{Bmatrix} l_1 & l_2 & L \\ \frac{1}{2} & \frac{1}{2} & S \\ j_1 & j_2 & J \end{Bmatrix}$$

$$\times \left|\left(l_1 \tfrac{1}{2}\right) j_1 \left(l_2 \tfrac{1}{2}\right) j_2; JM\right\rangle \,, \tag{1.137}$$

a relation that gives the $(jj) \rightarrow (LS)$ basis transformation.

1.9 Classical Limit of Wigner 3j-Symbols

It is now possible to construct a classical (in the limit of large angular momenta) model (Brussaard, Tolhoek 1957, Brussaard 1967) for angular momentum coupling and thus also for the Wigner 3j- (and similarly for the 6j-, 9j-, 3nj-) symbol. We make use of the fact that in quantum mechanics it is only possible to specify both the length and the projection on a quantization axis of the angular momentum. Therefore, a precessing vector model results where for constant precession velocity the azimuthal angle has a constant probability distribution. Since the Clebsch-Gordan coefficients denote the expansion coefficients in an orthonormal basis, the square can be interpreted as a probability. Thus, for the uncoupled representation where $\boldsymbol{J}_1^2$, J_{1z}, $\boldsymbol{J}_2^2$ and J_{2z} are the commuting operators, the coefficients $|\langle j_1 m_1, j_2 m_2 | jm\rangle|^2$ denote the probability that in a state with fixed $(j_1 m_1)$ and $(j_2 m_2)$, a given value of (j, m) will result with j expressing the length of the angular momentum vector (correct only for large values of j), (Fig. 1.5). Similarly, $|\langle j_1 m_1 j_2 m_2 | jm\rangle|^2$ (Fig. 1.5) can be interpreted, for the coupled basis where eigenstates of the operators $\boldsymbol{J}^2$, J_z, $\boldsymbol{J}_1^2$, $\boldsymbol{J}_2^2$ are considered, as the probability that for given (j, m) the values m_1 and m_2 will result as projection quantum numbers relating to the angular momenta j_1 and j_2, respectively. One can even calculate this

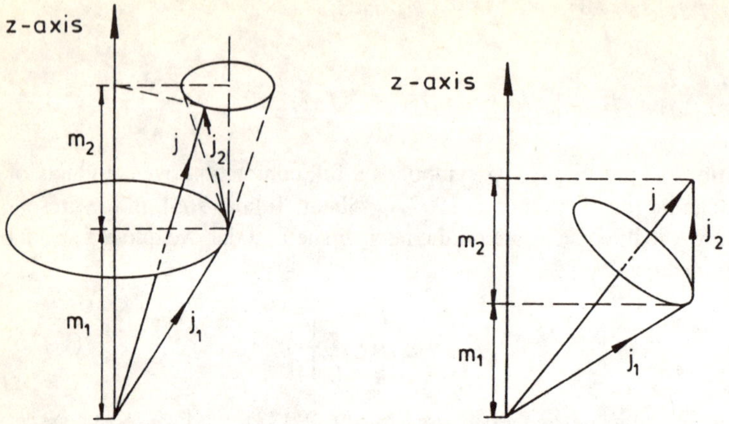

Fig. 1.5. Graphical representation of two angular momenta $\boldsymbol{j}_1$ and $\boldsymbol{j}_2$, shown as vectors that make a precession around the z-axis with constant angular velocity (vector model). Using the addition to a momentum $\boldsymbol{j} = \boldsymbol{j}_1 + \boldsymbol{j}_2$, the probability of obtaining a given value for the length j, given fixed m_1 and m_2 values, is given by the Clebsch-Gordan coefficient squared $|\langle j_1m_1, j_2m_2|jm >|^2$. If the two vectors $\boldsymbol{j}_1$ and $\boldsymbol{j}_2$ are coupled to form the total angular momentum $\boldsymbol{j}$ (which is a constant of motion), the two vectors will make a precession around the direction of $\boldsymbol{j}$. For fixed value of the length of $\boldsymbol{j}$ and projection m, the projections m_1 and m_2 can be obtained again as a probability distribution given by the Clebsch-Gordan coefficient squared $| < j_1m_1, j_2m_2|jm > |^2$

distribution in both cases from probability considerations (Edmonds 1957 gives an explicit calculation). Extending the above arguments, classical models can also be constructed for interpreting higher $3n - j$ symbols (Brussaard 1967).

Short Overview of Angular Momentum Coupling Formulas

One-particle central force motion-orbital angular momentum

l_x, l_y, l_z : differential operators

$$
\begin{aligned}
\left[\boldsymbol{l}^2, l_i\right] &= 0 \\
\left[l_i, l_j\right] &= \mathrm{i}\varepsilon_{ijk}\,\hbar l_k \\
\left[l_+, l_-\right] &= 2\,\hbar l_z \\
\left[l_z, l_+\right] &= \hbar l_+ \\
\left[l_z, l_-\right] &= -\,\hbar l_-
\end{aligned}
$$

$$l_\pm|lm\rangle = \hbar\big(l(l+1) - m(m\pm 1)\big)^{1/2}|l, m\pm 1\rangle\ .$$

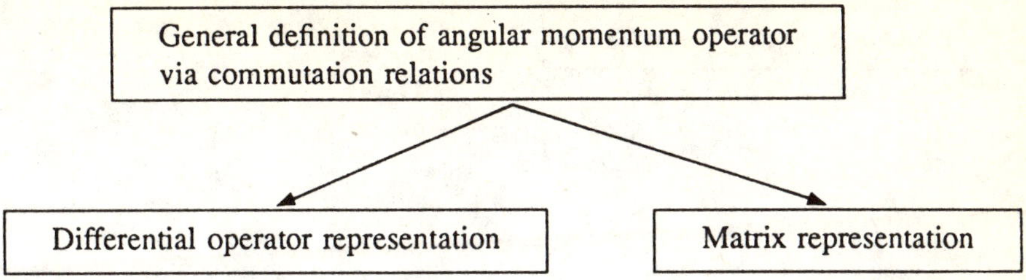

> General definition of angular momentum operator via commutation relations

> Differential operator representation

> Matrix representation

$l = 0, 1, 2, \ldots$

$j = 0, \frac{1}{2}, 1, \frac{3}{2}, \ldots$

> Total angular momentum

$j = l + s$

$J = J_1 + J_2$

$\left\{ J_1^2, J_{1z}, J_2^2, J_{2z} \right\}$ and $\left\{ J^2, J_z, J_1^2, J_2^2 \right\}$

$|j_1 j_2; jm\rangle = \sum_{m_1, m_2} \langle j_1 m_1, j_2 m_2 | jm \rangle | j_1 m_1 \rangle | j_2 m_2 \rangle$

$|j_1 m_1\rangle | j_2 m_2 > = \sum_{j,m} \langle j_1 m_1, j_2 m_2 | jm \rangle | j_1 j_2; jm \rangle \ .$

> Three angular momentum systems

$J = J_1 + J_2 + J_3$

$\left\{ J_1^2, J_{1z}, J_2^2, J_{2z}, J_3^2, J_{3z} \right\} \rightarrow |j_1 m_1\rangle | j_2 m_2\rangle | j_3 m_3\rangle$

$\left\{ J^2, J_z, J_1^2, J_2^2, J_3^2, J_{12}^2 \right\} \rightarrow |(j_1 j_2) J_{12} j_3; JM\rangle$

$\left\{ J^2, J_z, J_1^2, J_2^2, J_3^2, J_{13}^2 \right\} \rightarrow |(j_1 j_3) J_{13} j_2; JM\rangle$

$\left\{ J^2, J_z, J_1^2, J_2^2, J_3^2, J_{23}^2 \right\} \rightarrow |j_1 (j_2 j_3) J_{23}; JM\rangle \ .$

Recoupling Wigner $6j$-symbol

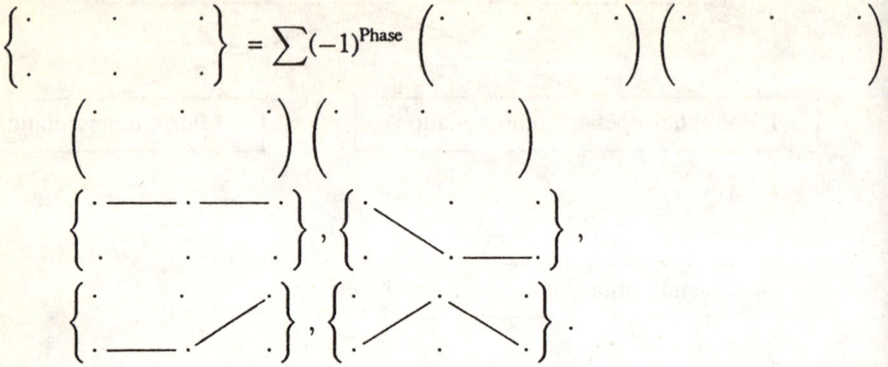

$$\begin{Bmatrix} \cdot & \cdot & \cdot \\ \cdot & \cdot & \cdot \end{Bmatrix} = \sum (-1)^{\text{Phase}} \begin{pmatrix} \cdot & \cdot & \cdot \\ & & \end{pmatrix} \begin{pmatrix} \cdot & \cdot & \cdot \\ & & \end{pmatrix} \begin{pmatrix} \cdot & \cdot & \cdot \\ & & \end{pmatrix} \begin{pmatrix} \cdot & \cdot & \cdot \\ & & \end{pmatrix}$$

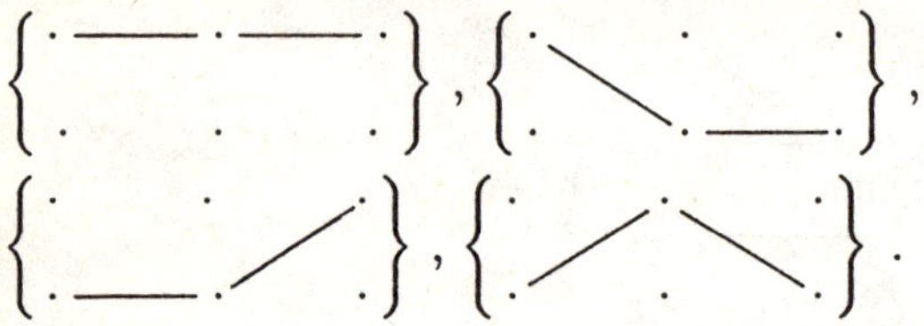

Notation:

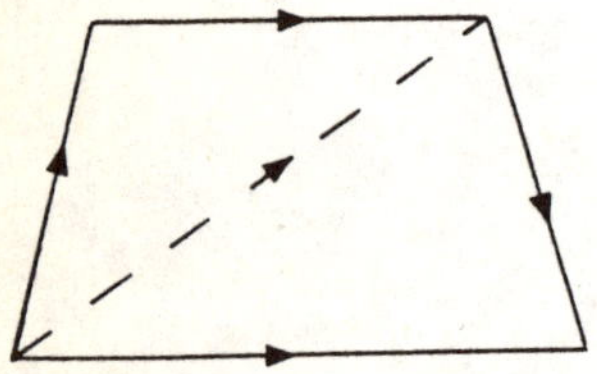

Four angular momentum systems

$$\boldsymbol{J} = \boldsymbol{J}_1 + \boldsymbol{J}_2 + \boldsymbol{J}_3 + \boldsymbol{J}_4$$

$$\left\{ \boldsymbol{J}_1^2,\ J_{1z},\ \boldsymbol{J}_2^2,\ J_{2z},\ \boldsymbol{J}_3^2,\ J_{3z},\ \boldsymbol{J}_4^2,\ J_{4z} \right\}$$

Basis states $\rightarrow |j_1 m_1\rangle |j_2 m_2\rangle |j_3 m_3\rangle |j_4 m_4\rangle$

$$\left\{ \boldsymbol{J}^2,\ J_z,\ \boldsymbol{J}_1^2,\ \boldsymbol{J}_2^2,\ \boldsymbol{J}_3^2,\ \hat{\boldsymbol{J}}_4^2 \ \begin{matrix}\text{two intermediate} \\ \text{angular momenta}\end{matrix} \right\} \text{e.g. } \boldsymbol{J}_{12}^2,\ \boldsymbol{J}_{34}^2$$

Basis states $\rightarrow |(j_1 j_2) J_{12} (j_3 j_4) J_{34}; JM\rangle$.

Recoupling-Wigner $9j$-symbol

$$\begin{Bmatrix} \cdot & \cdot & \cdot \\ \cdot & \cdot & \cdot \\ \cdot & \cdot & \cdot \end{Bmatrix} = \sum \left(\quad \right) \left(\quad \right) \left(\quad \right) \left(\quad \right) \left(\quad \right) \left(\quad \right) .$$

Notation:

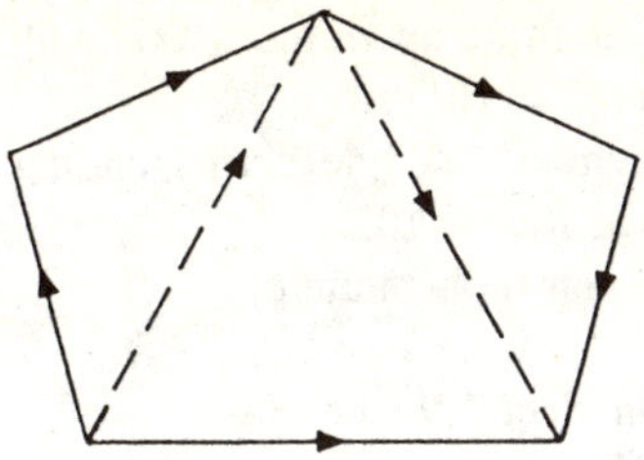

Problems

1.1 Prove the relation that the square of the angular momentum operator $\boldsymbol{l}^2$ can be written as (see (1.10, 11, 12))

$$\boldsymbol{l}^2 = \boldsymbol{r}^2\boldsymbol{p}^2 + h^2\frac{\partial}{\partial r}\left(r^2\frac{\partial}{\partial r}\right) .$$

1.2 Show that the operator $\boldsymbol{J} = \boldsymbol{l} + \boldsymbol{s}$, where $\boldsymbol{l}$ describes the orbital angular momentum operator and $\boldsymbol{s}$ the intrinsic spin angular momentum operator for a spin 1/2 particle, constitutes an angular momentum operator.

1.3 Derive an explicit form of the $2P_{3/2}, m = +1/2$ wave function in terms of the spin and orbital angular momentum wave functions.

1.4 Show that the total angular momentum operator $\boldsymbol{J}^2$ for a nucleon (obtained) by coupling the orbital and intrinsic spin angular momentum operators can be diagonalized in the basis $|lm_l\rangle|1/2m_s\rangle$. Determine the eigenvalues and show that the corresponding eigenfunctions are also eigenfunctions of the Hamiltonian $H = H_0 + a\boldsymbol{l}\cdot\boldsymbol{s}$ with

$$H_0 = \frac{\boldsymbol{p}^2}{2m} + \frac{1}{2}m\omega^2r^2 .$$

1.5 Prove the orthogonality relations for the Clebsch-Gordan coefficients (see (1.96)).

1.6 Show that the recoupling of (1.135) indeed leads to a Wigner $9-j$ symbol (with no *extra* phase factor).

1.7 Show that the recoupling coefficients $\langle(j_1j_2)J_{12}j_3; JM|j_1(j_2j_3)J_{23}; JM\rangle$, which describe the transformation between states with different coupling order in systems with three angular momenta, are independent of M.

1.8 Determine the relative weights for the $S = 0$ and $S = 1$ intrinsic spin components in the $|(1d_{5/2})^2; J = 2)$ two particle wave function.

1.9 Discuss the classical limit of the Wigner $6-j$ symbols, according to the methods of Sect. 1.9.

*1.10 Calculate the probability density $P(j)$ (i.e. the probability that the length of $\boldsymbol{j}$ lies between j and $j + dj$ is $P(j)dj$) if we suppose that, according to the upper part of Fig. 1.5, $\boldsymbol{j}_1$ rotates at a constant rate about the z-axis with respect to $\boldsymbol{j}_2$. $P(j)$ is then inversely proportional to dj/dt.

1.11 Prove the relation between the Wigner $3j$-symbol and the corresponding Clebsch-Gordan coefficient, as expressed in equation 1.102.

1.12 Determine the matrix representation of the angular momentum operators $\boldsymbol{J}_x$, $\boldsymbol{J}_y$ and $\boldsymbol{J}_z$ for angular momentum 3/2.

*1.13 Discuss, according to the method outlined in Sect. 1.9, the classical limit for the Wigner $6j$-symbol. Construct the graphical representation similar to Fig. 1.5 for the coupling of two angular momenta.

2. Rotations in Quantum Mechanics

2.1 Rotation of a Scalar Field-Rotation Group $O(3)$

In this section as well as in the rest of Chap. 2, we shall study the relationships between the angular momentum operator and rotation of a physical system described by a given wave function in more detail. We note that the angular momentum operator can act as a generator for rotations of a general scalar function.

We consider the change in a scalar function field $f(\boldsymbol{r})$ because of a rotation of this function field with respect to a fixed coordinate (x, y, z) system. This is an active point of view characterizing the transformations. For the particular case of a rotation about the z-axis through an angle $\delta\varphi$ (we denote this as a vector, defined in the sense of the rotation, perpendicular to the plane in which the rotation goes, Fig. 2.1).

We then know that a new function field $F(\boldsymbol{r})$ is defined, but with the constraint that

$$F(\boldsymbol{r}+\boldsymbol{a}) = f(\boldsymbol{r}) , \quad \text{or} \tag{2.1}$$

$$\begin{aligned} F(\boldsymbol{r}) &= f(\boldsymbol{r}-\boldsymbol{a}) \\ &= f(\boldsymbol{r}) - \boldsymbol{a}\cdot\nabla f(\boldsymbol{r}) + \dots . \end{aligned} \tag{2.2}$$

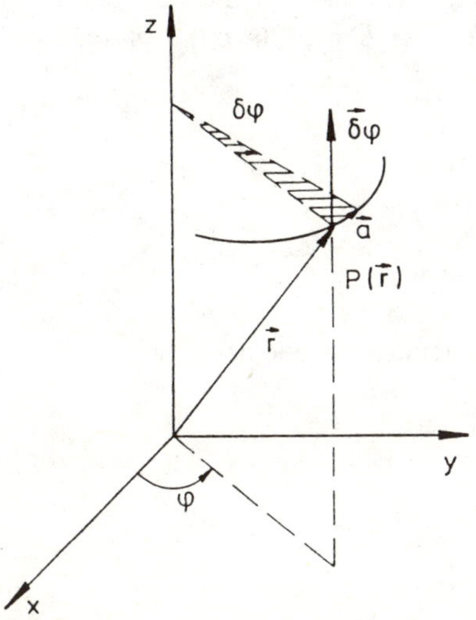

Fig. 2.1. Transformation of a point $P(\boldsymbol{r})$ under a rotation, in anti-clockwise direction, over an infinitesimal angle $\delta\varphi$. The resulting rotation is characterized by a vector $\delta\boldsymbol{\varphi}$ (perpendicular on the rotation plane in the sense of the rotation). The displacement vector $\boldsymbol{a}$ then becomes (infinitesimal quantities) $\boldsymbol{a} = \delta\boldsymbol{\varphi}\times\boldsymbol{r}$

The change of the function field in the same physical point $\boldsymbol{r}$ is expressed by

$$\begin{aligned}\delta f(\boldsymbol{r}) &= F(\boldsymbol{r}) - f(\boldsymbol{r}) \\ &= -\delta\boldsymbol{\varphi} \times \boldsymbol{r} \cdot \boldsymbol{\nabla} f(\boldsymbol{r}) + \ldots \\ &= -\delta\boldsymbol{\varphi} \cdot (\boldsymbol{r} \times \boldsymbol{\nabla}) f(\boldsymbol{r}) + \ldots \quad \text{or}\end{aligned}$$

$$\delta f(\boldsymbol{r}) = \frac{-\mathrm{i}}{\hbar}\delta\boldsymbol{\varphi} \cdot \boldsymbol{L} f(\boldsymbol{r}) \,, \tag{2.3}$$

up to lowest order in the infinitesimal angle $|\delta\boldsymbol{\varphi}|$. For a finite angle $\boldsymbol{\varphi}$ (which can be divided into n infinitesimal rotations $\delta\boldsymbol{\varphi} = \boldsymbol{\varphi}/n$) one gets

$$\begin{aligned}F(\boldsymbol{r}) &= \lim_{n\to\infty}\left(1 - \frac{\mathrm{i}}{\hbar}\frac{\boldsymbol{\varphi}}{n} \cdot \boldsymbol{L}\right)^n f(\boldsymbol{r}) \\ &= \mathrm{e}^{-\mathrm{i}/\hbar\boldsymbol{\varphi}\cdot\boldsymbol{L}} f(\boldsymbol{r}) \,. \end{aligned} \tag{2.4}$$

We generally denote the operator that transforms the old scalar field $f(\boldsymbol{r})$ into the new function field $F(\boldsymbol{r})$ in the same physical point by U_R, with

$$U_R = \mathrm{e}^{-\mathrm{i}/\hbar\boldsymbol{\varphi}\cdot\boldsymbol{L}} \,. \tag{2.5}$$

[Note that although the derivation was given for a rotation $\delta\varphi$ around the z-axis, (2.4, 5) hold generally for any rotation of an angle $|\boldsymbol{\varphi}|$ around a unit vector $\mathbf{1}_n$ specifying the rotational plane (Edmonds 1957, Brink, Satchler 1962, Rose, Brink 1967)].

The derivation, although slightly formal, can be illustrated in the example of a temperature field $T(x, y, z)$ (Fig. 2.2) where subsequent planes parallel to the (z, y) plane are characterized by increasing temperature when approaching the $x = 0$ point from the point at $x = a$ (0° → 100°). If we rotate the physical system, and thus the temperature field, a new field $T'(x, y, z)$ at every point is defined. We know, however, that for a given physical point $P(\boldsymbol{r})$, the new function field at the new coordinates has to be equal to the old function field at the old coordinates or,

$$T'(x', y', z') = T(x, y, z) \,. \tag{2.6}$$

As another example, we use the field

$$f(\boldsymbol{r}) = a\left(x^2 - y^2\right) . \tag{2.7}$$

When rotating the function field over 45° anticlockwise one has

$$\begin{aligned}x' &= x\cos 45° - y\sin 45° \,, \\ y' &= x\sin 45° + y\cos 45° \,, \end{aligned} \tag{2.8}$$

and the new function $F(\boldsymbol{r})$ becomes

$$F(\boldsymbol{r}) = 2axy \,. \tag{2.9}$$

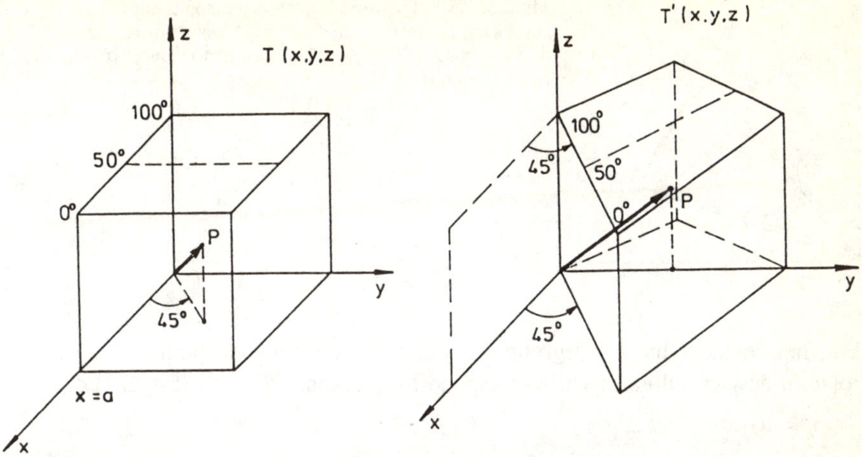

Fig. 2.2. Illustration of the transformation properties of a scalar field [here we take a temperature field, expressed by $T(x, y, z)$] under an *active* rotation of the system over an angle of 45° in anti-clockwise direction around the z-axis. In the figure, we present the system as a "temperature" cube with surfaces of given temperature: 100°–50°–0°. After rotation, a new temperature field $T'(x, y, z)$ results which describes the same physical system in the original system of axes (x, y, z)

Consider a point $P(x = 3, y = 2)$ with a function value $f(\boldsymbol{r}) = 5a$. The new coordinates become $x' = \sqrt{2}/2$; $y' = 5\sqrt{2}/2$ and the new function value at the new point becomes $F(\boldsymbol{r}') = 5a$, too, illustrating the above discussion. It is even possible to derive the new function form $F(\boldsymbol{r})$ by using the more formal definition of (2.5) applied to the present situation (see problem set).

The above discussion in deriving the rotation operator in an active image U_R has used the fact that the rotation induced was related to orbital angular momentum. A more general angular momentum operator $\boldsymbol{J}$ similarly induces a rotation operator

$$U_R = \mathrm{e}^{-\mathrm{i}/\hbar \boldsymbol{\alpha} \cdot \boldsymbol{J}} , \tag{2.10}$$

for a rotation about an angle $|\boldsymbol{\alpha}|$ around an axis defined by a unit vector $\mathbf{1}_n$ (Flügge 1974).

There thus exists a close relationship between the angular momentum operators (generators) and the rotations of a function field. Using some simple group theoretical elements this can even be more transparent (Wigner 1959, Hamermesh 1962, Goldstein 1980, Gilmore 1974).

The group $O(3)$ describes the orthogonal transformations in three-dimensional space. In defining a group structure we need the following rules for its elements to hold

$ab = c$	(product rule)
$(ab)c = a(bc)$	(associativity)
$ea = ae = a$	(a unit element e exists)
$aa^{-1} = a^{-1}a = 1$	(the inverse element a^{-1} exists) .

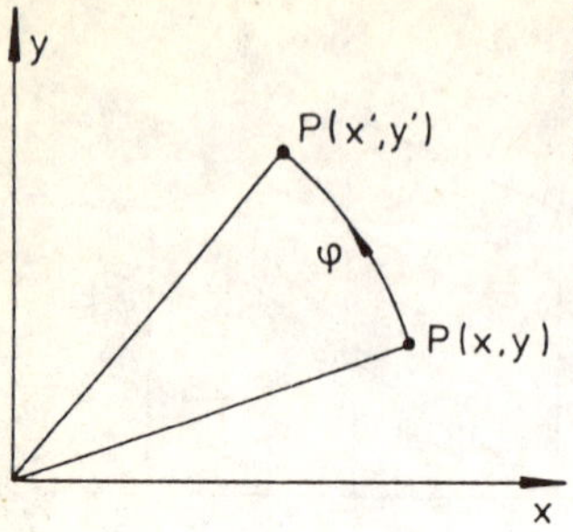

Fig. 2.3. Two-dimensional rotation group, characterized by one parameter, the rotation angle φ which transforms a point $P(x, y)$ into the point $P'(x', y')$. This operation characterizes the S0(2) rotation group

We first consider the $O(2)$ group $(0 \leq \varphi \leq 2\pi)$. With each point $P(x, y)$, after rotation new coordinates will correspond to the point $P(x', y')$ (Fig. 2.3) or

$$\begin{aligned} x' &= a_{11}x + a_{12}y \, , \\ y' &= a_{21}x + a_{22}y \, . \end{aligned} \tag{2.11}$$

Invariance of the length during rotation imposes

$$\begin{aligned} x'^2 + y'^2 &= \left(a_{11}^2 + a_{21}^2\right)x^2 + \left(a_{12}^2 + a_{22}^2\right)y^2 + 2\left(a_{11}a_{12} + a_{21}a_{22}\right)xy \\ &= x^2 + y^2 \, . \end{aligned} \tag{2.12}$$

Thus it follows that

$$a_{11}^2 + a_{21}^2 = 1 \, , \quad a_{12}^2 + a_{22}^2 = 1 \, , \quad a_{11}a_{12} + a_{21}a_{22} = 0 \, , \tag{2.13}$$

with, as a solution,

$$\begin{aligned} x' &= x \cos\varphi - y \sin\varphi \\ y' &= x \sin\varphi + y \cos\varphi \, . \end{aligned} \tag{2.14}$$

For the three-dimensional rotation group $O(3)$, a similar method can be used. Then, $\underline{9}$ parameters a_{ij} show up with $\underline{6}$ conditions, leaving $\underline{3}$ free parameters that can be chosen as the Euler-angles specifying the rotation $\theta \equiv (\theta_1, \theta_2, \theta_3)$ (Goldstein 1980). In general, for the rotation group $O(n)$ one has $r = \frac{1}{2}n(n-1)$ independent parameters.

For the rotation group $O(3)$, where the orbital angular momentum $\boldsymbol{L}(L_x, L_y, L_z)$ generates the rotations, the commutation relations

$$\begin{aligned} &\left[L_x, L_y\right] = i\hbar L_z \qquad \text{(and cyclic permutations)} \, , \\ &\left[\boldsymbol{L}^2, L_i\right] = 0 \, , \end{aligned} \tag{2.15}$$

hold. These operators obey the structure of a Lie algebra for which one generally has

$$\left[X_a, X_b\right] = \sum_c C_{ab}^c X_c \, , \tag{2.16}$$

with X_a, X_b, and X_c the generators of the Lie algebra and C_{ab}^c the structure constants. If all $C_{ab}^c = 0$ for *all* a, b, c one obtains an Abelian Lie algebra.

Invariant operators, (also called Casimir operators or Casimir invariants of the Lie-group) satisfy the condition

$$[C, X_a] = 0 \quad \text{for all} \quad a \; . \tag{2.17}$$

The rotation group in three-dimensional space generated by $\boldsymbol{L}$, forms a Lie group with $\boldsymbol{L}^2$ as the Casimir invariant operator [see (Iachello 1980, 1983) for an elementary discussion].

2.2 General Groups of Transformations

According to the definitions in Sect. 2.1, a group of general transformations can be defined. The set of $n \times n$ square matrices A also form a group under certain conditions:

- the set is closed,
- the rules of matrix multiplication guarantee associativity,
- the identity is the $n \times n$ unit matrix 1,
- the matrices are non-singular and thus A^{-1} exists.

Matrix groups can be finite or infinite, discrete or continuous and can be defined over the field of real (R), complex (C), ... numbers. Variables in the real field will be denoted by $\boldsymbol{x} \equiv (x_1, x_2, \ldots, x_i, \ldots, x_n)$ and in the complex field by $Z \equiv (z_1, z_2, \ldots, z_i, \ldots, z_n)$.

In the list below, we summarize the main matrix properties related to the original matrix A:

$A = \tilde{A}$	symmetric ,	$A = -A^*$	imaginary ,
$A = -\tilde{A}$	skew symmetric ,	$A = A^+$	hermitian ,
$\tilde{A}A = 1$	orthogonal ,	$A = -A^+$	skew hermitian ,
$A = A^*$	real ,	$A^+A = 1$	unitary .

Continuous groups are among the more important groups with many applications in physics (Iachello 1983).

i) The general linear group $GL(n, C)$ defines the most general linear transformation, characterized by $2n^2$ real parameters. If we restrict ourselves to real transformations we get $GL(n, R)$ with only n^2 elements and clearly

$$GL(n, C) \supset GL(n, R) \; . \tag{2.18}$$

ii) The special linear group of transformations has the extra condition that $\det|A| = +1$, and so we obtain $SL(n, C)$ with $2n^2 - 2$ parameters (real). For the real group, we get $SL(n, R)$ with $n^2 - 1$ parameters and

$$GL(n, C) \supset SL(n, C) \supset SL(n, R) \; . \tag{2.19}$$

iii) The unitary group $U(n, C)$ is a linear, complex transformation that keeps the "length" $\sum |z_i|^2$ invariant. The condition is also $A^+A = 1$ or

$$\sum_k a_{ik}a^*_{jk} = \delta_{ij} , \tag{2.20}$$

and hence $|a_{i,j}|^2 \leq 1$. The domain of n^2 parameters for $U(n)$ is bounded and closed (compact group). One could also define a transformation that leaves the quantity

$$-\sum_{i=1}^{p} |z_i|^2 + \sum_{p+1}^{p+q} |z_i|^2 = \text{invariant} , \tag{2.21}$$

defining a group $U(p, q)$. This group is non-compact. We clearly have

$$\begin{aligned} &GL(p+q, C) \supset U(p, q) , \\ &GL(n, C) \supset U(n) . \end{aligned} \tag{2.22}$$

iv) The special unitary groups now have, relative to iii), also det$|A| = +1$, leaving $n^2 - 1$ parameters for $SU(n)$. We can, similarly, construct the special unitary $SU(p, q)$ groups.

v) Orthogonal groups form an $n(n-1)$ parameter group that leave $\sum z_i^2$ invariant ($A\tilde{A} = 1$) and are denoted by $O(n, C)$ and have det$|A| = \pm 1$. The real, orthogonal groups $O(n, R)$ leave the quantity $\sum x_i^2$ invariant, and the particular subset with det$A = +1$ are the special orthogonal transformations $SO(n, R)$ or, in short, $SO(n)$. As with the unitary groups we can also define the non-compact orthogonal groups $SO(p, q)$.

We will not discuss the symplectic group.

As an example of a non-compact group, we discuss the 1 + 1 dimensional Lorentz group (Iachello 1983)

$$\begin{aligned} x' &= \gamma x - \gamma\beta(ct) \\ ct' &= -\gamma\beta x + \gamma(ct) , \end{aligned} \tag{2.23}$$

with

$$\beta = v/c \quad \text{and} \quad \gamma = \left[1 - (v/c)^2\right]^{-1/2} . \tag{2.24}$$

Here, the invariant quantity is

$$x^2 - c^2t^2 = x'^2 - c^2t'^2 . \tag{2.25}$$

The range of parameters is

$$\begin{aligned} &1 < \gamma < \infty \\ &-\infty < \beta\gamma < +\infty \\ &\gamma^2 - \gamma^2\beta^2 = 1 \end{aligned} \tag{2.26}$$

and the transformation (2 × 2) matrix A is expressed as (Fig. 2.4)

$$A = \begin{pmatrix} \cosh\theta & -\sinh\theta \\ -\sinh\theta & \cosh\theta \end{pmatrix} \quad [\text{with } \theta = \text{argth}(v/c)] . \tag{2.27}$$

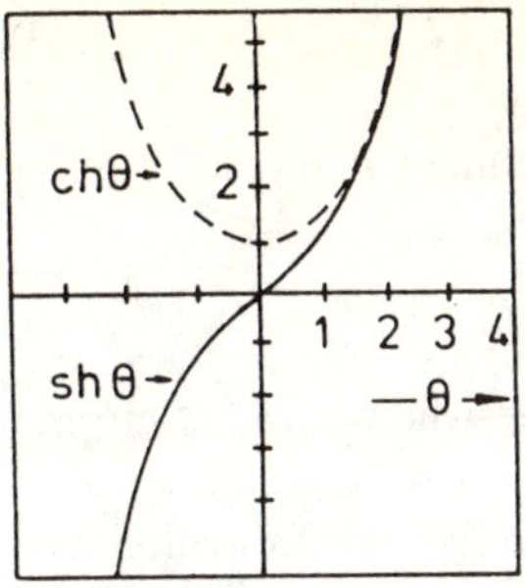

Fig. 2.4. Illustration of the variation of the elements of the transformation matrix [A] in (2.27), describing the 0(1,1) group. The $\cosh\theta$ and $\sinh\theta$ functions are illustrated

2.3 Representations of the Rotation Operator

2.3.1 The Wigner D-Matrices

Representations of the general rotation operator of (2.10) are formed by a set of square $n \times n$ matrices that follow the same group rules as the rotation operator U_R (Wigner 1959). A very well known representation is formed by the matrix elements of U_R in the basis spanned by the eigenvectors of $\boldsymbol{J}^2$ and Jz, i.e., the states $|j,m\rangle$.

The new, rotated state vector obtained by acting on $|j,m\rangle$ with the operator is called $|j,m\rangle'$, or

$$|j,m\rangle' = U_R|j,m\rangle \ . \tag{2.28}$$

Inserting the full set of basis states $|j,m'\rangle$, we get the result that

$$|j,m\rangle' = \sum_{m'} |j,m'\rangle\langle j,m'|U_R|j,m\rangle \ . \tag{2.29}$$

The $(2j+1)\times(2j+1)$ matrices $\langle j,m'|\mathrm{e}^{-\mathrm{i}/\hbar\alpha\cdot \boldsymbol{J}}|j,m\rangle$ are called the representation matrices of the rotation operator U_R, and are denoted by $D^{(j)}_{m',m}(R)$, the Wigner D-matrices.

$$D^{(j)}_{m',m}(R) \equiv \langle j,m'|\mathrm{e}^{-\mathrm{i}/\hbar\alpha\cdot \boldsymbol{J}}|j,m\rangle \ . \tag{2.30}$$

For a rotation around the z-axis, the D-matrix reduces to a diagonal matrix,

$$D^{(j)}_{m',m}(R) = \mathrm{e}^{-im\alpha}\delta_{m',m} \ . \tag{2.31}$$

The representation matrices now form a group with the same group structure as the rotation group $SO(3)$.

As an example, we can obtain the transformation properties of the spherical harmonics $Y_l^m(\theta,\varphi)$. By using the same method as in (2.29) we obtain

$$\left(Y_l^m(\theta,\varphi)\right)' = \sum_{m'} D^{(l)}_{m',m}(R) Y_l^{m'}(\theta,\varphi) \ . \tag{2.32}$$

Since the new functions $(Y_l^m(\theta,\varphi))'$ will be identical with $Y_l^m(R^{-1}\boldsymbol{r})$, we can write that

$$Y_l^m(\theta',\varphi') = \sum_{m'} D^{(l)}_{m',m}(R) Y_l^{m'}(\theta,\varphi)\,, \tag{2.33}$$

and the inverse transformation (since the D-matrices are unitary) reads

$$Y_l^m(\theta,\varphi) = \sum_{m'} D^{(l)^*}_{m,m'}(R) Y_l^{m'}(\theta',\varphi')\,. \tag{2.34}$$

Here (θ',φ') are the coordinates corresponding to $R^{-1}\boldsymbol{r}$ whereas (θ,φ) are the coordinates corresponding to $\boldsymbol{r}$.

One can work this out for $\boldsymbol{Y}_1$ and thereby determine the transformation matrices for the coordinates (x,y,z) of a given point $P(x,y,z)$ using the form of the $D^{(1)}(R)$ Wigner matrices.

2.3.2 The Group $SU(2)$-Relation with $SO(3)$

According to Sect. 2.2, the group of complex transformations that leave the expression $|u|^2 + |v|^2$ invariant

$$\begin{aligned} u' &= a_{11}u + a_{12}v \\ v' &= a_{21}u + a_{22}v\,, \end{aligned} \tag{2.35}$$

form a $U(2)$ group. The transformation has 8 real parameters. In matrix form (2.35) can be rewritten

$$A = \begin{pmatrix} a_{11} & a_{12} \\ a_{21} & a_{22} \end{pmatrix}\,, \quad A^+ = \begin{pmatrix} a_{11}^* & a_{21}^* \\ a_{12}^* & a_{22}^* \end{pmatrix}\,. \tag{2.36}$$

The conditions $A^+A = 1$ written in explicit form lead to the conditions

$$\begin{aligned} \begin{pmatrix} a_{11}^* & a_{21}^* \\ a_{12}^* & a_{22}^* \end{pmatrix} \begin{pmatrix} a_{11} & a_{12} \\ a_{21} & a_{22} \end{pmatrix} &= \begin{pmatrix} a_{11}^*a_{11} + a_{21}^*a_{21} & a_{11}^*a_{12} + a_{21}^*a_{22} \\ a_{12}^*a_{11} + a_{22}^*a_{21} & a_{12}^*a_{12} + a_{22}^*a_{22} \end{pmatrix} \\ &= \begin{pmatrix} 1 & 0 \\ 0 & 1 \end{pmatrix}\,, \end{aligned} \tag{2.37}$$

or

$$\begin{aligned} a_{11}^*a_{11} + a_{21}^*a_{21} &= 1\,, \\ a_{11}^*a_{12} + a_{21}^*a_{22} &= 0\,, \\ a_{12}^*a_{11} + a_{22}^*a_{21} &= 0\,, \\ a_{12}^*a_{12} + a_{22}^*a_{22} &= 1\,. \end{aligned} \tag{2.38}$$

This forms the $U(2)$ group. Now, 4 conditions reduce the set of 8 real parameters to 4. A further condition $\det A = 1$ gives the special group $SU(2)$ with

$$a_{11}a_{22} - a_{12}a_{21} = 1\,. \tag{2.39}$$

Now, only 3 real parameters remain. The transformation acts on the state vectors in an abstract space of spinors. If we use the notation

$$\tilde{u}' = \begin{pmatrix} u' \\ v' \end{pmatrix}\;; \quad \tilde{u} = \begin{pmatrix} u \\ v \end{pmatrix}\,, \tag{2.40}$$

then one has

$$\tilde{u}' = A\tilde{u} . \quad (2.41)$$

The $SU(2)$ 3-parameter group and the parameters of the $SO(3)$ group are related. This relation can be illustrated in some more detail. Using the substitution

$$x_1 = u^2 , \quad x_2 = uv , \quad x_3 = v^2 , \quad (2.42)$$

we study how (x_1, x_2, x_3) transform under the $SU(2)$ transformation matrix A. The conditions (2.38, 39) lead to a simplified notation for A as

$$A = \begin{pmatrix} a_{11} & a_{12} \\ -a_{12}^* & a_{11}^* \end{pmatrix} . \quad (2.43)$$

The transformation for (x_1, x_2, x_3) becomes

$$\begin{aligned} x_1' &= u'^2 = a_{11}^2 x_1 + 2a_{11}a_{12}x_2 + a_{12}^2 x_3 \\ x_2' &= u'v' = -a_{11}a_{12}^* x_1 + (a_{11}a_{11}^* - a_{12}a_{12}^*)x_2 + a_{11}^* a_{12}^* x_3 \\ x_3' &= v'^2 = a_{12}^{*2} x_1 - 2a_{11}^* a_{12}^* x_2 + a_{11}^{*2} x_3 . \end{aligned} \quad (2.44)$$

Using the linear combinations

$$x = (x_1 - x_3)/2 ; \quad y = (x_1 + x_3)/2i ; \quad z = x_2 , \quad (2.45)$$

the $SU(2)$ transformation induces the following transformation of the coordinates of the point $P(x, y, z) \to P(x', y', z')$

$$\begin{aligned} x' &= \frac{1}{2}\left(a_{11}^2 - a_{12}^{*2} - a_{12}^2 + a_{11}^{*2}\right)x + \frac{\mathrm{i}}{2}\left(a_{11}^2 - a_{12}^{*2} + a_{12}^2 - a_{11}^{*2}\right)y \\ &\quad + \left(a_{11}a_{12} + a_{11}^* a_{12}^*\right)z , \\ y' &= -\frac{\mathrm{i}}{2}\left(a_{11}^2 + a_{12}^{*2} - a_{12}^2 - a_{11}^{*2}\right)x + \frac{1}{2}\left(a_{11}^2 + a_{12}^{*2} + a_{12}^2 + a_{11}^{*2}\right)y \\ &\quad - \mathrm{i}\left(a_{11}a_{12} - a_{11}^* a_{12}^*\right)z , \\ z' &= -\left(a_{11}^* a_{12} + a_{11}a_{12}^*\right)x + \mathrm{i}\left(a_{11}^* a_{12} - a_{11}a_{12}^*\right)y + \left(a_{11}a_{11}^* - a_{12}a_{12}^*\right)z , \end{aligned}$$

with

$$x'^2 + y'^2 + z'^2 = x^2 + y^2 + z^2 . \quad (2.46)$$

Therefore we can construct for any $SU(2)$ matrix A, a corresponding $SO(3)$ matrix B.

We consider some special cases:

i) Rotation $R(\alpha, 0, 0)$ around the z-axis, corresponding to a diagonal $SU(2)$ matrix with $a_{12} = a_{12}^* = 0$:

$$\begin{pmatrix} e^{i\alpha/2} & 0 \\ 0 & e^{-i\alpha/2} \end{pmatrix} \Rightarrow \begin{pmatrix} \cos\alpha & -\sin\alpha & 0 \\ \sin\alpha & \cos\alpha & 0 \\ 0 & 0 & 1 \end{pmatrix} . \tag{2.47}$$

ii) Rotation $R(0,\beta,0)$ through β around the y-axis, corresponding to a choice of a real $SU(2)$ matrix:

$$\begin{pmatrix} \cos\beta/2 & \sin\beta/2 \\ -\sin\beta/2 & \cos\beta/2 \end{pmatrix} \Rightarrow \begin{pmatrix} \cos\beta & 0 & \sin\beta \\ 0 & 1 & 0 \\ -\sin\beta & 0 & \cos\beta \end{pmatrix} . \tag{2.48}$$

iii) The generalized matrix $R(\alpha,\beta,\gamma)$ leads to the $SU(2)$ matrix:

$$\begin{pmatrix} \cos\beta/2\, e^{i/2(\alpha+\gamma)} & \sin\beta/2\, e^{i/2(\alpha-\gamma)} \\ -\sin\beta/2\, e^{i/2(\gamma-\alpha)} & \cos\beta/2\, e^{-i/2(\alpha+\gamma)} \end{pmatrix} , \tag{2.49}$$

and one can (tediously) construct the related $SO(3)$ matrix $B(\alpha,\beta,\gamma)$.

We use the relations

$$\begin{aligned} R(\alpha,\beta,\gamma) &= R(\alpha)R(\beta)R(\gamma) \\ SU(2;\alpha\beta\gamma) &= SU(2;\alpha)SU(2;\beta)SU(2;\gamma) \end{aligned} . \tag{2.50}$$

Consider the rotation $R(0,0,0)$ with

$$\begin{pmatrix} 1 & 0 \\ 0 & 1 \end{pmatrix} \Rightarrow 1 , \tag{2.51}$$

and the rotation $R(0,2\pi,0)$ with

$$\begin{pmatrix} -1 & 0 \\ 0 & -1 \end{pmatrix} \Rightarrow 1 . \tag{2.52}$$

Thus, there is no one-to-one correspondence between the elements of the $SU(2)$ and $SO(3)$ matrices. *Two* elements of $SU(2)$ correspond to *one* element of $SO(3)$. This correspondence is called a homomorphic mapping (Hamermesh 1962, Wybourne 1974, Parikh 1978).

2.3.3 Application: Geometric Interpretation of Intrinsic Spin $\frac{1}{2}$

Electrons and protons are spin $\frac{1}{2}$ fermions. They contain a property that was called: "... a peculiar type of double-valuedness, not describable in classical terms" (Pauli 1925). This double-valuedness was illustrated experimentally in the Stern-Gerlach experiments.

The spin eigenvectors have been represented in an abstract two-dimensional space by using column matrices with two rows. In general

$$\chi = \begin{pmatrix} \chi_+ \\ \chi_- \end{pmatrix} , \tag{2.53}$$

where χ_+ and χ_- are amplitudes, and can also be written as

$\chi = \chi_+\alpha + \chi_-\beta$, where

$$\alpha \equiv \begin{pmatrix} 1 \\ 0 \end{pmatrix}, \quad \beta \equiv \begin{pmatrix} 0 \\ 1 \end{pmatrix}. \tag{2.54}$$

We have defined the spin $\frac{1}{2}$ angular momentum operators (Pauli-matrices) as

$$s_x = (\hbar/2)\sigma_x; \quad s_y = (\hbar/2)\sigma_y; \quad s_z = (\hbar/2)\sigma_z. \tag{2.55}$$

Intrinsic spin, described by using the abstract state vectors, has no, or only limited, relation to the more intuitive notion of spin as a vector pointing in a certain direction. This "classical" correspondence to an angular momentum vector operator can be expressed in a vector model by s_x, s_y, s_z and the expectation values for the different spin $\frac{1}{2}$ eigenvectors $\langle s_x\rangle$, $\langle s_y\rangle$, $\langle s_z\rangle$.

We can now put a link between the abstract rotations and geometric picture in the following way: we calculate the expectation values of s_x, s_y, s_z, for a spin $\frac{1}{2}$ eigenvector rotated over a general set of angles (α, β, γ) expressed via the D-matrices. One then has

$$|\tfrac{1}{2}, m_s\rangle' = \sum_{m_s'} D^{(1/2)}_{m_s', m_s}(R)|\tfrac{1}{2}, m_s'\rangle, \tag{2.56}$$

with

$$D^{(1/2)}_{m_s', m_s} = \begin{pmatrix} \cos\beta/2\,\mathrm{e}^{-\mathrm{i}/2(\alpha+\gamma)} & -\sin\beta/2\,\mathrm{e}^{-\mathrm{i}/2(\alpha-\gamma)} \\ \sin\beta/2\,\mathrm{e}^{-\mathrm{i}/2(\gamma-\alpha)} & \cos\beta/2\,\mathrm{e}^{\mathrm{i}/2(\alpha+\gamma)} \end{pmatrix}. \tag{2.57}$$

The calculation leads to

$$\begin{aligned} \langle s_x\rangle_{1/2,1/2} &= (\hbar/2)\sin\beta\cos\alpha \\ \langle s_y\rangle_{1/2,1/2} &= (\hbar/2)\sin\beta\sin\alpha \\ \langle s_z\rangle_{1/2,1/2} &= (\hbar/2)\cos\beta, \end{aligned} \tag{2.58}$$

and

$$\begin{aligned} \langle s_x\rangle_{1/2,-1/2} &= -(\hbar/2)\sin\beta\cos\alpha \\ \langle s_y\rangle_{1/2,-1/2} &= -(\hbar/2)\sin\beta\sin\alpha \\ \langle s_z\rangle_{1/2,-1/2} &= -(\hbar/2)\cos\beta. \end{aligned} \tag{2.59}$$

Thus, we find the "classical" vector via the expectation values of the components of the spin $\frac{1}{2}$ angular momentum operators in the spin $\frac{1}{2}$ and $-\frac{1}{2}$ eigenstates (Fig. 2.5).

For the particular choice $\alpha = \gamma = 0°$,

$$\begin{aligned} |\tfrac{1}{2}, \tfrac{1}{2}\rangle' &= \cos\beta/2|\tfrac{1}{2}, \tfrac{1}{2}\rangle + \sin\beta/2|\tfrac{1}{2}, -\tfrac{1}{2}\rangle \\ |\tfrac{1}{2}, -\tfrac{1}{2}\rangle' &= -\sin\beta/2|\tfrac{1}{2}, \tfrac{1}{2}\rangle + \cos\beta/2|\tfrac{1}{2}, -\tfrac{1}{2}\rangle. \end{aligned} \tag{2.60}$$

For $\beta = 2\pi$, one has

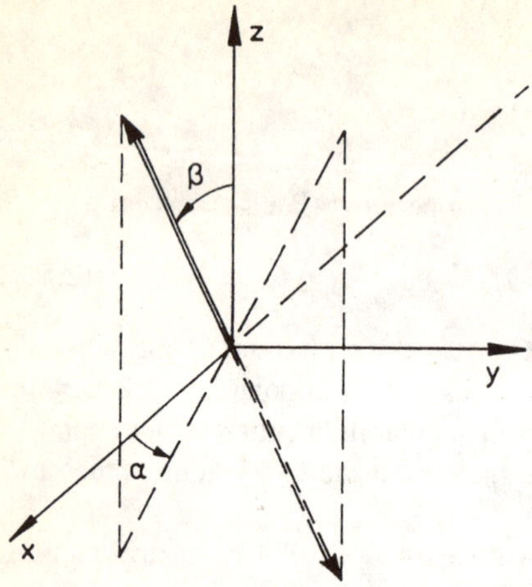

Fig. 2.5. Semi-classical representation of the spin vector in a cartesian system. The classical vector is given in the two possible states (double arrow) corresponding to the two possible orientations of the spin $1/2$ angular momentum vector. The projections on the x-, y- and z-axis given in terms of the rotation angles α, β are given by (2.58) and (2.59), α denoting a rotation around the z-axis, β a rotation around the y-axis

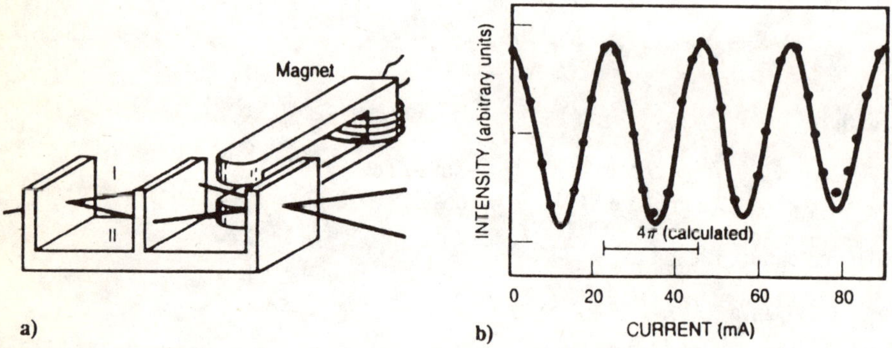

Fig. 2.6. Demonstration of spinor rotation for a spin $1/2$ system: (**a**) Schematic diagram of a neutron interferometer. The phase shift in path I is produced by inserting a magnetic field as shown. (**b**) The magnetic field induces a rotation of the neutron spin (wave function) and changes its phase, a rotation of 4π produces a phase shift of 2π [from (Werner 1980)]

$$\begin{aligned} |\tfrac{1}{2}, \tfrac{1}{2}\rangle' &= -|\tfrac{1}{2}, \tfrac{1}{2}\rangle \\ |\tfrac{1}{2}, -\tfrac{1}{2}\rangle' &= -|\tfrac{1}{2}, -\tfrac{1}{2}\rangle \,, \end{aligned} \tag{2.61}$$

and only for $\beta = 4\pi$ does the state coincide with the original state. This phase change has been observed experimentally in neutron interferometry [Fig. 2.6 (Werner 1980)].

2.4 Product Representations and Irreducibility

In Sect. 2.3, we have constructed the Wigner $D^{(j)}_{m',m}(R)$ representation matrices for a general rotation of the system $R(\alpha,\beta,\gamma)$. These D-matrices transform the eigenstates $|j,m\rangle$ under a rotation of the physical system. Thus we can write for the transformed eigenstates of a system $|j_1,m_1\rangle$

$$|j_1,m_1\rangle' = \sum_{m'_1} D^{(j_1)}_{m'_1,m_1}(R)|j_1,m'_1\rangle \,. \tag{2.62}$$

A second set of eigenstates for system 2 transform according to:

$$|j_2,m_2\rangle' = \sum_{m'_2} D^{(j_2)}_{m'_2,m_2}(R)|j_2,m'_2\rangle \,. \tag{2.63}$$

The product states $|j_1m_1\rangle|j_2m_2\rangle$ then transform as

$$\big(|j_1m_1\rangle|j_2m_2\rangle\big)' = \sum_{m'_1,m'_2} D^{(j_1)}_{m'_1,m_1}(R)D^{(j_2)}_{m'_2,m_2}(R)|j_1,m'_1\rangle|j_2,m'_2\rangle \tag{2.64}$$

$$= \sum_{m'_1,m'_2} D^{(j_1\times j_2)}_{m'_1m'_2;m_1m_2}(R)|j_1,m'_1\rangle|j_2,m'_2\rangle \,, \tag{2.65}$$

where $D^{(j_1\times j_2)}(R)$ is the product matrix describing the product representations. It is a direct product of D-matrices, also denoted by the notation $A\otimes B$ (Hamermesh 1962).

We will work this out in more detail for the particular situation of two spin $\frac{1}{2}$ particles: the eigenvectors are

$$\begin{pmatrix} u_1 \\ v_1 \end{pmatrix} \quad \text{and} \quad \begin{pmatrix} u_2 \\ v_2 \end{pmatrix} ,$$

respectively, and the product states are

$$\begin{pmatrix} u_1u_2 \\ v_1u_2 \\ u_1v_2 \\ v_1v_2 \end{pmatrix} \equiv \begin{pmatrix} |\frac{1}{2},\frac{1}{2}\rangle & |\frac{1}{2},\frac{1}{2}\rangle \\ |\frac{1}{2},-\frac{1}{2}\rangle & |\frac{1}{2},\frac{1}{2}\rangle \\ |\frac{1}{2},\frac{1}{2}\rangle & |\frac{1}{2},-\frac{1}{2}\rangle \\ |\frac{1}{2},-\frac{1}{2}\rangle & |\frac{1}{2},-\frac{1}{2}\rangle \end{pmatrix} \,. \tag{2.66}$$

The product transformation matrix $D^{(1/2)}\otimes D^{(1/2)}$ becomes

$$D^{(1/2)}\otimes D^{(1/2)} = \begin{pmatrix} a_{11}a_{11} & a_{11}a_{12} & a_{12}a_{11} & a_{12}a_{12} \\ a_{11}a_{21} & a_{11}a_{22} & a_{12}a_{21} & a_{12}a_{22} \\ a_{21}a_{11} & a_{21}a_{12} & a_{22}a_{11} & a_{22}a_{12} \\ a_{21}a_{21} & a_{21}a_{22} & a_{22}a_{21} & a_{22}a_{22} \end{pmatrix} \,. \tag{2.67}$$

Consider now the eigenstates

$$\begin{aligned}
\varphi_{0,0} &\equiv |\tfrac{1}{2}, \tfrac{1}{2}; 00\rangle = \frac{1}{\sqrt{2}}(u_1 v_2 - v_1 u_2) \\
\varphi_{1,1} &\equiv |\tfrac{1}{2}, \tfrac{1}{2}; 11\rangle = u_1 u_2 \\
\varphi_{1,0} &\equiv |\tfrac{1}{2}, \tfrac{1}{2}; 10\rangle = \frac{1}{\sqrt{2}}(u_1 v_2 + v_1 u_2) \\
\varphi_{1,-1} &\equiv |\tfrac{1}{2}, \tfrac{1}{2}; 1-1\rangle = v_1 v_2 \,.
\end{aligned} \tag{2.68}$$

The transformation of the eigenvectors will also modify the product transformation matrix. In shorthand notation: if $\bar{\varphi}$ describes the new basis and Z is the matrix describing the transformation (2.68), we have

$$\bar{\varphi} = Z \cdot \varphi \,, \tag{2.69}$$

and the general transformation (2.65)

$$\varphi' = A \cdot \varphi \,, \tag{2.70}$$

can be rewritten in the new basis as

$$Z\varphi' = ZAZ^{-1}Z \cdot \varphi \,, \tag{2.71}$$

or

$$\bar{\varphi}' = (Z \cdot A \cdot Z^{-1})\bar{\varphi} \,. \tag{2.72}$$

In the particular situation of two spin $\frac{1}{2}$ particles, via (2.68) the matrix Z becomes

$$Z = \begin{pmatrix} 0 & -\frac{1}{\sqrt{2}} & \frac{1}{\sqrt{2}} & 0 \\ 1 & 0 & 0 & 0 \\ 0 & \frac{1}{\sqrt{2}} & \frac{1}{\sqrt{2}} & 0 \\ 0 & 0 & 0 & 1 \end{pmatrix} . \tag{2.73}$$

The product matrix ZAZ^{-1} becomes (in the new basis) in shorthand notation

$$\begin{pmatrix} \varphi_{0,0} \\ \varphi_{1,1} \\ \varphi_{1,0} \\ \varphi_{1,-1} \end{pmatrix}' = \left(\begin{array}{c|ccc} b & 0 & 0 & 0 \\ \hline 0 & c_{1,1} & c_{1,0} & c_{1,-1} \\ 0 & c_{0,1} & c_{0,0} & c_{0,-1} \\ 0 & c_{-1,1} & c_{-1,0} & c_{-1,-1} \end{array}\right) \begin{pmatrix} \varphi_{0,0} \\ \varphi_{1,1} \\ \varphi_{1,0} \\ \varphi_{1,-1} \end{pmatrix} . \tag{2.74}$$

The product transformation $D^{(1/2)} \otimes D^{(1/2)}$ is in this case reducible in a spin 1 and spin 0 system. In the more general case of a product of representation matrices for particles with angular momentum j_1 and j_2 one obtains

$$D^{(j_1)} \otimes D^{(j_2)} = D^{(j_1+j_2)} \oplus D^{(j_1+j_2-1)} \oplus \ldots \oplus D^{(|j_1-j_2|)} \,, \tag{2.75}$$

and the product transformation is reducible in blocks, corresponding to angular momentum $J = j_1 + j_2,\ j_1 + j_2 - 1,\ \ldots\ |j_1 - j_2|$, and can be depicted diagrammatically as in Fig. 2.7.

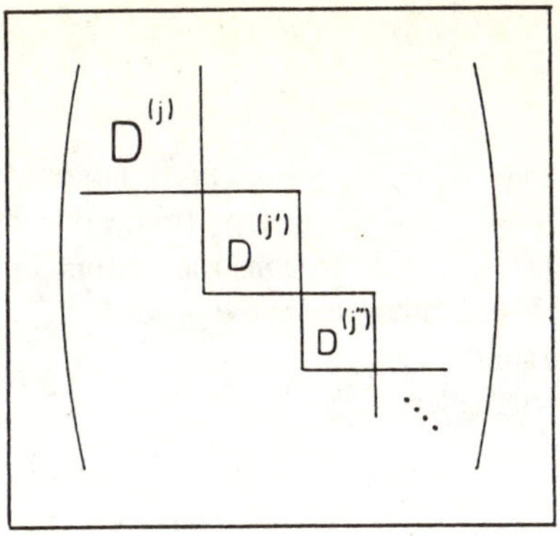

Fig. 2.7. Diagrammatic illustration of the reducible parts of the transformation Wigner D-matrix for the product system of the angular momentum eigenvectors $|j_1m_1\rangle|j_2m_2\rangle$. The blocks present the different irreducible parts, corresponding to the momenta j with $|j_1 - j_2| \leq j \leq j_1 + j_2$

The detailed product elements $D^{(j)}_{m',m}(R)$ are related to the separate elements of the separate D-matrices via the Clebsch-Gordan series

$$D^{(j_1)}_{m_1',m_1}(R)D^{(j_2)}_{m_2',m_2}(R) = \sum_{j=|j_1-j_2|}^{j_1+j_2} \sum_{m',m} \langle j_1m_1, j_2m_2|jm\rangle \times \langle j_1m_1', j_2m_2'|jm'\rangle D^{(j)}_{m',m}(R) \, . \tag{2.76}$$

A proof of (2.76) can be found in Hamermesh 1962, Edmonds 1957.

2.5 Cartesian Tensors, Spherical Tensors, Irreducible Tensors

In considering a vector $\boldsymbol{X}(x_1, x_2, x_3)$ with the above cartesian coordinates, under a genuine rotation of the vector the new coordinates can be expressed as

$$\boldsymbol{X}' = [A]\boldsymbol{X} \quad \text{or} \quad x_i' = \sum_k A_{i,k}x_k \, . \tag{2.77}$$

The components $A_{i,k}$ constitute a real, orthogonal matrix with determinant +1. If we now consider two vectors $\boldsymbol{X}(x_1, x_2, x_3)$ and $\boldsymbol{Y}(y_1, y_2, y_3)$, under a rotation of the vectors the product quantities $x_k y_l$ will transform like

$$x_i'y_j' = \sum_{k,l} A_{i,k}A_{j,l}x_k y_l \, . \tag{2.78}$$

It is also possible to have a non-separable quantity F_{kl} that transforms like the product $x_k y_l$, or

$$F_{ij}' = \sum_{k,l} A_{i,k}A_{j,l}F_{kl} \, . \tag{2.79}$$

The quantities $x_k y_l$, F_{kl}, ... transform like the components of a cartesian tensor of rank $\underline{2}$. A more general tensor (cartesian) of rank r will then transform as

$$\underbrace{F'_{i,j,k,\dots}}_{r} = \underbrace{\sum A_{i,l} A_{j,m} A_{k,n,\dots}}_{r} \underbrace{F_{l,m,n,\dots}}_{r} . \tag{2.80}$$

In general, the object F is reducible with respect to the group of orthogonal transformations $SO(3)$, indicating that it is possible by using appropriate linear combinations of the cartesian tensor components $F_{l,m,n,\dots}$, to construct a number of contributions that transform independently of each other under the group $SO(3)$ (Brink, Satchler 1962, Rose, Brink 1967, Weissbluth 1978).

For a cartesian tensor of rank 2, we can write (identity)

$$F_{i,k} \equiv S^{(0)}_{i,k} + A_{i,k} + \tau_{i,k} , \tag{2.81}$$

with

$$\tau_{i,k} \equiv \tfrac{1}{3}\delta_{i,k}\mathrm{Tr}(F_{i,k}) , \tag{2.82}$$

$$A_{i,k} \equiv \tfrac{1}{2}\left(F_{i,k} - F_{k,i}\right) , \tag{2.83}$$

$$S^{(0)}_{i,k} = \tfrac{1}{2}\left(F_{i,k} + F_{k,i}\right) - \tau_{i,k} , \tag{2.84}$$

with $\tau_{i,k}$ an invariant quantity, $A_{i,k}$ the components of a vector and $S^{(0)}_{i,k}$ the five components of a symmetric tensor of rank 2.

If, as an example, we take the 9 components $x_i x_j$ $(xx, xy, \dots, zz)$, we can construct

$$\begin{aligned} \tau_{i,k} &= \tfrac{1}{3}r^2\delta_{i,k} \\ A_{i,k} &= \tfrac{1}{2}(yz - zy) , \dots \\ S^{(0)}_{i,k} &= \tfrac{1}{2}(yz + zy) - \tfrac{1}{3}r^2\delta_{i,k} , \dots , \end{aligned} \tag{2.85}$$

where $\tau_{i,k}$ is proportional to the length of the scalar product $\boldsymbol{r} \cdot \boldsymbol{r}$, $A_{i,k}$ form the three independent components of the vector product $\boldsymbol{r} \times \boldsymbol{r}$ and the $S^{(0)}_{i,k}$ form the five independent components of the symmetric tensor $\boldsymbol{rr} - \frac{1}{3}r^2\delta_{i,k}$ with zero diagonal sum. It can be shown that in this example the quantities $\tau_{i,k}$, $A_{i,k}$ and $S^{(0)}_{i,k}$ transform under rotation according to the spherical harmonics of order 0, 1 and 2, i.e., Y_0, Y_1^{μ} and Y_2^{μ}, respectively (Weissbluth 1978).

The quantities $\tau_{i,k}$, $A_{i,k}$ and $S^{(0)}_{i,k}$ transform among themselves (see e.g. Weissbluth 1978) and are therefore called "irreducible" tensors of rank 0, 1 and 2. In general, an irreducible spherical tensor of rank k is a set of $2k+1$ independent components $T^{(k)}_{\kappa}$ (with $-k \le \kappa \le +k$) that transform under rotation as the spherical harmonics of the same rank Y_k^{κ} do. Thus we obtain the transformation law

$$T^{(k)\prime}_{\kappa} = \sum_{\kappa'} D^{(k)}_{\kappa',\kappa}(R) T^{(k)}_{\kappa'} . \tag{2.86}$$

The quantities $S^{(k)}_{\kappa} \equiv (-1)^{\kappa} T^{(k)+}_{-\kappa}$ also form an irreducible tensor of rank k (Brink, Satchler 1962, Edmonds 1957), a fact that can be proved by using the relation for the D-matrices

$$D^{(k)*}_{\kappa',\kappa}(R) = (-1)^{\kappa-\kappa'} D^{(k)}_{-\kappa',-\kappa}(R) \, . \tag{2.87}$$

For example, for the radius vector $\boldsymbol{r}$ of a point P we have the cartesian coordinates $\boldsymbol{r}(x, y, z)$ that form a cartesian tensor of rank 1. The spherical components are $\boldsymbol{r}(r_{+1}, r_{-1}, r_0)$, with

$$\begin{aligned} r_{+1} &= -\frac{1}{\sqrt{2}}(x + \mathrm{i}y) \\ r_{-1} &= \frac{1}{\sqrt{2}}(x - \mathrm{i}y) \\ r_0 &= z \, , \end{aligned} \tag{2.88}$$

form a spherical tensor of rank 1 and are proportional to Y_1^{+1}, Y_1^{-1} and Y_1^0, respectively.

2.6 Tensor Product

By using the Clebsch-Gordan coefficients just as in angular momentum coupling, spherical tensors of higher rank can be constructed in a systematic way. This "building"-up process goes as (Brink, Satchler 1962, Edmonds 1957, de-Shalit, Talmi 1963)

$$T^{(k_3)}_{\kappa_3} = \sum_{\kappa_1,\kappa_2} \langle k_1\kappa_1, k_2\kappa_2 | k_3\kappa_3 \rangle T^{(k_1)}_{\kappa_1} T^{(k_2)}_{\kappa_2} \, , \tag{2.89}$$

or, in a short-hand notation

$$T^{(k_3)}_{\kappa_3} = \left[\boldsymbol{T}^{(k_1)} \otimes \boldsymbol{T}^{(k_2)} \right]^{(k_3)}_{\kappa_3} \, . \tag{2.90}$$

One can prove that the $2k_3 + 1$ components of $T^{(k_3)}_{\kappa_3}$ form the components of a spherical tensor of rank k_3 if the quantities $\boldsymbol{T}^{(k_1)}$ and $\boldsymbol{T}^{(k_2)}$ are spherical tensors of rank k_1 and k_2, respectively.

The relation (2.90) can also be inverted to give

$$T^{(k_1)}_{\kappa_1} T^{(k_2)}_{\kappa_2} = \sum_{k_3,\kappa_3} \langle k_1\kappa_1, k_2\kappa_2 | k_3\kappa_3 \rangle T^{(k_3)}_{\kappa_3} \, . \tag{2.91}$$

As an example, it is possible to show that the tensor product of the vectors $\boldsymbol{r}$ and $\boldsymbol{p}$, coupled to a tensor of rank 1, becomes proportional to the angular momentum vector $\boldsymbol{l}$. The exact relation is

$$\boldsymbol{l} \equiv -\mathrm{i}\sqrt{2} \left[\boldsymbol{r} \otimes \boldsymbol{p} \right]^{(1)} \, . \tag{2.92}$$

A particular case is the tensor product of rank 0, thereby forming a scalar (invariant) quantity. One therefore often uses the notation of a scalar product (de-Shalit, Talmi 1963)

$$\boldsymbol{T}^{(k)} \cdot \boldsymbol{U}^{(k)} \equiv (-1)^k \hat{k} \left[\boldsymbol{T}^{(k)} \otimes \boldsymbol{U}^{(k)}\right]^{(0)} . \tag{2.93}$$

By using the definition in (2.89) one then gets

$$\boldsymbol{T}^{(k)} \cdot \boldsymbol{U}^{(k)} = \sum_{\kappa} (-1)^{\kappa} T_{\kappa}^{(k)} U_{-\kappa}^{(k)} . \tag{2.94}$$

2.7 Spherical Tensor Operators: The Wigner-Eckart Theorem

Up to now, we have studied the transformation properties of wave functions, state vectors or, more generally, of tensor quantities. The operators used in quantum mechanics that relate to observables can be classified according to their transformation properties: we speak of "scalar, vector,... tensor of rank k" operators. They are the key objects that appear when calculating measurable quantities that can be expressed as matrix elements of a given spherical tensor operator acting between eigenstates of the angular momentum $\boldsymbol{J}^2$ and J_z:

$$\langle \alpha' j' m' | T_{\kappa}^{(k)} | \alpha j m \rangle = \int \psi^*_{\alpha' j' m'}(\boldsymbol{x}) T_{\kappa}^{(k)}(\boldsymbol{x}) \psi_{\alpha j m}(\boldsymbol{x})\, d\boldsymbol{x} , \tag{2.95}$$

where at the right hand side, $\boldsymbol{x}$ denotes *all* coordinates that are present in the wave function and in the operator.

We now consider the general state vector $|\psi\rangle$. Under a rotation (characterized by the rotation operator U_R), the new state vector will be given by $|\psi\rangle'$, via

$$|\psi\rangle' = U_R |\psi\rangle .$$

The operator $\boldsymbol{A}(A_i)$ is now called a vector operator if the matrix elements transform as the components of a vector, i.e.,

$$\begin{aligned} {}'\langle \psi | A_i | \psi \rangle' &= \sum_j R_{i,j} \langle \psi | A_j | \psi \rangle , \quad \text{or} \\ \langle \psi | U_R^{+} A_i U_R | \psi \rangle &= \sum_j R_{i,j} \langle \psi | A_j | \psi \rangle . \end{aligned} \tag{2.96}$$

This should hold for any $|\psi\rangle$ vector, so we obtain, formally,

$$\begin{aligned} U_R^{+} A_i U_R &= \sum_j R_{i,j} A_j , \quad \text{or} \\ U_R A_i U_R^{+} &= \sum_j R_{j,i} A_j . \end{aligned} \tag{2.97}$$

We now call the $2k + 1$ components $T_{\kappa}^{(k)}$ a spherical tensor operator if, likewise, the components $T_{\kappa}^{(k)}$ transform according to the spherical harmonics of rank k, thus

$$T_{\kappa}^{(k)\prime} = U_R T_{\kappa}^{(k)} U_R^{+} = \sum_{\kappa'} D_{\kappa',\kappa}^{(k)}(R) T_{\kappa'}^{(k)} . \tag{2.98}$$

We can now construct the following schematic drawing, illustrating the effect of a transformation of coordinates, described through the operator U_R, on both the state vectors $|\alpha jm\rangle$ and the spherical tensor operator $T_\kappa^{(k)}$. If the action of the spherical tensor operator $T_\kappa^{(k)}$ on the state vector $|\alpha jm\rangle$ produces a new state vector $|\overline{\alpha jm}\rangle$, i.e.

$$|\overline{\alpha jm}\rangle = T_\kappa^{(k)}|\alpha jm\rangle \,, \tag{2.99}$$

then the transformed state vectors and the transformed spherical tensor operator obey a similar relation, i.e.

$$|\alpha jm\rangle' = T_\kappa^{(k)\prime}|\alpha jm\rangle' \,. \tag{2.100}$$

The various relations (2.99), (2.100) are depicted in Fig. 2.8.

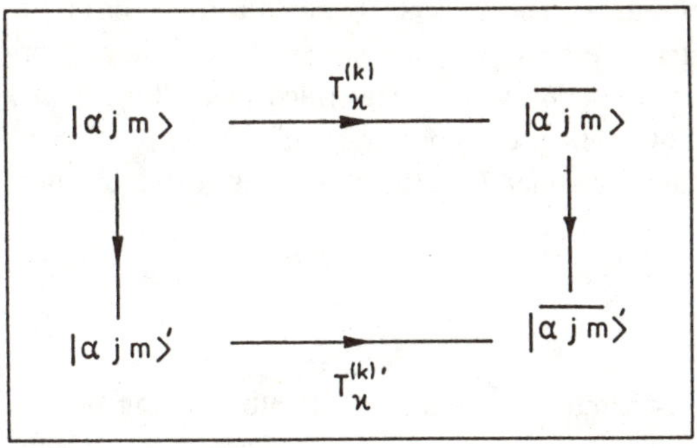

Fig. 2.8. Relations between the effects of acting on an angular momentum state vector $|\alpha jm\rangle$ with a spherical tensor operator $T_\kappa^{(k)}$ and/or a rotation operator U_R

When calculating the matrix elements of spherical tensor operator components $T_\kappa^{(k)}$, the Wigner-Eckart theorem allows a separation in the part that only depends on the projection quantum numbers (called the geometrical part) and another part that depends on, e.g., the radial properties (and angular momentum properties) of the operator and of the state vectors (Eckart 1930). The latter is called a *reduced* matrix element defined by

$$\langle \alpha jm|T_\kappa^{(k)}|\alpha' j' m'\rangle = (-1)^{j-m}\begin{pmatrix} j & k & j' \\ -m & \kappa & m' \end{pmatrix}\langle \alpha j\|\boldsymbol{T}^{(k)}\|\alpha' j'\rangle \,. \tag{2.101}$$

We give no proof here [see e.g. (Brussaard, Glaudemans 1977)]. Expression (2.101) can also be rewritten as

$$\langle \alpha jm|T_\kappa^{(k)}|\alpha' j' m'\rangle = (\hat{j})^{-1}(-1)^{2k}\langle j'm', k\kappa|jm\rangle\langle \alpha j\|\boldsymbol{T}^{(k)}\|\alpha' j'\rangle \,. \tag{2.102}$$

[Here, and in further discussions, we use $\hat{j}$ as a shorthand notation for $(2j+1)^{1/2}$.] We illustrate this by calculating the reduced matrix element $\langle j\|\boldsymbol{j}\|j\rangle$. We know that

$$\begin{aligned} m = \langle jm|j_z|jm\rangle &= (-1)^{j-m}\begin{pmatrix} j & 1 & j \\ -m & \kappa & m \end{pmatrix}\langle j\|\boldsymbol{j}\|j\rangle \\ &= (-1)^{2(j-m)}m\big(j(j+1)(2j+1)\big)^{-1/2}\langle j\|\boldsymbol{j}\|j\rangle \,, \quad \text{or} \end{aligned}$$

$$\langle j \| \boldsymbol{j} \| j \rangle = \left(j(j+1)(2j+1) \right)^{1/2} . \tag{2.103}$$

Likewise, one can prove that

$$\langle j \| \mathbb{1} \| j \rangle = \hat{j} , \tag{2.104}$$

where $\mathbb{1}$ is the unit-operator.

2.8 Calculation of Matrix Elements

In the general program of calculating matrix elements of spherical tensor operators, one deals with composite wave functions and operators, i.e., one handles many-body nuclear wave functions and operators symmetrized in all the participating particles. It should be possible to reduce such matrix elements to their basic entities which are the reduced matrix elements of one-body spherical tensor operators and the wave functions of a single system. The rules that allow such a reduction use techniques of angular momentum coupling and recoupling (Racah 1942a,b, 1943, 1949, 1951). We shall consider two distinct cases to which more general situations can always be reduced.

2.8.1 Reduction Rule I

We consider the tensor operator as a tensor product of operators acting on two independent subsystems

$$T_\kappa^{(k)}(1,2) = \left[\boldsymbol{T}^{(k_1)}(1) \otimes \boldsymbol{T}^{(k_2)}(2) \right]_\kappa^{(k)} , \tag{2.105}$$

denoted by 1 and 2, respectively. Here 1 stands for all coordinates characterizing subsystem 1, i.e., $\boldsymbol{r}_1$, $\boldsymbol{\sigma}_1$, ... In the same way, the state vectors will be the angular momentum coupled eigenvectors of the two subsystems denoted by $|\alpha_1 j_1, \alpha_2 j_2; JM\rangle$.

The reduced matrix element can then be reduced to the separate reduced matrix elements and some recoupling coefficient. In particular one gets

$$\begin{aligned} &\langle \alpha_1 j_1, \alpha_2 j_2; J \| \boldsymbol{T}^{(k)}(1,2) \| \alpha_1' j_1', \alpha_2' j_2'; J' \rangle \\ &= \hat{J}\hat{J}'\hat{k} \begin{Bmatrix} j_1 & j_2 & J \\ j_1' & j_2' & J' \\ k_1 & k_2 & k \end{Bmatrix} \langle \alpha_1 j_1 \| \boldsymbol{T}^{(k_1)} \| \alpha_1' j_1' \rangle \langle \alpha_2 j_2 \| \boldsymbol{T}^{(k_2)} \| \alpha_2' j_2' \rangle . \end{aligned} \tag{2.106}$$

The way to obtain the above result is outlined in Appendix C. This method can be used for the reduction of a matrix element into its basic ingredients. We now consider some particular cases of the type (2.106).

i) If the operator $\boldsymbol{T}^{(k)}(1,2)$ reduces to a scalar product (i.e., $k = 0$), or

$$\boldsymbol{T}^{(k)}(1) \cdot \boldsymbol{U}^{(k)}(2) ,$$

the reduction formula (2.106) leads to

$$\langle \alpha_1 j_1, \alpha_2 j_2; J \| \boldsymbol{T}^{(k)}(1) \cdot \boldsymbol{U}^{(k)}(2) \| \alpha_1' j_1', \alpha_2' j_2'; J' \rangle$$
$$= (-1)^{j_2+j_1'+J} \hat{J} \begin{Bmatrix} j_1 & j_2 & J \\ j_2' & j_1' & k \end{Bmatrix} \langle \alpha_1 j_1 \| \boldsymbol{T}^{(k)} \| \alpha_1' j_1' \rangle \langle \alpha_2 j_2 \| \boldsymbol{U}^{(k)} \| \alpha_2' j_2' \rangle \delta_{JJ'} \,. \tag{2.107}$$

ii) If one of the operators is a unit operator, this means that the product tensor operator is

$$T_\kappa^{(k)} = \left[\boldsymbol{T}^{(k)}(1) \otimes \mathbb{1}\right]_\kappa^{(k)} \,, \quad \text{or}$$
$$T_\kappa^{(k)} = \left[\mathbb{1} \otimes \boldsymbol{T}^{(k)}(2)\right]_\kappa^{(k)} \,. \tag{2.108}$$

Matrix elements can also be easily reduced with the result (we give some extra intermediate steps in working this out)

$$\langle \alpha_1 j_1, \alpha_2 j_2; J \| \boldsymbol{T}^{(k)}(1) \| \alpha_1' j_1', \alpha_2' j_2'; J' \rangle$$
$$= \langle \alpha_1 j_1, \alpha_2 j_2; J \| \left[\boldsymbol{T}^{(k)}(1) \otimes \mathbb{1}\right] \| \alpha_1' j_1', \alpha_2' j_2'; J' \rangle$$
$$= \hat{J}\hat{J}'\hat{k} \begin{Bmatrix} j_1 & j_2 & J \\ j_1' & j_2' & J' \\ k & 0 & k \end{Bmatrix} \langle \alpha_1 j_1 \| \boldsymbol{T}^{(k)} \| \alpha_1' j_1' \rangle \langle \alpha_2 j_2 \| \mathbb{1} \| \alpha_2' j_2' \rangle$$
$$= \hat{J}\hat{J}'(-1)^{j_1+j_2+J'+k} \begin{Bmatrix} j_1 & j_2 & J \\ J' & k & j_1' \end{Bmatrix} \langle \alpha_1 j_1 \| \boldsymbol{T}^{(k)} \| \alpha_1' j_1' \rangle \delta_{j_2 j_2'} \delta_{\alpha_2 \alpha_2'} \,. \tag{2.109}$$

We can similarly calculate the case when the tensor operator only acts on the coordinates of the second system. We only give the result:

$$\langle \alpha_1 j_1, \alpha_2 j_2; J \| \boldsymbol{T}^{(k)}(2) \| \alpha_1' j_1', \alpha_2' j_2'; J' \rangle$$
$$= \hat{J}\hat{J}'(-1)^{j_1+j_2'+J+k} \begin{Bmatrix} j_1 & j_2 & J \\ k & J' & j_2' \end{Bmatrix} \langle \alpha_2 j_2 \| \boldsymbol{T}^{(k)} \| \alpha_2' j_2' \rangle \delta_{j_1 j_1'} \delta_{\alpha_1 \alpha_1'} \,. \tag{2.110}$$

2.8.2 Reduction Rule II

In contrast to rule I, it is also possible to construct a composite tensor operator that acts on a single system, giving

$$T_\kappa^{(k)}(1) = \left[\boldsymbol{T}^{(k_1)}(1) \otimes \boldsymbol{T}^{(k_2)}(1)\right]_\kappa^{(k)} \,. \tag{2.111}$$

Now the state vector is *not* a product state but only a single eigenstate $|\alpha j m\rangle$. In a way that is analogous to the method of rule I, we can derive the reduced matrix element as (Brussaard, Glaudemans 1977)

$$\langle \alpha j \| \boldsymbol{T}^{(k)}(1) \| \alpha' j' \rangle = (-1)^{j+j'+k} \hat{k}$$
$$\times \sum_{\alpha'', j''} \begin{Bmatrix} k_1 & k_2 & k \\ j' & j & j'' \end{Bmatrix} \langle \alpha j \| \boldsymbol{T}^{(k_1)} \| \alpha'' j'' \rangle \langle \alpha'' j'' \| \boldsymbol{T}^{(k_2)} \| \alpha' j' \rangle \,. \tag{2.112}$$

By using the above rules, any matrix element can be reduced into a small number of basic one-body reduced matrix elements. The Hamiltonian, being a scalar operator (tensor operator of rank $k = 0$) can quite often be decomposed into a sum over multiple operators

$$H(1,2) = \sum_{\lambda} \boldsymbol{Y}_{\lambda}(1) \cdot \boldsymbol{Y}_{\lambda}(2) f(r_1, r_2) , \tag{2.113}$$

and matrix elements can be calculated by using the reduction rule I. The operators appearing in the electromagnetic one-body operators can mainly be reduced to products of the basic matrix elements

$$\langle l \| \boldsymbol{l} \| l \rangle , \quad \langle l \| \boldsymbol{Y}_{\lambda} \| l \rangle , \quad \langle \tfrac{1}{2} \| \boldsymbol{s} \| \tfrac{1}{2} \rangle , \tag{2.114}$$

from which all more complicated cases can be built.

2.9 Summary of Chaps. 1 and 2

Although short overviews, recapitulating the main results to be used later, have been given at the end of Chaps. 1 and 2, it is perhaps advisable to restate the basic content and philosophy.

Before concentrating on the nuclear shell-model where residual interactions, as well as independent particle motion, are to be treated in detail, we feel it necessary to bring together the essential techniques of angular momentum geometry.

In Chap. 1, we discussed the angular momentum operator properties, as well as the techniques necessary to couple many angular momenta together to a given total value of the angular momentum. Since the final nuclear wave functions are characterized by given J, M values (at least for spherical nuclei) resulting from the A nucleons, such coupling techniques need to be familiarized. We develop coupling schemes and various recoupling schemes for up to four angular momentum systems in quite some technical detail. More complex systems can then be derived from the methods worked out here, or can be studied from the references cited on angular momentum theory.

The tensor operators are discussed in Chap. 2 since, in conjunction with the properties of angular momentum eigenvectors, the necessary matrix elements have to be calculated. The Wigner-Eckart theorem, as well as methods for combining spherical tensor operators to a higher rank spherical tensor operator are discussed, and the various reduction rules to evaluate even highly complex matrix elements in terms of its elementary building block reduced matrix elements are worked out.

Thereby, the reader should be in a position to "attack" the nuclear shell model at a level far beyond the more general and intuitive discussions often presented in many text books.

Short Overview of Rotation Properties, Tensor Operators, Matrix Elements

Rotation of scalar function field

$$x' = Rx$$

$$f'(x) = U_R f(x) \to U_R = \mathrm{e}^{-\mathrm{i}/\hbar \boldsymbol{\varphi}\cdot\boldsymbol{L}} \; .$$

Representations of $O(3)$ and $SU(2)$

$$|jm\rangle' = U_R |jm\rangle$$

$$|lm\rangle' = \sum_{m'} D^l_{m',m}(R)|lm\rangle$$

$$l = 0,\ 1,\ 2,\ \ldots \; .$$

$SU(2)$ is homomorphic with $SO(3)$

$$|jm\rangle' = \sum_{m'} D^{(j)}_{m',m}(R)|jm\rangle$$

$$j = 0,\ \tfrac{1}{2},\ 1,\ \tfrac{3}{2},\ 2,\ \ldots \; .$$

Product representations: Clebsch-Gordan series

$$D^{(j_1)} \otimes D^{(j_2)} = \sum_{J=|j_1-j_2|}^{j_1+j_2} D^{(J)}$$

$$D^{(j_1)}_{m_1',m_1}(R) D^{(j_2)}_{m_2',m_2}(R) = \sum_{J=|j_1-j_2|}^{j_1+j_2} \sum_{M,M'} \langle j_1 m_1, j_2 m_2 | JM\rangle \times \langle j_1 m_1', j_2 m_2' | JM'\rangle D^{(J)}_{M',M}(R)$$

Tensors, Spherical tensors, Irreducible tensors

$$F'_{\underbrace{i,j,k,}_{r}\cdots} = \sum_{l,m,n,\ldots} \underbrace{A_{i,l}A_{j,m}A_{k,n}}_{r} \cdots F_{\underbrace{l,m,n,}_{r}\cdots} \; .$$

Cartesian:

$\boldsymbol{r}(x, y, z)$

$\boldsymbol{p}(p_x, p_y, p_z)$

$\boldsymbol{l}(l_x, l_y, l_z)$

products $x_i x_j,\ p_i p_j,\ x_i p_j$: reducible .

Spherical:

$$T_{\kappa}^{(k)\prime} = \sum_{\kappa'} D_{\kappa',\kappa}^{(k)}(R) T_{\kappa'}^{(k)}$$

$$\boldsymbol{r}\left(r_{+1}, r_{-1}, r_0\right) \text{ with } r_{\pm 1} = \mp\frac{1}{\sqrt{2}}(x \pm iy),\ r_0 = z\ .$$

Tensor product of spherical tensors

$$T_{\kappa_3}^{(k_3)} = \sum_{\kappa_1,\kappa_2} \langle k_1\kappa_1, k_2\kappa_2|k_3\kappa_3\rangle T_{\kappa_1}^{(k_1)} T_{\kappa_2}^{(k_2)}$$

$$\boldsymbol{T}^{(k)} \cdot \boldsymbol{U}^{(k)} = (-1)^k \hat{k} \left[\boldsymbol{T}^{(k)} \otimes \boldsymbol{U}^{(k)}\right]^{(0)} : \text{ Scalar product}\ .$$

Wigner-Eckart theorem

$$\langle \alpha j m|T_{\kappa}^{(k)}|\alpha' j' m'\rangle = (-1)^{j-m} \begin{pmatrix} j & k & j' \\ -m & \kappa & m' \end{pmatrix} \langle \alpha j\|\boldsymbol{T}^{(k)}\|\alpha' j'\rangle\ .$$

Reduction rules

– rule I *qquad*(see text)
– rule II (see text) .

Problems

*2.1 Derive the transformed function of $F(x,y) = a(x^2 - y^2)$ as $F'(x,y) = 2axy$ (the new function), using the formal definition

$$F'(x,y) = \exp\left[-\frac{\mathrm{i}}{\hbar}\varphi_z L_z\right] F(x,y)\ ,$$

for a rotation of $\varphi_z = 45°$ around the z-axis.

2.2 Derive (2.10) i.e. show that the rotation operator for a general angular momentum operator $\boldsymbol{J}$ and a rotation over an angle α around an axis, specified by the unit vector $1\!\!1_\alpha$, becomes

$$U_R(\alpha) = \exp\left[-\frac{\mathrm{i}}{\hbar} 1\!\!1_\alpha \cdot \boldsymbol{J}\right]\ .$$

*2.3 Derive the transformation matrix acting on the coordinates of a point $P(x,y)$ for a rotation around the z-axis over an angle φ, using the transformation properties of the spherical harmonics $Y_1^M(\hat{\boldsymbol{r}})$ as given in (2.33).

2.4 Show, by explicit calculation, that the transformation matrix in (2.74), for a system of two spin-1/2 particles decouples into an $S = 0$ and $S = 1$ part, when coupling the two spin-1/2 systems.

2.5 Determine the representation 2×2 matrix $\exp(-\mathrm{i}\beta\sigma_y/\hbar)$ in the space spanned by the spin 1/2 eigenvectors.

2.6 Given are the D-functions for spin-1/2. Study the effect of rotation of 180°, 360°, 720° around the x-axis concerning the transformation properties of the spin-1/2 eigenvectors $\chi_m^{1/2}$, (see figure). Discuss your results.

Table P.1. The Wigner $D^{1/2}_{m',m}(\alpha,\beta,\gamma)$ rotation matrix representation for the basis of the intrinsic spin $s = 1/2$ states

m' \ m	$+1/2$	$-1/2$
$+1/2$	$\mathrm{e}^{-\mathrm{i}\alpha/2}\cos(\beta/2)\mathrm{e}^{-\mathrm{i}\gamma/2}$	$-\mathrm{e}^{-\mathrm{i}\alpha/2}\sin(\beta/2)\mathrm{e}^{\mathrm{i}\gamma/2}$
$-1/2$	$\mathrm{e}^{\mathrm{i}\alpha/2}\sin(\beta/2)\mathrm{e}^{-\mathrm{i}\gamma/2}$	$\mathrm{e}^{\mathrm{i}\alpha/2}\cos(\beta/2)\mathrm{e}^{\mathrm{i}\gamma/2}$

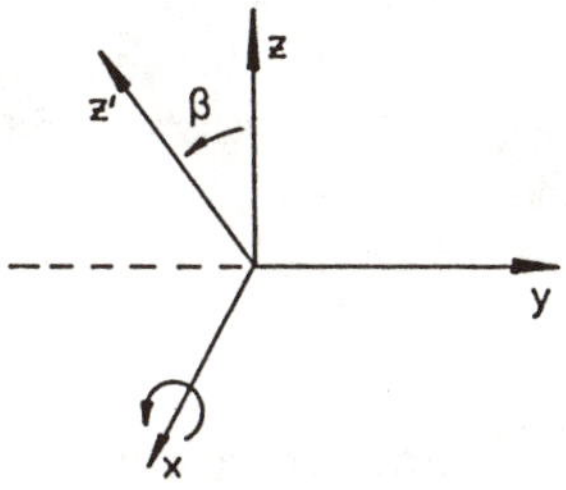

2.7 Prove, as an example of a tensor product, for the rank 1 tensor $\boldsymbol{r}^{(1)}$ and $\boldsymbol{p}^{(1)}$, that $\boldsymbol{l}^{(1)} = -\mathrm{i}\sqrt{2}[\boldsymbol{r}^{(1)} \otimes \boldsymbol{p}^{(1)}1^{(1)}$.

2.8 Derive the reduction rule II (2.112) using the methods of Sect. 2.8.1.

*2.9 If the $2k+1$ components $T_\kappa^{(k)}$ form a spherical tensor of rank k, prove that the following commutation relations with the angular momentum operator components exists:

(i) $\left[J_\pm, T_\kappa^{(k)}\right] = \sqrt{(k \mp \kappa)(k \pm \kappa + 1)}\, T_{\kappa\pm 1}^{(k)}$

(ii) $\left[J_z, T_\kappa^{(k)}\right] = \kappa T_\kappa^{(k)}$.

2.10 Making use of the set of reduction rules I, prove that the spin-orbit coupling matrix element $\boldsymbol{l}.\boldsymbol{s}$ can be evaluated as

$$\langle (l\,1/2)j, m|\boldsymbol{l}.\boldsymbol{s}|(l\,1/2)j, m\rangle = 1/2.\left[j(j+1) - l(l+1) - 3/4\right] .$$

3. The Nuclear Shell Model

3.1 One-particle Excitations

3.1.1 Introduction

The basic assumption in the nuclear shell model is that, to first order, each nucleon (proton or neutron) is moving in an independent way in an average field. This is not so, a priori, since the nucleus constitutes an A-body problem interacting via the nucleon-nucleon force in the nuclear medium. It is clear from the very beginning that this nucleon-nucleon force will be different from the free nucleon-nucleon interaction (Bohr, Mottelson 1969). As was already expressed in the introduction to this book, the non-relativistic form of the nucleon-nucleon interaction behaves as shown in Fig. 3.1. At large separations $|\boldsymbol{r}_i - \boldsymbol{r}_j| \cong 1.5 - 2\,\text{fm}$, the force behaves according to a one-pion exchange potential (OPEP) which has an analytic dependence on $r = |\boldsymbol{r}_i - \boldsymbol{r}_j|$ of (Bohr, Mottelson 1969)

$$V(r) = -\frac{e^{-\mu r}}{\mu r}\left(1 + \frac{3}{\mu r} + \frac{3}{(\mu r)^2}\right) . \tag{3.1}$$

For small distances the attractive part turns over and becomes repulsive at distances $r < 0.5\,\text{fm}$; this is the hard-core potential. At such short distances the energy for interactions between nucleons becomes so high that non-relativistic treatments are no longer justified. In this region, exchange of more pions or heavy mesons is needed. It is probably even more correct to go to QCD where nucleons are

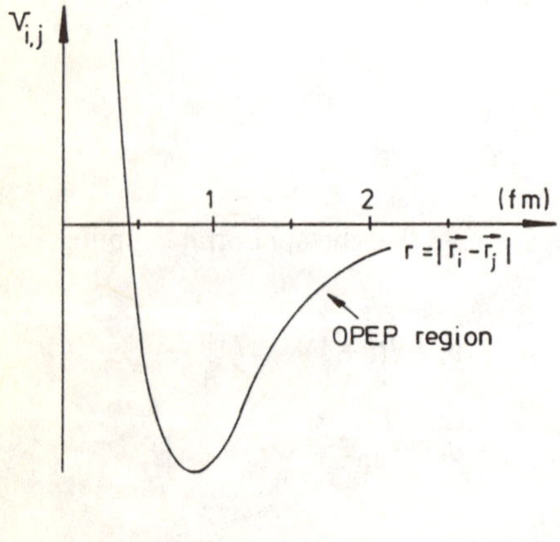

Fig. 3.1. Schematic illustration of the nucleon two-body interaction $V_{i,j}$ as a function of the nucleon separation $r = |\boldsymbol{r}_i - \boldsymbol{r}_j|$. For large separation ($r \cong 1.5 - 2fm$), the OPEP tail results. For short distances, a short range repulsive core shows up

considered as being obtained from their quark constituents and so, a nucleon-nucleon interaction process can be better depicted, as in Fig. 3.2.

One of the most unexpected features is still the very large nuclear mean free path in the nuclear medium (Fig. 3.3).

In the study of the nuclear structure observed at low excitation energy ($E_x <$ 8 MeV), two important aspects show up:

- how to handle (even in a non-relativistic way) the A-nucleon problem,
- how to describe the nuclear average field starting from the nucleon-nucleon force $V(|\boldsymbol{r}_i - \boldsymbol{r}_j|)$ between free nucleons.

In the atomic case, a shell structure was shown to exist by N. Bohr. Starting from an average Coulomb field $V(r) = -Z\,e^2/r$, the corresponding one-electron Schrödinger equation can be solved and the atomic orbits studied in detail. According to the Pauli principle, only one fermion particle can be in a specific quantum state defined by the radialquantum number (n), and the orbital (l), total (j) and

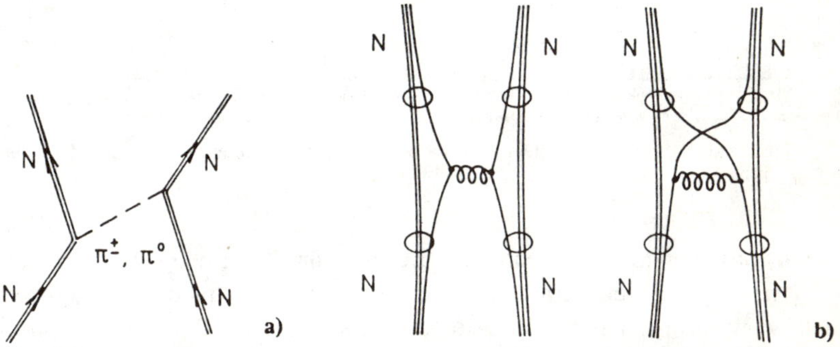

Fig. 3.2. The nucleon-nucleon interaction **(a)** on the level of QHD (quantum hadro-dynamics) where the force is mediated by the exchange of pions ($\pi^{\pm}, \pi^0$) between the interacting nucleons, or **(b)** on the level of QCD (quantum chromo-dynamics) where a gluon is exchanged between one of the three quarks constituting the nucleon. The left-hand side diagram is zero due to the colour selection rules, the right-hand diagram contributes since colour selection rules are obeyed

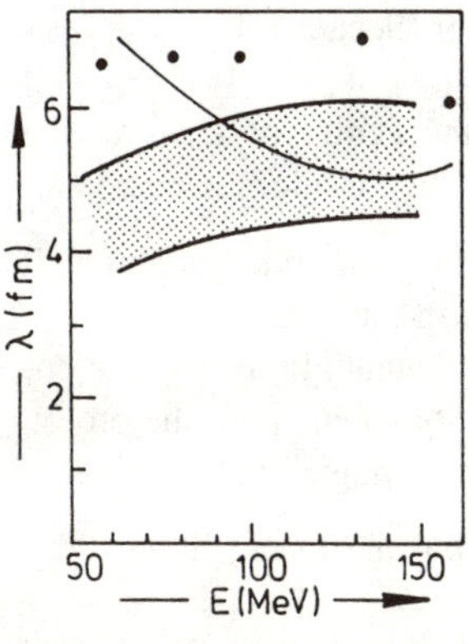

Fig. 3.3. The empirical mean free path ($\lambda(fm)$). The shaded band denotes the range of values determined from reaction cross sections for Ca, Zr and Pb in (Nadasen 1981), the solid line indicates the representative values selected in (Bohr 1969) and the data points are determined from optical potential fits for ^{208}Pb in (Nadasen 1981) [taken from (Negele 1983)]

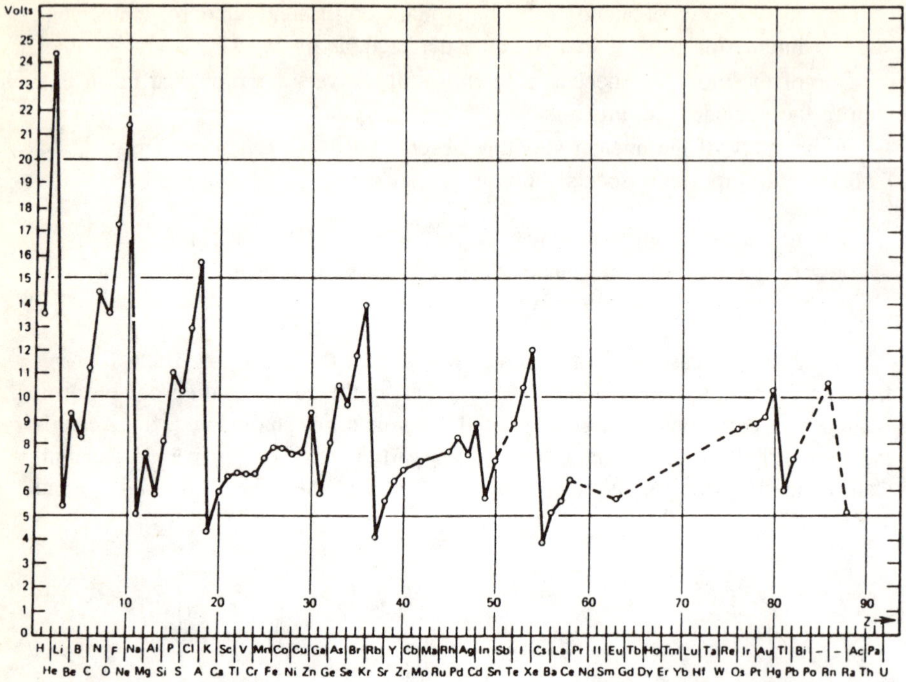

Fig. 3.4. Dependence of the ionization potential of the neutral atom on the atomic number Z [taken from (Herzberg 1944)]

magnetic (m) of the angular momentum. For each j, the $2j+1$ magnetic substates with $-j \leq m \leq j$ are degenerate and form a given shell structure. A number of subshells now form a major shell. Atoms with a major closed-shell configuration form configurations that are particularly stable against losing the last electron (Figure 3.4 shows the ionization potentials of the elements).

In the nucleus a similar description seems to be possible. However, a number of distinct differences to the atomic case arise:

i) The nuclear mean field is very different from the Coulomb potential. Moreover, strong spin-orbit coupling is shown to exist in nuclei (Sect. 3.1.3).
ii) In the nucleus both protons and neutrons are present.
iii) There is no preferential central point other than the center of mass in the nucleus in contrast to the atomic field generated by the atomic nucleus.

Because of the above conditions the shell structure in the atomic nucleus will be very different from the corresponding shell structure in the atom.

We now quote a number of nuclear properties that unambiguously point towards nuclear shell structure and increased nuclear stability when either the proton number (Z) and/or the neutron number (N) has a certain "magic" value.

i) Deviations of the nuclear mass (binding energy) from the mean, liquid drop value exist (Fig. 3.5).

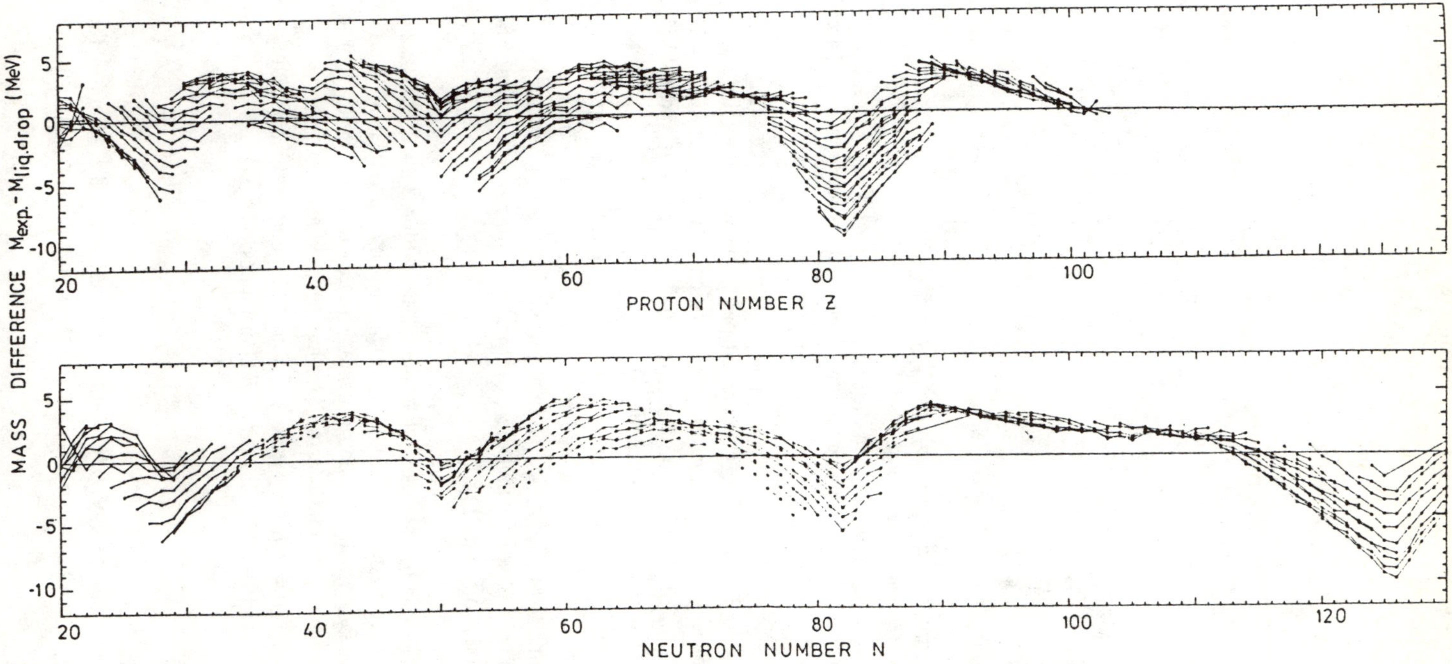

Fig. 3.5. Deviations of nuclear masses from their mean (liquid drop) values, and this as a function of neutron and proton number [taken from (Myers 1966)]

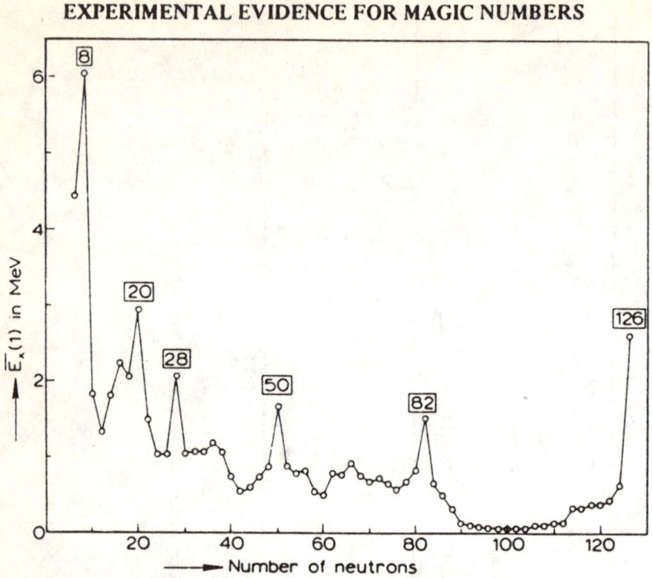

Fig. 3.6. The magic numbers, demonstrated by the excitation energy for the first excited state in doubly-even nuclei (mainly a $J^\pi = 2^+$ level) plotted as a function of the neutron number N [taken from (Brussaard 1977)]

ii) Since nucleons couple into $J^\pi = 0^+$ coupled pairs, the way to excite nuclei (the excitation energy of the first excited state which is most often a $J^\pi = 2^+$ state) as a function of neutron number (Fig. 3.6) again correlates very well with the shell closures as obtained under i).

iii) Specific tests result when, in one-nucleon transfer reactions (pick-up or stripping), a nucleon is taken out or added to a nucleus with given A (Z, N) nucleon constitution. In the case of adding a proton to $^{208}_{82}\mathrm{Pb}_{126}$ via a $(^3\mathrm{He}, d)$ reaction, it is clearly observed that the extra proton is placed in very specific nuclear shell model orbitals (Fig. 3.7).

The stable nucleon configurations so determined are N (or Z) = 2, 8, 20, 28, 50, 82, 126, (...). These numbers can now be explained by starting from a one-body Schrödinger equation using a central average (and attractive) field $U(r)$ to which a strong spin-orbit interaction term $\zeta \boldsymbol{l} \cdot \boldsymbol{s}$ has been added.

Suppose that $\varphi_a(\boldsymbol{r})$ $(a = n_a, l_a, j_a, m_a, \ldots)$ are solutions to the Schrödinger one-body equation (here $\boldsymbol{r}$ is a notation for *all* coordinates, $\boldsymbol{r} \equiv \boldsymbol{r}, \boldsymbol{\sigma}, \ldots$)

$$[T + U(r)]\varphi_a(\boldsymbol{r}) = \varepsilon_a \varphi_a(\boldsymbol{r}) \,. \tag{3.2}$$

Here T describes the kinetic energy, $U(r)$ the average field (Sect. 3.1.4) and ε_a the single-particle energy (recall that we do not write the spin and charge coordinates explicitly).

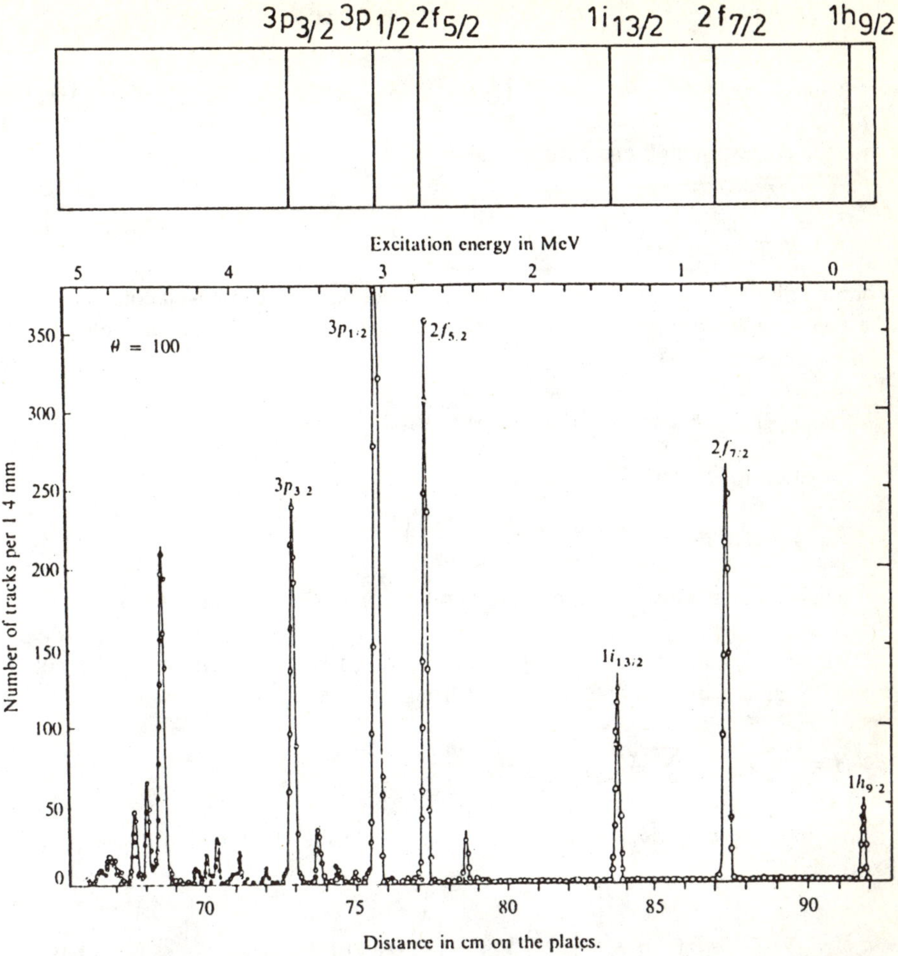

Fig. 3.7. Single-particle states in ^{209}Bi, obtained from the pick-up reaction ^{208}Pb (^{3}He,d)^{209}Bi. At the angle of $\theta = 100°$, the $1h_{9/2}$ ground-state level is less populated than the excited states but at other angles, this situation can become reversed. Above the actual spectrum, the proton single-particle states are drawn as an illustration [taken from (Mottelson 1967)]

Orthogonality demands that

$$\int \varphi_a^*(\boldsymbol{r})\varphi_b(\boldsymbol{r})\, d\boldsymbol{r} = \delta_{ab} \,. \tag{3.3}$$

The model Hamiltonian for the A nucleons (taken as independent particles) can then be written as

$$H_0 = \sum_{i=1}^{A} \big(T_i + U(r_i)\big) = \sum_{i=1}^{A} h_0(i) \,. \tag{3.4}$$

The eigenfunctions of H_0 are now of the product type

$$\Psi_{a_1,a_2,\ldots,a_A}(\boldsymbol{r}_1,\boldsymbol{r}_2,\ldots,\boldsymbol{r}_A)=\prod_{i=1}^{A}\varphi_{a_i}(\boldsymbol{r}_i)\,, \tag{3.5}$$

with the corresponding energy eigenvalue

$$E_0=\sum_{i=1}^{A}\varepsilon_{a_i}\,. \tag{3.6}$$

For a number of identical nucleons, the wave function (3.5) is not well constructed: the Pauli (exclusion) principle is not fulfilled. The correct wave function, e.g., for two particles becomes

$$\Psi_{a_1,a_2}(\boldsymbol{r}_1,\boldsymbol{r}_2)=\frac{1}{\sqrt{2}}\left(\varphi_{a_1}(\boldsymbol{r}_1)\varphi_{a_2}(\boldsymbol{r}_2)-\varphi_{a_1}(\boldsymbol{r}_2)\varphi_{a_2}(\boldsymbol{r}_1)\right), \tag{3.7}$$

or, rewritten in (Slater) determinant form

$$\Psi_{a_1,a_2}(\boldsymbol{r}_1,\boldsymbol{r}_2)=\frac{1}{\sqrt{2}}\begin{pmatrix}\varphi_{a_1}(\boldsymbol{r}_1) & \varphi_{a_1}(\boldsymbol{r}_2)\\ \varphi_{a_2}(\boldsymbol{r}_1) & \varphi_{a_2}(\boldsymbol{r}_2)\end{pmatrix}. \tag{3.8}$$

For an A-nucleon wave function, the generalization of (3.8) becomes an A-particle Slater determinant.

We should note that the average field expressed by the potential $U(r)$ is not explicitly given. In fact, one has to start from the A-nucleon Hamiltonian

$$H=\sum_{i=1}^{A}T_i+\frac{1}{2}\sum_{i,j=1}^{A}V_{i,j}\,, \tag{3.9}$$

(restricting to two-body interactions only), to write the Hamiltonian that can be expressed

$$H=\sum_{i=1}^{A}\left[T_i+U(r_i)\right]+\left(\frac{1}{2}\sum_{i,j=1}^{A}V_{i,j}-\sum_{i=1}^{A}U(r_i)\right) \tag{3.10}$$

$$\begin{aligned}&=H_0+H_{\text{res}}\\&=\sum_{i=1}^{A}h_0(i)+H_{\text{res}}\,,\end{aligned} \tag{3.11}$$

where H_0 describes the motion of A nucleons, independent of each other in the same average field. The smaller the effect of H_{res}, the better the assumption of an average, independent field becomes. The method of determining $U(r)$, starting from a known $V_{i,j}$ and a Slater determinant A-nucleon wave function that is a good approximation to the total ground state wave function for the full Hamiltonian H, is carried out by using the Hartree-Fock method (Sect. 3.1.4). Before that, however, we study the independent nucleon motion with a harmonic oscillator potential $U(r)=\frac{1}{2}m\omega^2r^2$.

3.1.2 The Radial Equation and the Single-particle Spectrum: the Harmonic Oscillator in the Shell Model

If we start from the central, one-body problem discussed in Chap. 1 (1.14–18) the total wave function can be written as

$$\begin{aligned}\varphi(\boldsymbol{r}) &= R(r)Y(\theta,\varphi)\,,\\ &= \frac{u(r)}{r}Y(\theta,\varphi)\,.\end{aligned}\tag{3.12}$$

The equation governing the radial motion is then

$$-\frac{\hbar^2}{2m}\frac{d^2u(r)}{dr^2}+\left[l(l+1)\hbar^2/2mr^2+U(r)\right]u(r)=Eu(r)\,.\tag{3.13}$$

In studying bound states ($E<0$), conditions have to be imposed on the radial solution $u(r)$:

$$\begin{aligned}&u(r)\underset{r\to\infty}{\longrightarrow}0\,,\\ &u(0)=0\,.\end{aligned}\tag{3.14}$$

The normalization of the radial wave function leads to the integrals

$$\int_0^\infty R^2(r)\,dr=\int_0^\infty u^2(r)\,dr=1\,.\tag{3.15}$$

Starting now from the harmonic oscillator potential $U(r)=\frac{1}{2}m\omega^2r^2$, we obtain the radial Laguerre equation, with the solution (Abramowitz, Stegun 1964)

$$u_{kl}(r)=N_{k,l}\cdot r^{l+1}\,\mathrm{e}^{-\nu r^2}\,L_k^{l+1/2}\left(2\nu r^2\right)\,,\tag{3.16}$$

with $N_{k,l}$ a normalization factor, $\nu=m\omega/2\hbar$ the oscillator frequency and the Laguerre polynomial, given by (Abramowitz, Stegun 1964)

$$L_k^{l+1/2}(x)=\sum_{k'=0}^{k}a_{k'}^{l}(-1)^{k'}x^{k'}\,.\tag{3.17}$$

A number of radial solutions are illustrated for $Z=82$, $N=126$ for the neutron motion. Although the wave functions in Fig. 3.8 are calculated for a more realistic potential, a Woods-Saxon potential (Blomqvist, Wahlborn 1960), the overall behavior is the same as for the harmonic oscillator potential.

The energy eigenvalues corresponding with the eigenfunctions (3.16) are given by

$$E=\hbar\omega\left(2k+l+\tfrac{3}{2}\right)=\hbar\omega\left(N+\tfrac{3}{2}\right)\,,\tag{3.18}$$

with

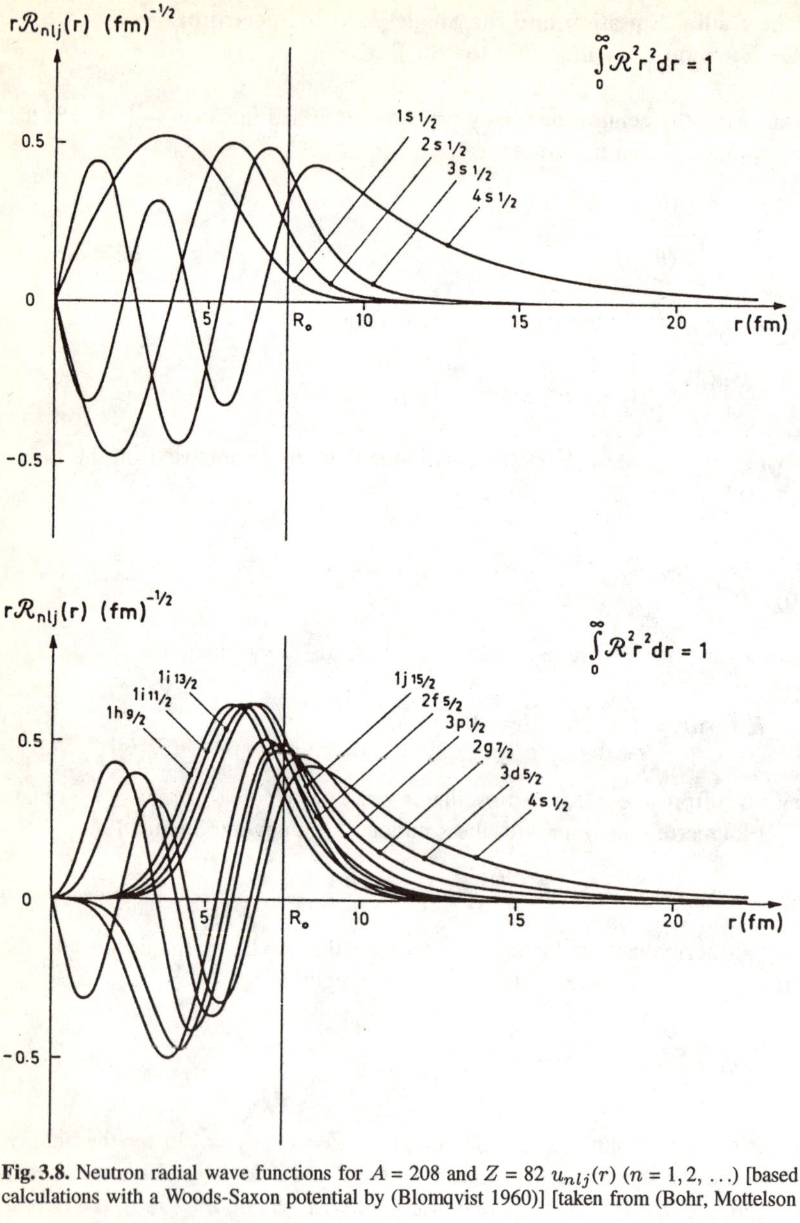

Fig. 3.8. Neutron radial wave functions for $A = 208$ and $Z = 82$ $u_{nlj}(r)$ $(n = 1, 2, \ldots)$ [based on the calculations with a Woods-Saxon potential by (Blomqvist 1960)] [taken from (Bohr, Mottelson 1969)]

$$\begin{aligned} &N = 0, 1, 2, && \text{(major oscillator quantum number)}, \\ &l = N, N-2, \ldots, 1 \text{ or } 0 && \text{(orbital quantum number)}, \\ &k = (N-l)/2 && \text{(radial quantum number)}. \end{aligned} \tag{3.19}$$

Thus, the spectrum of eigenvalues presents a large number of degenerate (l, k) quantum numbers corresponding to a fixed N major oscillator quantum number (Fig. 3.9).

The radial quantum number more often used (de-Shalit, Talmi 1963) is related to k via

$$n = k + 1 = (N - l + 2)/2 , \tag{3.20}$$

and expresses the number of nodes of the radial wave function in the interval $(0, \infty)$ including the node at the origin (excluding the one at infinity).

We now give a number of interesting properties of the Laguerre polynomials that allow for an elegant calculation of the normalization factor $N_{k,l}$ (Abramowitz, Stegun 1964)

$$\int_0^\infty z^a \,\mathrm{e}^{-z} L_k^a(z) L_{k'}^a(z)\, dz = \delta_{kk'} \cdot \Gamma(k + a + 1)^3/k! . \tag{3.21}$$

We use (3.21) in calculating the norm $N_{k,l}$ by putting $z = 2\nu r^2$, $r = (z/2\nu)^{1/2}$ and $dr = dz/[2(2\nu z)^{1/2}]$ and evaluate the integral

$$\begin{aligned} &N_{k,l}^2 \int_0^\infty r^{2l+2}\,\mathrm{e}^{-2\nu r^2} \left[L_k^{l+1/2}(2\nu r^2)\right]^2 dr \\ &\quad = N_{k,l}^2 \int_0^\infty z^{l+1/2}/(2(2\nu)^{l+3/2})\mathrm{e}^{-z} \left[L_k^{l+1/2}(z)\right]^2 dz \end{aligned}$$

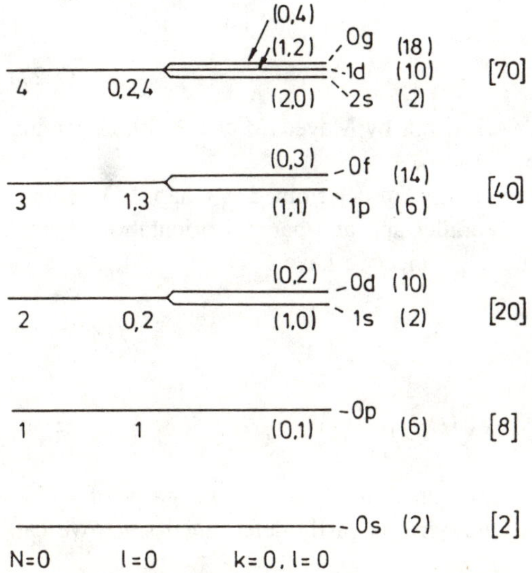

Fig. 3.9. Illustration of the degenerate harmonic oscillator energy spectrum up to $N = 4$. Besides the major shell quantum number (N), the (k, l) degeneracies are drawn explicitly. Partial and cumulative occupation numbers are given in round and squared brackets, respectively

$$= N_{k,l}^2/(2(2\nu)^{l+3/2}) \cdot \Gamma\left(l+\tfrac{3}{2}+k\right)^3/k! = 1 \quad \text{or}$$

$$N_{k,l} = \left(2(2\nu)^{l+3/2}k!/\Gamma(l+k+3/2)^3\right)^{1/2} . \tag{3.22}$$

Summarizing some properties:

$$\begin{aligned} &\Gamma(z+1) = z\Gamma(z), \quad \Gamma\left(\tfrac{1}{2}\right) = \sqrt{\pi}, \\ &L_0^a(z) = \Gamma(a+1), \quad L_0^{1/2}(z) = \sqrt{\pi}/2 . \end{aligned} \tag{3.23}$$

The radial wave functions can be determined as follows, since

$$L_k^a(z) = \Gamma(a+k+1)/k!\, e^z z^{-a} \frac{d^k}{dz^k}\left(z^{a+k} e^{-z}\right) , \tag{3.24}$$

(k = entire). As an example, we illustrate the above equation for $L_1^{1/2}(z)$:

$$L_1^{1/2}(z) = \tfrac{3}{2} \cdot \frac{\sqrt{\pi}}{2}\left(\tfrac{3}{2} - z\right) . \tag{3.25}$$

A number of often used expressions relating to the calculation of radial integrals are

$$\begin{aligned} &\int_0^\infty x^{2n} e^{-px^2} dx = (2n-1)!!/\left(2(2p)^n\right)\sqrt{\pi/p} \quad (p > 0) , \\ &\int_0^\infty x^{2n+1} e^{-px^2} dx = n!/2p^{n+1} \quad (p > 0) . \end{aligned} \tag{3.26}$$

Having determined the solutions to the radial equation for a single-particle harmonic oscillator potential, we observe a large degeneracy in the orbitals. Moreover, the nucleon numbers that form the stable configurations are N, Z = 2, 8, 20, 40, 70 and do not agree with the experimental numbers. There is a strong spin-orbit term $\zeta(r)\,\boldsymbol{l} \cdot \boldsymbol{s}$ that is modifying the single-particle spectrum in the right direction. We shall work this out in more detail. The original Hamiltonian h_0 becomes

$$h = h_0 + \zeta(r)\boldsymbol{l} \cdot \boldsymbol{s} . \tag{3.27}$$

The consequences were originally worked out by Mayer (Mayer 1949, 1950) and Haxel, Jensen and Suess (Haxel et al. 1949).

The single-particle wave functions that were determined in Chap. 1 are eigenfunctions of h_0. Moreover, both the parallel and anti-parallel orientations correspond to the same energy eigenvalue $\varepsilon_{nlj}^{(0)}$ since we have

$$\langle nlj, m|h_0|nlj, m\rangle = \varepsilon_{nlj}^{(0)} , \tag{3.28}$$

with

$$\langle \boldsymbol{r}, \boldsymbol{\sigma}|nlj, m\rangle = \frac{u_{nl}(r)}{r}\left[\boldsymbol{Y}_l(\theta,\varphi) \otimes \boldsymbol{\chi}^{1/2}(\boldsymbol{\sigma})\right]_m^{(j)} , \tag{3.29}$$

the $(l, \frac{1}{2})j$ coupled single-particle wave function. By using the basis of (3.29) the energy correction for the spin-orbital term is easily determined since we can express the spin-orbit term $\zeta(r)\boldsymbol{l} \cdot \boldsymbol{s}$ as

$$\zeta(r)\tfrac{1}{2}\left(\boldsymbol{j}^2 - \boldsymbol{l}^2 - \boldsymbol{s}^2\right) . \tag{3.30}$$

Thus we obtain

$$\varepsilon_{nlj} = \varepsilon_{nlj}^{(0)} + \Delta\varepsilon_{nlj} , \tag{3.31}$$

with

$$\Delta\varepsilon_{nlj} = \langle nlj, m|\zeta(r)\boldsymbol{l}\cdot\boldsymbol{s}|nlj, m\rangle , \tag{3.32}$$

or

$$\Delta\varepsilon_{nlj} = \frac{D}{2}\left[j(j+1) - l(l+1) - \tfrac{3}{4}\right] . \tag{3.33}$$

We define

$$D = \int u_{nl}^2(r)\zeta(r)\,dr . \tag{3.34}$$

This gives rise to a spin-orbit splitting of

$$\begin{aligned}\Delta\varepsilon_{nl\,j=l+1/2} &= \frac{D}{2}\cdot l \\ \Delta\varepsilon_{nl\,j=l-1/2} &= -\frac{D}{2}(l+1) ,\end{aligned} \tag{3.35}$$

as illustrated in Fig. 3.10b.

A much used form of $\zeta(r)$ is the derivative of the average $U(r)$ potential and is shown for a Woods-Saxon potential in Fig. 3.10a. One can thus express $\zeta(r)$ as

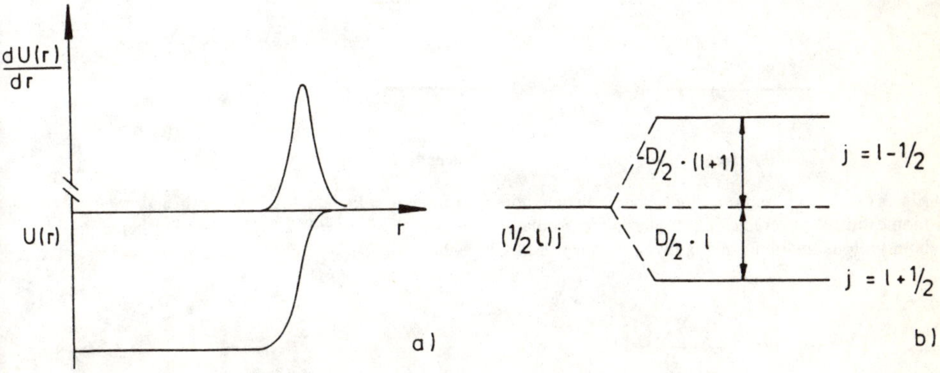

Fig. 3.10. (**a**) We draw the possible radial form for the spin-orbit strength function $\zeta(r)$ as determined by the derivative of a Woods-Saxon potential, described by (3.36). This function $\zeta(r)$ peaks at the nuclear surface. (**b**) The spin-orbit splitting between $j = l \pm \frac{1}{2}$ partners, according to (3.35), using the factor D, with $D \equiv \int u_{nl}^2(r)\zeta(r)dr$ [see (3.34)]

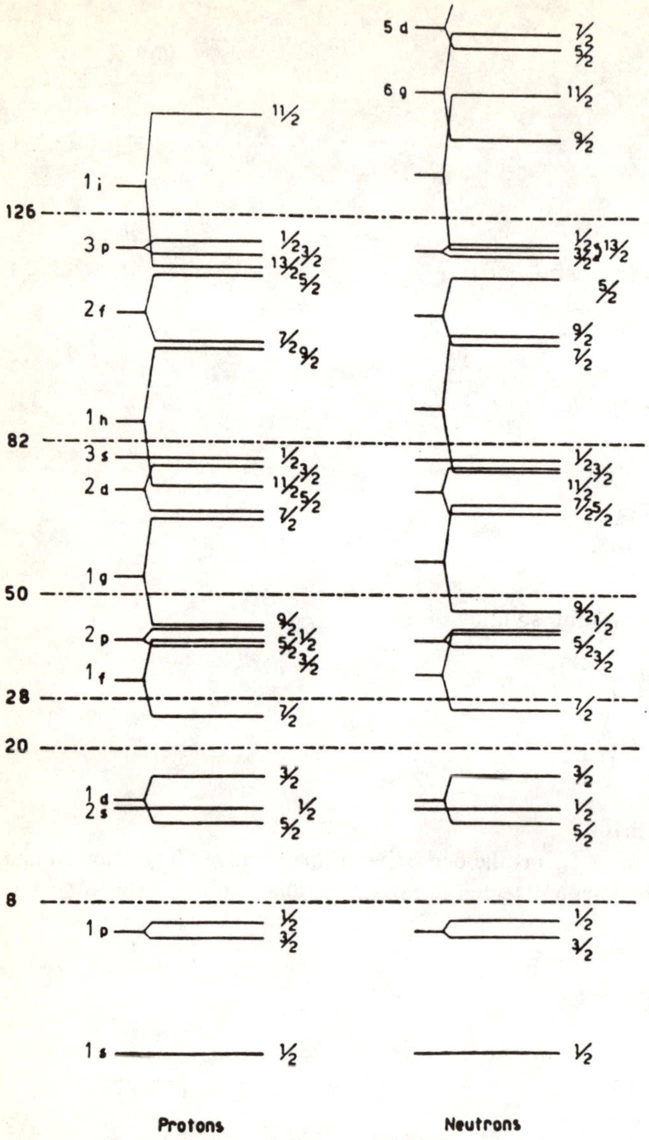

Fig. 3.11. A full single-particle spectrum, including terms that split both the spin-orbit and angular momentum degeneracies in the harmonic oscillator case of Fig. 3.9 ($n = 1, 2, \ldots$). A level scheme for both protons and neutrons is given [taken from (Klingenberg 1952)]

$$\zeta(r) = V_{ls} \cdot r_0^2 \cdot \frac{1}{r} \frac{\partial U(r)}{\partial r} \,. \tag{3.36}$$

A full single-particle spectrum, including a term proportional to l^2 for splitting the remaining degeneracies on (k, l) for given N in addition to the spin-orbit interaction, is illustrated in Fig. 3.11. This figure gives a general idea of

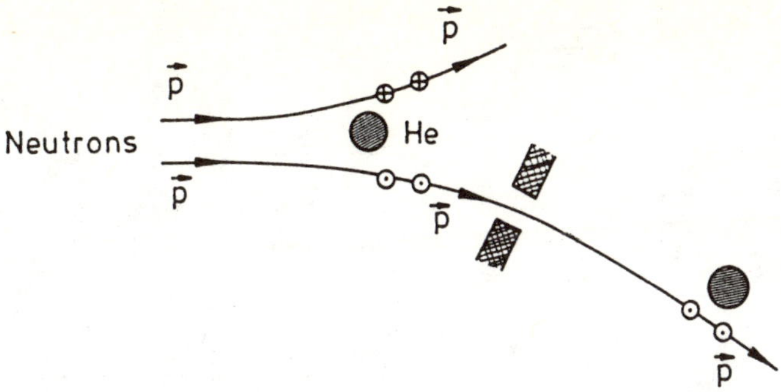

Fig. 3.12. Classical illustration of the existence of a spin-orbit interaction. A beam of unpolarized neutrons is directed to ^{4}He scattering nuclei. If the coefficient multiplying the spin-orbit interaction is such as to favour parallel orientation, then neutrons passing on the lower side have their spin pointing *out* of the plane ($\odot$), the neutrons on the upper side have their spin pointing *into* the plane ($\otimes$). By collimating the beam, a polarized beam is selected. For a 100% polarized beam and scattering on a second ^{4}He nucleus, *all* neutrons should pass on the lower side. This asymmetry is indeed observed in realistic cases. This experiment is an example of the polarizer-analyzer set-up

the nucleon shell structure with, this time, the correct "magic" numbers N, $Z = 2, 8, 20, 28, 50, 82, 126, \ldots$.

Besides the correct reproduction of the stable nucleon configurations, there exists experimental evidence for a spin-orbit term in the nucleon-nucleon interaction as observed from nucleon-nucleus scattering (Bohr, Mottelson 1969). In a first collision of neutrons or protons with He nuclei of energy between 5 and 15 MeV, the beam is partially polarized. Classically, we can present the scattering in the following way (Fig. 3.12). The potential that an incoming nucleon feels in the nucleus is $U(r) + \zeta(r)\boldsymbol{l} \cdot \boldsymbol{s}$ with a negative value of V_{ls} in (3.36). Thus the parallel orientation is favored by the scattering process. The nucleons scattered to the right now have their spin preferentially in the forward direction; the ones scattered to the left have their spin in the opposite direction. Using a diaphragm, a partial polarization of the beam is obtained. By now using a second collision process, right-left asymmetry is observed because of the partial polarization of parallel spin-orbital orientation in the incoming beam. Although the above discussion is somewhat too simplified, the basic outcome is indeed observed and so unambiguously indicates a term in the nuclear potential proportional to $\boldsymbol{l} \cdot \boldsymbol{s}$.

3.1.3 Illustrative Examples of Energy Spectra

In the next two figures, we illustrate for both light nuclei (^{16}O region) and heavy nuclei (^{208}Pb region) both the proton and neutron particle and hole excited states at relatively low excitation energy (Figs. 3.13, 14). Besides some extra states, the concept of single-particle motion in an average field following the structure obtained in Sect. 3.1.2 is indeed very well realized. Also, it is clear that in the heavier ^{208}Pb nucleus the average field is better determined compared to ^{16}O where only 16 particles are present.

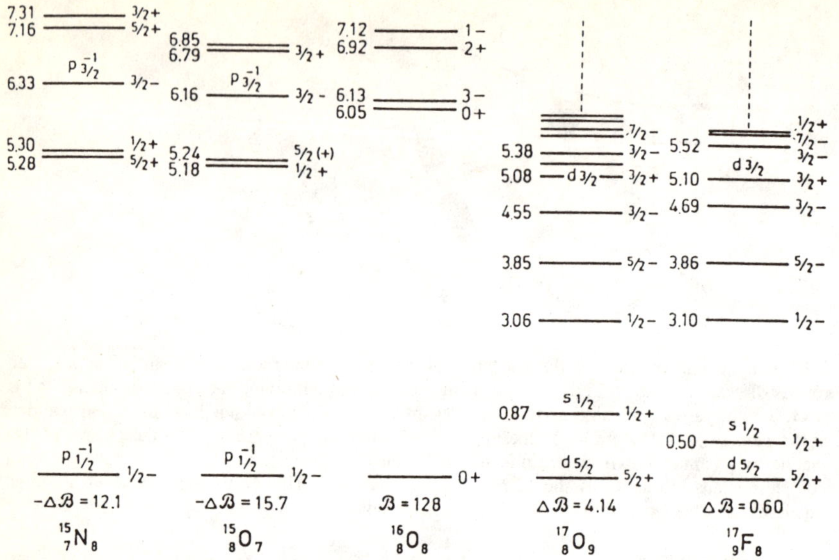

Fig. 3.13. Illustration of both proton and neutron single-particle and single-hole states around ^{16}O. On each nucleus, the binding energy difference ΔB, relative to ^{16}O is given. The levels where the major single-particle (single-hole) character is concentrated are given with the quantum numbers (lj) although these quantities are not good quantum numbers in general (only J^{π} is). The level scheme is taken from (Bohr, Mottelson 1969)

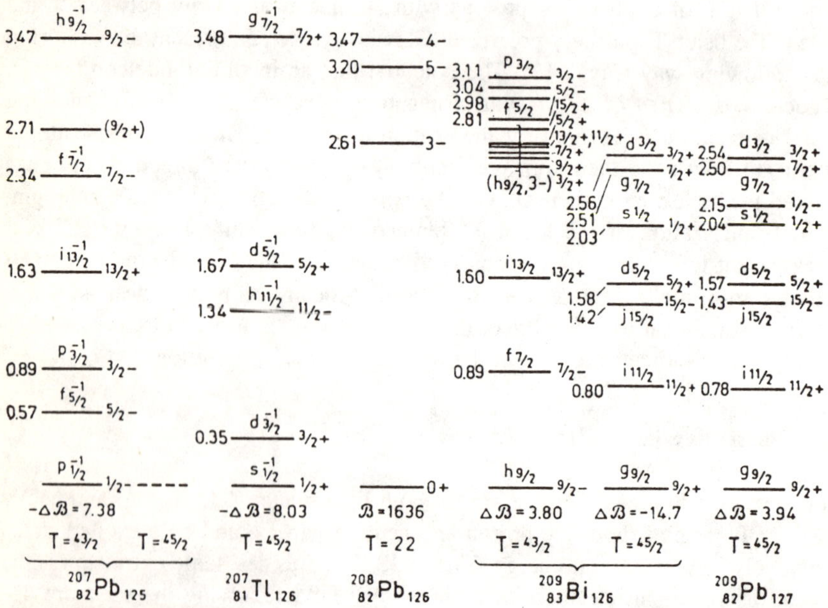

Fig. 3.14. See caption to Fig. 3.13, but now for the proton and neutron single-particle and single-hole states around the nucleus ^{208}Pb. Details on the origin of the experimental data can be found in the Nuclear Data Sheets for the appropriate nuclei [Figure taken from (Bohr, Mottelson 1969)]

We show, moreover, the variation of the single-particle states as a function of nucleon number for the neutron bound states (Fig. 3.15), as well as an excerpt for ^{208}Pb (Fig. 3.16) around the $Z = 82$ and $N = 126$ closed shells. [The calculated levels result from a Woods-Saxon potential as studied by Blomqvist and Wahlborn (Blomqvist, Wahlborn 1960).]

We now discuss how the average field $U(r)$, used before in a rather phenomenological way, can be determined from a microscopic starting point (Hartree-Fock).

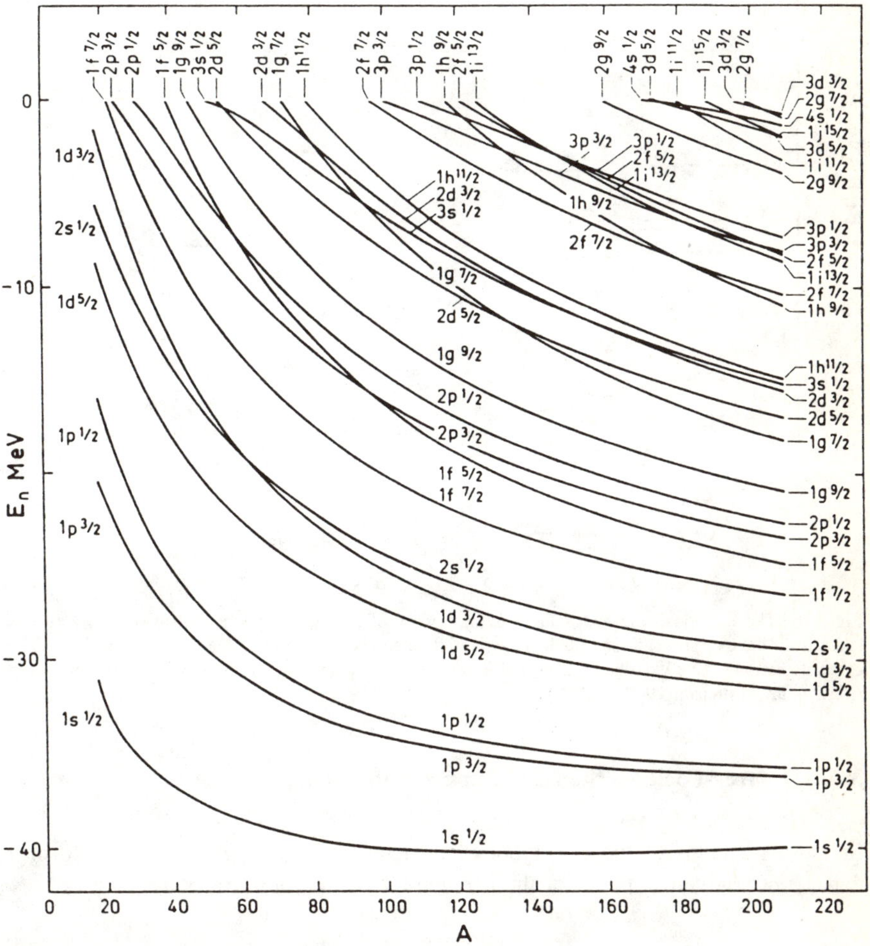

Fig. 3.15. Energies of neutron orbitals, calculated by C.J. Veje as quoted in (Bohr 1969). Use has been made of a Woods-Saxon potential $U = Vf(r) + V_{ls}(\boldsymbol{l} \cdot \boldsymbol{s})r_0^2(1/r)(d/dr)f(r)$, with $f(r)$ having a Woods-Saxon shape $[1 + \exp(r - R_0/a)]^{-1}$, and $R_0 = r_0 A^{1/3}$ ($r_0 = 1.27\,fm$) and $a = 0.67\,fm$. The potentials V and V_{ls} are given as $V = (-51 + 33((N - Z)/A))$ MeV, $V_{ls} = -0.44$ V, [taken from (Bohr, Mottelson 1969)]

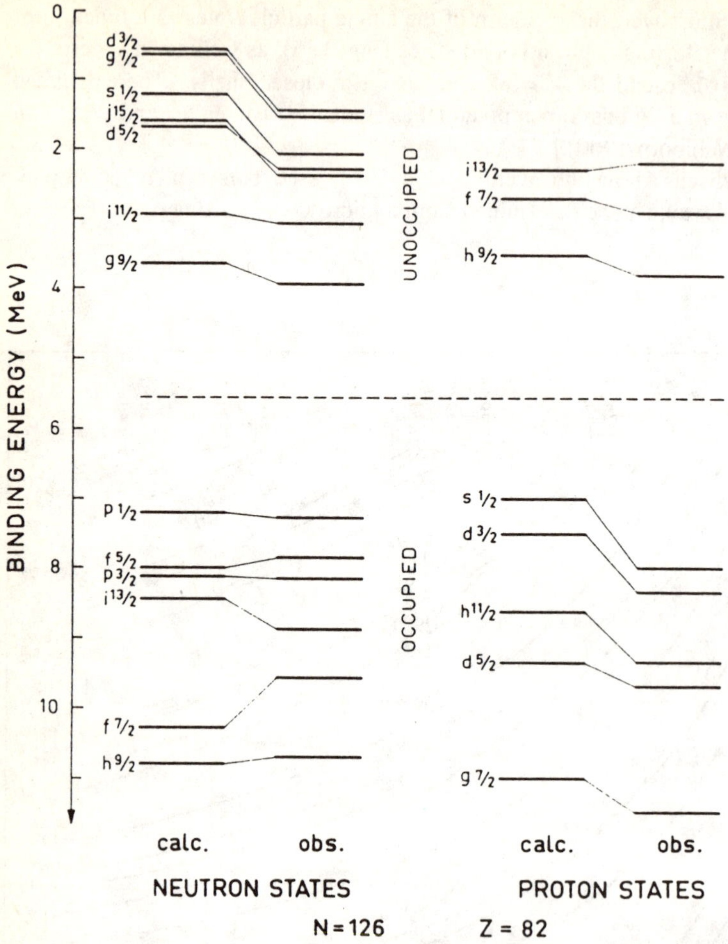

Fig. 3.16. The empirical values for the binding energy of a single nucleon with respect to ^{208}Pb as taken from the experimental one-nucleon separation energies (see the quantities ΔB in Fig. 3.14). The calculated values have been taken from J. Blomqvist and S. Wahlborn (Blomqvist 1960) [Figure taken from (Bohr, Mottelson 1969)]

3.1.4 Hartree-Fock Methods: A Simple Approach

Suppose nucleons fill up a number of nucleon orbitals $\varphi_a(\boldsymbol{r})$ such as to form a density $\varrho(\boldsymbol{r})$ given in terms of the occupied single-particle states as

$$\varrho(\boldsymbol{r}) = \sum_{b \in F} \varphi_b^*(\boldsymbol{r}) \varphi_b(\boldsymbol{r}) \,. \tag{3.37}$$

Then the potential at a point $\boldsymbol{r}'$, generated because of the nucleon-nucleon two-body interaction $V(\boldsymbol{r}, \boldsymbol{r}')$ reads

$$U_H(\boldsymbol{r}') = \sum_{b\in F} \int \varphi_b^*(\boldsymbol{r})V(\boldsymbol{r},\boldsymbol{r}')\varphi_b(\boldsymbol{r})\,d\boldsymbol{r}\,. \tag{3.38}$$

We denote by $U_H(\boldsymbol{r}')$, the Hartree term neglecting exchange effects, and this term is used in the case of atoms. Within the atomic nucleus, $U_H(\boldsymbol{r}')$ is the direct term of the potential affecting the nucleon motion in the nucleus. The more correct, single-particle Schrödinger equation for the orbital $\varphi_i(\boldsymbol{r})$ now becomes

$$\begin{aligned} &-\frac{\hbar^2}{2m}\Delta\varphi_i(\boldsymbol{r}) + \sum_{b\in F}\int \varphi_b^*(\boldsymbol{r}')V(\boldsymbol{r},\boldsymbol{r}')\varphi_b(\boldsymbol{r}')d\boldsymbol{r}'\cdot\varphi_i(\boldsymbol{r}) \\ &\quad -\sum_{b\in F}\int \varphi_b^*(\boldsymbol{r}')V(\boldsymbol{r},\boldsymbol{r}')\varphi_b(\boldsymbol{r})\varphi_i(\boldsymbol{r}')\,d\boldsymbol{r}' = \varepsilon_i\varphi_i(\boldsymbol{r})\,. \end{aligned} \tag{3.39}$$

The second contribution on the left hand side takes into account the antisymmetry for two identical nucleons, one moving in the orbital $\varphi_b(\boldsymbol{r}')$, the other in the orbital $\varphi_i(\boldsymbol{r})$. The product wave function $\varphi_b(\boldsymbol{r}')\varphi_i(\boldsymbol{r})$ has to be replaced by $\varphi_b(\boldsymbol{r}')\varphi_i(\boldsymbol{r}) - \varphi_b(\boldsymbol{r})\varphi_i(\boldsymbol{r}')$.

The above Hartree-Fock equations [since for every $\varphi_i(\boldsymbol{r})$ an analogous differential equation is obtained, all coupling via the potential terms] can be written in shorthand form as

$$-\frac{\hbar^2}{2m}\Delta\varphi_i(\boldsymbol{r}) + U_H(\boldsymbol{r})\varphi_i(\boldsymbol{r}) - \int U_F(\boldsymbol{r},\boldsymbol{r}')\varphi_i(\boldsymbol{r}')\,d\boldsymbol{r}' = \varepsilon_i\varphi_i(\boldsymbol{r})\,, \tag{3.40}$$

with

$$\begin{aligned} U_H(\boldsymbol{r}) &= \sum_{b\in F}\int \varphi_b^*(\boldsymbol{r}')V(\boldsymbol{r},\boldsymbol{r}')\varphi_b(\boldsymbol{r}')\,d\boldsymbol{r}'\,, \\ U_F(\boldsymbol{r},\boldsymbol{r}') &= \sum_{b\in F}\varphi_b^*(\boldsymbol{r}')V(\boldsymbol{r},\boldsymbol{r}')\varphi_b(\boldsymbol{r})\,. \end{aligned} \tag{3.41}$$

The iterative Hartree-Fock method now starts from an initial guess of the average field, or of the wave functions, starting from the knowledge of $V(\boldsymbol{r},\boldsymbol{r}')$ to solve the coupled equations (3.40) in order to determine a better value for $U_H(\boldsymbol{r})$ and $U_F(\boldsymbol{r},\boldsymbol{r}')$, the $\varphi_i(\boldsymbol{r})$ and ε_i. One can thus proceed in this way until convergence in the above quantities results. Schematically, one has

$$\begin{array}{cccccc} U_{H(F)}^{(0)}(\boldsymbol{r}) & & U_{H(F)}^{(1)}(\boldsymbol{r}) & & U_{H(F)}^{(2)}(\boldsymbol{r}) & \ldots \\ \downarrow & \nearrow & \downarrow & \nearrow & \downarrow & \nearrow \\ \varphi_i^{(0)}(\boldsymbol{r}) & & \varphi_i^{(1)}(\boldsymbol{r}) & & \varphi_i^{(2)}(\boldsymbol{r}) & \\ \varepsilon_i^{(0)} & & \varepsilon_i^{(1)} & & \varepsilon_i^{(2)} & . \end{array} \tag{3.42}$$

At the end, a final field $U_H(\boldsymbol{r})$, wave function $\varphi_i(\boldsymbol{r})$, single-particle energy ε_i is obtained. It is now possible to prove that with the wave functions so obtained, by calculating the energy expectation value

$$E_{HF} = \langle \Psi_{a_1,a_2,\ldots,a_A}(\boldsymbol{r}_1, \ldots, \boldsymbol{r}_A)|H|\Psi_{a_1,a_2,\ldots,a_A}(\boldsymbol{r}_1, \ldots, \boldsymbol{r}_A)\rangle , \tag{3.43}$$

with the $\Psi_{a_1,a_2,\ldots,a_A}(\boldsymbol{r}_1, \ldots, \boldsymbol{r}_A)$ defined in (3.5), the minimal value is obtained.

In the present discussion, one assumes that the original one and two-body Hamiltonian (3.9) does *not* contain strong short-range correlations nor density dependent two-body interactions. In those cases, the variational aspect of Hartree-Fock theory becomes lost (Ring, Schuck 1980, de-Shalit, Feshbach 1974). Detailed discussions on the determination of the Hartree-Fock energy and on variational approaches to the energy of an interacting many-body system can be found in (Ring, Schuck 1980, Irvine 1972). Since it is not our aim to devote an extensive treatment to those aspects of the nuclear A-body system, we refer the reader to the above references. The situation with two-body density dependent interactions will be discussed in some detail in Chap. 8.

There now exist many alternative methods in addition to the Hartree-Fock Schrödinger equations to derive this condition. We do not go into detail about these aspects. We only mention that an often used method to determine the "best" wave functions $\varphi_i(\boldsymbol{r})$ is to expand in a harmonic oscillator basis

$$\varphi_i(\boldsymbol{r}) = \sum_k a_k^i \varphi_k^{\text{h.o.}}(\boldsymbol{r}) , \tag{3.44}$$

after which the coefficients a_k^i are determined so that the total energy E_{HF} (3.43) is minimized.

The Hartree-Fock wave functions (and potential) indeed give a firm basis to the independent particle shell-model approach for the study of nuclear excited states. Besides the determination of the single-particle energies that can be compared with the data and show good overall agreement throughout the nuclear mass table (see Fig. 3.16 for a comparison in ^{208}Pb), nuclear densities [charge densities $\varrho^c(\boldsymbol{r})$] have also been determined and compared in detail. We do this for ^{16}O–^{208}Pb (Fig. 3.17a) and compare in Fig. 3.17b their nuclear matter densities $\varrho^m(r)$ with that for the corresponding densities of nuclear matter. Calculations have been carried out using Skyrme type of effective interactions (see also Chap. 8) but other interactions give very similar results. Very recently, using the difference of charge densities as obtained via (e, e') scattering experiments at Saclay for ^{206}Pb-^{205}Tl, (Cavedon et al. 1982, Frois et al. 1983, Doe 1983)

$$\begin{aligned} \varrho(^{206}\text{Pb}) - \varrho(^{205}\text{Tl}) &= \sum_{b\in F} |\varphi_b(\boldsymbol{r})|^2(^{206}\text{Pb}) - \sum_{b\in F} |\varphi_b(\boldsymbol{r})|^2(^{205}\text{Tl}) \\ &= |\varphi_{3s^2_{1/2}}(\boldsymbol{r})|^2 , \end{aligned} \tag{3.45}$$

the shape of the $3s_{1/2}$ orbit could unambiguously be determined (Fig. 3.18). This gives a sound basis for the independent motion of nucleons as a very good picture of the nuclear A-body problem. Thus this is a good point to study more detailed features related to the residual nucleon-nucleon interactions remaining outside the average field (H_{res}). We shall study the two-particle and three-particle systems

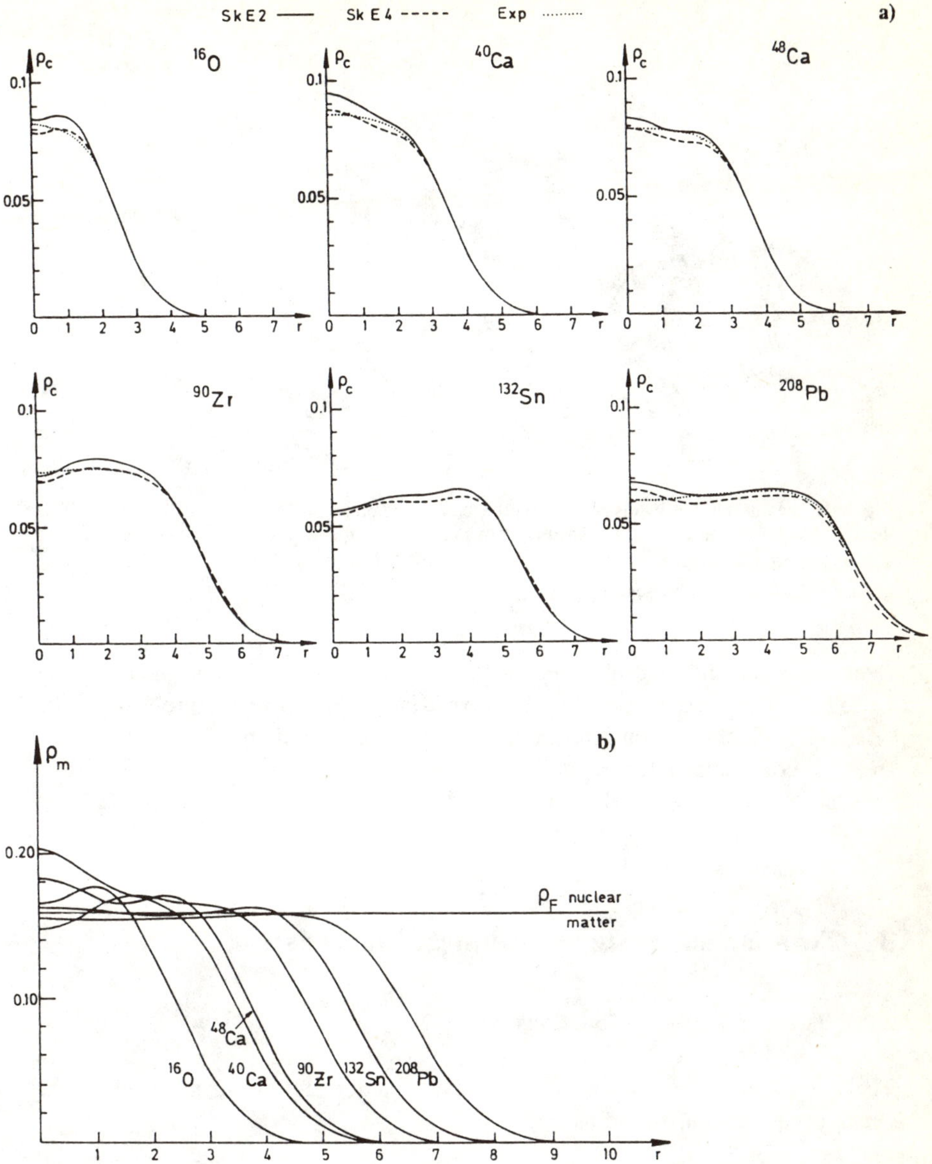

Fig. 3.17. (**a**) Charge densities for the magic nuclei ^{16}O, ^{40}Ca, ^{48}Ca, ^{90}Zr, ^{132}Sn and ^{208}Pb. The theoretical curves correspond to the effective interactions SkE2 and SkE4, respectively (see Chap. 8 for a detailed discussion of the extended Skyrme forces) and are compared with the data. The units for ϱ_c are efm^{-3}, with the radius r in fm. (**b**) Combined nuclear matter densities ϱ_m (fm^{-3}) for the above set of doubly-closed shell nuclei. Nuclear matter density ϱ_F (nuclear matter) is given as a "measure" for comparison

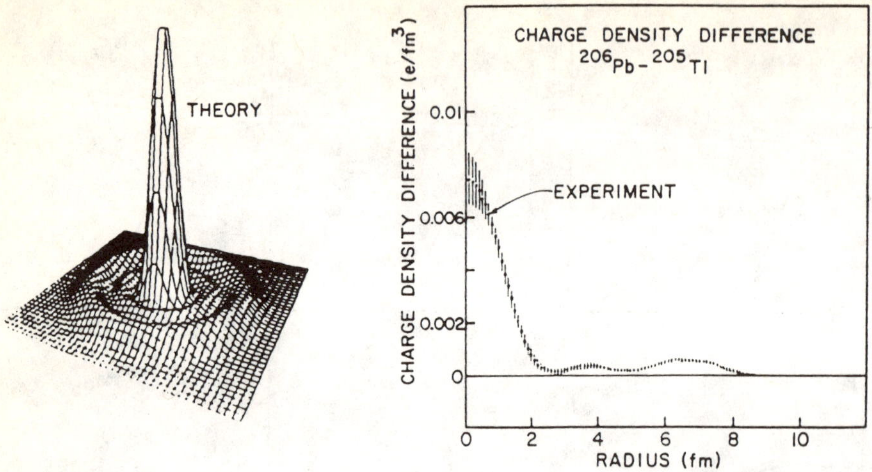

Fig. 3.18. The nuclear density distribution for the least bound proton in ^{206}Pb. The shell-model predicts the last ($3s_{1/2}$) proton in ^{206}Pb to have a sharp maximum at the centre, as shown on the left-hand side. On the right-hand side the nuclear charge density difference ϱ_c (^{206}Pb)$-\varrho_c$ (^{205}Tl) = $\varphi^2_{3s_{1/2}}(r)$ is given [taken from (Frois 1983) and Doe 1983)]

(identical nucleons) and also address the proton-neutron systems, incorporating isospin into the discussion. We also give some attention to the problem of the effective nucleon-nucleon force acting *in* the nuclear medium as compared to the free nucleon-nucleon force. In Chap. 8, we shall discuss a fully self-consistent version of the shell-model approach and show the state-of-the-art possibilities using present day high speed computers.

3.2 Two-particle Systems: Identical Nucleons

3.2.1 Two-particle Wavefunctions

The two-particle angular momentum wave function, following the methods of Chap. 1, can be constructed as

$$\psi\big(j_1(1)j_2(2); JM\big)\ . \tag{3.46}$$

(φ will always be the notation for a single-particle wave function, ψ and Ψ for composite wave functions). In what follows we denote with $1 \equiv \boldsymbol{r}_1, \boldsymbol{\sigma}_1, \ldots$ all coordinates of particle 1, and j_1 is a notation for all quantum numbers necessary to specify the single-particle state in a unique way $j_1 \equiv n_1, l_1, j_1$.

In describing the full Hamiltonian for a nucleus formed by a closed shell system described by H_0 and two extra identical valence nucleons described by

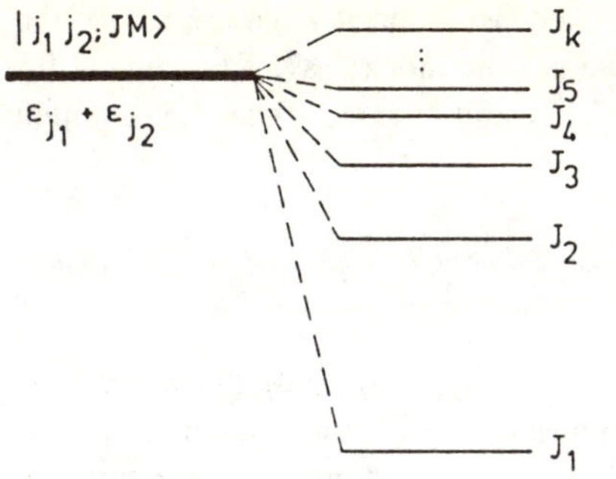

Fig. 3.19. Splitting of the two-particle states $\psi(j_1 j_2; JM)_{\mathrm{nas}}$ with $|j_1 - j_2| \leq J \leq j_1 + j_2$ through the residual nucleon-nucleon interaction $V_{1,2}$. The unperturbed energy for the configurations, $\varepsilon_{j_1} + \varepsilon_{j_2}$ (all J) is given on the left-hand side. The energy splitting is drawn schematically

the single-particle Hamiltonian $h_0(i)$ with $V(\boldsymbol{r}_1, \boldsymbol{r}_2)$ as a two-body interaction, the wave functions (3.46) are eigenfunctions of

$$H = H_0 + \sum_{i=1}^{2} h_0(i) , \tag{3.47}$$

with energy eigenvalue $E_0 + \varepsilon_{j_1} + \varepsilon_{j_2}$. The energy shift induced by the residual interaction now becomes

$$\Delta E(j_1 j_2; J) \equiv \langle j_1 j_2; JM | V(\boldsymbol{r}_1, \boldsymbol{r}_2) | j_1 j_2; JM \rangle , \tag{3.48}$$

and will split the degeneracy in J for the $(j_1 j_2)J$ multiplet of states (Fig. 3.19). In the remaining discussion, we always shall discard the core-part H_0 and only consider the energy of the valence nucleons. We also recall that the wave function (3.46) stands for

$$\psi(j_1(1) j_2(2); JM) = \sum_{m_1, m_2} \langle j_1 m_1, j_2 m_2 | JM \rangle \times \varphi_{j_1 m_1}(1) \varphi_{j_2 m_2}(2) . \tag{3.49}$$

In the case of identical particles ($p - p$; $n - n$) the wave function (3.46) needs to be made explicitly antisymmetric under the interchange of *all* coordinates of the two nucleons, however. We consider separately the cases (i) $j_1 \neq j_2$ and (ii) $j_1 = j_2$. In the discussion below, when expressing and evaluating matrix elements we shall always use the Dirac bra-ket notation and an index to show which matrix elements are calculated. We also use the notation

as : antisymmetrized wave function,
nas : normalized, antisymmetrized wave function,
no index : non-antisymmetrized wave function.

i) $j_1 \neq j_2$. We construct

$$\psi_{\mathrm{as}}(j_1 j_2; JM) = N \sum_{m_1, m_2} \langle j_1 m_1, j_2 m_2 | JM \rangle \times \left[\varphi_{j_1 m_1}(1) \varphi_{j_2 m_2}(2) - \varphi_{j_1 m_1}(2) \varphi_{j_2 m_2}(1) \right] . \tag{3.50}$$

Because of the explicit antisymmetrization, there is no need anymore to use the coordinates in the wave function on which we put the index "as". The demand for a normalized wave function results in $N = 1/\sqrt{2}$. Equation (3.50) can be rewritten as

$$\psi_{\text{nas}}(j_1 j_2; JM) = \frac{1}{\sqrt{2}} \left[\psi(j_1 j_2; JM) - (-1)^{j_1+j_2-J}\psi(j_2 j_1; JM)\right] . \qquad (3.51)$$

On the right hand side, we can also leave out coordinates by using the convention of always coupling the quantum state for particle 1 with the quantum state of particle 2 in *that* order reading from left to right! (standard convention for carrying out Racah-Algebra within the nuclear shell-model) (de-Shalit, Talmi 1963).

ii) $j_1 = j_2$. We now construct, as above,

$$\begin{aligned}\psi_{\text{nas}}(j^2; JM) &= N' \sum_{m_1,m_2} \langle jm_1, jm_2|JM\rangle \\ &\times \left[\varphi_{jm_1}(1)\varphi_{jm_2}(2) - \varphi_{jm_2}(1)\varphi_{jm_1}(2)\right] \\ &= N' \sum_{m_1,m_2} \left[\langle jm_1, jm_2|JM\rangle - \langle jm_2, jm_1|JM\rangle\right]\varphi_{jm_1}(1)\varphi_{jm_2}(2) \\ &= N'(1-(-1)^{2j-J}) \sum_{m_1,m_2} \langle jm_1, jm_2|JM\rangle\varphi_{jm_1}(1)\varphi_{jm_2}(2) . \end{aligned} \qquad (3.52)$$

Thus, only even J values are obtained ($J = 0, 2, \ldots, 2j - 1$) and $N' = \frac{1}{2}$.

As an example, we give the possible states for

- the $1d_{5/2}\, 1d_{3/2}$ two-particle states

$$\psi_{\text{as}}(1d_{5/2}\, 1d_{3/2}; JM) \qquad J = 1, 2, 3, 4 ,$$

- the $1d_{5/2}\, 1d_{5/2}$ two-particle states

$$\psi_{\text{as}}((1d_{5/2})^2; JM) \qquad J = 0, 2, 4 .$$

The two-body matrix elements including the residual interaction can be written as ($j_1 \neq j_2$)

$$\begin{aligned}\Delta E(j_1 j_2, J) &\equiv \langle j_1 j_2; JM|V_{12}|j_1 j_2; JM\rangle_{\text{nas}} \\ &= \langle j_1 j_2; JM|V_{12}|j_1 j_2; JM\rangle \\ &\quad - (-1)^{j_1+j_2-J}\langle j_1 j_2; JM|V_{12}|j_2 j_1; JM\rangle , \end{aligned} \qquad (3.53)$$

by assuming $V_{12} = V_{21}$ [where $V_{12} \equiv V(\boldsymbol{r}_1, \boldsymbol{r}_2)$] and if $j_1 = j_2$, one gets

$$\Delta E(j^2, J) \equiv \langle j^2; JM|V_{12}|j^2; JM\rangle . \qquad (3.54)$$

Before discussing methods to evaluate the matrix elements (3.53, 54) by using the two-particle wave functions as constructed above, we first discuss in some detail methods to get a better understanding of the effective interaction $V(\boldsymbol{r}_1, \boldsymbol{r}_2)$ itself.

3.2.2 Two-particle Residual Interaction

The problem of an appropriate choice for the two-nucleon interaction, especially when nucleons are surrounded by other nucleons (the nuclear medium) is a difficult one (Brown 1964, Brown, Kuo 1967). It does not have a unique answer since the "effective" force will be dependent on the particular model space that one considers in handling low-lying excited states. Still, a number of avenues have been followed in the past and we will discuss some of the more effective ones.

a) Effective Two-Body Matrix Elements

Within this approach, no attempt is made to pin down the (radial) shape of the interaction itself. Rather, the two-body matrix elements, together with the single-particle energies, are taken as "free" parameters for use in the Schrödinger equation to describe the nuclear excited states (Brussaard, Glaudemans 1977). Thus as parameters for a given model space one has

$$\varepsilon_{j_i}\,, \qquad \langle j_1 j_2; JM|V_{12}|j_3 j_4; JM\rangle\,.$$

If one takes as an example the full *sd* space (in order to discuss nuclei between ^{16}O and ^{40}Ca), the parameters are the two relative energies $\varepsilon_{2s_{1/2}} - \varepsilon_{1d_{5/2}}$, $\varepsilon_{2s_{1/2}} - \varepsilon_{1d_{3/2}}$ and the 63 two-body interaction matrix elements using as two-particle configurations all two particle states $|j_1 j_2; JM\rangle$ with $j_1, j_2 \in (2s_{1/2}, 1d_{3/2}, 1d_{5/2})$. Since in the Schrödinger equation for a given nucleus $A(Z, N)$ with n valence nucleons (protons and neutrons) $0 \le n \le 24$, the eigenvalues $E_{J_i^\pi}$ are functions of the essential parameters given above,

$$H\Psi_{J_i^\pi}(n) = E_{J_i^\pi}\Psi_{J_i^\pi}(n)\,, \tag{3.55}$$

one can use iterative least-squares methods to get convergence to a final set of two-body matrix elements and relative single-particle energies (Fig. 3.20). This method has been in use for a long time, and with the advent of high-speed computers has been applied to a large variety of nuclei. In particular the Utrecht group (Brussaard, Glaudemans 1977) and Wildenthal-Brown (Wildenthal 1976, 1985) have studied *p* shell nuclei and, more recently, the full *sd* shell using a single set of two-body matrix elements. We illustrate this by some examples for ^{27}Al, ^{28}Si, ^{29}Si in the middle of the *sd* shell for the excited states and for the full *sd* shell including binding energies (two-neutron separation energies (Figs. 3.21, 22).

Eventually, of course, one should try to compare and understand the fitted values with two-body matrix elements determined by other methods.

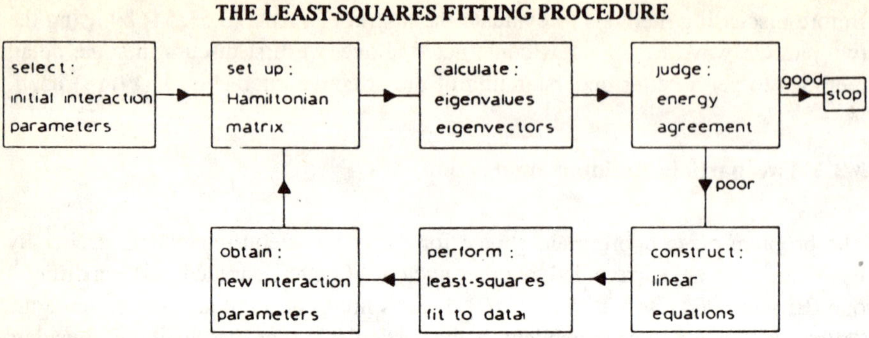

Fig. 3.20. Illustration of the various steps necessary in order to deduce an effective interaction (the two-body matrix elements $\langle j_1 j_2; J|V_{1,2}|j_3 j_4; J\rangle$ and single-particle energies ε_{j_i}) along the method discussed in Sect. 3.2.2a. Thereby a fit to experimental excitation energies and/or electromagnetic properties can be imposed [taken from (Brussaard 1977)]

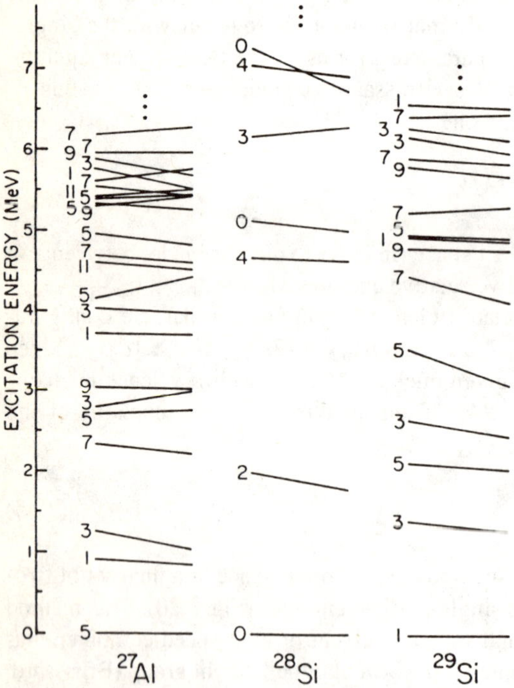

Fig. 3.21. Excited states for ^{27}Al, ^{28}Si and ^{29}Si. The ground-state energies are set equal in the figure. The left-hand ending of each line contains the calculated number, the right-hand point the corresponding experimental number. In odd-mass nuclei, the level is characterized by the value $2J$. Known negative parity states have been left out from the data. All other data are given (up to the upper energy considered here) [taken from (Wildenthal 1985)]

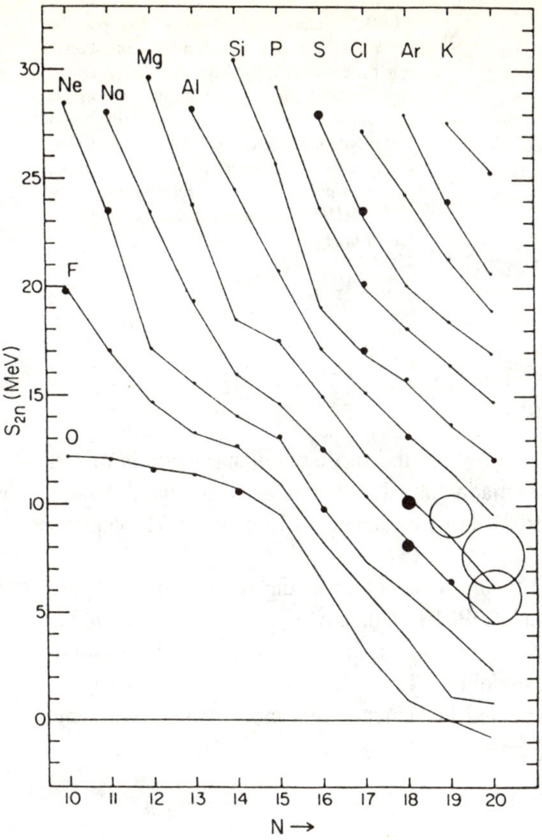

Fig. 3.22. Calculated and measured two-neutron separation energies S_{2n} along sd-isotope chains. The lines connect the theoretical points. The data are indicated by dot, solid or open circles. The diameters and their placement relative to the lines indicate the *magnitude and directions* of the differences $S_{2n}(\text{th}) - S_{2n}(\text{exp})$ [taken from (Wildenthal 1985)]

b) Realistic Interactions

In a completely different method, one tries to start from the free nucleon-nucleon interaction and to incorporate the necessary modifications to obtain the appropriate nuclear two-body interaction matrix elements. Pioneering work in this direction has been carried out by Brown and Kuo (Brown, Kuo 1967, Kuo, Brown 1966, Kuo et al. 1966).

Here we shall outline briefly how such "realistic" forces are determined, forces that are obtained by fitting to the free nucleon-nucleon scattering observables [phase shifts in all possible reaction channels $\delta(lS)\mathcal{J}$ (E_{lab}), polarization data, ...]. We illustrate in Fig. 3.23 those basic data in the *pp*, *nn*, *pn* case (isospin $T = 1$ and $T = 0$ channels) for channels denoted by ${}^{2S+1}(l)\mathcal{J}$ quantum numbers (l: relative orbital angular momentum, S: relative intrinsic spin, $\mathcal{J}$: total relative angular momentum) (see also Fig. 3.24).

In general one describes processes up to $E_{\text{lab}}\lambda 350$ MeV. Thus the most important partial waves are $l = 0, 1, 2$ S, P and D-waves (MacGregor et al. 1968a, b, Wright et al. 1967). At higher energies the concept of a non-relativistic scalar

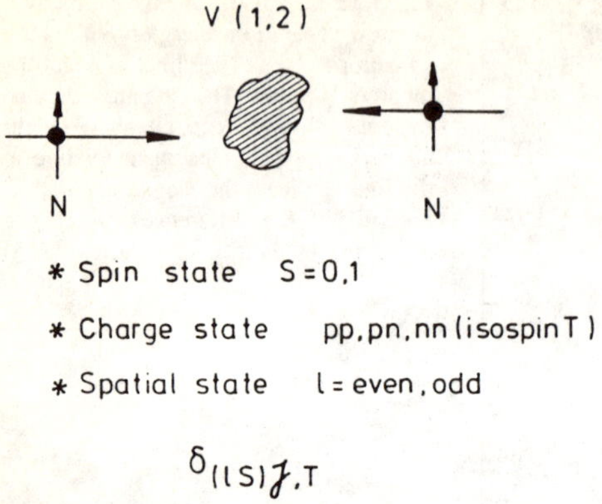

Fig. 3.23. Characterization of the possible channels in a free nucleon-nucleon scattering process $(lS)\mathcal{J}, T$ where $l, S, \mathcal{J}$ and T denote the relative angular momentum, the total intrinsic spin, the total relative angular momentum and isospin (see Sect. 3.4), respectively. The process can be described (for non-polarized quantities) by the phase shifts $\delta(lS)\mathcal{J}, T$ as a function of the scattering energy E_{lab}

potential loses its precise definition. Some of the most popular potentials of this "realistic" type are the hard-core Hamada-Johnston potential (Hamada, Johnston 1962), the Reid soft-core (Reid 1968) and the Tabakin (Tabakin 1964) velocity dependent potentials.

One starts from a potential with analytic structure in the radial part that goes into the one-pion exchange potential (OPEP) at distances $|\boldsymbol{r}_1 - \boldsymbol{r}_2| \cong 2fm$ but with strengths that have to be fitted so as to describe the data as well as possible. These general shapes, also dependent on intrinsic spin orientation, charge of the interacting particles, etc., are dictated by general invariance principles (Ring, Schuck 1980):

i) under exchange of the coordinates,
ii) translation
iii) Galilean
iv) space reflection
v) time reversal
vi) rotational, in coordinate space
vii) rotational, in charge space.

Thus for local forces not depending on the velocity, the central force is the most important part:

$$\begin{aligned} V_C(1,2) = {} & V_0(r) + V_\sigma(r)\boldsymbol{\sigma}_1 \cdot \boldsymbol{\sigma}_2 + V_\tau(r)\boldsymbol{\tau}_1 \cdot \boldsymbol{\tau}_2 \\ & + V_{\sigma\tau}(r)\boldsymbol{\sigma}_1 \cdot \boldsymbol{\sigma}_2 \boldsymbol{\tau}_1 \cdot \boldsymbol{\tau}_2 \,. \end{aligned} \tag{3.56}$$

(We shall discuss isospin where the $\boldsymbol{\tau}$ operators occur in detail in Sect. 3.4.)

The remaining local part is of tensor character and given by

$$V_T(1,2) = \left[V_{T_0}(r) + V_{T_\tau}(r)\boldsymbol{\tau}_1 \cdot \boldsymbol{\tau}_2\right] S_{12}$$

with

$$S_{12} = \frac{3}{r^2}\left(\boldsymbol{\sigma}_1 \cdot \boldsymbol{r}\right)\left(\boldsymbol{\sigma}_2 \cdot \boldsymbol{r}\right) - \boldsymbol{\sigma}_1 \cdot \boldsymbol{\sigma}_2 \,. \tag{3.57}$$

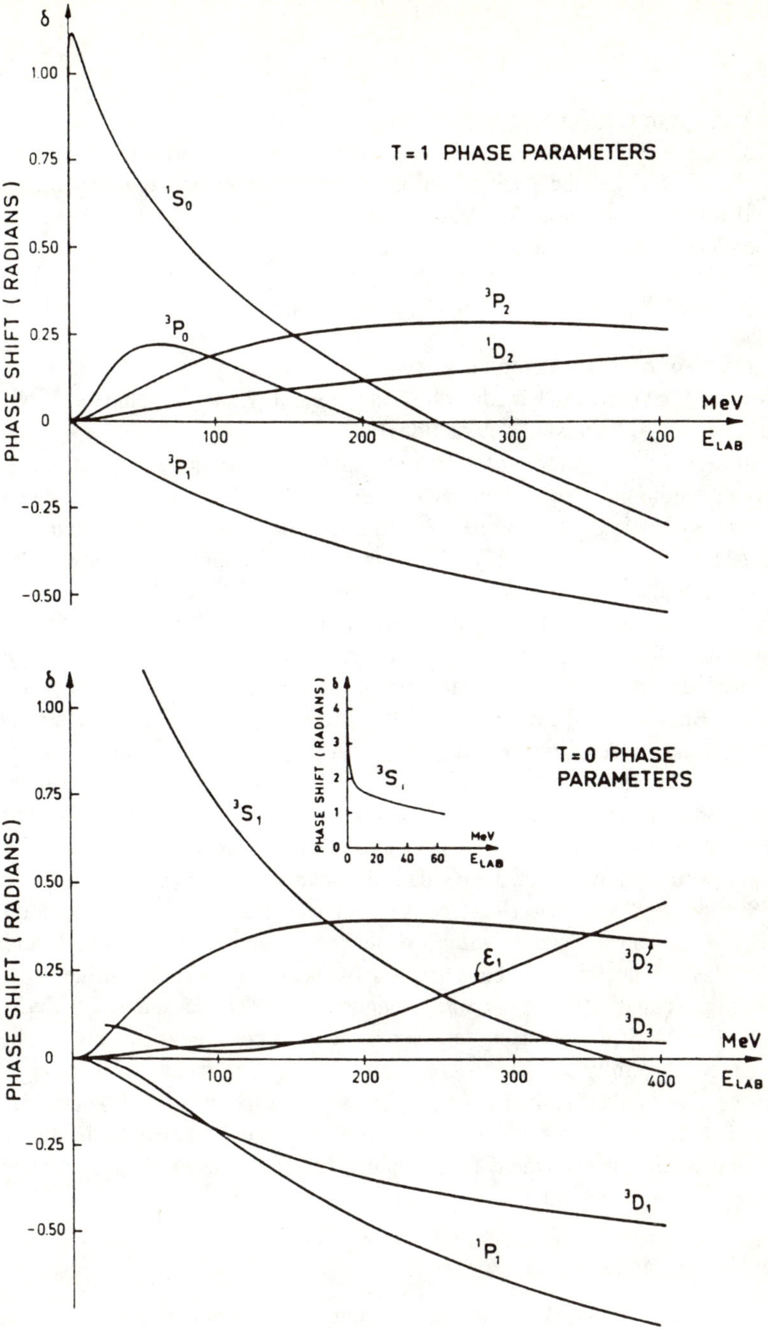

Fig. 3.24. The nuclear phase parameters for both the $T = 1$ and $T = 0$ channels and this for S, P and D channels, up to a laboratory energy of $E_{\text{lab}} = 400\,\text{MeV}$. The low-energy behavior (insert) is determined by the effective-range parameters, the data for $E > 24$ have been taken from (Arndt 1966). Phase shifts are given in radians [taken from (Bohr, Mottelson 1969)]

The important non-local term is the spin-orbit term

$$V_{LS}(1,2) = V_{LS}(r)\boldsymbol{l} \cdot \boldsymbol{S} , \tag{3.58}$$

and a quadratic spin-orbit can also be added.

In all situations the radial shape is not determined by invariance principles. The idea of Yukawa that the nucleon-nucleon interaction mainly derives from a meson field theory (Yukawa 1935) leads to the Yukawa-shape as a fundamental radial dependence given by the form

$$V_{\text{Yukawa}}(r) = \mathrm{e}^{-\mu r}/\mu r , \tag{3.59}$$

where $1/\mu \equiv \hbar/m\pi c$ is the Compton wavelength of the pion.

We show in Fig. 3.25 the Hamada-Johnston potential (Hamada, Johnston 1962) as a good example of "realistic" interactions.

Even though the evaluation of two-body matrix elements of these interactions becomes quite involved, they have been studied in detail (Kuo, Brown 1966, Brown, Kuo 1967). We compare in Fig. 3.26 the Tabakin matrix elements $\langle (1g_{7/2})^2; JM|V_{\text{Tabakin}}|(1g_{7/2})^2; JM\rangle$ with other matrix elements which we discuss below. One immediately observes that these "realistic" two-body matrix elements can not be used as such in shell model calculations. Only after bringing in the nuclear medium effects (a process called renormalization of the force) can one perform a serious comparison. We refer the reader to Kuo and Brown for such studies (Kuo, Brown 1966, Brown, Kuo 1967, Kuo 1974). In using realistic interactions in order to describe the nuclear many-body system, a number of extra complications arise in comparison to the effective and schematic interactions. Since most bare nucleon-nucleon forces contain an infinite, or at least a strong repulsive core, perturbation theory and Hartree-Fock methods do not yield even a good first approximation (Sect. 3.1.4). The Brueckner method (Brueckner et al. 1955, 1958, 1960, 1962, Brown 1964) consists in replacing the nucleon-nucleon potential by a G-matrix. Nuclei with few nucleons outside closed shells and doubly-closed shell nuclei have been treated by using Brueckner-Hartree-Fock methods (Ring, Schuck 1980, Brussaard, Glaudemans 1977, de-Shalit, Feshbach 1974, Brown 1964). Still, the strong tensor force causes considerable problems in the development of a fully self-consistent microscopic theory of nuclear structure.

Along the line of the "realistic" forces, in recent years fully field-theoretical forces have been constructed [the Paris potential (Lacombe 1975, 1980), the Bonn potential (Machleidt 1987)], where the exchange of the π-meson but also higher mesons (ω, ϱ, ...) is involved.

c) Schematic Interactions

In contrast to the complicated process of bringing the free nucleon-nucleon force into the nucleus, simple forces have been used that are immediately useful in applications into a given mass region. One defines a simple radial shape leading to a numerically rather simple calculation and determines the strength(s) so that

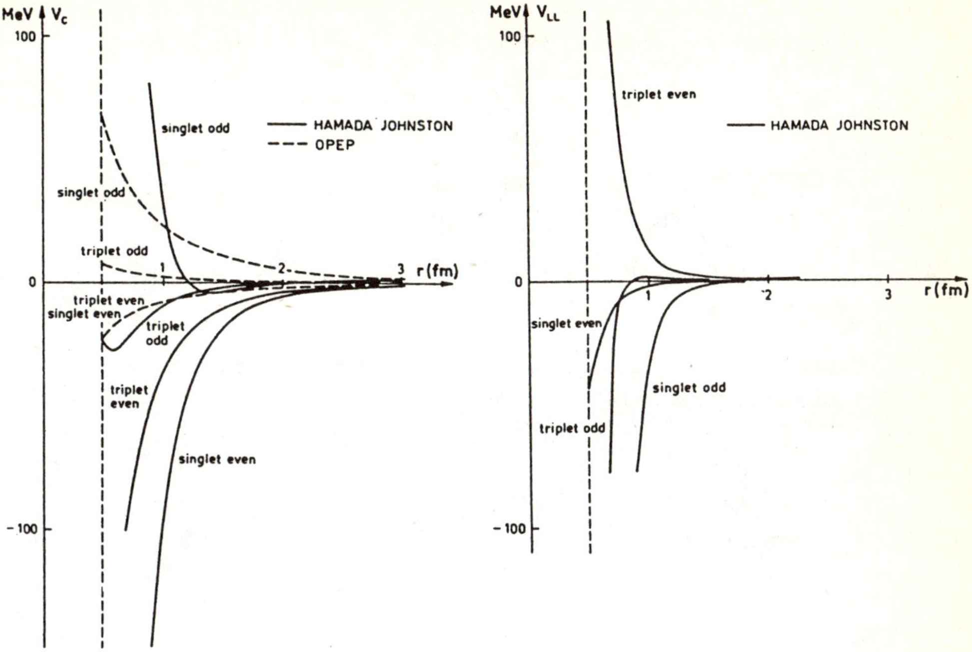

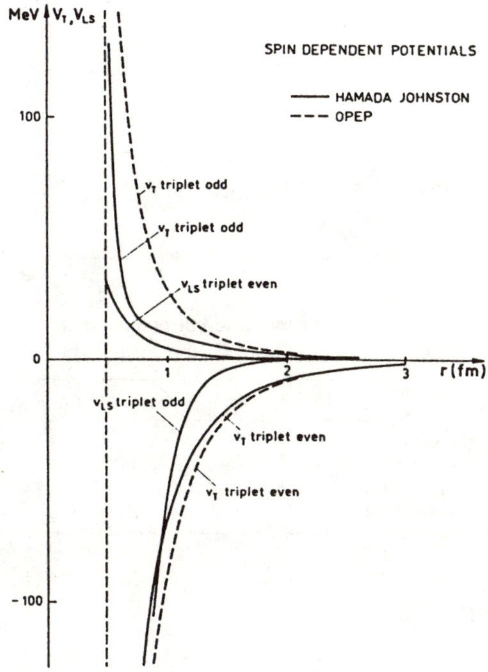

Fig. 3.25. The "realistic" nucleon-nucleon potentials as obtained from the analysis of T. Hamada and I.D. Johnston (Hamada 1962) are illustrated for both the central spin-orbit, tensor and quadratic spin-orbit parts [see (3.56–58)]. The dotted potentials (OPEP) correspond to the one-pion exchange potential. We give on p. 84 a box with the details of the Hamada-Johnston potential [taken from (Bohr, Mottelson 1969)]

The Hamada-Johnston Potential

$$V = V_C(r) + V_T(r)S_{12} + V_{LS}(r)\boldsymbol{l}\cdot\boldsymbol{S} + V_{LL}(r)L_{12} ,$$

with

$$S_{12} \equiv \frac{3}{r^2}(\boldsymbol{\sigma}_1\cdot\boldsymbol{r})\,(\boldsymbol{\sigma}_2\cdot\boldsymbol{r}) - \boldsymbol{\sigma}_1\cdot\boldsymbol{\sigma}_2 ,$$
$$L_{12} \equiv (\boldsymbol{\sigma}_1\cdot\boldsymbol{\sigma}_2)\boldsymbol{l}^2 - \tfrac{1}{2}\left[(\boldsymbol{\sigma}_1\cdot\boldsymbol{l})\,(\boldsymbol{\sigma}_2\cdot\boldsymbol{l}) + (\boldsymbol{\sigma}_2\cdot\boldsymbol{l})\,(\boldsymbol{\sigma}_1\cdot\boldsymbol{l})\right] ,$$
$$\equiv \left(\delta_{l,J} + \boldsymbol{\sigma}_1\cdot\boldsymbol{\sigma}_2\right)\boldsymbol{l}^2 - (\boldsymbol{l}\cdot\boldsymbol{S})^2 .$$

The radial functions are, at large distances, restricted by the condition of approaching the OPEP.

$$V_C(r) = v_0\left(\boldsymbol{\tau}_1\cdot\boldsymbol{\tau}_2\right)\,\left(\boldsymbol{\sigma}_1\cdot\boldsymbol{\sigma}_2\right)Y(x)\left[1 + a_C Y(x) + b_C Y^2(x)\right] ,$$
$$V_T(r) = v_0\left(\boldsymbol{\tau}_1\cdot\boldsymbol{\tau}_2\right)\,\left(\boldsymbol{\sigma}_1\cdot\boldsymbol{\sigma}_2\right)Z(x)\left[1 + a_T Y(x) + b_T Y^2(x)\right] ,$$
$$V_{LS}(r) = g_{LS} v_0 Y^2(x)\left[1 + b_{LS} Y(x)\right] ,$$
$$V_{LL}(r) = g_{LL} v_0 \frac{Z(x)}{x^2}\left[1 + a_{LL} Y(x) + b_{LL} Y^2(x)\right] ,$$
$$v_0 = \frac{1}{3}\frac{f^2}{\hbar c} m_\pi c^2 = 3.65\,\mathrm{MeV} ,$$
$$x = \left(m_\pi c\right)/\hbar\cdot r = r/1.43\,\mathrm{fm} ,$$
$$Y(x) = \frac{1}{x}\exp(-x) ,$$
$$Z(x) = \left(1 + \frac{3}{x} + \frac{3}{x^2}\right)\cdot Y(x) .$$

In addition, an infinite repulsion at the radius $c = 0.49$ fm ($x_c = 0.343$), is assumed.

The optimum, adjusted parameters are given in the table.

	Singlet even	Triplet even	Singlet odd	Triplet odd
a_C	8.7	6.0	−8.0	−9.07
b_C	10.6	−1.0	12.0	3.38
a_T	−.	−0.5	−.	−1.29
b_T	−.	0.2	−.	0.55
g_{LS}	−.	2.77	−.	7.36
b_{LS}	−.	−0.1	−.	−7.1
g_{LL}	−0.033	0.1	−0.1	−0.033
a_{LL}	0.2	1.8	2.0	−7.3
b_{LL}	−0.2	−0.4	6.0	6.9

Thc values of the different potentials at the hard core $r = c$ have been determined as

	V_c	V_T	V_{LS}	V_{LL}
Singlet, even	−1460	–	–	−42
Triplet,even	−207	−642	34	668
Singlet, odd	2371	–	–	−6683
Triplet, odd	−23	173	−1570	−1087

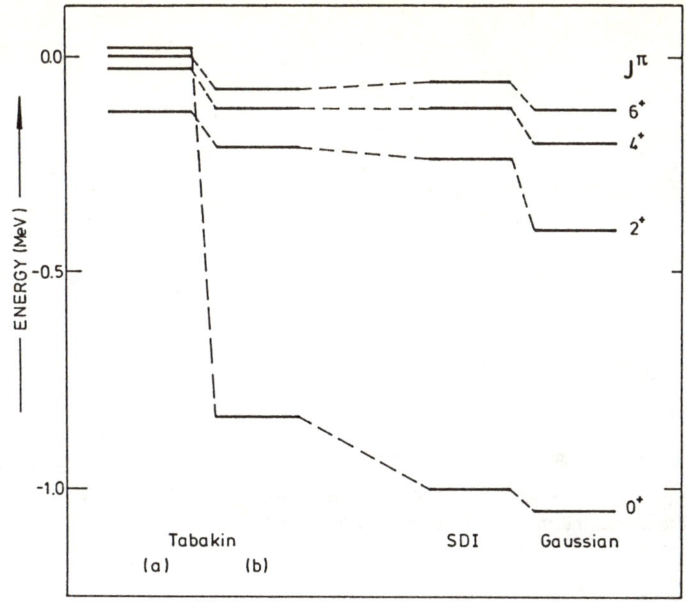

Fig. 3.26. Comparison of two-body matrix elements for the $(1g_{7/2})^2 J$ configuration according to a number of different interactions. We compare the surface-delta interaction (SDI) (see Appendix D), the Gaussian interaction and the realistic, velocity-dependent Tabakin interaction (Tabakin 1964). (i) For the Tabakin force, we give the total, bare matrix elements (*a*) and the Tabakin matrix elements corresponding to the $^1S_0$ channel only (*b*). (ii) The Gaussian interaction of Fig. 3.27, with the projection operators of (3.64) i.e. $V(r) = V_0(P_S + tP_T)$, (with $V_0 = -35\,\text{MeV}$ and $t = +0.2$). (iii) A SDI interaction, such that for the force $V_{\text{SDI}}(r) = -4\pi A'_T \delta(\boldsymbol{r}_1 - \boldsymbol{r}_2)\delta(r_1 - R_0)$, the product $A_T = A'_T C(R_0) = 0.25\,\text{MeV}$ with $C(R_0) \equiv R^4_{1g_{7/2}}(R_0) R_0^2$

the nuclear structure can be well described. Eventually, the invariance principles discussed above are also incorporated to suit certain specific purposes.

Often used potentials for $V(1,2)$ are the Yukawa potential, $e^{-\mu r}/\mu r$, square-well potential, the Gaussian potential, $e^{-\mu r^2}$ and δ or surface-δ (SDI) interactions (Brussaard, Glaudemans 1977): $\delta(r)$ or $\delta(r)\delta(r_1 - R_0)$.

We give a combined illustration of various potentials in Fig. 3.27. In the brief Appendix D, we point out that the SDI interaction is actually much better founded than one would think and relates to properties of the nucleon-nucleon free scattering process.

The schematic interactions all illustrate the short-range aspects of the nucleon-nucleon interaction in the nucleus, with the δ-function form as an extreme case. The latter form is particularly interesting to study the main features of the nuclear two-body interaction matrix since one can work out most results in analytic form when using the single-particle wave functions constructed in Sect. 3.1.

In addition specific combinations of the intrinsic spin and charge properties often occur, e.g.,

$$P_\sigma \equiv \tfrac{1}{2}\left(1 + \boldsymbol{\sigma}_1 \cdot \boldsymbol{\sigma}_2\right) . \tag{3.60}$$

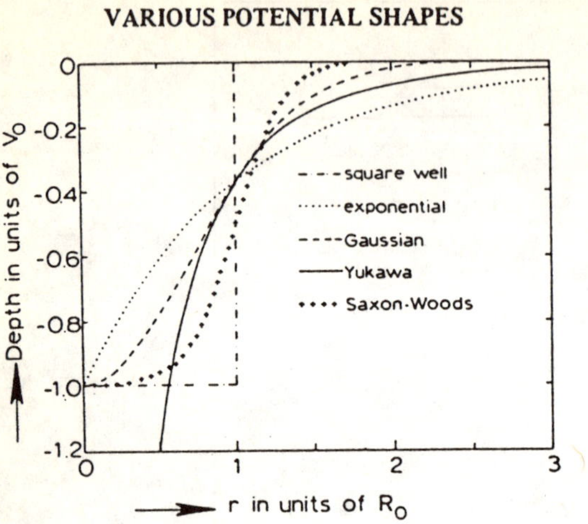

Fig. 3.27. Plots of the potentials, given below. In addition, the Woods-Saxon potential with a diffuseness parameter $a = 0.7$ is also shown [taken from (Brussaard, Glaudemans 1977)]

Square well	$V(r) = -V_0 \quad r \leq R_0$, $\quad\;\; = 0 \quad\;\; r > R_0$,
Exponential potential	$V(r) = -V_0 \exp(-r/R_0)$,
Gaussian potential	$V(r) = -V_0 \exp(-r^2/R_0^2)$,
Yukawa potential	$V(r) = -V_0 \exp(-r/R_0)/(r/R_0)$.

This form is called the Bartlett term and is interesting for the following reason. If we add the intrinsic spins of the interacting nucleons

$$S = s_1 + s_2 \,, \tag{3.61}$$

we have

$$\sigma_1 \cdot \sigma_2 = 2\left[S^2 - s_1^2 - s_2^2\right] \,, \tag{3.62}$$

or for expectation value for an $S = 0$ or $S = 1$ state

$$\langle \sigma_1 \cdot \sigma_2 \rangle_S = 2S(S+1) - 3 \,. \tag{3.63}$$

Thus, the operator P_σ is an exchange operator since

$$P_\sigma \psi_{S=1} = \psi_{S=1} \,,$$
$$P_\sigma \psi_{S=0} = -\psi_{S=0} \,,$$

and the two spin coordinates appear in a symmetric way in the $S = 1$ (parallel) state and in an antisymmetric way in the $S = 0$ (antiparallel) state. From the P_σ's, projection operators for the $S = 1$ (triplet) and $S = 0$ (singlet) two-nucleon state can be defined:

$$\begin{aligned} P_S &\equiv \Pi_{S=0} = \tfrac{1}{2}\left(1 - P_\sigma\right) \,, \\ P_T &\equiv \Pi_{S=1} = \tfrac{1}{2}\left(1 + P_\sigma\right) \,. \end{aligned} \tag{3.64}$$

Similarly, an operator can be defined in charge space for the isospin operators, giving the Heisenberg term $P_\tau = \frac{1}{2}(1 + \boldsymbol{\tau}_1 \cdot \boldsymbol{\tau}_2)$. Moreover, the Majorana term which interchanges the spatial coordinates $\boldsymbol{r}_1 - \boldsymbol{r}_2 \rightarrow -(\boldsymbol{r}_1 - \boldsymbol{r}_2)$, induced by the operator P_r can be introduced.

3.2.3 Calculation of Two-Body Matrix Elements

a) Central Interactions

Starting from a general central interaction $V(|\boldsymbol{r}_1 - \boldsymbol{r}_2|)$ one can expand it within a complete set of functions (e.g., with the Legendre polynomials). Thus, we can express the interaction as

$$V(|\boldsymbol{r}_1 - \boldsymbol{r}_2|) = \sum_{k=0}^{\infty} v_k(r_1, r_2) P_k(\cos\theta_{12}) , \tag{3.65}$$

with

$$P_k(\cos\theta_{12}) = \sum_{\kappa} 4\pi/(2k+1) Y_k^{\kappa^*}(\Omega_1) Y_k^{\kappa}(\Omega_2) . \tag{3.66}$$

From the normalized, antisymmetric two-particle wave functions constructed in Sect. 3.2.1, we obtain for the diagonal matrix element

$$\begin{aligned}\langle j_1 j_2; JM | V_{12} | j_1 j_2; JM \rangle_{\text{nas}} &= \langle j_1 j_2; JM | V_{12} | j_1 j_2; JM \rangle_{\text{dir}} \\ &\quad - (-1)^{j_1+j_2-J} \langle j_1 j_2; JM | V_{12} | j_2 j_1; JM \rangle_{\text{exch}} ,\end{aligned} \tag{3.67}$$

with a direct and an exchange term. We carry out the calculation in some detail, for both the direct and the exchange term:

The direct term can be written

$$\langle j_1 j_2; JM | V_{12} | j_1 j_2; JM \rangle_{\text{dir}} = \sum_k f_k F^k , \tag{3.68}$$

with

$$f_k = 4\pi/(2k+1) \langle j_1 j_2; JM | \boldsymbol{Y}_k(\Omega_1) \cdot \boldsymbol{Y}_k(\Omega_2) | j_1 j_2; JM \rangle , \tag{3.69}$$

and

$$\begin{aligned}F^k &= F^k(n_1 l_1, n_2 l_2) \\ &= \int |u_{n_1 l_1}(r_1) u_{n_2 l_2}(r_2)|^2 v_k(r_1, r_2)\, dr_1\, dr_2 .\end{aligned} \tag{3.70}$$

The expression for f_k can be evaluated using the reduction rule I from Chap. 2, and yields

$$\begin{aligned}f_k = &4\pi/(2k+1) \cdot (-1)^{j_2+j_1+J} \begin{Bmatrix} j_1 & j_2 & J \\ j_2 & j_1 & k \end{Bmatrix} \\ &\times \langle j_1 \| \boldsymbol{Y}_k \| j_1 \rangle \langle j_2 \| \boldsymbol{Y}_k \| j_2 \rangle .\end{aligned} \tag{3.71}$$

For the exchange matrix element we get, analogously

$$\langle j_1 j_2; JM|V_{12}|j_2 j_1; JM\rangle_{\text{exch}} = \sum_{k=0}^{\infty} g_k G^k \,, \tag{3.72}$$

with

$$g_k = 4\pi/(2k+1) \cdot (-1)^{1+J} \begin{Bmatrix} j_1 & j_2 & J \\ j_1 & j_2 & k \end{Bmatrix} \times \langle j_1\|\boldsymbol{Y}_k\|j_2\rangle\langle j_2\|\boldsymbol{Y}_k\|j_1\rangle \,, \tag{3.73}$$

and

$$G^k = G^k(n_1 l_1, n_2 l_2) \equiv \int u_{n_1 l_1}(r_1) u_{n_2 l_2}(r_2) u_{n_1 l_1}(r_2) \times u_{n_2 l_2}(r_1) v_k(r_1, r_2)\, dr_1\, dr_2 \,. \tag{3.74}$$

Thus the total matrix element becomes for

i) $j_1 \neq j_2$

$$\Delta E_{j_1 j_2, J} = \sum_k f_k F^k - (-1)^{j_1+j_2-J} \sum_k g_k G^k \,, \tag{3.75}$$

and

ii) $j_1 = j_2$

$$\Delta E_{j^2, J} = \sum_k f_k F^k \,. \tag{3.76}$$

For a general central interaction when the $v_k(r_1, r_2)$ are determined, in most cases one has to evaluate the Slater integrals G^k and F^k numerically. If one chooses a simpler schematic force like the $\delta(\boldsymbol{r}_1 - \boldsymbol{r}_2)$ or the SDI $\delta(\boldsymbol{r}_1 - \boldsymbol{r}_2)\delta(r_1 - R_0)$ force, the integrals simplify significantly.

In Appendix F, we bring together the necessary Racah algebra expressions needed to carry out the calculation of the $\langle j\|\boldsymbol{Y}_k\|j'\rangle$ reduced matrix elements as well as the sums in (3.75, 76). We derive in Appendix E the multipole expansion for a zero-range $\delta(\boldsymbol{r}_1 - \boldsymbol{r}_2)$ interaction.

The result is

$$\delta(\boldsymbol{r}_1 - \boldsymbol{r}_2) = \sum_k \delta(r_1 - r_2)/(r_1 r_2) \cdot (2k+1)/4\pi \cdot P_k(\cos\theta_{12}) \,, \tag{3.77}$$

and so we easily get the $v_k(r_1, r_2)$ coefficient as

$$v_k(r_1, r_2) = \delta(r_1 - r_2)/(r_1 r_2)(2k+1)/4\pi \,. \tag{3.78}$$

Thus, the above radial integrals F^k and G^k reduce to

$$
\begin{aligned}
F^k &= \frac{2k+1}{4\pi}\int_0^\infty \frac{1}{r^2}\Big[u_{n_1l_1}(r)u_{n_2l_2}(r)\Big]^2 dr = (2k+1)F^0 \,, \\
G^k &= \frac{2k+1}{4\pi}\int_0^\infty \frac{1}{r^2}\Big[u_{n_1l_1}(r)u_{n_2l_2}(r)\Big]^2 dr = (2k+1)F^0 \,.
\end{aligned} \tag{3.79}
$$

Furthermore, we use the explicit expression for the reduced matrix element $\langle j\|\boldsymbol{Y}_k\|j\rangle$, which becomes (Appendix F) after using the reduction rule I of Chap. 2

$$
\begin{aligned}
\langle(1/2l)j\|\boldsymbol{Y}_k\|(1/2l')j'\rangle &= (-1)^{j-1/2}\hat{j}\hat{j}'\hat{k}/\sqrt{4\pi} \\
&\times \begin{pmatrix} j & k & j' \\ -\frac{1}{2} & 0 & \frac{1}{2} \end{pmatrix} \frac{1}{2}\left(1+(-1)^{l+l'+k}\right) .
\end{aligned} \tag{3.80}
$$

Combining the above result with (3.71) we get the explicit form of the f_k term

$$
\begin{aligned}
f_k &= (-1)^{2(j_1+j_2)+J-1}(2j_1+1)(2j_2+1)\tfrac{1}{2}\left(1+(-1)^k\right) \\
&\times \begin{Bmatrix} j_1 & j_2 & J \\ j_2 & j_1 & k \end{Bmatrix} \begin{pmatrix} j_1 & k & j_1 \\ -\frac{1}{2} & 0 & \frac{1}{2} \end{pmatrix} \begin{pmatrix} j_2 & k & j_2 \\ -\frac{1}{2} & 0 & \frac{1}{2} \end{pmatrix} ,
\end{aligned} \tag{3.81}
$$

and the total direct contribution reads (3.79)

$$
\begin{aligned}
\sum_k f_k F^k &= F^0(2j_1+1)(2j_2+1)\sum_k (-1)^{J-1}\frac{1}{2}\left(1+(-1)^k\right) \\
&\times (2k+1)\begin{Bmatrix} j_1 & j_2 & J \\ j_2 & j_1 & k \end{Bmatrix} \begin{pmatrix} j_1 & k & j_1 \\ -\frac{1}{2} & 0 & \frac{1}{2} \end{pmatrix} \begin{pmatrix} j_2 & k & j_2 \\ -\frac{1}{2} & 0 & \frac{1}{2} \end{pmatrix} .
\end{aligned} \tag{3.82}
$$

Using the expressions of Appendix F where

$$
\begin{aligned}
&\sum_k (2k+1)\begin{Bmatrix} j_a & j_b & J \\ j_d & j_c & k \end{Bmatrix} \begin{pmatrix} j_a & k & j_c \\ -\frac{1}{2} & 0 & \frac{1}{2} \end{pmatrix} \begin{pmatrix} j_b & k & j_d \\ -\frac{1}{2} & 0 & \frac{1}{2} \end{pmatrix} \\
&= (-1)^{j_b+j_d+J}\begin{pmatrix} j_b & j_a & J \\ -\frac{1}{2} & -\frac{1}{2} & 1 \end{pmatrix} \begin{pmatrix} j_c & j_d & J \\ \frac{1}{2} & \frac{1}{2} & -1 \end{pmatrix} ,
\end{aligned} \tag{3.83}
$$

$$
\begin{aligned}
&\sum_k (2k+1)(-1)^k\begin{Bmatrix} j_a & j_b & J \\ j_d & j_c & k \end{Bmatrix} \begin{pmatrix} j_a & k & j_c \\ -\frac{1}{2} & 0 & \frac{1}{2} \end{pmatrix} \begin{pmatrix} j_b & k & j_d \\ -\frac{1}{2} & 0 & \frac{1}{2} \end{pmatrix} \\
&= (-1)^{J-2j_d}\begin{pmatrix} j_b & j_a & J \\ \frac{1}{2} & -\frac{1}{2} & 0 \end{pmatrix} \begin{pmatrix} j_c & j_d & J \\ \frac{1}{2} & -\frac{1}{2} & 0 \end{pmatrix} ,
\end{aligned} \tag{3.84}
$$

we get the direct term

$$
\begin{aligned}
\langle j_1j_2;JM|V_{12}|j_1j_2;JM\rangle_{\text{dir}} &= F^0(2j_1+1)(2j_2+1)\tfrac{1}{2} \\
&\times \left[\begin{pmatrix} j_1 & j_2 & J \\ \frac{1}{2} & -\frac{1}{2} & 0 \end{pmatrix}^2 + \begin{pmatrix} j_1 & j_2 & J \\ \frac{1}{2} & \frac{1}{2} & -1 \end{pmatrix}^2\right] .
\end{aligned} \tag{3.85}
$$

In a completely analogous way one can calculate the exchange term on which we give some intermediate results, i.e.,

$$g_k = (-1)^{J+j_1+j_2} \begin{Bmatrix} j_1 & j_2 & J \\ j_1 & j_2 & k \end{Bmatrix} (2j_1+1)(2j_2+1) \times \begin{pmatrix} j_1 & k & j_2 \\ -\frac{1}{2} & 0 & \frac{1}{2} \end{pmatrix} \begin{pmatrix} j_2 & k & j_1 \\ -\frac{1}{2} & 0 & \frac{1}{2} \end{pmatrix} \frac{1}{2}\left(1+(-1)^{l_1+l_2+k}\right), \quad (3.86)$$

with the exchange matrix element

$$\sum_k g_k G^k = F^0 (2j_1+1)(2j_2+1)(-1)^{J+j_1+j_2} \times \sum_k \tfrac{1}{2}\left(1+(-1)^{l_1+l_2+k}\right) \begin{Bmatrix} j_1 & j_2 & J \\ j_1 & j_2 & k \end{Bmatrix} \times \begin{pmatrix} j_1 & k & j_2 \\ -\frac{1}{2} & 0 & \frac{1}{2} \end{pmatrix} \begin{pmatrix} j_2 & k & j_1 \\ -\frac{1}{2} & 0 & \frac{1}{2} \end{pmatrix} (2k+1)\,. \quad (3.87)$$

Again using (3.83, 84), the exchange matrix element finally becomes

$$\langle j_1 j_2; JM|V_{12}|j_2 j_1; JM\rangle_{\text{exch}} = F^0 (2j_1+1)(2j_2+1) \times (-1)^{j_1+j_2+J} \frac{1}{2}\left[\begin{pmatrix} j_1 & j_2 & J \\ \frac{1}{2} & \frac{1}{2} & -1 \end{pmatrix}^2 - (-1)^{l_1+l_2+J} \begin{pmatrix} j_2 & j_1 & J \\ \frac{1}{2} & -\frac{1}{2} & 0 \end{pmatrix}^2 \right]. \quad (3.88)$$

The sum gives now when $j_1 \neq j_2$,

$$\Delta E_{j_1 j_2, J} = F^0 (2j_1+1)(2j_2+1) \begin{pmatrix} j_1 & j_2 & J \\ \frac{1}{2} & -\frac{1}{2} & 0 \end{pmatrix}^2 \times \left(1+(-1)^{l_1+l_2+J}\right)/2\,. \quad (3.89)$$

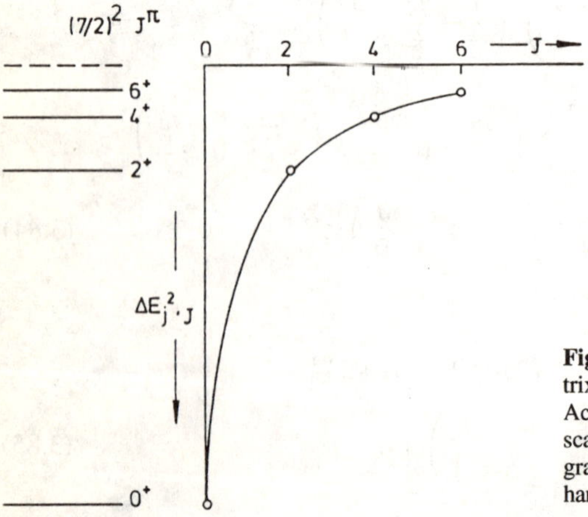

Fig. 3.28. Illustration of the two-particle matrix elements $\Delta E_{j^2,J}$ for the $j=\frac{7}{2}$ orbital. According to (3.90), the matrix elements are scaled with the corresponding Slater integrals $4F^0$. We also construct (on the right-hand side) the $\Delta E_{j^2,J}$ vs. J plot

Table 3.1. The two-body matrix elements of (3.90) and this for a $j = \frac{7}{2}$ particle. The matrix elements are expressed in units $4F^0$ where F^0 denotes the corresponding Slater integral for the $j = \frac{7}{2}$ particle using a δ-residual interaction

J	$\begin{pmatrix} j & j & J \\ \frac{1}{2} & -\frac{1}{2} & 0 \end{pmatrix}$	$\Delta E_{j^2,J}/4F^0$
0	$-\sqrt{\frac{1}{8}}$	1
2	$\sqrt{\frac{5}{3.7.8}}$	$\frac{5}{21} = 0.238$
4	$-\sqrt{\frac{9}{8.7.11}}$	$\frac{9}{77} = 0.117$
6	$\sqrt{\frac{25}{3.8.11.13}}$	$\frac{25}{429} = 0.058$

Using the same method as discussed above, for the specific case of $j_1 = j_2$, one obtains the result

$$\Delta E_{j^2,J} = \frac{F^0}{2}(2j+1)^2 \begin{pmatrix} j & j & J \\ \frac{1}{2} & -\frac{1}{2} & 0 \end{pmatrix}^2 . \tag{3.90}$$

As an example, we show this in Fig. 3.28 for a $j = 7/2$ particle, combining to a $J^\pi = 0^+$, 2^+, 4^+ and 6^+ state (Table 3.1).

b) Multipole Expansion

Instead of using the multipole summation relations from appendix F, as worked out explicitly in (3.83) and (3.84), the various multipole contributions to a particular two-body matrix element can be studied in more detail.

This sum over various multipoles, as depicted in (3.75) and (3.76), is restricted due to the angular momentum coupling conditions implied by the Wigner $6j$-symbol and Wigner $3j$-symbols which appear in (3.81) and (3.86). For the particular situation where $j_1 = j_2 = j$ i.e. evaluating diagonal two-body matrix elements, the above conditions imply that $0 \leq k \leq 2j$, with only even k values contributing. The specific angular momentum (J) dependence in each give multipole contribution only comes from the Wigner $6j$-symbol $\begin{Bmatrix} j & j & J \\ j & j & k \end{Bmatrix}$ and has the following interesting properties:

i) for $k = 0$ (the monopole part), a constant contribution for all J-values results,
ii) for $k = 2$, the Wigner $6j$-symbol results in a quadratic expression in $J(J+1)$ (see also Fig. 3.29),
iii) for high-k values (for $k \to \infty$), the contribution for the different J values is very small compared to the $J = 0$ values. Therefore, it is often stated that the pairing properties in a given two-body force result from the highest multipole contributions.

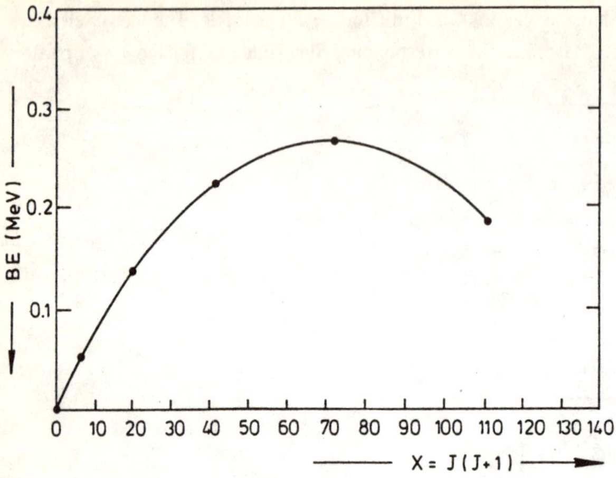

Fig. 3.29. Spin (J) dependence for the quadrupole ($k = 2$) component in the two-body matrix element for a $\langle j^2; JM|V|j^2; JM\rangle$ configuration with, in particular, $j = 1h_{11/2}$

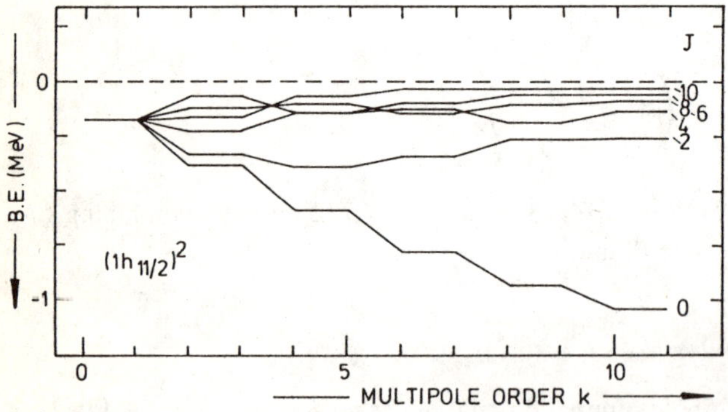

Fig. 3.30. Multipole decomposition ($0 \leq k \leq 10$) for the two-body matrix element $\langle(1h_{11/2})^2; JM|V|(1h_{11/2})^2; JM\rangle$, using a zero-range interaction

A full multipole decomposition is illustrated in Fig. 3.30 for a $(1h_{11/2})^2 J$ configuration. Here, only the multipoles $k = 0, 2, \ldots, 10$ contribute, with $k = 0$ giving the constant, monopole shift, $k = 2$ the quadrupole "parabolic" $J(J + 1)$ dependence and with the high-k values ($k = 8, 10$) exhibiting the pairing properties in the two-body matrix elements $\langle(1h_{11/2})^2; JM|V|(1h_{11/2})^2; JM\rangle$.

c) Examples

In Chap. 9 we provide a computer program that calculates the two-body matrix elements in order to evaluate the δ-function matrix elements. The Slater integrals for the δ-interaction F^0 are calculated fully analytically. They are given together with the code.

We compare the two-body matrix elements for several nuclei:

i) ^{18}O (Fig. 3.31): The $(1d_{5/2})^2 J$ matrix elements are compared with the experimental spectrum. The agreement is much improved when we also take more configurations into account ($2s_{1/2}$, $1d_{3/2}$) as well as the interactions among the different configurations. This we shall discuss in detail in Sect. 3.2.4.

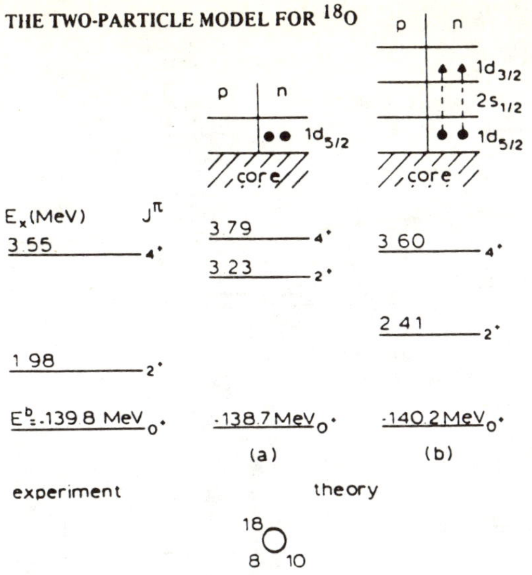

Fig. 3.31. A comparison between theory and experiment for both the binding energy and the excited states in ^{18}O (**a**) the spectrum for the neutron $(1d_{5/2})^2$ configuration only and (**b**) the spectrum for two neutrons in the full $(1d_{5/2}\,2s_{1/2}\,1d_{3/2})$ space. In both cases, a MSDI interaction was used to calculate the spectra. The Modified SDI (MSDI) differs from the regular SDI force in the addition of terms $V_{MSDI} = V_{SDI} + B\tau_1 \cdot \tau_2$. For obtaining the spectra in (**a**) and (**b**), the values $A_{T=1} = B = 25$ MeV/A were used [taken from (Brussaard, Glaudemans 1977)]

ii) ^{210}Po (Fig. 3.32): Here we compare the effect of the possible configurations on the description of the energy spectrum and the effect of the model space on the strength of the two-body MSDI interaction [*M*odified *S*urface *D*elta *I*nteraction (Brussaard, Glaudemans 1977)]. In one case, we only take the $(1h_{9/2})^2 J$, $(1h_{9/2}\,2f_{7/2})J$ and $(1h_{9/2}\,1i_{13/2})J$ configurations into account, whereas in the other case, all two-body configurations within the full $(1h_{9/2}, 2f_{7/2}, 1i_{13/2})$ space are taken into account.

iii) We also make a comparison for the $N = 82$ single-closed shell nuclei ^{134}Te-^{146}Gd (Fig. 3.33) where a two-quasi particle calculation was carried out. Full details of the quasi-particle excitations and the pairing degree of freedom will be discussed in Chap. 7. However, here one can clearly observe the typical features in the 0^+-2^+-4^+-6^+ separation that show the short range correlations in the nucleon-nucleon residual interaction.

d) Semi-Classical Interpretation

In comparing two-body matrix elements for many different mass regions, it is interesting to compare the matrix elements relative to the average matrix element (Fig. 3.34)

$$\bar{V} = \sum_J (2J+1)\langle (j_1)^2; JM|V_{12}|(j_1)^2; JM\rangle \Big/ \sum_J (2J+1) \,. \tag{3.91}$$

It is also possible to plot the matrix elements

$$\langle (j_1)^2; JM|V_{12}|(j_1)^2; JM\rangle / \bar{V} \,, \tag{3.92}$$

versus the angle between the classical orbits for the two valence nucleons θ_{12}, which is defined as

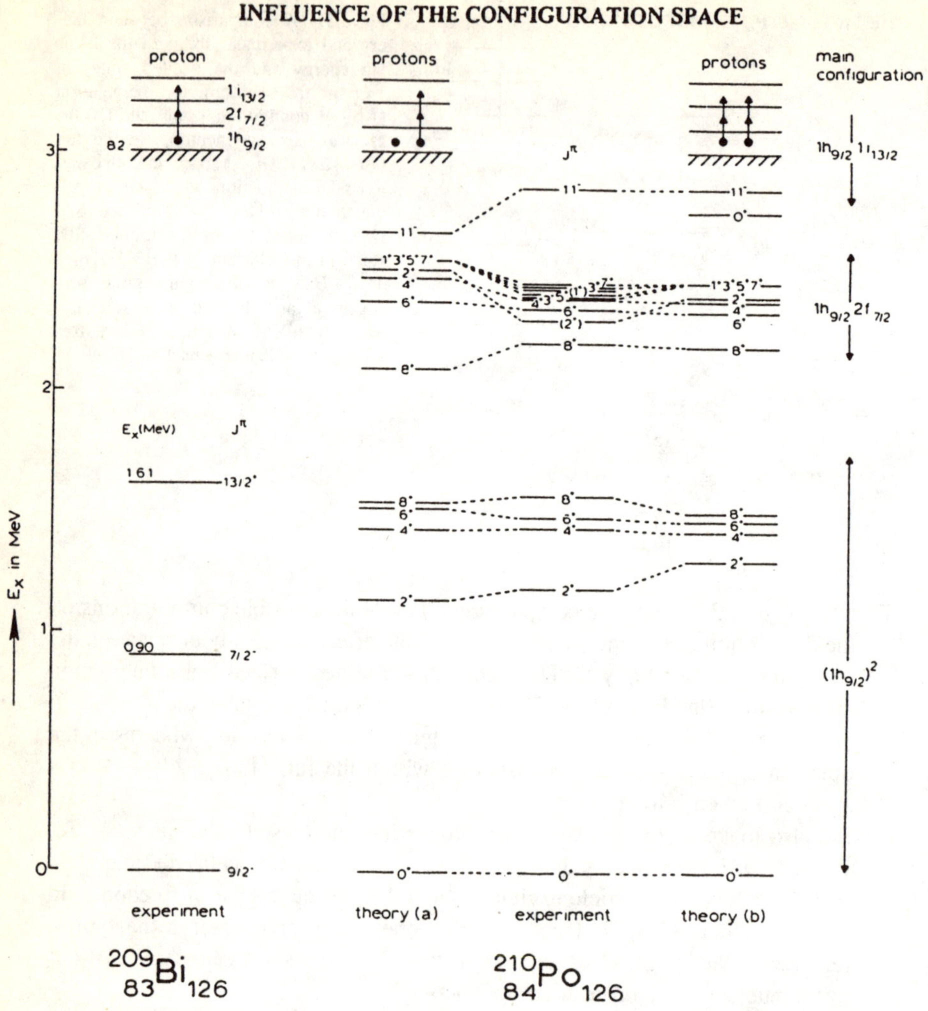

Fig. 3.32. The effect of the possible configurations on the spectrum and, in particular, on the MSDI parameter $A_{T=1}$ is illustrated for $^{210}_{84}Po_{126}$ (see also Sect. 3.2.4 for a detailed discussion on configuration mixing). In case (**a**) only the configurations $(1h_{9/2})^2$, $(1h_{9/2}\,2f_{7/2})$ and $(1h_{9/2}\,1i_{13/2})$ are considered with a value of $A_{T=1} = 0.33$ MeV. In case (**b**), both protons can be in the full space $(1h_{9/2}\,2f_{7/2}\,1i_{13/2})$ with now an optimum value of $A_{T=1} = 0.17$ MeV [taken from (Brussaard, Glaudemans 1977)]

$$\cos\theta_{12} = \frac{\boldsymbol{j}_1 \cdot \boldsymbol{j}_2}{|\boldsymbol{j}_1||\boldsymbol{j}_2|} = \frac{\boldsymbol{J}^2 - \boldsymbol{j}_1^2 - \boldsymbol{j}_2^2}{2|\boldsymbol{j}_1||\boldsymbol{j}_2|} \; . \tag{3.93}$$

This results in

$$\langle\cos\theta_{12}\rangle = \frac{J(J+1) - j_1(j_1+1) - j_2(j_2+1)}{2\sqrt{j_1(j_1+1)j_2(j_2+1)}} \; . \tag{3.94}$$

In such a way one measures the overlap of the orbitals. For $j_1 \cong j_2$ (large values) one obtains approximately $\theta_{12} \cong 0°$ for $J = j_1 + j_2$ and $\theta_{12} \cong 180°$ for $J = 0$

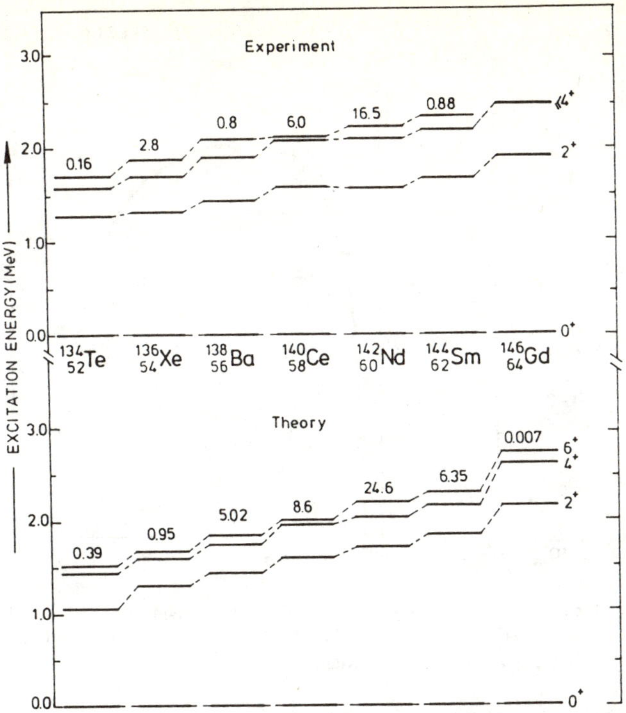

Fig. 3.33. We illustrate, for the $N = 82$ single closed-shell nuclei, the energy spectra in even-even nuclei $_{52}$Te, $_{54}$Xe, $_{56}$Ba, $_{58}$Ce, $_{60}$Nd, $_{62}$Sm and $_{64}$Gd. We show the lowest 2^+, 4^+, 6^+ levels that are mainly formed from proton 2 quasi-particle ($2qp$) excitations (see Chap. 7 for a discussion on the theory of pairing among n identical particles). The lifetime ($T_{1/2}$) for the 6^+ level is indicated in units μs

(Fig. 3.35). The empirical results for a number of configurations $(1g_{9/2})^2$, $(1f_{7/2})^2$, $(2d_{3/2})^2$, $(2g_{9/2})^2$ do follow a single universal curve with the largest attractive matrix element at $\theta_{12} \cong 180°$ (antiparallel spins) (Fig. 3.34). It indeed follows that a δ-interaction force, due to its short-range attractive characteristics, explains this experimental feature rather well.

e) Moshinsky Transformation Method

In many cases of a central residual interaction depending only on the relative nucleon separation $r = |\boldsymbol{r}_1 - \boldsymbol{r}_2|$, a method exists to separate the relative from the center-of-mass coordinates when calculating the two-body matrix elements. When using harmonic oscillator radial wave functions, this method is extremely interesting.

If we consider harmonic oscillator potentials for the interacting nucleons, using the transformation

$$\boldsymbol{r} = \boldsymbol{r}_1 - \boldsymbol{r}_2 , \quad \boldsymbol{R} = \tfrac{1}{2}\left(\boldsymbol{r}_1 + \boldsymbol{r}_2\right) , \tag{3.95}$$

one can rewrite

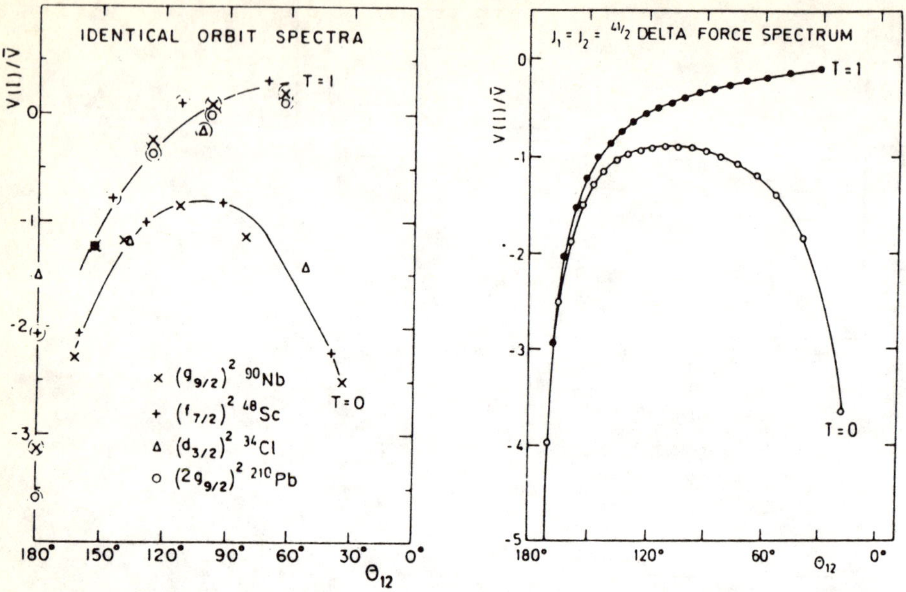

Fig. 3.34. The relative matrix elements $\langle j^2; JM|V_{1,2}|j^2; JM\rangle/\bar{V}$ [with $\bar{V}$ defined via (3.91) as a function of the overlap angle θ_{12} [see (3.94)]. Experimental values from (Anantaraman 1971) and for a pure δ-force spectrum with $j = \frac{41}{2}$ are given [taken from (Ring 1980)]

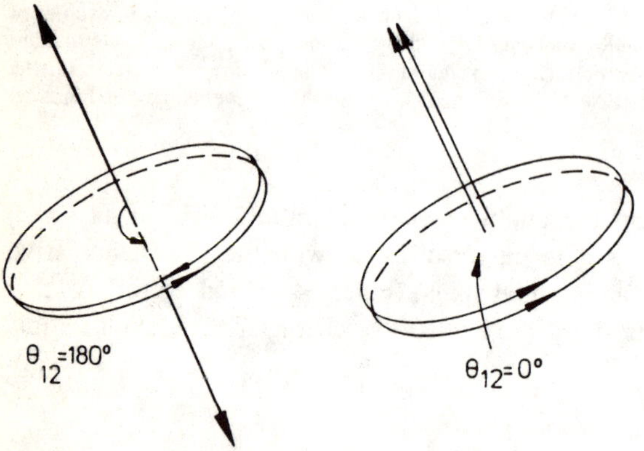

Fig. 3.35. The classical orbits for two particles coupled in the parallel ($\theta_{12} = 0°$) and antiparallel ($\theta_{12} = 180°$) way. The motion is, in a schematic way, presented by the arrow within the plane, perpendicular to the angular momentum vectors

$$\frac{1}{2}m\omega^2\left(r_1^2 + r_2^2\right), \tag{3.96}$$

as

$$\frac{1}{2}\left(\frac{m}{2}\right)\omega^2 r^2 + \frac{1}{2}(2m)\omega^2 R^2 . \tag{3.97}$$

Similarly, by transforming the momenta to a relative ($\boldsymbol{p}$) and center-of-mass total ($\boldsymbol{P}$) momentum, one can also rewrite the kinetic energy part such that

$$\frac{1}{2m}\left(\boldsymbol{p}_1^2 + \boldsymbol{p}_2^2\right) , \tag{3.98}$$

becomes

$$\tfrac{1}{2}\boldsymbol{p}^2/(m/2) + \tfrac{1}{2}\boldsymbol{P}^2/(2m) . \tag{3.99}$$

Thus, the total Hamiltonian describing the unperturbed motion of two nucleons in a harmonic oscillator potential becomes

$$\begin{aligned} H &= H_{\text{ho}}(1) + H_{\text{ho}}(2) = H_{\text{rel}} + H_{\text{com}} \\ &= \boldsymbol{p}^2/m + \frac{1}{4}m\omega^2 r^2 + \frac{\boldsymbol{P}^2}{4m} + m\omega^2 R^2 , \end{aligned} \tag{3.100}$$

indicating that in the relative harmonic oscillator a mass of $m/2$ appears and for the center-of-mass motion a mass of $2m$.

The wave function describing the unperturbed motion of the two nucleons can be written in Dirac ket notation

$$|n_1 l_1 m_1\rangle |n_2 l_2 m_2\rangle , \tag{3.101}$$

with

$$\langle \boldsymbol{r}|nlm\rangle = R_{nl}(r) Y_l^m(\theta,\varphi) , \tag{3.102}$$

and

$$R_{nl}(r) = N_{n,l} \cdot r^l \, \mathrm{e}^{-\nu r^2} L_{n-1}^{l+1/2}(2\nu r^2) . \tag{3.103}$$

Here $\nu = m\omega/2\hbar$ and $n = k + 1$.

Since the total Hamiltonian is also equal to the sum of a relative $(m/2)$ and center-of-mass $(2m)$ oscillator potential, the total two-particle wave function can also be written as

$$|nlm\rangle |N\Lambda M_\Lambda\rangle , \tag{3.104}$$

where (n, l, m) and (N, Λ, M_Λ) describe the quantum numbers of the relative and center-of-mass wave function. It is possible to transform between the two equivalent sets of basis functions (3.101, 104) by using the Moshinsky transformation (Brody, Moshinsky 1960). One writes

$$\begin{aligned} &|n_1 l_1, n_2 l_2; LM\rangle \\ &\quad = \sum_{n,l,N,\Lambda} \langle nl, N\Lambda; L|n_1 l_1, n_2 l_2; L\rangle |nl, N\Lambda; LM\rangle . \end{aligned} \tag{3.105}$$

One can easily prove that the transformation coefficients are independent of the projection quantum number M. During the transformation, a number of conservation laws hold:

$$\text{(i)} \quad \boldsymbol{l}_1 + \boldsymbol{l}_2 = \boldsymbol{L} = \boldsymbol{l} + \boldsymbol{\Lambda} , \tag{3.106}$$

$$\text{(ii)} \quad E_{n_1 l_1} + E_{n_2 l_2} = E_{nl} + E_{N\Lambda} , \tag{3.107}$$

or

$$2n_1 + l_1 + 2n_2 + l_2 = 2n + l + 2N + \Lambda \ . \tag{3.108}$$

We illustrate some of these transformation coefficients in Table 3.2. Applying this Moshinsky transformation to the calculation of the two-body matrix elements with a central force $V(r)$, we can evaluate

$$\begin{aligned} M &\equiv \langle n_1 l_1, n_2 l_2; LM | V(r) | n_1' l_1', n_2' l_2'; LM \rangle \\ &= \sum_{\substack{n,l,N,\Lambda \\ n',l',N',\Lambda'}} \langle n_1 l_1, n_2 l_2; L | nl, N\Lambda; L \rangle \\ &\quad \times \langle n_1' l_1', n_2' l_2'; L | n'l', N'\Lambda'; L \rangle \\ &\quad \langle nl, N\Lambda; LM | V(r) | n'l', N'\Lambda'; LM \rangle \ . \end{aligned} \tag{3.109}$$

In the radial integral part on the right-hand side of (3.109), the center-of-mass part of the wave function drops out; we get orthogonality conditions $\delta_{NN'}\delta_{\Lambda\Lambda'}$ and an extra factor $\delta_{ll'}$, since in the relative wave function $|nlm\rangle$, the angular part described by the $Y_l^m(\theta, \varphi)$ spherical harmonics, also drops out (to show this part of the calculation explicitly is left as an exercise). One then obtains

$$\begin{aligned} M = &\sum_{n,n',l,N,\Lambda} \langle n_1 l_1, n_2 l_2; L | nl, N\Lambda; L \rangle \\ &\times \langle n_1' l_1', n_2' l_2'; L | n'l, N\Lambda; L \rangle \\ &\times \langle nl | V(r) | n'l \rangle \ . \end{aligned} \tag{3.110}$$

Here, the radial integral $\langle nl | V(r) | n'l \rangle$ can be put explicitly, in coordinate space, as the integral

$$R \equiv \int R_{nl}(r) V(r) R_{n'l}(r) r^2 \, dr \ . \tag{3.111}$$

Using the explicit form of the Laguerre polynomials

$$L_{n-1}^{l+1/2}(x) = \sum_{k=0}^{n-1} a_k (-1)^k x^k \ , \tag{3.112}$$

and a Gaussian type of residual interaction

$$V(r) = V_0 \, \mathrm{e}^{-r^2/r_0^2} \ , \tag{3.113}$$

the radial integral can be evaluated as the double sum (note that $\nu \rightarrow \nu_{\mathrm{rel}} = \nu/2$)

$$\begin{aligned} R = N_{n,l} N_{n',l} &\sum_{k,k'=0}^{(n-1)(n'-1)} (-1)^{k+k'} a_k a_{k'} \\ &\times \int r^{2l} \, \mathrm{e}^{-\nu r^2} \, \mathrm{e}^{-r^2/r_0^2} \left(\nu r^2\right)^{k+k'} r^2 \, dr \ . \end{aligned} \tag{3.114}$$

Table 3.2. Illustration of transformation brackets from the product harmonic oscillator wave functions to a product basis of relative times centre-of-mass (c.o.m.) oscillator wave functions. The table lists coefficients $\langle n_1 l_1, n_2 l_2;\ L|nl, N\Lambda;\ L\rangle$ [taken from (Brody 1960)] with $n_1 = n_2 = 0$. Note that in the tables of Brody and Moshinsky, the total orbital angular momentum is denoted by λ, the c.o.m. angular momentum by L and that the radial quantum number n takes the values n = 0, 1, 2, ...

l_1	l_2	λ	n	l	N	L	ϱ	$\langle\|\rangle$	τ
0	0	0	0	0	0	0		1.00000000	1
0	1	1	0	0	0	1	1	0.70710678	
			0	1	0	0		−0.70710677	2
0	2	2	0	0	0	2	2	0.49999999	
			0	1	0	1		−0.70710679	
			0	2	0	0		0.49999999	3
0	3	3	0	0	0	3	3	0.35355340	
			0	1	0	2		−0.61237245	
			0	2	0	1		0.61237245	
			0	3	0	0		−0.35355340	4
0	4	4	0	0	0	4	4	0.24999999	
			0	1	0	3		−0.50000000	
			0	2	0	2		0.61237245	
			0	3	0	1		−0.50000000	
			0	4	0	0		0.24999999	5
0	5	5	0	0	0	5	5	0.17677670	
			0	1	0	4		−0.39528471	
			0	2	0	3		0.55901700	
			0	3	0	2		−0.55901699	
			0	4	0	1		0.39528471	
			0	5	0	0		−0.17677670	6
0	6	6	0	0	0	6	6	0.12500000	
			0	1	0	5		−0.30618623	
			0	2	0	4		0.48412292	
			0	3	0	3		−0.55901699	
			0	4	0	2		0.48412292	
			0	5	0	1		−0.30618622	
			0	6	0	0		0.12500000	
1	1	0	0	0	1	0	2	0.70710679	
			0	1	0	1		0.00000000	
			1	0	0	0		−0.70710679	3
1	1	1	0	1	0	1	2	0.99999999	1
1	1	2	0	0	0	2	2	0.70710678	
			0	1	0	1		0.00000000	
			0	2	0	0		−0.70710678	3
1	2	1	0	0	1	1	3	0.40824829	
			0	1	0	2		0.23570226	
			0	1	1	0		−0.52704628	
			0	2	0	1		0.23570226	
			1	0	0	1		−0.52704628	
			1	1	0	0		0.40824829	6
1	2	2	0	1	0	2	3	0.70710676	
			0	2	0	1		−0.70710676	2

l_1	l_2	λ	n	l	N	L	ϱ	$\langle\|\rangle$	τ
1	2	3	0	0	0	3	3	0.61237245	
			0	1	0	2		−0.35355340	
			0	2	0	1		−0.35355340	
			0	3	0	0		0.61237245	4
1	3	2	0	0	1	2	4	0.27386128	
			0	1	0	3		0.20000000	
			0	1	1	1		−0.45825757	
			0	2	0	2		0.00000000	
			0	2	1	0		0.41833000	
			0	3	0	1		−0.20000000	
			1	0	0	2		−0.41833000	
			1	1	0	1		0.45825757	
			1	2	0	0		−0.27386128	9
1	3	3	0	1	0	3	4	0.50000000	
			0	2	0	2		−0.70710677	
			0	3	0	1		0.50000000	3
1	3	4	0	0	0	4	4	0.50000001	
			0	1	0	3		−0.49999999	
			0	2	0	2		−0.00000001	
			0	3	0	1		0.49999999	
			0	4	0	0		−0.50000001	5
1	4	3	0	0	1	3	5	0.18898224	
			0	1	0	4		0.15152288	
			0	1	1	2		−0.37115375	
			0	2	0	3		−0.05976143	
			0	2	1	1		0.43915504	
			0	3	0	2		−0.05976143	
			0	3	1	0		−0.32732684	
			0	4	0	1		0.15152288	
			1	0	0	3		−0.32732684	
			1	1	0	2		0.43915504	
			1	2	0	1		−0.37115375	
			1	3	0	0		0.18898224	12
1	4	4	0	1	0	4	5	0.35355339	
			0	2	0	3		−0.61237243	
			0	3	0	2		0.61237243	
			0	4	0	1		−0.35355339	4
1	4	5	0	0	0	5	5	0.39528471	
			0	1	0	4		−0.53033006	
			0	2	0	3		0.25000000	
			0	3	0	2		0.24999999	
			0	4	0	1		−0.53033005	
			0	5	0	0		0.39528471	6
1	5	4	0	0	1	4	6	0.13176157	
			0	1	0	5		0.11111111	
			0	1	1	3		−0.29133579	
			0	2	0	4		−0.07273929	
			0	2	1	2		0.40458680	
			0	3	0	3		0.00000000	

l_1	l_2	λ	n	l	N	L	ϱ	$\langle\vert\rangle$	τ
			0	3	1	1		−0.39086798	
			0	4	0	2		0.07273929	
			0	4	1	0		0.25230420	
			0	5	0	1		−0.11111111	
			1	0	0	4		−0.25230420	
			1	1	0	3		0.39086798	
			1	2	0	2		−0.40458680	
			1	3	0	1		0.29133579	
			1	4	0	0		−0.13176157	15
1	5	5	0	1	0	5	6	0.25000000	
			0	2	0	4		−0.50000000	
			0	3	0	3		0.61237245	
			0	4	0	2		−0.50000001	
			0	5	0	1		0.25000000	5

We define the dimensionless coordinate $x \equiv \nu^{1/2} r$, and obtain

$$R = N_{n,l} N_{n',l} \sum_p a_p \int x^{2p}\, e^{-x^2}\, e^{-x^2/(\nu r_0^2)} x^2\, dx \,, \tag{3.115}$$

where $l \leq p \leq l + n + n' - 2$, $a_p = a_k a_{k'} (-1)^{k+k'} (\nu)^{-(l+3/2)}$.

Combining all intermediate results, the matrix element M can be written in compact form

$$\begin{aligned} M = \sum_{\substack{n,n',l,\\ N,\Lambda}} & \langle n_1 l_1, n_2 l_2; L | nl, N\Lambda; L\rangle \langle n_1' l_1', n_2' l_2'; L | n'l, N\Lambda; L\rangle \\ & \times N_{n,l} N_{n',l} \sum_p I_p \frac{a_p}{2} \Gamma\left(p + \tfrac{3}{2}\right) , \end{aligned} \tag{3.116}$$

and where the Talmi integrals defined as I_p

$$I_p = \frac{2}{\Gamma\left(p + \frac{3}{2}\right)} \int x^{2(p+1)}\, e^{-(1+\lambda^2)x^2}\, dx \,, \tag{3.117}$$

with $\lambda = (\nu^{1/2} r_0)^{-1}$, characterize the interaction. This Talmi integral can be evaluated in closed form and gives

$$I_p = \left(1 + \lambda^2\right)^{-p-3/2} . \tag{3.118}$$

In Fig. 3.36 we plot the Talmi integrals I_p for the Gaussian interaction. This method is very general, and Talmi integrals for, e.g., the Yukawa shape $e^{-\mu r}/\mu r$, the Coulomb shape $1/r$, or any other force depending on $r = |\boldsymbol{r}_1 - \boldsymbol{r}_2|$ can be obtained and evaluated (Talmi 1952).

In the above discussion we concentrated on the orbital part of the wave function. In actual situations, we calculate the two-body matrix elements

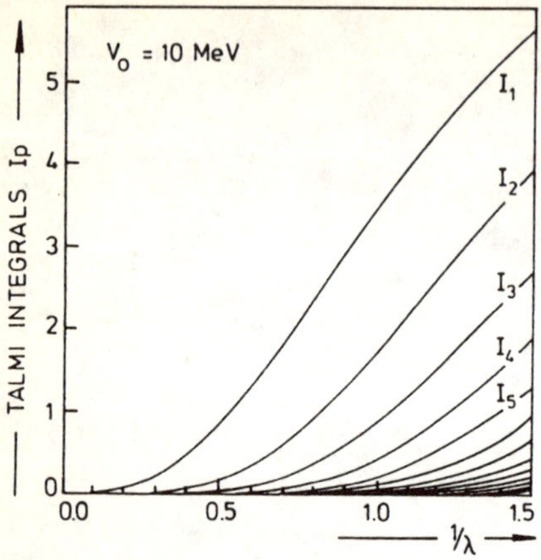

Fig. 3.36. The nuclear Talmi integrals [see (3.117) and (3.118)] for a residual interaction with a Gaussian shape. The abcissa is indicated by $1/\lambda$ where $\lambda = (\nu^{1/2} r_0)^{-1}$ with ν the harmonic oscillator parameter ($\nu = m\omega/2\hbar$) and r_0 the Gaussian shape parameter of $V(r) = V_0 \exp(-r^2/r_0^2)$. The values $I_1, \ldots, I_{13}$ are shown (in descending order). The strength of the Gaussian interaction $V_0 = 10\,\text{MeV}$ is taken

$\langle j_1 j_2; JM | V_{12} | j_3 j_4; JM \rangle$ where $j_1, \ldots, j_4$ are notations for a full single-particle wave function, containing besides the orbital part also the intrinsic spin part in a coupled form $(l\frac{1}{2})j$. In the problem set there is an exercise to go from the above two-body matrix elements into the relative integrals via the Moshinsky transformation brackets.

3.2.4 Configuration Mixing: Model Space and Model Interaction

In many cases when we consider nuclei with just two valence nucleons outside closed shells, it is not possible to single out one orbital j. Usually a number of valence shells are present in which the two nucleons can, in principle, move.

Let us consider the case of ^{18}O with two neutrons outside the ^{16}O core. In the most simple approach the two neutrons move in the energetically most favored orbital, i.e., in the $1d_{5/2}$ orbital (Fig. 3.37). Thus we can only form the $(2d_{5/2})^2\, 0^+$, 2^+, 4^+ configurations and then determine the strength of the residual interaction V_{12} so that the theoretical 0^+-2^+-4^+ spacing reproduces the experimental spacing as well as possible (Fig. 3.31). The next step is to consider the full sd model space with many more configurations for each J^π value. For the $J^\pi = 0^+$ state, we have three configurations, i.e., the $(2d_{5/2})^2 0^+$, $(3s_{1/2})^2 0^+$ and the $(2d_{3/2})^2 0^+$ configurations. In the latter situation, the strength of the residual interaction V_{12} will be different from the first model space where only the $(2d_{5/2})^2 0^+$ state was considered. Thus one generally concludes that the strength of the residual interaction depends on the model space chosen, that is, $V_{12} = V_{12}$ (model space) such that the larger the model space is, the smaller V_{12} will become in order to get a similar overall agreement. In the larger model spaces, one will, in general, be able to describe the observed properties in the nucleus better than with the smaller model spaces. This argument only relates to effective forces using a given form, i.e., a Gaussian interaction, an (M)SDI interaction, etc. for which only the strength parameter determines the overall magnitude of the two-body matrix elements in a given finite dimensional model space. Thus one should not extrapolate to the full

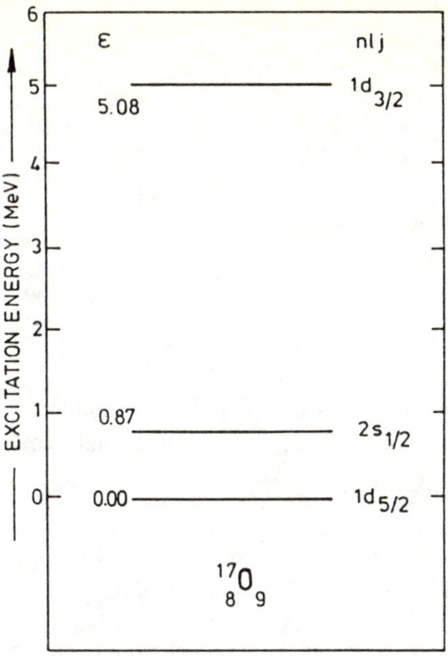

Fig. 3.37. The neutron single-particle energies in $^{17}_{8}O_9$ (relative to the $1d_{5/2}$ orbital) for $2s_{1/2}$ and $1d_{3/2}$ orbitals. The energies are taken from the experimental spectrum in ^{17}O

(infinite dimensional initial) configuration space in which the bare nucleon-nucleon force would be acting.

Thus, in the case of ^{18}O where the $3s_{1/2}$ and $2d_{5/2}$ orbitals separate from the higher-lying $2d_{3/2}$ orbital, for the model spaces one has

$$
\begin{aligned}
J^\pi &= 0^+ \to \left(1d_{5/2}\right)^2_{0^+}, \ \left(3s_{1/2}\right)^2_{0^+} \\
J^\pi &= 2^+ \to \left(1d_{5/2}\right)^2_{2^+}, \ \left(1d_{5/2}2s_{1/2}\right)_{2^+} \\
J^\pi &= 3^+ \to \left(1d_{5/2}2s_{1/2}\right)_{3^+} \\
J^\pi &= 4^+ \to \left(1d_{5/2}\right)^2_{4^+} .
\end{aligned}
$$

The energy eigenvalues for the $J^\pi = 0^+$ states, for example, will be the corresponding eigenvalues for the eigenstates to the Hamiltonian

$$
\begin{aligned}
H &= H_0 + H_{\mathrm{res}} , \\
&= \sum_{i=1}^{2} h_0(i) + V_{12} ,
\end{aligned}
\tag{3.119}
$$

where the core energy corresponding to the closed shell system E_0 is taken as the reference value.

The wave functions will, in general, be linear combinations of the possible basis functions. This means that for $J^\pi = 0^+$ we will get two eigenfunctions

$$|\Psi_{0^+;1}\rangle = \sum_{k=1}^{n} a_{k,1}|\psi_k^{(0)};0^+\rangle ,$$

$$|\Psi_{0^+;2}\rangle = \sum_{k=1}^{n} a_{k,2}|\psi_k^{(0)};0^+\rangle , \tag{3.120}$$

where for the particular case of ^{18}O we define

$$|\psi_1^{(0)};0^+\rangle \equiv |(1d_{5/2})^2;0^+\rangle ,$$
$$|\psi_2^{(0)};0^+\rangle \equiv |(2s_{1/2})^2;0^+\rangle .$$

Before turning back to the particular case of ^{18}O, we make the method more general. If the basis set is denoted by $|\psi_k^{(0)}\rangle (k = 1, 2, \ldots, n)$, the total wave function can be expanded as

$$|\Psi_p\rangle = \sum_{k=1}^{n} a_{kp}|\psi_k^{(0)}\rangle . \tag{3.121}$$

The coefficients a_{kp} have to be determined by solving the Schrödinger equation for $|\Psi_p\rangle$, or

$$H|\Psi_p\rangle = E_p|\Psi_p\rangle . \tag{3.122}$$

In explicit form this becomes [using the Hamiltonian of (3.119)]

$$\left(H_0 + H_{\text{res}}\right) \sum_{k=1}^{n} a_{kp}|\psi_k^{(0)}\rangle = E_p \sum_{k=1}^{n} a_{kp}|\psi_k^{(0)}\rangle , \tag{3.123}$$

or

$$\sum_{k=1}^{n} \langle\psi_l^{(0)}|H_0 + H_{\text{res}}|\psi_k^{(0)}\rangle a_{kp} = E_p a_{lp} . \tag{3.124}$$

Since the basis functions $|\psi_k^{(0)}\rangle$ correspond to eigenfunctions of H_0 with eigenvalues (unperturbed energies) $E_k^{(0)}$, we can rewrite (3.124) in shorthand form as

$$\sum_{k=1}^{n} H_{lk} a_{kp} = E_p a_{lp} , \tag{3.125}$$

with

$$H_{lk} \equiv E_k^{(0)}\delta_{lk} + \langle\psi_l^{(0)}|H_{\text{res}}|\psi_k^{(0)}\rangle . \tag{3.126}$$

The eigenvalue equation becomes a matrix equation

$$[H][A] = [E][A] . \tag{3.127}$$

This forms a secular equation for the eigenvalues E_p which are determined from

$$\begin{vmatrix} H_{11}-E_p & H_{12} & \dots H_{1n} \\ H_{21} & H_{22}-E_p & \dots H_{2n} \\ \vdots & \ddots & \vdots \\ H_{n1} & \dots & H_{nn}-E_p \end{vmatrix} = 0\,. \tag{3.128}$$

This is a nth degree equation for the n-roots E_p ($p = 1, 2, \dots, n$). Substitution of each value of E_p separately in (3.125) gives a set of linear equations that can be solved for the coefficients a_{kp}. The wave functions $|\Psi_p\rangle$ can be orthonormalized since

$$\sum_{k=1}^{n} a_{kp} a_{kp'} = \delta_{pp'}\,. \tag{3.129}$$

From (3.125) it now follows that

$$\sum_{l,k=1}^{n} a_{lp'} H_{lk} a_{kp} = E_p \delta_{pp'}\,, \tag{3.130}$$

or, in matrix form

$$[\widetilde{A}][H][A] = [E]\,, \tag{3.131}$$

with

$$[\widetilde{A}] = [A]^{-1}\,.$$

Equation (3.131) indicates a similarity transformation to a new basis that makes $[H]$ diagonal and thus produces the n energy eigenvalues. In practical situations (n large), this process needs high-speed computers. A number of algorithms exist for $[H]$ (hermitian, real in most cases) matrix diagonalization which we do not discuss here (Wilkinson 1965): the Jacobi method (small n or $n\lambda 50$), the Householder method ($50 < n < 200$), the Lanczos algorithm ($n\gamma 1000$, requiring the calculation of only a small number of eigenvalues, normally the lowest lying ones). In cases where the non-diagonal matrix elements $|H_{ij}|$ are of the order of the unperturbed energy differences $|E_i^{(0)} - E_j^{(0)}|$, large configuration mixing will result and the final energy eigenvalues E_p can be very different from the unperturbed spectrum of eigenvalues $E_p^{(0)}$. If, on the other hand, the $|H_{ij}|$ are small compared to $|E_i^{(0)} - E_j^{(0)}|$, energy shifts will be small and even perturbation theory might be applied.

Now to make these general considerations more specific, we discuss the case of $J^\pi = 0^+$ levels in ^{18}O for the $(1d_{5/2}2s_{1/2})$ model space. As shown before, the model space reduces to a two-dimensional space $n = 2$ (3.120) and the 2×2 energy matrix can be written out as

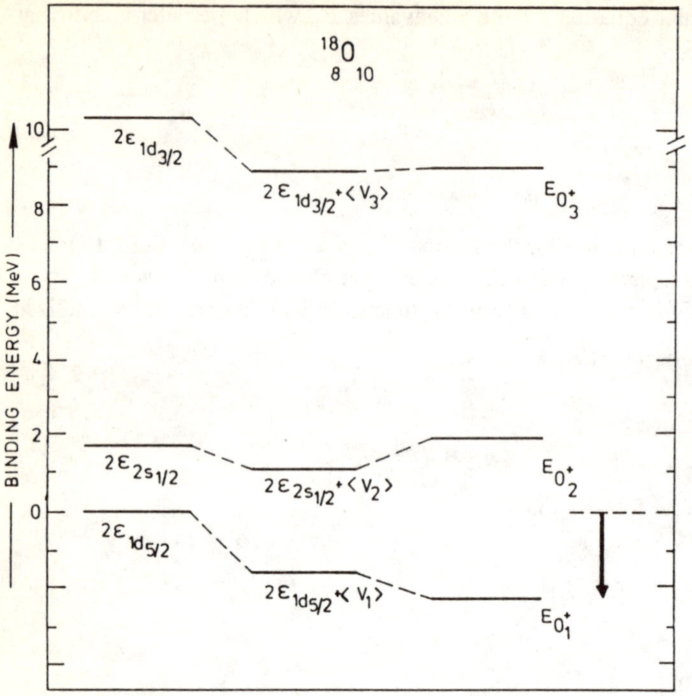

Fig. 3.38. The solution of the secular equation for 0^+ states in ^{18}O, shown in a diagrammatic way. On the extreme left we show the unperturbed two-neutron single-particle energies $2\varepsilon_j$. In the middle part the diagonal interaction matrix elements $\langle j^2; J^\pi = 0^+|V_{1,2}|j^2; J^\pi = 0^+\rangle$ for the three configurations are added and denoted by $\langle V_1\rangle$, $\langle V_2\rangle$, $\langle V_3\rangle$, respectively. On the right-hand side, the resulting energy eigenvalues $E_{0_i^+}$ $(i = 1, 2, 3)$ from diagonalizing the energy matrix (3.132) are shown

$$H = \begin{bmatrix} 2\varepsilon_{1d_{5/2}} + \langle (1d_{5/2})^2; 0^+|V_{12}|(1d_{5/2})^2; 0^+\rangle & \langle (1d_{5/2})^2; 0^+|V_{12}|(2s_{1/2})^2; 0^+\rangle \\ \langle (2s_{1/2})^2; 0^+|V_{12}|(1d_{5/2})^2; 0^+\rangle & 2\varepsilon_{2s_{1/2}} + \langle (2s_{1/2})^2; 0^+|V_{12}|(2s_{1/2})^2; 0^+\rangle \end{bmatrix} \tag{3.132}$$

These diagonal elements yield the first correction to the unperturbed single-particle energies $2\varepsilon_{1d_{5/2}}$ and $2\varepsilon_{2s_{1/2}}$, respectively (the diagonal two-body interaction matrix elements H_{11} and H_{22}, Fig. 3.38). The energy matrix is Hermitian (for a real matrix this means symmetric and $H_{12} = H_{21}$) and, in shorthand notation gives the secular equation

$$\begin{bmatrix} H_{11} - \lambda & H_{12} \\ H_{12} & H_{22} - \lambda \end{bmatrix} = 0 . \tag{3.133}$$

We can solve this easily since we get a quadratic equation in λ

$$\lambda^2 - \lambda(H_{11} + H_{22}) - H_{12}^2 + H_{11}H_{22} = 0 \tag{3.134}$$

with the roots

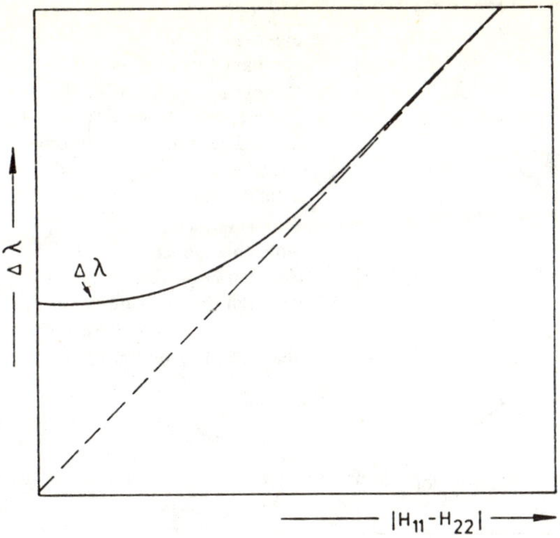

Fig. 3.39. The variation of the difference of eigenvalues $\Delta\lambda \equiv \lambda_+ - \lambda_-$ for the two-level model of (3.133) and (3.136) as a function of the energy difference between the unperturbed states $\Delta H \equiv |H_{11} - H_{22}|$

$$\lambda_\pm = \frac{H_{11} + H_{22}}{2} \pm \frac{1}{2}\left[(H_{11} - H_{22})^2 + 4H_{12}^2\right]^{1/2} . \tag{3.135}$$

The difference $\Delta\lambda \equiv \lambda_+ - \lambda_-$ then becomes

$$\Delta\lambda = \left[(H_{11} - H_{22})^2 + 4H_{12}^2\right]^{1/2} , \tag{3.136}$$

and is shown in Fig. 3.39. Even for $H_{11} = H_{22}$, the degenerate situation for the two basis states, a difference of $\Delta\lambda = 2H_{12}$ results. It is as if the two levels are repelled over a distance of H_{12}. Thus $2H_{12}$ is the minimal energy difference. In the limit of $|H_{11} - H_{22}| \gg |H_{12}|$, the energy difference $\Delta\lambda$ becomes asymptotically equal to $\Delta H \equiv H_{11} - H_{22}$.

The equation for $\lambda_\pm$ given in (3.135) is interesting with respect to perturbation theory.

i) If we consider the case $|H_{11} - H_{22}| \gg |H_{12}|$, then we see that one can obtain by expanding the square root around $H_{11} = H_{22}$

$$\begin{aligned} \lambda_1 &= H_{11} + \frac{H_{12}^2}{H_{11} - H_{22}} + \dots , \\ \lambda_2 &= H_{22} + \frac{H_{12}^2}{H_{22} - H_{11}} + \dots . \end{aligned} \tag{3.137}$$

[We use the expansion $(1 + x)^{1/2} \cong 1 + \frac{1}{2}x + \dots$.)

ii) One can show that if the perturbation expansion does not converge easily, one has to sum the full perturbation series up to infinity. The final result of this sum can be shown to be equal to the square root expression of (3.135) (see Problem set).

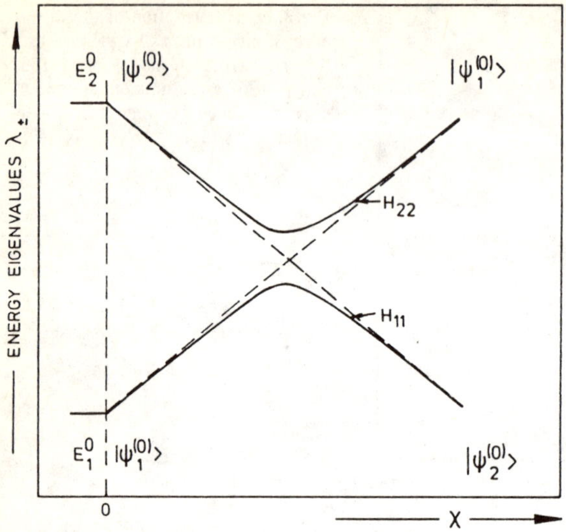

Fig. 3.40. The variation of eigenvalues $\lambda_\pm$ [see (3.135)] obtained from the two-level model as a function of the parameter χ which describes the variation of the unperturbed energies H_{11}, H_{22}. We take a linear variation $H_{11} = E_1^{(0)} + \chi a$, $H_{22} = E_2^{(0)} - \chi b$ with an interaction matrix element H_{12}. The variation (through the crossing zone) of the wave function character is also indicated on the figure

It is also interesting to study (3.135) with a constant interaction matrix element H_{12} when the unperturbed energies H_{11}, H_{22} vary linearly, i.e., $H_{11} = E_1^{(0)} + \chi a$ and $H_{22} = E_2^{(0)} - \chi b$ (Fig. 3.40). There will be a crossing point for the unperturbed energies at a certain value of $\chi = \chi_{\text{crossing}}$. However, the eigenvalues E_1, E_2 will first approach the crossing but then change direction (no-crossing rule). The wave functions are also interesting. First of all, we study the wave functions analytically (the coefficients a_{kp}).

We get

$$\begin{aligned} |\Psi_1\rangle &= a_{11}|\psi_1^{(0)}\rangle + a_{21}|\psi_2^{(0)}\rangle \\ |\Psi_2\rangle &= a_{12}|\psi_1^{(0)}\rangle + a_{22}|\psi_2^{(0)}\rangle \,. \end{aligned} \tag{3.138}$$

If we use one of the eigenvalues, say λ_1, the coefficients follow from

$$\left(H_{11} - \lambda_1\right)a_{11} + H_{12}a_{21} = 0 \,, \tag{3.139}$$

or

$$\frac{a_{11}}{a_{21}} = \frac{-H_{12}}{H_{11} - \lambda_1} \,. \tag{3.140}$$

The normalizing condition $a_{11}^2 + a_{21}^2 = 1$ then gives

$$a_{21} = \left(1/\left(1 + \left(H_{12}/\left(H_{11} - \lambda\right)\right)^2\right)\right)^{1/2} , \tag{3.141}$$

and similar results for the other coefficients. In the situation that $H_{11} = H_{22}$, the absolute values of the coefficients a_{11}, a_{12}, a_{21} and a_{22} all have absolute value $1/\sqrt{2}$. These coefficients then also determine the wave functions of Fig. 3.41 at the crossing point. One can see in Fig. 3.40 that for the case of $\chi = 0$, one has

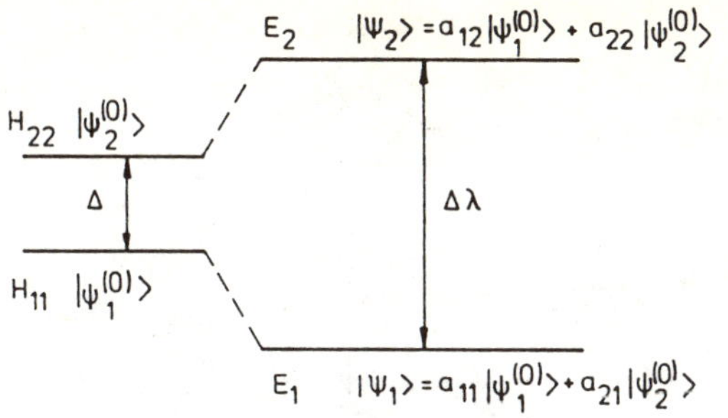

Fig. 3.41. The change in energy separation $\Delta\,(\Delta \equiv H_{22} - H_{11})$ to $\Delta\lambda\,(\equiv E_2 - E_1)$ from unperturbed to perturbed spectrum in a two-level model. The specific form of the wave functions $|\Psi_1\rangle$ and $|\Psi_2\rangle$ is also given using the basis functions $|\psi_1^{(0)}\rangle$, $|\psi_2^{(0)}\rangle$

$$E_1 \simeq E_1^{(0)} \quad \text{and} \quad |\Psi_1\rangle \simeq |\psi_1^{(0)}\rangle \,,$$
$$E_2 \simeq E_2^{(0)} \quad \text{and} \quad |\Psi_2\rangle \simeq |\psi_2^{(0)}\rangle \,.$$

On the other hand, after the level crossing and in the region where again $|H_{11} - H_{22}| \gg |H_{12}|$, one has

$$E_1 \simeq H_{22} \quad \text{and} \quad |\Psi_1\rangle \simeq |\psi_2^{(0)}\rangle \,,$$
$$E_2 \simeq H_{11} \quad \text{and} \quad |\Psi_2\rangle \simeq |\psi_1^{(0)}\rangle \,,$$

so that one can conclude that the "character" of the states has been interchanged in the crossing region, although the levels never actually cross!

In the realistic situations of ^{18}O and ^{210}Po (Figs. 3.31 and 3.32) (Brussaard, Glaudemans 1977), the results are now the results of configuration mixing: in ^{18}O (the $1d_{5/2}$ and full $1d_{5/2}$, $2s_{1/2}$, $1d_{3/2}$ spaces, respectively) and for ^{210}Po [using the $(1h_{9/2})^2$, $1h_{9/2}\,2f_{7/2}$, $1h_{9/2}\,1i_{13/2}$ and the full $(1h_{9/2}, 2f_{7/2}, 1i_{13/2})$ spaces, respectively]. In the case of ^{18}O, one observes a net improvement for the larger model space when comparing with the data. Also, for ^{210}Po some improvement can be observed for the larger model space. In both cases (^{18}O, ^{210}Po) one needs to reduce the MSDI force strength when going to the larger space.

The above method of configuration mixing was discussed and worked out in detail for the two-level model (as applied) to the lowest 0^+ levels in ^{18}O. If we restrict ourselves to the full sd-shell, the complete two-neutron configuration space can be constructed for all possible J^π values, as was done for the 0^+ state shown in Fig. 3.38. According to the specific neutron single-particle spacings as deduced from the experimental spectrum of ^{17}O, we construct the full unperturbed two-particle spectrum on the extreme left-hand side of Fig. 3.42. At the same time (second column of levels), all possible J^π values are shown. Subsequently, the various diagonal two-body matrix elements (using a zero-range interaction) $\langle j_1 j_2; JM|V|j_1 j_2; JM\rangle$, which already split the original degeneracy to a large extent, are added. The results of diagonalizing the various energy matrices ($J^\pi = 0^+, 1^+, \ldots, 4^+$) are given separately and, finally, at the extreme right-hand side, the full theoretical two-particle spectrum for ^{18}O is obtained. The global binding energy of the 0 ground-state configuration is also indicated, relative to the unperturbed energy of $2\varepsilon_{1d_{5/2}}$. In this figure, the full details of how such a calculation within a two-particle space is carried out, is clearly illustrated. The above method can now also be applied to any other nucleus containing either a two-particle or a two-hole configuration.

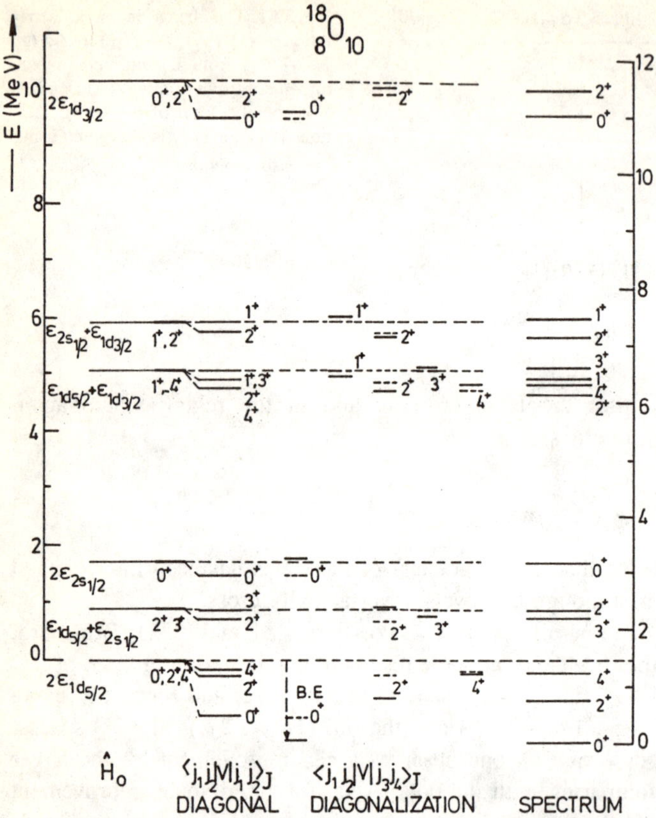

Fig. 3.42. Energy spectrum for ^{18}O with two neutrons occupying the full (sd) model space. Besides the unperturbed energy spectrum and the various diagonal contributions, the results of diagonalizing the variou $J^\pi = 0^+, \ldots, 4^+$ matrices as well as the total spectrum are illustrated

3.3 Three-particle Systems and Beyond

It is our purpose to construct fully antisymmetric and normalized many-particle wave functions in order to be able, using the methods as outlined in Sect. 3.2, to obtain energy spectra and wave functions for the many-particle nuclear Schrödinger equation. We first start from the three-particle system to show the general methods before extending to the n-particle system.

3.3.1 Three-particle Wave Functions

First of all, we consider three-particle systems with the nucleons moving in the orbitals j_a, j_b and j_c [remember we use the notation $j \equiv (n, l, j)$].

i) We consider the simplest situation of $j_a \neq j_b \neq j_c$. The antisymmetrized, normalized wave function is constructed in a straightforward way as

$$\mathcal{N}\sum_P (-1)^P \psi\big((j_a(1)j_b(2))J_{12}, j_c(3); JM\big) \ , \tag{3.142}$$

where $\mathcal{N}$ is the normalization and $\sum$ means a sum over all permutations of the particle coordinates 1, 2 and 3 over the orbitals j_a, j_b and j_c with $(-1)^P = +1$ for an even permutation of $(1,2,3)$ and -1 for an odd permutation of $(1,2,3)$. So, (3.142) becomes

$$\begin{aligned}\frac{1}{\sqrt{6}}\Big\{ &\psi\big((j_a(1)j_b(2))J_{12}, j_c(3); JM\big) - \psi\big((j_a(2)j_b(1))J_{12}, j_c(3); JM\big) \\ &+ \psi\big((j_a(2)j_b(3))J_{12}, j_c(1); JM\big) - \psi\big((j_a(3)j_b(2))J_{12}, j_c(1); JM\big) \\ &+ \psi\big((j_a(3)j_b(1))J_{12}, j_c(2); JM\big) - \psi\big((j_a(1)j_b(3))J_{12}, j_c(2); JM\big)\Big\} \ .\end{aligned} \tag{3.143}$$

ii) We consider the case $(j_a = j_b = j) \neq j_c$. In this case, the two-particle wave function where j_a and j_b in (3.143) show up now contains two identical orbitals j. Thus, the six terms are reduced to only three terms because of the symmetry condition

$$\begin{aligned}&\psi\big((j(2)j(1))J_{12}, j_c(3); JM\big) \\ &\quad = (-1)^{2j-J_{12}}\psi\big((j(1)j(2))J_{12}, j_c(3); JM\big) \ ,\end{aligned} \tag{3.144}$$

with J_{12} even and $(-1)^{2j} = -1$. Thus the total wave function of (3.143) reduces to

$$\begin{aligned}\frac{1}{\sqrt{3}}\Big\{&\psi\big((j^2(12))J_{12}, j_c(3); JM\big) - \psi\big((j^2(13))J_{12}, j_c(2); JM\big) \\ &+ \psi\big((j^2(23))J_{12}, j_c(1); JM\big)\Big\} \ .\end{aligned} \tag{3.145}$$

iii) The case where all three angular momenta become equal $j_a = j_b = j_c = j$ is the most interesting one. It is possible to recouple the angular momenta in (3.145) such that the particle coordinates in the three terms always come in the sequence 1, 2, 3 and in increasing order. In this case, the standard rules of angular momentum algebra when calculating matrix elements are applicable. We try to bring the second and third term in (3.145) in the order of the first term by recoupling

$$\begin{aligned}&\psi\big((j^2(23))J_{12}, j(1); JM\big) \\ &\quad = \sum_{J'_{12}} (-1)^{J_{12}} \hat{J}_{12}\hat{J}'_{12} \begin{Bmatrix} j & j & J_{12} \\ J & j & J'_{12} \end{Bmatrix} \psi\big((j^2(12))J'_{12}, j(3); JM\big) \ ,\end{aligned}$$

and

$$\begin{aligned}&\psi\big((j^2(13))J_{12}, j(2); JM\big) \\ &\quad = \sum_{J'_{12}} (-1)^{J_{12}+J'_{12}+1} \hat{J}_{12}\hat{J}'_{12} \begin{Bmatrix} j & j & J_{12} \\ J & j & J'_{12} \end{Bmatrix} \\ &\qquad \times \psi\big((j^2(12))J'_{12}, j(3); JM\big) \ .\end{aligned} \tag{3.146}$$

By bringing these terms together with the first term in (3.145) we always have the particles in the order $(1,2,3)$. Thus we can actually leave out the nucleon coordinates. The state thus constructed is an antisymmetric state of three particles moving in a j-orbital and coupled to angular momentum J, denoted as $\psi(j^3; JM)_{\mathrm{as}}$. Written explicitly,

$$\psi(j^3; JM)_{\mathrm{nas}} = \mathcal{N}' \sum_{J'_{12}\ (\mathrm{even})} \left[\delta_{J_{12}J'_{12}} + 2\hat{J}_{12}\hat{J}'_{12} \times \begin{Bmatrix} j & j & J_{12} \\ J & j & J'_{12} \end{Bmatrix} \right] \psi\left((j^2(12)) J'_{12}, j(3); JM \right) . \qquad (3.147)$$

The expansion on the right-hand side goes over all states J'_{12} where one couples a third particle $j(3)$ to an antisymmetric state of two particles in a single j-shell $j^2(12)J'_{12}$. The latter basis, a basis of three-particle wave functions antisymmetrized in particle coordinates 1 and 2 but not in particle coordinate 3, is much larger than the basis of three-particle states antisymmetric in all three particle coordinates. In (3.147) one writes the $\psi(j^3; JM)_{\mathrm{nas}}$ by calculating the overlap with all states $\psi((j^2(12))J_{12}, j(3); JM)$, i.e., we project onto the subspace $\psi(j^3; JM)_{\mathrm{nas}}$. The expansion (projection) coefficients [...] do not form a unitary transformation, therefore a special notation is used for the coefficients:

$$\psi(j^3; JM)_{\mathrm{nas}} = \sum_{\substack{J_1 \\ \mathrm{even}}} \left[j^2(J_1)jJ| \} j^3 J \right] \psi\left(j^2(J_1), j; JM \right) , \qquad (3.148)$$

and the coefficient $[j^2(J_1)jJ|\}j^3J]$ is the cfp coefficient (coefficient of fractional parentage) giving the projection of the state $\psi(j^3; JM)_{\mathrm{nas}}$ on the basis $\psi(j^2(J_1)j; JM)$. In the latter wave functions we leave out of the right-hand side the nucleon coordinates since they are ordered in the sequence $1, 2, 3$.

As an example we discuss the case of the $(d_{5/2})^3 J$ configurations.

In making the two-particle configurations first, we have $(d_{5/2})^2\ 0^+, 2^+, 4^+$ configurations. Next we couple a third $d_{5/2}$ particle by just using angular momentum coupling, thereby constructing the basis $|(d_{5/2})^2 J_{12}(d_{5/2}); JM\rangle$, i.e., we have

$$\begin{aligned} &|(d_{5/2})^2_{0^+}, d_{5/2}; 5/2^+\rangle \\ &|(d_{5/2})^2_{2^+}, d_{5/2}; 1/2^+, \ldots, 9/2^+\rangle \ . \\ &|(d_{5/2})^2_{4^+}, d_{5/2}; 3/2^+, \ldots, 13/2^+\rangle \end{aligned}$$

This space has 12 states. Now using (3.147) to explicitly calculate the three-particle cfp's for the $(d_{5/2})^3 J$ configurations, for all J_{12} intermediate values one keeps just 3 fully antisymmetric states with $J^\pi = 3/2^+, 5/2^+, 9/2^+$. If we call the first space A (12 states) and the latter space B (3 states), then B is entirely within space A and the states in B can be expanded using the projection coefficients of the fully antisymmetric states on the basis spanning space A (Fig. 3.43). Making this explicit for $(d_{5/2})^3 J$, we have

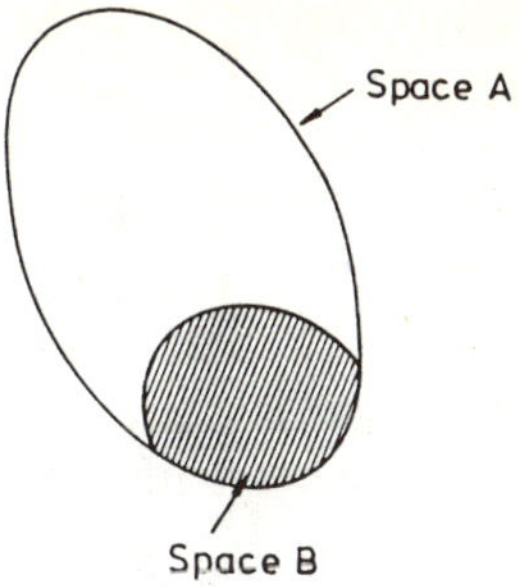

Fig. 3.43. Example for the construction of antisymmetric three-particle states in the $(d_{5/2})^3$ configuration. We indicate space A: the larger space of states $|(d_{5/2})^2 J_1, (d_{5/2}); JM\rangle$ antisymmetric in two-particles only and the smaller space B: the space $|(d_{5/2})^3; JM\rangle$ of antisymmetrized three-particle configurations. Space (B) is a subspace of space (A) and can be obtained via a projection

$$\begin{aligned}|(d_{5/2})^3;\ J = \tfrac{5}{2}\rangle = &- \frac{\sqrt{2}}{3}|(d_{5/2})^2_{0^+},\ d_{5/2};\ 5/2^+\rangle \\ &+ \frac{\sqrt{5}}{3\sqrt{2}}|(d_{5/2})^2_{2^+},\ d_{5/2};\ 5/2^+\rangle \\ &+ \frac{1}{\sqrt{2}}|(d_{5/2})^2_{4^+},\ d_{5/2};\ 5/2^+\rangle\ ,\end{aligned}$$

$$\begin{aligned}|(d_{5/2})^3;\ J = \tfrac{3}{2}\rangle = &- \frac{\sqrt{5}}{\sqrt{7}}|(d_{5/2})^2_{2^+},\ d_{5/2};\ 3/2^+\rangle \\ &+ \frac{\sqrt{2}}{\sqrt{7}}|(d_{5/2})^2_{4^+},\ d_{5/2};\ 3/2^+\rangle\ ,\end{aligned}$$

$$\begin{aligned}|(d_{5/2})^3;\ J = \tfrac{9}{2}\rangle = &\frac{\sqrt{3}}{\sqrt{14}}|(d_{5/2})^2_{2^+},\ d_{5/2};\ 9/2^+\rangle \\ &- \frac{\sqrt{11}}{\sqrt{14}}|(d_{5/2})^2_{4^+},\ d_{5/2};\ 9/2^+\rangle\ .\end{aligned}$$

In describing the fully antisymmetric three-particle states as in (3.148), in most cases more than just one J state can be made from the j^3 configuration: an extra quantum number α will be needed to characterize the state uniquely.

If only *one* state can be formed, α is not necessary. In the latter case, the wave function should be unique (except for an eventual overall phase factor) and independent of the initial two-particle spin J_{12} [see (3.147)]. In making the discussion somewhat more detailed, we first normalize the states $\psi(j^3; JM)$ leading to normalized cfp coefficients

$$\begin{aligned}\left[j^2(J_1)jJ|\}j^3J\right] = &\Big[\delta_{J_1,J_{12}} + 2\hat{J}_1\hat{J}_{12} \\ &\left\{\begin{matrix} j & j & J_1 \\ J & j & J_{12}\end{matrix}\right\}\Big]\left[3 + 6(2J_{12}+1)\left\{\begin{matrix} j & j & J_{12} \\ J & j & J_{12}\end{matrix}\right\}\right]^{-1/2}\ . \qquad (3.149)\end{aligned}$$

In the case $J = j$, we can start from the choice $J_{12} = 0$. For $j \le \frac{7}{2}$ there is only one antisymmetric state $\psi(j^3;\ J = j)$ (by construction). For $j > \frac{7}{2}$, this is no longer the case.

If we now put $J_{12} = 0$, and fill in the explicit values of the Wigner $6j$-symbols, we get

$$\left[j^2(0)j\, J = j|\}j^3 J = j\right] = \left((2j-1)/\left(3(2j+1)\right)\right)^{1/2},$$
$$\left[j^2(J_1)j\, J = j|\}j^3 J = j\right] = -2\left((2J_1+1)/\left(3(2j-1)(2j+1)\right)\right)^{1/2}, \qquad (3.150)$$

with $J_1 > 0$, even.

For the case of $(d_{5/2})^3$, these expressions (3.150) reduce to the wave function

$$\begin{aligned}\psi\left(\left(d_{5/2}\right)^3(J_{12}=0),\, J=\tfrac{5}{2}\right) =& \frac{\sqrt{2}}{3}\psi\left(\left(d_{5/2}\right)^2 J_1 = 0,\, d_{5/2};\, J=\tfrac{5}{2}\right)\\ &- \frac{1}{3}\frac{\sqrt{5}}{\sqrt{2}}\psi\left(\left(d_{5/2}\right)^2 J_1 = 2,\, d_{5/2};\, J=\tfrac{5}{2}\right)\\ &- \frac{1}{\sqrt{2}}\psi\left(\left(d_{5/2}\right)^2 J_1 = 4,\, d_{5/2};\, J=\tfrac{5}{2}\right). \qquad (3.151)\end{aligned}$$

Here, we include $J_{12} = 0$ on the left-hand side to point out that we have antisymmetrized starting from the $J_{12} = 0$ two-particle state. One can now verify that starting from other states $J_{12} = 2, 4$ leads to the same wave function (3.151) multiplied by a phase of -1. This will no longer be the case for $j = \frac{9}{2}$ in the $(9/2)^3$ configuration where different wave functions (cfp coefficients) result when starting from $J_{12} = 0$ or from $J_{12} = 2$. In these cases, one has to construct orthonormalized states explicitly.

3.3.2 Extension to n-particle Wave Functions

Using the discussion in Sect. 3.3.1, by using the building up principle one can construct consecutively more complicated states, ending up with n-particle states

$$\begin{aligned}\psi(j^n\alpha; JM)_{\text{nas}} =& \sum_{\alpha_1, J_1}\left[j^{n-1}(\alpha_1 J_1)jJ|\}j^n\alpha J\right]\\ &\times \psi\left(j^{n-1}(\alpha_1 J_1)j; JM\right), \qquad (3.152)\end{aligned}$$

where cfp coefficients are used to project from the fully antisymmetric n-particle wave functions onto the space of wave functions antisymmetric in the first $(n-1)$ particles, coupled to the nth particle with the use of angular momentum coupling. Thus, this is a $2 \to 3 \to \ldots n-1 \to n$ building up principle. One can also carry out the process by using two-particle cfp coefficients where the $\psi(j^n\alpha; JM)_{\text{nas}}$ wave functions are constructed from an antisymmetrized $(n-2)$ and an antisymmetrized 2 particle wave function. This leads to an expression of the form

$$\begin{aligned}\psi(j^n\alpha; JM)_{\text{nas}} =& \sum_{\alpha_1, J_1, J_2}\left[j^{n-2}(\alpha_1 J_1)j^2(J_2)J|\}j^n\alpha J\right]\\ &\times \psi\left(j^{n-2}(\alpha_1 J_1)j^2(J_2); JM\right). \qquad (3.153)\end{aligned}$$

Here the $n \to n-2$ cfp can be expressed in terms of $n \to n-1$ and $n-1 \to n-2$ one particle cfp coefficients (see problem set) with as a result

$$\left[j^{n-2}(\alpha_1 J_1)j^2(J_2)J|\}j^n\alpha J\right] = \sum_{\alpha_1', J_1'} \left[j^{n-2}(\alpha_1 J_1)jJ_1'|\}j^{n-1}\alpha_1' J_1'\right]$$
$$\times \left[j^{n-1}(\alpha_1' J_1')jJ|\}j^n\alpha J\right] \hat{J}_2 \hat{J}_1'(-1)^{2j+J+J_1} \begin{Bmatrix} J_1 & j & J_1' \\ j & J & J_2 \end{Bmatrix} . \tag{3.154}$$

Using the above wave functions, we are now in a position (using the reduction rules I, II of Chap. 2 to calculate matrix elements of spherical tensor operators) to evaluate all matrix elements necessary to set up a n-particle energy matrix and, later on, to calculate one-body operator expectation or transition matrix elements to test the wave functions against observable nuclear properties (electromagnetic moments and transition rates, beta-decay transition probabilities, etc.).

i) One-Body Matrix Elements. Here, the general one-body operator for an n-particle system reads

$$F_\kappa^{(k)} = \sum_{i=1}^{n} f_\kappa^{(k)}(i) , \tag{3.155}$$

where i denotes the particle coordinates $(\boldsymbol{r}_i, \boldsymbol{\sigma}_i, \ldots)$. We wish to evaluate the reduced n-particle matrix element

$$\langle j^n \alpha J \| \boldsymbol{F}^{(k)} \| j^n \alpha' J' \rangle \tag{3.156}$$

which is related to the normal matrix element $\langle j^n\alpha; JM|F_\kappa^{(k)}|j^n\alpha'; J'M'\rangle$ via the Wigner-Eckart theorem. Since $F_\kappa^{(k)}$ is a sum over n particles and since $\psi(j^n\alpha; JM)$ and $\psi(j^n\alpha'; J'M')$ are antisymmetrized wave functions, one can write that

$$\langle j^n\alpha; JM|f_\kappa^{(k)}(1)|j^n\alpha'; J'M'\rangle + \langle j^n\alpha; JM|f_\kappa^{(k)}(2)|j^n\alpha'; J'M'\rangle$$
$$+ \ldots + \langle j^n\alpha; JM|f_\kappa^{(k)}(n)|j^n\alpha'; J'M'\rangle$$
$$= n\langle j^n\alpha; JM|f_\kappa^{(k)}(n)|j^n\alpha'; J'M'\rangle . \tag{3.157}$$

Using the $n \to n-1$ cfp coefficients, one can separate the $(n-1)$ particle part out of the wave function and after some simple algebra obtain for the reduced matrix element of (3.156)

$$\langle j^n\alpha J\|\boldsymbol{F}^{(k)}\|j^n\alpha' J'\rangle = n \sum_{\alpha_1, J_1} \left[j^{n-1}(\alpha_1 J_1)jJ|\}j^n\alpha J\right]$$
$$\times \left[j^{n-1}(\alpha_1 J_1)jJ'|\}j^n\alpha' J'\right](-1)^{J_1+j+J+k}\hat{J}\hat{J}'$$
$$\times \begin{Bmatrix} j & J & J_1 \\ J' & j & k \end{Bmatrix} \langle j\|\boldsymbol{f}^{(k)}\|j\rangle . \tag{3.158}$$

Thus, the n-particle matrix elements are expressed in terms of the one-body matrix elements $\langle j\|\boldsymbol{f}^{(k)}\|j\rangle$ only. This is an important result which we shall discuss later in some detail.

ii) Two-Body Matrix Elements. Here, a general two-body operator is

$$V = \sum_{i<k=1}^{n} V(i,k) \,.$$

The nuclear two-body interaction has this structure and we shall need these operators in evaluating the energy for an n-particle system $\langle j^n\alpha; JM|V|j^n\alpha; JM\rangle$ or calculating the matrix elements needed to solve the n-particle secular equation.

Using similar arguments as under (i), since the two-body operator $V = V(1,2) + V(1,3) + \ldots V(2,3) + \ldots V(n-1,n)$, is symmetric in the coordinates of the interacting nucleons and since the wave functions are antisymmetric in the interchange of the coordinates of any two nucleons, one has

$$\langle j^n\alpha; JM|V|j^n\alpha; JM\rangle = \frac{n(n-1)}{2}\langle j^n\alpha; JM|V(n-1,n)|j^n\alpha; JM\rangle \,. \tag{3.159}$$

Now, using the $n \to n-2$ cfp coefficients, (3.159) can be reduced to

$$\langle j^n\alpha; JM|V|j^n\alpha; JM\rangle = \frac{n(n-1)}{2} \sum_{\alpha_2 J_2, J'} \left[j^{n-2}(\alpha_2 J_2) j^2(J')J|\}j^n\alpha J\right]^2 \times \langle j^2; J'M'|V|j^2; J'M'\rangle \,, \tag{3.160}$$

where a reduction of the n-particle interaction matrix element is obtained as a linear combination of two-body matrix elements only. If we now use the total Hamiltonian for the n-particle system

$$H = H_0 + H_{\text{res}} = \sum_{i=1}^{n} h_0(i) + \sum_{i<j=1}^{n} V(i,j) \,, \tag{3.161}$$

the expectation value of the energy becomes

$$\langle j^n\alpha; JM|H|j^n\alpha; JM\rangle = n\varepsilon_j + \frac{n(n-1)}{2} \sum_{\alpha_2 J_2, J'} \left[j^{n-2}(\alpha_2 J_2) j^2(J')J|\}j^n\alpha J\right]^2 \langle j^2; J'M'|V|j^2; J'M'\rangle \,. \tag{3.162}$$

As we conclude this section, before discussing some applications, it becomes clear that once we know the "basic" constituents: the single-particle matrix elements $\varepsilon_j = \langle jm|h_0(i)|jm\rangle$, $\langle j\|\boldsymbol{f}^{(k)}\|j\rangle$ for a general operator $f_\kappa^{(k)}$ and the two-body interaction matrix elements $\langle j^2; J'M'|V|j^2; J'M'\rangle$, by using (3.158, 162), the related n-particle matrix elements can be obtained in order to study n-particle nucleon systems when the cfp coefficients are known. Thus one can test, e.g., the consistency of the nuclear observables in the $1f_{7/2}$ shell model orbital (excitation energies, transition rates): nuclei between ${}^{40}_{20}\text{Ca}_{20}$ and ${}^{56}_{28}\text{Ni}_{28}$ starting from a single set of two-body matrix elements $\langle (1f_{7/2})^2; JM|V|(1f_{7/2})^2; JM\rangle$ and the one-body matrix elements $\langle 1f_{7/2}\|\boldsymbol{f}^{(k)}\|1f_{7/2}\rangle$. This process is schematically given in Fig. 3.44.

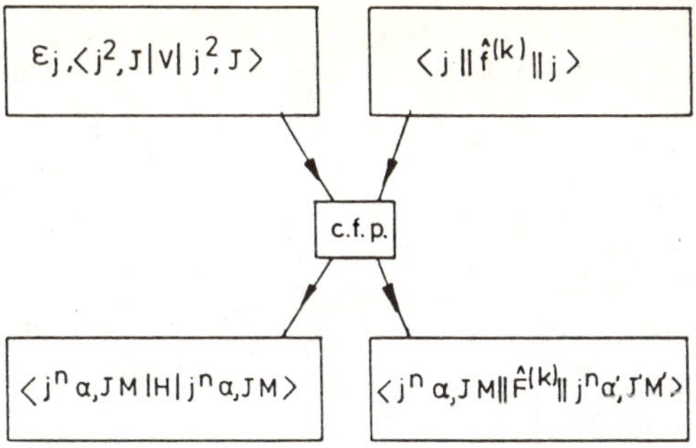

Fig. 3.44. Schematic illustration of how the n-particle properties (energy matrix elements and transition matrix elements) are related to the one- and two-body properties ε_j, $\langle j^2; JM|V_{1,2}|j^2; JM\rangle$ and $\langle j\|\boldsymbol{f}^{(k)}\|j\rangle$ using the cfp coefficients for a $n \to n-1 \to n-2 \to \ldots \to 2$ decomposition

iii) Fitting One- and Two-Body Matrix Elements. In the light of the above discussion, it is clear that (see (3.159)) two-body matrix elements in a j^n configuration depend quadratically on n, and that one-body matrix elements (see (3.158)) in a j^n configuration only depend linearly on the particle number n.

In order to deduce, using a fit between theoretical and experimental energies, two-body matrix elements and single-particle energies in a given mass region (the $1f_{7/2}$ shell for the Ca nuclei, the full sd-shell between ^{16}O and ^{40}Ca, ...) as precisely as possible, it is important to pay special attention to those energies in nuclei with a large number of valence nucleons n. On the contrary, if two-body matrix elements are fixed by the two-particle spectra in two-particle configuration only, even small deviations from the optimal values will 'propagate' through a multiplicative factor of n^2 in proceeding towards nuclei with a large number of valence nucleons outside of the closed shells.

3.3.3 Some Applications: Three-particle Systems

Starting from the discussion at the end of Sect. 3.3.2, we are now able to calculate spectra in nuclei having three particles outside a closed shell configuration. The first application, of course, is the case where only a single configuration j is important.

In this case,

$$\langle j^3\alpha; JM|H|j^3\alpha; JM\rangle = 3\varepsilon_j + 3\sum_{J'}\left[j^2(J')jJ|\}j^3J\right]^2 A_{J'} \,, \tag{3.163}$$

(with $A_{J'} \equiv \langle j^2; J'M|V|j^2; J'M\rangle$), the unperturbed energy $3\varepsilon_j$ can be left out as a constant energy shift. For a $(d_{5/2})^3$ configuration, one has $J = \frac{3}{2}, \frac{5}{2}, \frac{9}{2}$ and A_0, A_2 and A_4. The corresponding cfp coefficients are given in Table 3.3.

Using these cfp, we obtain

$$\begin{aligned} J &= \tfrac{5}{2} \quad \rightarrow \quad \tfrac{2}{3}A_0 + \tfrac{5}{6}A_2 + \tfrac{3}{2}A_4 \,, \\ J &= \tfrac{3}{2} \quad \rightarrow \quad \tfrac{15}{7}A_2 + \tfrac{6}{7}A_4 \,, \\ J &= \tfrac{9}{2} \quad \rightarrow \quad \tfrac{9}{14}A_2 + \tfrac{33}{14}A_4 \,. \end{aligned} \tag{3.164}$$

Table 3.3. The one-particle coefficients of fractional parentage (cfp) for the $(d_{5/2})^3$ configuration. So, we denote the coefficients as $[(d_{5/2})^2 J' \; d_{5/2} J|\}(d_{5/2})^3 J]$ and label the rows and columns by J and J', respectively

J \ J'	0	2	4
$\frac{5}{2}$	$-\frac{\sqrt{2}}{3}$	$\frac{\sqrt{5}}{(3\sqrt{2})}$	$\frac{1}{\sqrt{2}}$
$\frac{3}{2}$		$-\frac{\sqrt{5}}{\sqrt{7}}$	$\frac{\sqrt{2}}{\sqrt{7}}$
$\frac{9}{2}$		$\frac{\sqrt{3}}{\sqrt{14}}$	$-\frac{\sqrt{11}}{\sqrt{14}}$

Now, by using a residual δ-interaction the two-particle matrix elements can be obtained, using (3.90) as

$$A_0 = F_0'/2\,;\; A_2 = 8/35\left(F_0'/2\right)\,;\; A_4 = 2/21\left(F_0'/2\right)\,,$$

with

$$F_0' = (2j+1)F^0 = 6F^0\,.$$

This then leads to the interaction matrix elements

$$\begin{aligned} J &= \tfrac{5}{2} \;\rightarrow\; F_0'/2\,,\\ J &= \tfrac{3}{2} \;\rightarrow\; \tfrac{4}{7}\left(F_0'/2\right)\,,\\ J &= \tfrac{9}{2} \;\rightarrow\; \tfrac{13}{35}\left(F_0'/2\right)\,. \end{aligned} \tag{3.165}$$

The result is shown in Fig. 3.45.

It is now possible to use the expressions (3.164) in a slightly different way. If one knows the experimental 2^+ and 4^+ states and their excitation energy in the nucleus with a two-particle $(d_{5/2})^2$ configuration, then one can use (3.164), but

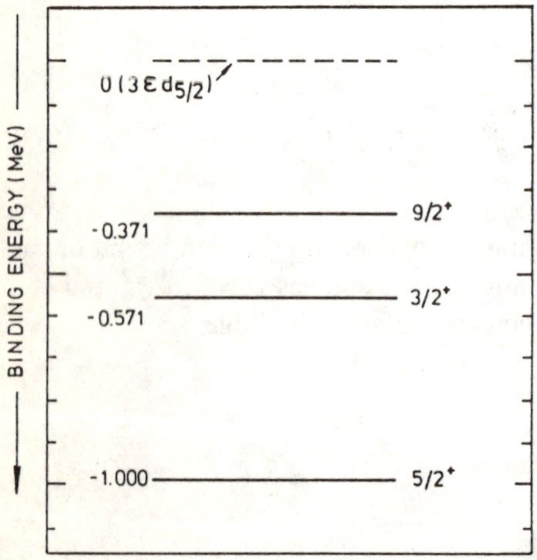

Fig. 3.45. The three-particle spectrum $(d_{5/2})^3$ J^π ($J^\pi = 5/2^+, 3/2^+, 9/2^+$ expressed in terms of the two-body properties. The equations (3.165) are used with F^0 the Slater integral for the δ-interaction. Relative to the unperturbed energy $3\varepsilon d_{5/2}$, the energies are given in units $F_0'/2$ [with $F_0' = (2j+1)F^0$]

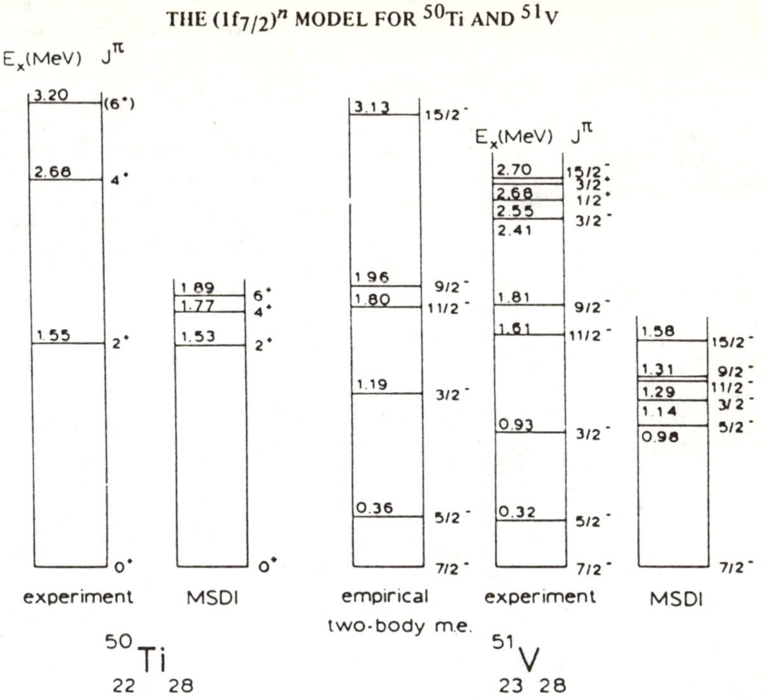

Fig. 3.46. The spectra of $^{50}_{22}\text{Ti}_{28}$ and $^{51}_{23}\text{V}_{28}$ for the configurations $(1f_{7/2})^2$ and $(1f_{7/2})^3$. Both the MSDI interaction (using a value of $A_{T=1} = 25/A\,\text{MeV}$) and the case of two-body empirical matrix elements (taken from the experimental spectrum of ^{50}Ti itself) are illustrated [taken from (Brussaard, Glaudemans 1977)]

now interpreting the matrix elements A_J as differences from matrix elements, i.e., $A_{J'} \equiv A_J - A_0$ (A_0 can not be determined from experiment but the $A_{J'}$ are the relative or excitation energies for the corresponding J states). Thus one obtains the relative energies in the three-particle case from the relative energies in the two-particle case (see problem set).

We illustrate this for the relation $^{50}_{22}\text{Ti}_{28} \rightarrow ^{51}_{23}\text{V}_{28}$ (Fig. 3.46) where (Brussaard, Glaudemans 1977)

i) We use theoretical values (MSDI-matrix elements $A_0, \ldots A_6$) obtained from a fit to the $^{50}_{22}$Ti spectrum in order to evaluate the $^{53}_{23}$V spectrum. Agreement is not very good.
ii) Now, using the empirical method where relative matrix elements $A'_2, \ldots A'_6$ are taken from the $^{50}_{22}$Ti experimental spectrum, agreement becomes rather good. This latter point proves that the spectra in ^{50}Ti-^{51}V are rather consistent with a single $1f_{7/2}$ shell model configuration using an empirical set of two-body matrix elements.

This discussion illustrates very well the methods discussed in Sect. 3.2.2.

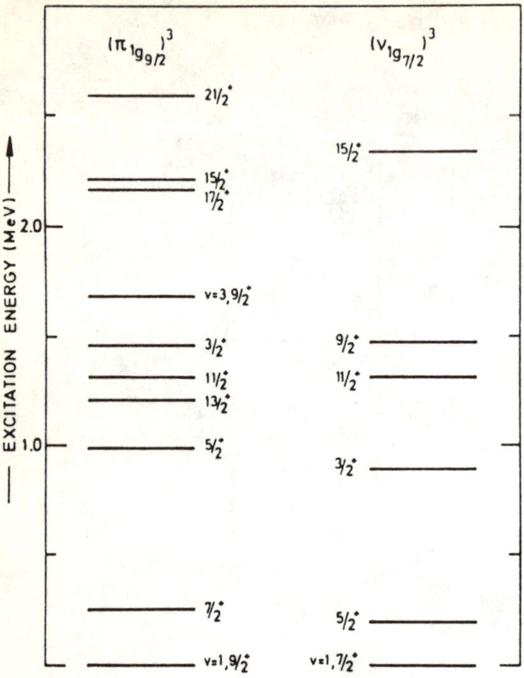

Fig. 3.47. We show the proton $(\pi 1g_{9/2})^3$ and neutron $(\nu 1g_{7/2})^3$ energy spectra with the proton and neutron two-body interaction determined as follows: the proton $(\pi 1g_{9/2})^2$ matrix elements are taken from the experimental spectrum of ${}^{88}_{40}Zr_{48}$ and the neutron $(\nu 1g_{7/2})^2$ matrix elements from the experimental spectrum of ${}^{134}_{52}Te_{82}$ [taken from (De Gelder 1980)]

Table 3.4. The cfp coefficients $[(f_{7/2})^2 J' \; f_{7/2} J | \} (f_{7/2})^3 J]$ for the three particle configurations. We denote the rows and columns by J and J', respectively [taken from (Brussaard, Glaudemans 1977)]

J	$J' = 0$	$J' = 2$	$J' = 4$	$J' = 6$
$\frac{3}{2}$	0	0.463	−0.886	0.000
$\frac{5}{2}$	0	0.782	0.246	−0.573
$\frac{7}{2}$	0.500	−0.373	−0.500	−0.601
$\frac{9}{2}$	0	0.321	−0.806	0.497
$\frac{11}{2}$	0	0.527	−0.444	−0.725
$\frac{15}{2}$	0	0.000	0.477	−0.879

We further illustrate three-particle spectra for some higher j orbitals: the proton $(1g_{9/2})^3$ spectrum and the neutron $(1g_{7/2})^3$ spectrum (Fig. 3.47), (Table 3.4) (De Gelder et al. 1980).

From all the above illustrations, it is clear that the lowest-lying state is always the $(j)^3 J = j$ state. This is easily explained since only for this $J = j$ state does the cfp for the intermediate two-particle 0^+ state occur. Now, the 0^+ two-particle matrix elements are by far the largest attractive matrix elements and still appear in the three-particle spectra via the cfp coefficients.

In the above situation, if we consider besides the $1f_{7/2}$ orbital also the relatively close-lying $2p_{3/2}$ and $1f_{5/2}$ orbitals, a more complicated situation results.

This is because as in the two-particle case with configuration mixing, we have to construct all three-particle configurations for given J^π values. For the present $(1f_{7/2}, 2p_{3/2}, 1f_{5/2})$ space with three particles one obtains

$$
\begin{array}{ll}
(1f_{7/2})^3 & J^\pi = 3/2^-\,,\ 5/2^-\,,\ 7/2^-\,,\ 9/2^-\,,\ 11/2^-\,,\ 15/2^- \\
(2p_{3/2})^3 & J^\pi = 3/2^- \\
(1f_{5/2})^3 & J^\pi = 3/2^-\,,\ 5/2^-\,,\ 9/2^- \\
(1f_{7/2})^2_{0^+}\,,\ 2p_{3/2} & J^\pi = 3/2^- \\
(1f_{7/2})^2_{2^+}\,,\ 2p_{3/2} & J^\pi = 1/2^-\,,\ 3/2^-\,,\ 5/2^-\,,\ 7/2^- \\
\vdots &
\end{array}
$$

In this manner one constructs the model space for each J^π, and using the techniques discussed in Sects. 3.2.4, 3.3 one can calculate the energy matrix in order to obtain the energy eigenvalues and the corresponding wave functions.

3.4 Non-identical Particle Systems: Isospin

Up to now we have considered only those situations where identical nucleons outside closed shells determine the nuclear structure. In many cases, however, both protons *and* neutrons are present outside the closed shells. Therefore, we would like to construct an extension of the methods outlined above so that the same general structure remains. We shall introduce the concept of "charge" quantum number or "isospin" quantum number, depending on which nuclei we are treating. Before constructing the nuclear wave functions including this new quantum number, we shall point out the evidence that exists in nuclear properties to introduce the concept of isospin as a valuable quantum number.

3.4.1 Isospin: Introduction and Concepts

Protons and neutrons have almost identical mass ($\Delta m/m \cong 1.4 \times 10^{-3}$). In addition, they show an almost identical behavior in their nuclear interactions. In 1932, Heisenberg proposed to consider protons and neutrons as two distinct forms of "nucleons", by using a double-valued variable called isospin that distinguishes between the "proton" state (described by a projection quantum number $t_z = -\frac{1}{2}$) and a "neutron" state (described by a projection quantum number $t_z = +\frac{1}{2}$) (Heisenberg 1932). The isospin (or isotopic spin) formalism can now be duplicated from the properties we studied in Chaps. 1, 2 on intrinsic spin ($\boldsymbol{s}$) of the proton and the neutron. We first present some of the experimental evidence for the equivalence of protons and neutrons in their nuclear interactions.

i) Low energy *np* scattering and *pp* scattering below $E < 5\,\text{MeV}$, after correcting for Coulomb effects, is equal within a few percent in the ^1S scattering channel (Arndt, MacGregor 1966, Wright et al. 1967, MacGregor et al. 1968a, b).

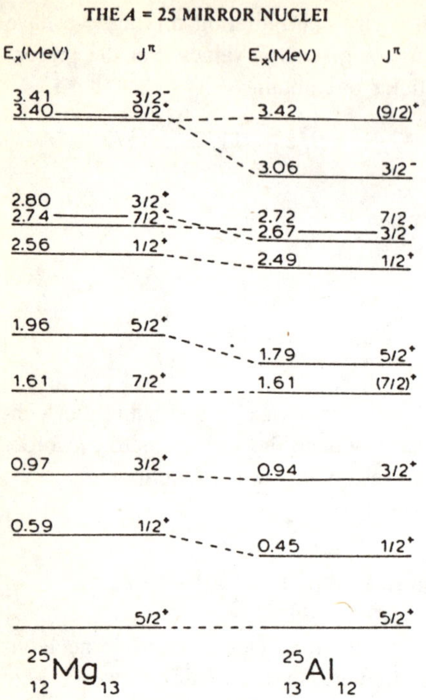

Fig. 3.48. A comparison of the level schemes of the $A = 25$ (^{25}Mg-^{25}Al) mirror nuclei shows the close similarity of the excitation energies for the states with identical J^{π} values [taken from (Brussaard, Glaudemans 1977)]

ii) Energy spectra in "mirror" nuclei are almost identical (Fig. 3.48). The small differences are both a consequence of the difference in the Coulomb interaction energy and of specific nuclear wave functions. From this observation one concluded that the exchange of protons and neutrons gives no modification of the nuclear interaction energy or, that the substitution $n - n \Leftrightarrow p - p$; $n - p \Leftrightarrow p - n$ does not modify the interaction energy. This observation implies the concept of *charge symmetry* in nuclear forces.

iii) Further information on how the $n - n$, $p - p$ forces relate to the $n - p$ force cannot be deduced from mirror nuclei. If we, therefore, study the triplet of nuclei, e.g., $^{30}_{14}\mathrm{Si}_{16}$, $^{30}_{15}\mathrm{P}_{15}$, $^{30}_{16}\mathrm{S}_{14}$, it is immediately clear that within a number of states ($0^+, 2^+$) (after correcting for Coulomb energies) the nuclear binding energies are equal in all three nuclei. From this observation a new characteristic of the nuclear forces can be deduced. Taking as a core $^{28}_{14}\mathrm{Si}_{14}$, the data show that the residual interaction energies due to $n-n$, $p-p$ and $n-p$ interactions are equal in a number of states (Fig. 3.49). The above leads to an even more stringent condition than obtained from (ii), i.e., *charge independence* in nuclear configurations that are possible in an $n-n$, $n-p$ and $p-p$ interacting system. A number of states in ^{30}P do not find a partner in the ^{30}Si, ^{30}S nuclei. This follows from the Pauli principle that excludes the realization of a number of configurations in identical nucleon systems ($n - n$, $p - p$) compared to the $n - p$ (non-identical nucleons) system. Thus the Pauli principle explains the large number of extra states in ^{30}P, as shown in Fig. 3.49.

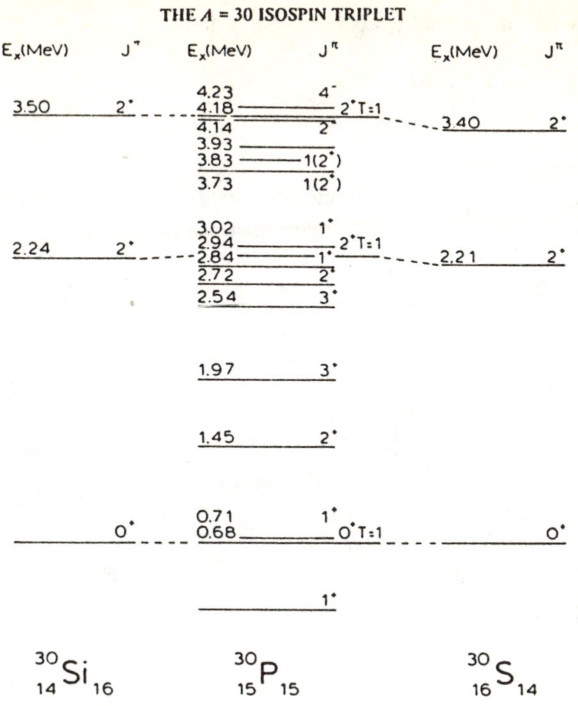

Fig. 3.49. The level spacings between the $T = 1$ isospin states in the mass $A = 30$ chain (^{30}Si-^{30}P-^{30}S) are very similar to the corresponding spacings in ^{30}Si, ^{30}P and ^{30}S. The states in ^{30}P where isospin is not given are the isospin $T = 0$ states [taken from (Brussaard, Glaudemans 1977)]

3.4.2 Isospin Formalism

a) General Properties of Isospin

We describe the neutron and proton by using a two-valued quantity just like the intrinsic spin, but now in isospin space:

$$\varphi_n(\boldsymbol{r}) = \varphi(\boldsymbol{r}) \begin{pmatrix} 1 \\ 0 \end{pmatrix} , \tag{3.166}$$

$$\varphi_p(\boldsymbol{r}) = \varphi(\boldsymbol{r}) \begin{pmatrix} 0 \\ 1 \end{pmatrix} , \tag{3.167}$$

where $\varphi(\boldsymbol{r})$ describes the spatial wave function which is the same for the proton and the neutron in the present case.

As for intrinsic spin, we can introduce Pauli isospin matrices $\boldsymbol{\tau}(\tau_x, \tau_y, \tau_z)$

$$\tau_x = \begin{pmatrix} 0 & 1 \\ 1 & 0 \end{pmatrix} ; \ \tau_y = \begin{pmatrix} 0 & -\mathrm{i} \\ \mathrm{i} & 0 \end{pmatrix} ; \ \tau_z = \begin{pmatrix} 1 & 0 \\ 0 & -1 \end{pmatrix} , \tag{3.168}$$

and define the isospin ($t = \frac{1}{2}$) operator as

$$\boldsymbol{t} = \frac{\boldsymbol{\tau}}{2} . \tag{3.169}$$

All other spin $\frac{1}{2}$ algebra results and properties can be used again: we have

$$\begin{aligned} &\left[t_x, t_y\right] = it_z, \text{ and cyclic permutations} , \\ &\left[\boldsymbol{t}^2, t_i\right] = 0 , \; i = x, y, z , \end{aligned} \tag{3.170}$$

and $\boldsymbol{t}^2$, acting on the isospin spinors, gives as an eigenvalue $t(t+1)$. The proton and neutron wave functions (3.166, 167) are now eigenfunctions of $\boldsymbol{t}^2$ and t_z

$$\begin{aligned} t_z \varphi_p(\boldsymbol{r}) &= -\tfrac{1}{2} \varphi_p(\boldsymbol{r}) , \\ t_z \varphi_n(\boldsymbol{r}) &= \tfrac{1}{2} \varphi_n(\boldsymbol{r}) . \end{aligned} \tag{3.171}$$

Now, the following relations hold

$$\begin{aligned} \tfrac{1}{2}\left(1 - \tau_z\right) \varphi_p(\boldsymbol{r}) &= \varphi_p(\boldsymbol{r}) , \\ \tfrac{1}{2}\left(1 + \tau_z\right) \varphi_p(\boldsymbol{r}) &= 0 , \\ \tfrac{1}{2}\left(1 - \tau_z\right) \varphi_n(\boldsymbol{r}) &= 0 , \\ \tfrac{1}{2}\left(1 + \tau_z\right) \varphi_n(\boldsymbol{r}) &= \varphi_n(\boldsymbol{r}) , \end{aligned} \tag{3.172}$$

and we can introduce the charge operator

$$\frac{Q}{e} \equiv \frac{1}{2}\left(1 - \tau_z\right) \quad \text{with} \quad \frac{Q}{e} = \begin{pmatrix} 0 & 0 \\ 0 & 1 \end{pmatrix} . \tag{3.173}$$

The ladder operator properties hold here, too, and change proton into neutron states (t_+) or neutron into proton states (t_-), respectively:

$$\begin{aligned} t_+ \varphi_p(\boldsymbol{r}) &= \varphi_n(\boldsymbol{r}) , \\ t_+ \varphi_n(\boldsymbol{r}) &= 0 , \\ t_- \varphi_n(\boldsymbol{r}) &= \varphi_p(\boldsymbol{r}) , \\ t_- \varphi_p(\boldsymbol{r}) &= 0 . \end{aligned} \tag{3.174}$$

Extending to a many-nucleon system, the total isospin operators are constructed according to the methods of adding angular momenta (Chap. 2) and we obtain

$$\begin{aligned} \boldsymbol{T} &= \sum_{i=1}^{A} \boldsymbol{t}_i \\ T_z &= \sum_{i=1}^{A} t_{z,i} . \end{aligned} \tag{3.175}$$

Acting on a many-nucleon eigenstate, the eigenvalues of $\boldsymbol{T}^2$ are $T(T+1)$ with $-T \leq T_z \leq +T$. The total isospin will be integer or half-integer according to whether A is even or odd. For each T value, there exists an isospin multiplet with $2T+1$ members, characterized by the z-component T_z since T_z varies from $-T$ to $+T$. The eigenvalue of T_z according to its definition in (3.175) becomes $\frac{1}{2}(N-Z)$.

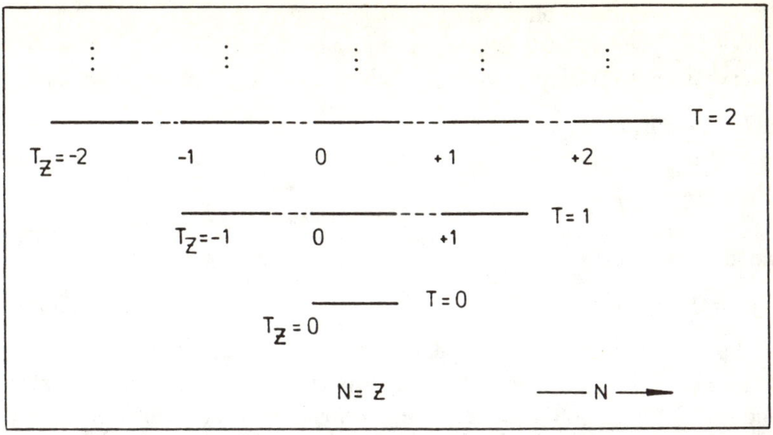

Fig. 3.50. Illustration of a series of levels in isotopic chains $T = 0, 1, 2, \ldots$. The variation with the neutron excess $N - Z = 2T_z$ is illustrated (isospin states in different nuclei of a given mass A chain) and, for a given nucleus (given T_z) of the different isospin states T ($|T_z| \leq T \leq A/2$) as a function of excitation energy

On the other hand, in a given nucleus $A(Z, N)$ with a given T_z value one can find a number of states with different isospin T according to the conditions

$$T = |T_z|\,,\; |T_z| + 1\,,\; \ldots\,,\; \frac{A}{2}\,.$$

This multiplet structure is illustrated in Fig. 3.50 for a number of nuclei. The different members of an isospin multiplet denoted by the eigenvector $|T, T_z\rangle$ are connected via the ladder operators

$$T_\pm = T_x \pm \mathrm{i}T_y = \sum_{i=1}^{A} t_\pm(i)\,, \tag{3.176}$$

and give as a result

$$T_\pm |T, T_z\rangle = \left[T(T+1) - T_z\left(T_z \pm 1\right)\right]^{1/2} |T, T_z \pm 1\rangle\,. \tag{3.177}$$

Consecutive action of $T_\pm$ changes T_z but does not affect the isospin T. This means that within an isospin multiplet the spin-spatial wave function remains constant; only the charge part changes as is expressed in the $|T, T_z\rangle$ eigenvectors.

We now try to express the charge symmetry and charge independence properties of the nuclear interactions in a more formal way.

i) Conservation of charge implies

$$\left[H, T_z\right] = 0\,, \tag{3.178}$$

ii) Charge independence implies that all members of a given isospin multiplet have the same energy, so that along a multiplet $E = E(T)$ only and no T_z-dependence shows up. Expressed in a mathematical way this demands that

$$\begin{aligned} H|T,T_z\rangle &= E(T)|T,T_z\rangle , \\ HT_+|T,T_z\rangle &= E(T)T_+|T,T_z\rangle , \\ HT_-|T,T_z\rangle &= E(T)T_-|T,T_z\rangle . \end{aligned} \tag{3.179}$$

The conditions (3.179) imply that one has the commutator relation

$$[H, T_\pm] = 0 , \tag{3.180}$$

indicating that H should be a scalar operator in isospin space. If we apply this to a Hamiltonian describing N neutrons and Z protons with nuclear residual interactions and the Coulomb interactions among protons, the Hamiltonian is written

$$\begin{aligned} H = &-\sum_{i=1}^{N} \frac{\hbar^2}{2m_n}\Delta(n) - \sum_{i=N+1}^{A} \frac{\hbar^2}{2m_p}\Delta(p) \\ &+ \frac{1}{2}\sum_{i,j=1}^{A} V(i,j) + \sum_{i<k=1}^{Z} \frac{e^2}{|\boldsymbol{r}_i - \boldsymbol{r}_j|} \end{aligned} \tag{3.181}$$

$$\begin{aligned} H \cong &-\sum_{i=1}^{A} \frac{\hbar^2}{2m}\Delta_i + \sum_{i=1}^{A} \frac{\hbar^2}{m^2} t_z(i) \cdot \Delta m \Delta_i + \frac{1}{2}\sum_{i,j=1}^{A} V(i,j) \\ &+ \sum_{i<j=1}^{A} \frac{e^2}{4|\boldsymbol{r}_i - \boldsymbol{r}_j|} - \sum_{i<j=1}^{A} \frac{e^2}{2}\left(t_z(i) + t_z(j)\right)/|\boldsymbol{r}_i - \boldsymbol{r}_j| \\ &+ \sum_{i<j=1}^{A} e^2 t_z(i) t_z(j)/|\boldsymbol{r}_i - \boldsymbol{r}_j| . \end{aligned} \tag{3.182}$$

In the second expression $m = (m_n + m_p)/2$ and Δm is the difference $\Delta m = |m_n - m| = |m_p - m|$. Here one observes that the Hamilton operator is not a scalar in isospin space but contains terms that are the $T_z = 0$ component of an isospin vector (rank 1) and of an isospin tensor (rank 2), indicating that the actual many nucleon Hamiltonian is not fully charge independent. In shorthand,

$$H = H_0^{(2)} + H_0^{(1)} + H_0^{(0)} , \tag{3.183}$$

in rank (2) (1) (0) with $T_z = 0$ in each case.

From this general Hamiltonian, an energy expression can easily be obtained as

$$\begin{aligned} \langle T,T_z|H|T,T_z\rangle = &(-1)^{T-T_z} \begin{pmatrix} T & 0 & T \\ -T_z & 0 & Tz \end{pmatrix} \langle T\|\boldsymbol{H}^{(0)}\|T\rangle \\ &+ (-1)^{T-T_z} \begin{pmatrix} T & 1 & T \\ -T_z & 0 & T_z \end{pmatrix} \langle T\|\boldsymbol{H}^{(1)}\|T\rangle \\ &+ (-1)^{T-T_z} \begin{pmatrix} T & 2 & T \\ -T_z & 0 & T_z \end{pmatrix} \langle T\|\boldsymbol{H}^{(2)}\|T\rangle . \end{aligned} \tag{3.184}$$

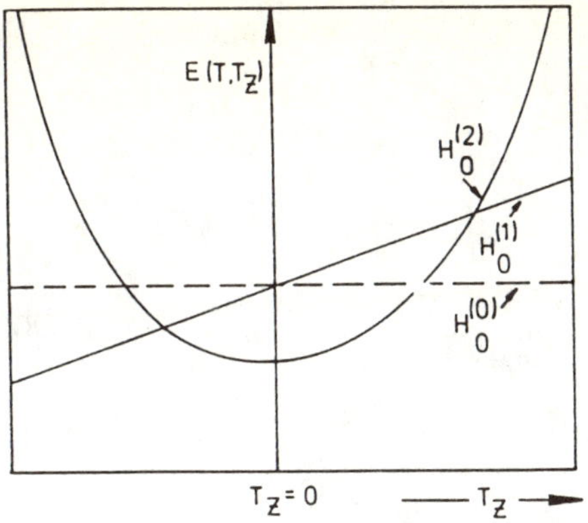

Fig. 3.51. Variation of the energy $E(T,T_z) \equiv \langle T,T_z|H|T,T_z\rangle$ with T_z ($= (N-Z)/2$). The different contributions from (3.184) are illustrated: the constant energy coming from the $H_0^{(0)}$ part (isoscalar contribution), the linear term coming from the $H_0^{(1)}$ (isovector) term and the quadratic term coming from the $H_0^{(2)}$ (tensor of rank 2) term. The dependence on T_z (for a given T) values is given by

$$\begin{pmatrix} T & 0 & T \\ -T_z & 0 & T_z \end{pmatrix}, \quad \begin{pmatrix} T & 1 & T \\ -T_z & 0 & T_z \end{pmatrix}, \quad \begin{pmatrix} T & 2 & T \\ -T_z & 0 & T_z \end{pmatrix}, \quad \text{respectively}$$

Here, the first term gives a constant value for given T, and expresses the charge independent part of H. The second term (coming from the mass difference in the kinetic energies of protons and neutrons and from the Coulomb term) induces a linear dependence on T_z, and the third term (Coulomb effect) introduces a quadratic T_z dependence. The total dependence is shown schematically in Fig. 3.51 for (T,T_z).

Taking into account the explicit values of the different Wigner $3j$-symbols appearing in (3.184), an energy formula results for the different $|T,T_z\rangle$ members of an isospin multiplet as

$$\begin{aligned} E(T,T_z) &= \langle T,T_z|H|T,T_z\rangle\,, \\ &= a(T) + b(T)T_z + c(T)T_z^2\,. \end{aligned} \tag{3.185}$$

The coefficients $a(T)$, $b(T)$ and $c(T)$ are specific for a given T-value and the full expression of $E(T,T_z)$ is a quadratic function in T_z. More examples of such simple energy or mass relations have also been obtained in other domains of physics (Gell-Mann (1962), Okubo (1962)) and will be discussed as realizations of a given dynamical symmetry that is present in the Hamiltonian describing the interacting fermion (boson) A-body system (see discussion in Chap. 7).

b) Isospin Wave Functions

We first construct the two-nucleon isospin wave functions by using

$$\begin{aligned} \varphi_n(\boldsymbol{r}) &= \zeta_{+1/2}^{1/2}\cdot\varphi(\boldsymbol{r}) \\ \varphi_p(\boldsymbol{r}) &= \zeta_{-1/2}^{1/2}\cdot\varphi(\boldsymbol{r})\,, \end{aligned} \tag{3.186}$$

with $\zeta_{t_z}^{1/2}$ the isospin spinors formally corresponding to the spin eigenvectors $\chi_{m_s}^{1/2}$.

For the two-nucleon isospin eigenvectors, we construct

$$\zeta\left(\tfrac{1}{2}\,\tfrac{1}{2};\,TT_z\right) = \sum_{t_z,t_z'}\langle\tfrac{1}{2}t_z,\,\tfrac{1}{2}t_z'|TT_z\rangle\zeta_{t_z}^{1/2}(1)\zeta_{t_z'}^{1/2}(2)\,, \tag{3.187}$$

or

$$
\begin{aligned}
\zeta\left(\tfrac{1}{2}\tfrac{1}{2};T=1,T_z=+1\right)&=\zeta^{1/2}_{+1/2}(1)\zeta^{1/2}_{+1/2}(2)\,,\\
\zeta\left(\tfrac{1}{2}\tfrac{1}{2};T=1,T_z=-1\right)&=\zeta^{1/2}_{-1/2}(1)\zeta^{1/2}_{-1/2}(2)\,,\\
\zeta\left(\tfrac{1}{2}\tfrac{1}{2};T=1,T_z=0\right)&=\frac{1}{\sqrt{2}}\zeta^{1/2}_{+1/2}(1)\zeta^{1/2}_{-1/2}(2)\\
&\quad+\frac{1}{\sqrt{2}}\zeta^{1/2}_{-1/2}(1)\zeta^{1/2}_{+1/2}(2)\,,\\
\zeta\left(\tfrac{1}{2}\tfrac{1}{2};T=0,T_z=0\right)&=\frac{1}{\sqrt{2}}\zeta^{1/2}_{+1/2}(1)\zeta^{1/2}_{-1/2}(2)\\
&\quad-\frac{1}{\sqrt{2}}\zeta^{1/2}_{-1/2}(1)\zeta^{1/2}_{+1/2}(2)\,,
\end{aligned}
\tag{3.188}
$$

written explicitly. The first gives a two-neutron, the second a two-proton wave function. The other two are linear combinations for a proton-neutron wave function, however, with a specific symmetry for the interchange of the quantum numbers of the nucleons: $T=1$ symmetric and $T=0$ antisymmetric.

Consider now the more general case of a proton moving in the orbital (j_a, m_a) and a neutron moving in the orbital (j_b, m_b). The wave function describing this particular situation reads

$$
\psi_{pn}(j_a j_b; JM) = \sum_{m_a,m_b} \langle j_a m_a, j_b m_b | JM\rangle \varphi_{j_a m_a}(\boldsymbol{r}_p)\varphi_{j_b m_b}(\boldsymbol{r}_n)\,. \tag{3.189}
$$

In the nuclear potential (in particular for light and for medium heavy nuclei) the situation with a proton moving in the orbital (j_b, m_b) and a neutron in the orbital (j_a, m_a) is almost degenerate in the former case. This tells us that the two configurations,

$$
\begin{aligned}
&\psi_{pn}(j_a j_b; JM)\,,\\
&\psi_{np}(j_a j_b; JM)\,,
\end{aligned}
\tag{3.190}
$$

are almost degenerate in energy. If we now diagonalize the residual proton-neutron interaction V_{pn}, the degeneracy in (3.190) will be lifted and we get two states with given symmetry $\psi^{\pm}_{pn}(j_a j_b; JM)$

$$
\begin{aligned}
\psi^{\pm}_{pn}(j_a j_b; JM) &= N \sum_{m_a,m_b} \langle j_a m_a, j_b m_b | JM\rangle\\
&\times \left\{\varphi_{j_a m_a}(\boldsymbol{r}_p)\varphi_{j_b m_b}(\boldsymbol{r}_n) \pm \varphi_{j_a m_a}(\boldsymbol{r}_n)\varphi_{j_b m_b}(\boldsymbol{r}_p)\right\}\,,
\end{aligned}
\tag{3.191}
$$

and we can also evaluate the energy shift between the states with different symmetry $\Delta E^{\pm}_J$ (Fig. 3.52). In (3.191), in contrast to (3.189) and (3.190), the index pn of the wave function $\Psi^{\pm}_{pn}(j_a j_b; JM)$ simply indicates that we are still describing a proton and a neutron, but the information regarding a precize localization in a given single-particle orbital is lost through the diagonalization process. One calls

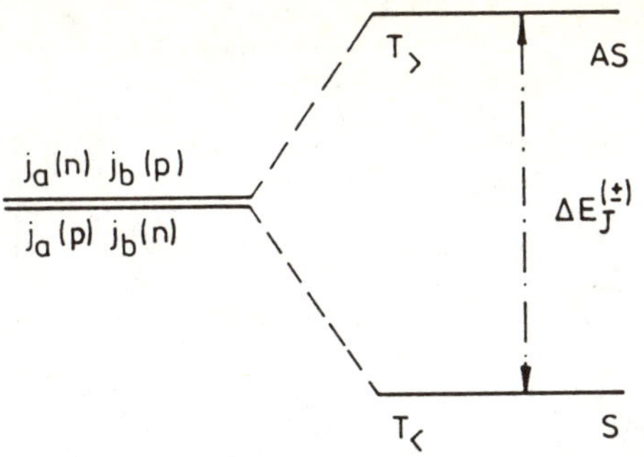

Fig. 3.52. Illustration of the mechanism of the proton-neutron interaction establishing the formation of two states with a specific spatial symmetry character. Starting from the *two* unperturbed configurations $\psi_{\rm pn}(j_a j_b; JM)$ and $\psi_{\rm np}(j_a j_b; JM)$ [see (3.190)], the proton-neutron force generates a symmetric (S) and antisymmetric (AS) state [see (3.191)]. The lower, symmetric state corresponds to the isospin $T_<$ (lower isospin, $T = 0$ for two-particle state) and the upper, antisymmetric state corresponds to the isospin $T_>$ (upper isospin, $T = 1$ for two-particle state). The energy separation is denoted by $\Delta E_J^{(\pm)}$

the low-lying state with the spatial symmetric wave function the $T_<$ state (lower isospin; $T = 0$ for a two-particle system) and the high-lying state with the spatial antisymmetric wave function the $T_>$ state (upper isospin; $T = 1$ for a two-particle system). Using now the two-particle isospin wave functions of (3.188), one can rewrite (3.191) as

$$\begin{aligned} N & \sum_{m_a, m_b} \langle j_a m_a, j_b m_b | JM \rangle \\ & \times \Big(\varphi_{j_a m_a}(1) \varphi_{j_b m_b}(2) \\ & \pm \varphi_{j_a m_a}(2) \varphi_{j_b m_b}(1) \Big) \zeta_{-1/2}^{1/2}(1) \zeta_{+1/2}^{1/2}(2) \, . \end{aligned} \tag{3.192}$$

If we now exchange the coordinates of particles 1 and 2 in (3.192), the physical content of the wave function remains identical. Thus we get extra information that can be removed by making linear combinations of (3.192), and (3.192) with $1 \Leftrightarrow 2$ interchanged. We make linear combinations such that the final wave functions obey a generalized exclusion principle: antisymmetry with respect to *all* (spatial, spin, isospin) coordinates. Remember that this is *not* an extra assumption but just a convenient formalism that allows us to handle $n - n$, $p - p$ and $n - p$ systems consistently. The above combinations are now constructed as (using $a \equiv j_a, m_a, \ldots$)

$$\begin{aligned} \psi_{\rm pn}^{+}\big(j_a j_b; JM\big) &= \frac{N}{\sqrt{2}} \sum \langle \ldots | \ldots \rangle \\ & \quad \times \big(\varphi_a(1)\varphi_b(2) + \varphi_a(2)\varphi_b(1)\big) \zeta\big(T = 0,\, T_z = 0\big) \, , \\ \psi_{\rm pn}^{-}\big(j_a j_b; JM\big) &= \frac{N}{\sqrt{2}} \sum \langle \ldots | \ldots \rangle \\ & \quad \times \big(\varphi_a(1)\varphi_b(2) - \varphi_a(2)\varphi_b(1)\big) \zeta\big(T = 1,\, T_z = 0\big) \, , \end{aligned} \tag{3.193}$$

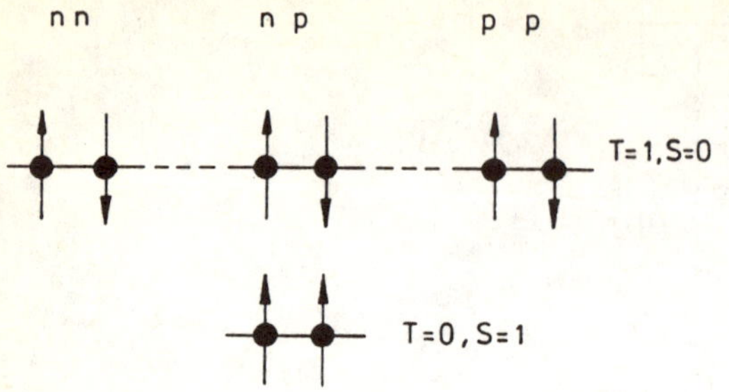

Fig. 3.53. Schematic illustration of possible combinations for the two-nucleon configurations nn, pp, np. The nn and pp only occur for $S = 0$, $T = 1$ (taking a symmetric orbital wave function, i.e. $l = 0$ wave). The $n - p$ (deuteron) occurs in both the $T = 1$, $S = 0$ and $T = 0$, $S = 1$ (lowest state) configurations

$$\psi_{\text{pp}}(j_a j_b; JM) = \frac{N}{\sqrt{2}} \sum \langle \ldots | \ldots \rangle$$
$$\times \left(\varphi_a(1)\varphi_b(2) - \varphi_a(2)\varphi_b(1)\right)\zeta\left(T = 1,\, T_z = -1\right) ,$$
$$\psi_{\text{nn}}(j_a j_b; JM) = \frac{N}{\sqrt{2}} \sum \langle \ldots | \ldots \rangle$$
$$\times \left(\varphi_a(1)\varphi_b(2) - \varphi_a(2)\varphi_b(1)\right)\zeta\left(T = 1,\, T_z = 1\right) .$$

In the above, we have either a symmetric spatial-spin function (S) and an antisymmetric isospin function $\zeta(T = 0, T_z = 0)$ or an antisymmetric spatial-spin function (AS) and a symmetric isospin function $\zeta(T = 1, T_z)$ (see Fig. 3.53 for application to the deuteron). Note once again that we could have studied nuclei without using the generalized Pauli principle with isospin, but simply by using the charge quantum numbers.

The outcome of constructing *all* basis configurations according to the proton-neutron valence numbers, is, however, in line with the isospin results.

To illustrate, we take the example of a light nucleus $^{42}_{21}\text{Sc}_{21}$ where we consider possible proton-neutron configurations (Fig. 3.54). Besides the $1f_{7/2}(p)\ 2p_{3/2}(n)$ basis state, we shall also have to take the $1f_{7/2}(n)\ 2p_{3/2}(p)$ basis state into account. In this small model space diagonalizing the nucleon-nucleon force, one ends up with two classes of states: one set with a symmetric wave function in the interchange of the charge coordinates, ($T = 1$) states that are high in energy, and another set with antisymmetric wave functions ($T = 0$) that are low in energy. Thus, if $\varepsilon_{1f_{7/2}}(p) = \varepsilon_{1f_{7/2}}(n)$, $\varepsilon_{2p_{3/2}}(p) = \varepsilon_{2p_{3/2}}(n)$ and all matrix elements are exactly charge independent, the coefficients in the S and AS combination will be equal to $1/\sqrt{2}$, thus again forming the isospin structure. Slight differences in the above conditions will induce some isospin mixing and thus, depending on whether one likes charge-quantum numbers or isospin, constitutes a convenient basis for obtaining the same physics.

If we now move to heavy nuclei with a neutron excess, the underlying principles in constructing isospin cannot be used easily since configurations $\psi_{\text{pn}}(j_a j_b; JM)$ and $\psi_{\text{np}}(j_a j_b; JM)$ are not at *all* degenerate any more: it costs a lot of energy to create the "exchanged" state and isospin can be heavily broken for low-lying states (Fig. 3.55). Here, one should better use a charge quantum number formalism. It is, however, possible to define certain states that correspond to a good isospin, i.e., configurations with two valence protons in orbitals that

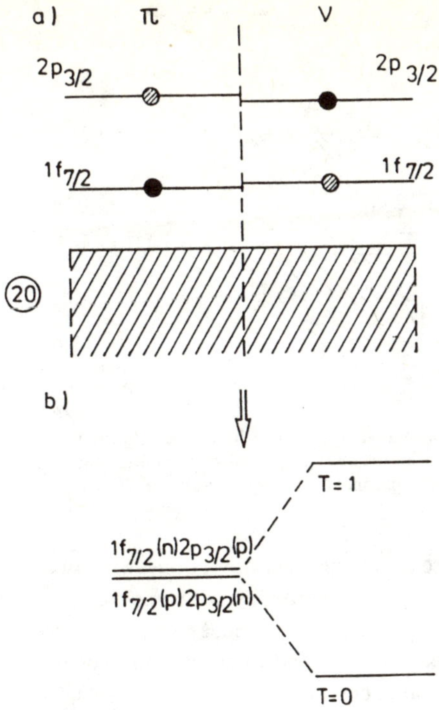

Fig. 3.54a, b. Illustration how, for a nucleus like ${}^{42}_{21}\mathrm{Sc}_{21}$, the isospin $T = 0$ and $T = 1$ states can be realized for the $1f_{7/2}\,2p_{3/2}$ configuration. (**a**) In the upper part we show the two possible, almost degenerate proton-neutron configurations $\pi 1f_{7/2}\ \nu 2p_{3/2}$ (•) and $\nu 1f_{7/2}\ \pi 2p_{3/2}$ (⊘) configurations. (**b**) In the lower part, we indicate how the two, nearly degenerate configurations form specific linear combinations when diagonalizing $V_{p,n}$. The lowest state, the $T = 0$ state, gives a symmetric spatial state, the upper one, the $T = 1$ state, corresponds to an antisymmetric spatial state

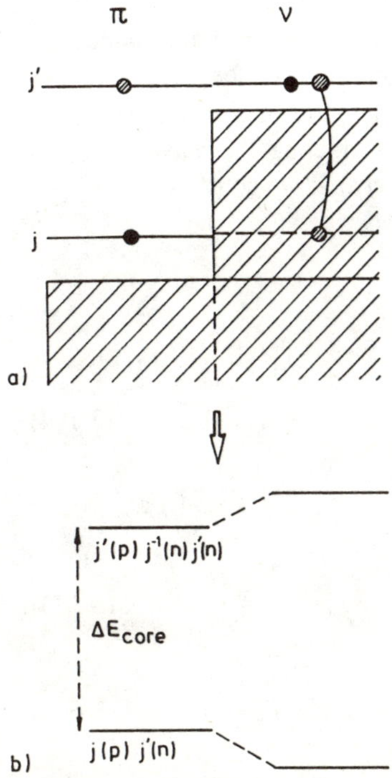

Fig. 3.55. Construction, as in Fig. 3.54, for proton-neutron configurations, but now for heavy nuclei with a large neutron excess (**a**). In the upper part, we have the $\pi j\ \nu j'$ configuration. Due to the neutron excess, the $\nu j\, \pi j'$ configuration does not exist, instead we have to make a core excited state $(\nu j)^{-1}(\nu j')\pi j'$. The latter state occurs at a much higher excitation energy compared with the former one. In the lower part (**b**) we illustrate these two states corresponding to a large energy difference ΔE_{core}. The residual interaction $V_{p,n}$ will only induce minor admixtures. Isospin is not needed to be introduced

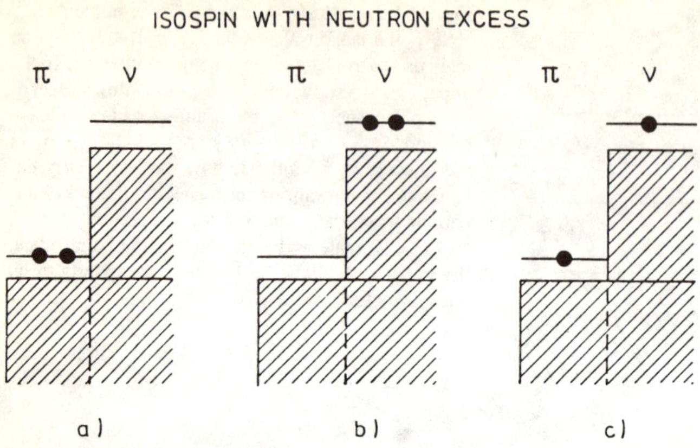

Fig. 3.56. States with fixed isospin $T = T_z$ for nuclei with a neutron excess. In the cases (**a**), (**b**) and (**c**), it is impossible to change a proton into a neutron (via the T_+ operator) since the corresponding neutron orbitals are fully occupied in the core. So, states with maximal isospin result with $T = T_z$

correspond to filled orbitals in the neutron excess core, configurations with two valence neutrons above the neutron excess and configurations with one valence proton (corresponding to a neutron excess orbital) and one valence neutron above the neutron excess (Fig. 3.56). The above states correspond to states with isospin obtained by coupling the isospin of the neutron excess system $T_{\text{core}} = (N - Z)/2$, to the isospin $T = 1$ of the two-particle system. In both cases, the isospin is *maximal* and corresponds to the projection T_z or $T = T_z$!

We shall illustrate the analogies and differences between a proton-neutron formalism versus isospin formalism in Chap. 5 for particle-hole excitations in ^{16}O.

To finish the present discussion, we consider the case of two-nucleons ($n - n$, $p - p$ or $p - n$) in a single j-shell when constructing the wave functions $\psi(j^2; JM, TT_z)$. Using the same method as in Sect. 3.2, we construct

$$\begin{aligned}\psi\left(j^2; JM, TT_z\right) &= N'\left(1 - (-1)^{2j-J+1-T}\right) \\ &\times \sum_{m,m'} \langle jm, jm'|JM\rangle \sum_{t_z,t'_z} \langle \tfrac{1}{2}t_z, \tfrac{1}{2}t'_z, |TT_z\rangle \\ &\times \varphi_{jm}(1)\varphi_{jm'}(2)\zeta^{1/2}_{t_z}(1)\zeta^{1/2}_{t'_z}(2)\,. \end{aligned} \tag{3.194}$$

These functions disappear except when $J + T =$ odd, or

$$\begin{aligned} &T = 1\,, \quad J = \text{even} = 0, 2, \ldots, 2j - 1\,, \\ &T = 0\,, \quad J = \text{odd} = 1, 3, \ldots, 2j\,, \end{aligned} \tag{3.195}$$

with, as an example, the $(1f_{7/2})^2$ ($T = 1$, $J = 0,2,4,6$) and ($T = 0$, $J = 1,3,5,7$) configurations.

3.4.3 Two-Body Matrix Elements with Isospin

Using the same methods as discussed in Sect. 3.2.3, one can evaluate two-body interaction matrix elements, now including isospin, with the result (using $\boldsymbol{l}+\boldsymbol{s}=\boldsymbol{j}$ coupling)

$$
\begin{aligned}
&\langle j_1 j_2; JM, TT_z|V|j_3 j_4; JM, TT_z\rangle = -A_0^{(T)}\hat{j}_1\,\hat{j}_2\,\hat{j}_3\,\hat{j}_4\,\cdot \\
&\left[\left(1+\delta_{j_1j_2}\right)\left(1+\delta_{j_3j_4}\right)\right]^{-1/2}\left(1+(-1)^{l_1+l_2+l_3+l_4}\right)/2\,\cdot \\
&\left[(-1)^{j_1+l_2+1/2}\begin{pmatrix} j_1 & j_2 & J \\ \frac{1}{2} & -\frac{1}{2} & 0\end{pmatrix}(-1)^{j_3+l_4+1/2}\begin{pmatrix} j_3 & j_4 & J \\ \frac{1}{2} & -\frac{1}{2} & 0\end{pmatrix}\right. \\
&\quad\left(1-(-1)^{J+T+l_3+l_4}\right)/2+\left(1+(-1)^T\right)/2(-1)^{j_1+j_2} \\
&\quad\left.\begin{pmatrix} j_1 & j_2 & J \\ \frac{1}{2} & \frac{1}{2} & -1\end{pmatrix}(-1)^{j_3+j_4}\begin{pmatrix} j_3 & j_4 & J \\ \frac{1}{2} & \frac{1}{2} & -1\end{pmatrix}\right]\,. \qquad (3.196)
\end{aligned}
$$

By using the above expressions we can also evaluate the matrix elements of the interaction $-4\pi V_0\delta(\boldsymbol{r}_1-\boldsymbol{r}_2)(1+\alpha\boldsymbol{\sigma}_1\cdot\boldsymbol{\sigma}_2)$ by the concordance

$$
\begin{aligned}
A_0^{(0)} &= (1+\alpha)A_0\,, \\
A_0^{(1)} &= (1-3\alpha)A_0\,,
\end{aligned} \qquad (3.197)
$$

with

$$
A_0 = V_0\int u_{n_1l_1}(r)u_{n_2l_2}(r)u_{n_3l_3}(r)u_{n_4l_4}(r)\frac{1}{r^2}dr\,.
$$

This is true because we have $S=0$, $T=1$ and $S=1$, $T=0$ since the δ-function only allows the spatially symmetric ($L=$ even) wave functions to give non-vanishing results.

One can verify that in the special case of $\alpha=0$, $j_1=j_3$, $j_2=j_4$, $j_1\neq j_2$, the result of Sect. 3.2.3 is reproduced.

Another case is for all $j_i=j$ with

$$
\begin{aligned}
\langle j^2; JM, TT_z|V|j^2; JM, TT_z\rangle &= -A_0^{(T)}(2j+1)^2\frac{1}{2}\left[\begin{pmatrix} j & j & J \\ \frac{1}{2} & -\frac{1}{2} & 0\end{pmatrix}^2\right. \\
&\left.\times\left(1-(-1)^{J+T}\right)/2+\begin{pmatrix} j & j & J \\ \frac{1}{2} & \frac{1}{2} & -1\end{pmatrix}^2\left(1+(-1)^T\right)/2\right]\,. \qquad (3.198)
\end{aligned}
$$

In studying the above example for $(1f_{7/2})^2_{T=0}$ states, we obtain the spectrum shown in Fig. 3.57. Here we see that the interaction is most attractive in the parallel and anti-parallel angular momenta for $T=0$ states. We can, moreover, compare the above theoretical results with the $^{42}_{21}$Sc nucleus where indeed, the $1f_{7/2}(p)$ $1f_{7/2}(n)$ case occurs. In Sect. 3.2.3, the results for $T=0$ for a δ-interaction with large j are illustrated and compare well with experimental data on odd-odd nuclei.

In Fig. 3.58 we compare the low-lying spectra for $^{42}_{20}\mathrm{Ca}_{22}$ with $^{42}_{21}\mathrm{Sc}_{21}$, and observe that although many more states occur in ^{42}Sc, for the $(1f_{7/2})^2_{T=1;\,J=0,2,4,6}$ states, very similar results occur because of charge independence.

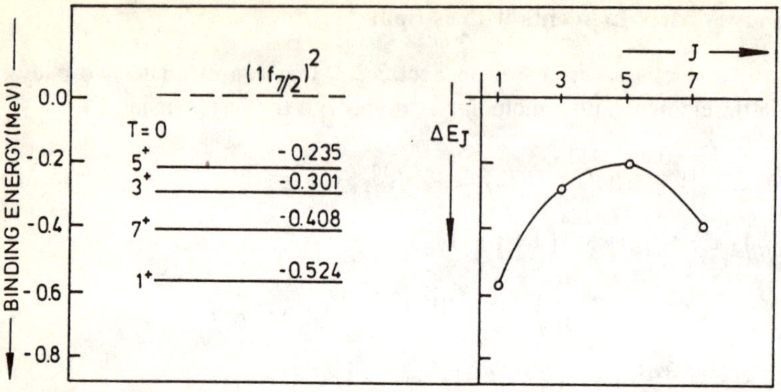

Fig. 3.57. Isospin $T = 0$ two-body matrix elements for two particles in the $1f_{7/2}$ orbital. The matrix elements are expressed as $\Delta E_J \equiv \langle j^2; JM, T = 0|V_{1,2}|j^2; JM, T = 0\rangle / 4A_0^{(0)}$. On the right-hand side, a corresponding ΔE_J vs. J plot has been made

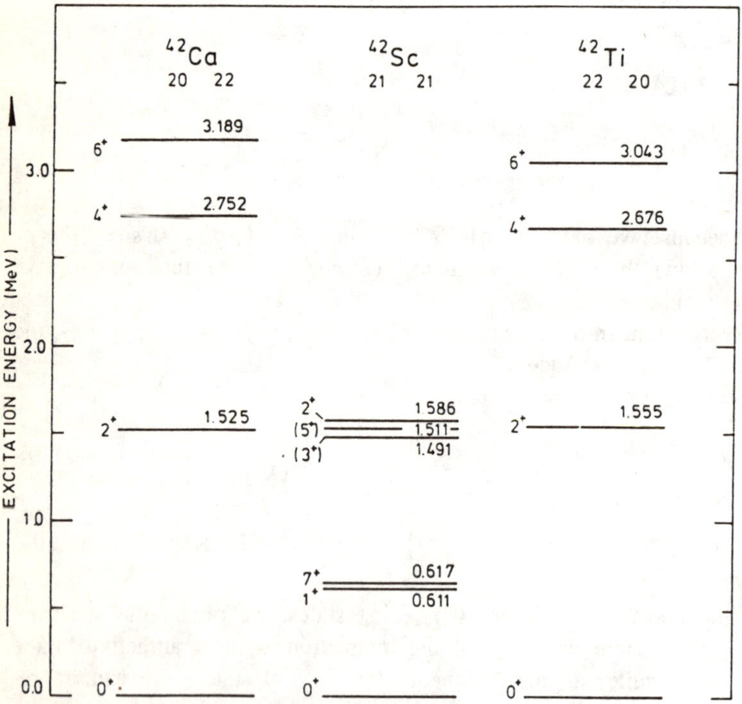

Fig. 3.58. Comparison of the data in ^{42}Ca, ^{42}Sc and ^{42}Ti. We compare the $T = 1$ and $T = 0$ $(1f_{7/2})^2$ states. The conditions of charge independence are well followed when comparing ^{42}Ca and ^{42}Ti. In ^{42}Sc, the extra states 1^+, 7^+, (3^+), (5^+) are also given. In the case of ^{42}Ca and ^{42}Ti, some low-lying levels $(0^+, 2^+)$ have been left out since they do not correspond to the $(1f_{7/2})^2$ configuration

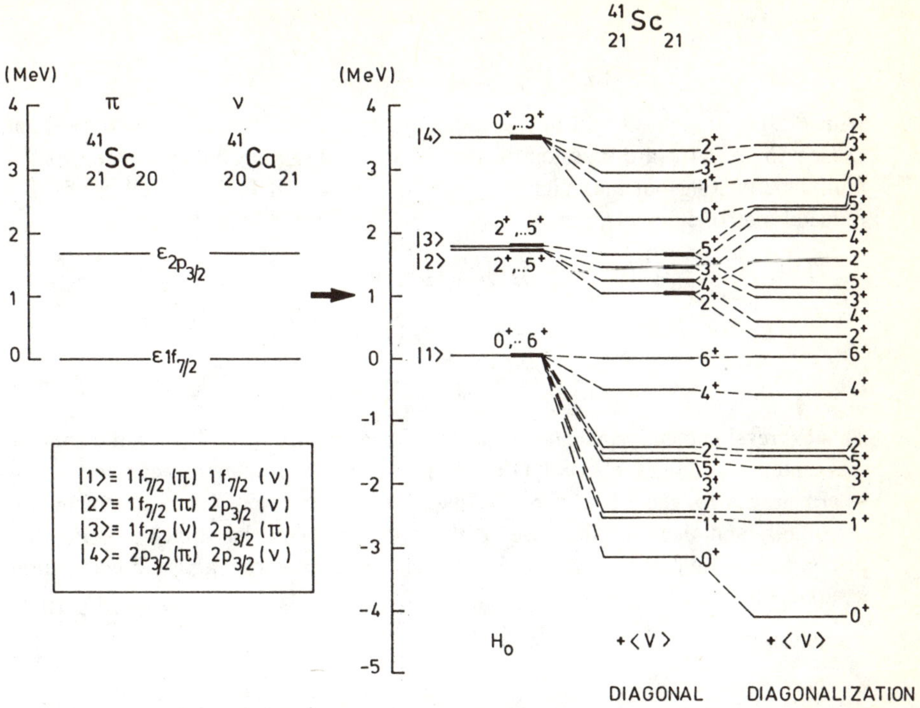

Fig. 3.59. Spectrum for the doubly-odd nucleus ^{42}Sc with a proton and a neutron occupying the $1f_{7/2}$ and $2p_{3/2}$ orbitals. On the left-hand side, the unperturbed energies are given. The energy spectrum is decomposed in to the unperturbed spectrum, the spectrum after adding diagonal terms and, finally, the full result after diagonalizing the residual interactions

Starting from a knowledge of the isospin $T = 0$ and $T = 1$ splitting that is obtained for the two-fold degeneracy resulting in a proton-neutron system, the full spectrum of the nucleus $^{42}_{21}\mathrm{Sc}_{21}$ can then be obtained when we constrain the separate proton and neutron single-particle space to the $1f_{7/2}$ and the $2p_{3/2}$ orbitals. Taking the single-particle energies to be degenerate (left-hand side of Fig. 3.59), the states combined from the $|1f_{7/2}(p)1f_{7/2}(n); JM\rangle$; $|1f_{7/2}(p)2p_{3/2}(n); JM\rangle$; $|1f_{7/2}(n)2p_{3/2}(p); JM\rangle$ and the $|2p_{3/2}(p)2p_{3/2}(n); JM\rangle$ configurations are given on the right-hand side. Then, subsequently adding the two-particle diagonal matrix elements and, on the extreme right, taking into account the results of diagonalization in the above model space, a rather good approximation to the full (J, T) structure of the energy spectrum in an odd-odd nucleus such as $^{42}_{21}\mathrm{Sc}_{21}$ can be reached.

The above particle-particle spectra in odd-odd nuclei are remarkably consistent with a parabolic behavior. This works throughout the whole nuclear mass region (In nuclei, $N = 81$ and $N = 83$ nuclei) and is known as the parabolic rule (Paar 1979, Van Maldeghem, Heyde 1985, Van Maldeghem 1988). Although somewhat outside the scope of the present discussion, it has been shown that such a parabolic dependence on spin J can be derived (Van Maldeghem 1988):

$$\Delta E(J) = \frac{-2\pi}{5}\theta(p)\theta(n)(-1)^{j_p+j_n+J}\begin{Bmatrix} j_p & j_n & J \\ j_n & j_p & 2 \end{Bmatrix}$$
$$\times \langle j_p\|\boldsymbol{Y}_2\|j_p\rangle\langle j_n\|\boldsymbol{Y}_2\|j_n\rangle\,, \qquad (3.199)$$

with $\theta(\omega) = u^2(\omega) - v^2(\omega)$ and $v^2(\omega)$, the occupation probability of the orbital ω (see Chap. 7 for a discussion on how to determine the orbit occupation probabilities). Writing out the Wigner $6j$-symbol in explicit form as well as the $\boldsymbol{Y}_2$ reduced matrix elements, one obtains for the "parabolic" rule

$$\Delta E(J) = -\tfrac{3}{4}\Big\{\Big[J(J+1) - j_p(j_p+1) - j_n(j_n+1)\Big]$$
$$+ \Big[J(J+1) - j_p(j_p+1) - j_n(j_n+1)\Big]^2\Big\}$$
$$\times \big((2j_p)(2j_n)(2j_p+2)(2j_n+2)\big)^{-1}\theta(p)\theta(n)\,. \qquad (3.200)$$

This expression clearly indicates a quadratic dependence of $\Delta E(J)$ on the angular momentum combination $J(J+1)$ of the proton-neutron multiplet members.

Examples for the $1g_{9/2}^{-1}(\pi) \times 1h_{9/2}(\nu)$ multiplet are first shown schematically (Fig. 3.60) and then for the case of the odd-odd In nuclei (Fig. 3.61). For the proton hole state one has $v^2_{1g_{9/2}} = 1$, whereas for the neutron, a partial occupation is included $\left(v^2_{1h_{11/2}} \neq 0, 1\right)$ and points towards using a neutron quasi-particle excitation (see Chap. 7 for more details).

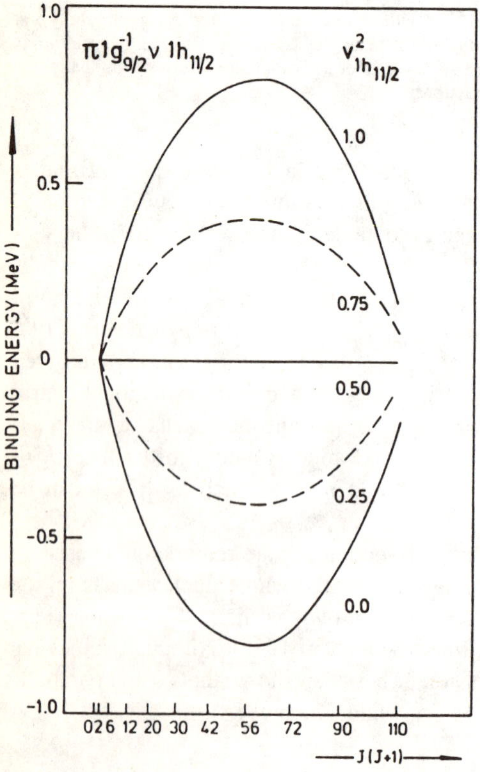

Fig. 3.60. Schematic illustration of the $[\pi(1g_{9/2})^{-1}\,\nu(1h_{11/2})]$ proton-neutron multiplet members, according to the parabolic rule of (3.200). It is the pairing factor (see Chap. 7) that causes a turn-over of the parabola from convex to concave in shape. Here, the notation $\pi(\nu)$ is used for proton (neutron) orbitals

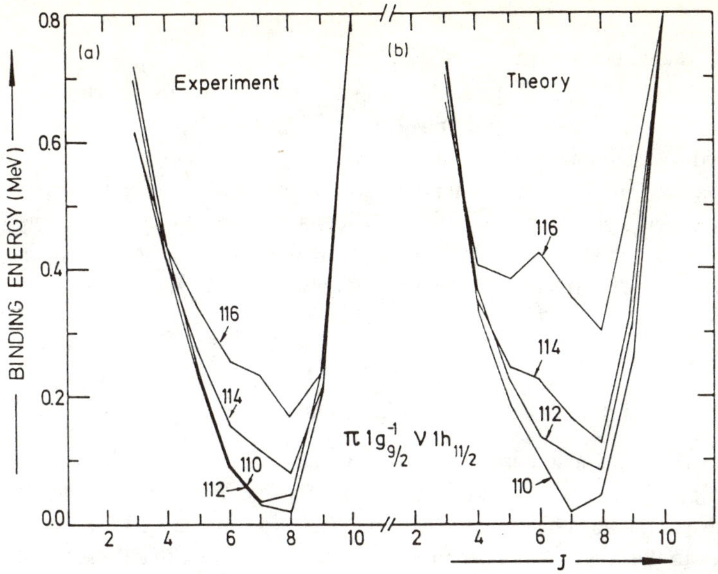

Fig. 3.61. (**a**) The experimental splitting of the $\left[\pi(1g_{9/2})^{-1}\nu(1h_{11/2})\right]$ multiplet members relative to the 10^- level as deduced from the level schemes in odd-odd In nuclei. (**b**) The same multiplet but now for the calculated spectra using a δ-force [taken from (Van Maldeghem 1985)]

3.5 Large-Scale Shell-Model Calculations

In this chapter we have learned how to study nuclei where the number of nucleons outside closed shells is not too large to allow for an exact shell model treatment. We have discussed methods of calculating the matrix elements needed to set up the secular equation for the energy eigenvalues both for identical and non-identical nucleons. We have also shown, once the wave functions are known, how to calculate one-body expectation values that determine the nuclear observables (half-lives of nuclear excited states, decay rates, moments, etc.). In particular, the evaluation of these observables for electric and magnetic transitions and moments will be outlined in Chap. 4.

Here we summarize the above procedure and philosophy for the study of a given nucleus (N, Z).

i) One determines the nearby closed shells given by $Z_{\rm cl}$ and $N_{\rm cl}$ so as to fix the number of valence protons and neutrons.
ii) The number of active particles becomes $n_p = (Z - Z_{\rm cl})$ and $n_n = (N - N_{\rm cl})$ where n_p and n_n are the valence number of proton (neutron) particles (or holes). Through (i) one can fix the single-particle orbitals $j_{p_1}, j_{p_2}, \ldots$ and $j_{n_1}, j_{n_2}, \ldots$ that determine the properties at low energy when constructing the model space.

iii) One constructs the model space and the configurations that span the space for each J^π value. The basis configurations are denoted by $|(j_{p_1} j_{p_2} \cdots j_{p_n})^{n_p} J_p, (j_{n_1} j_{n_2} \cdots j_{n_n})^{n_n} J_n; JM\rangle$ which is a shorthand notation for constructing the proton n_p particle state J_p multiplied by the neutron n_n particle state J_n, both coupled to total spin J^π. This basis has $n(J^\pi)$ basis configurations.

iv) Starting from the single-particle energies $\varepsilon_{j_{p_i}}, \varepsilon_{j_{n_j}}$ and the two-body matrix elements for identical and non-identical nucleons, one builds up the energy matrix $[H]$ determined by its elements H_{ij} and diagonalizes the $n \times n$ energy matrix. Thus one obtains the n energy eigenvalues and n corresponding eigenfunctions.

v) With the wave functions $\psi_i(J_i^{\pi_i})$ and $\psi_f(J_f^{\pi_f})$, we calculate all possible observables to compare with the data and so, in retrospect, can determine the wave functions and the basic input quantities that appear via point iv).

This method (i-v) sets a standard shell model calculation that is able to give a good description of a large body of nuclear observables. When many valence nucleons are present outside closed shells, we shall have to construct good approximation schemes to the more general shell model calculations that readily become unfeasible. Such approximation schemes will be discussed in Chap. 6 and Chap. 7 in detail.

With the advent of large and fast computing facilities, however, shell-model calculations using very big shell-model configuration spaces have become possible. One of the most thoroughly investigated regions in that respect is formed by the full sd shell-model region spanning the various nuclei between ^{16}O and ^{40}Ca. Here, a maximum of 24 particles can fill up the proton and neutron $1d_{5/2}$, $2s_{1/2}$ and $1d_{3/2}$ orbitals. For a given number of valence proton particles (n_π) and neutron particles (n_ν), either a proton-neutron or an isospin basis can be used to construct the energy matrices. Because the proton and neutron single-particle energies are not fully identical, one could start from a basis which is a product state of proton and neutron basis configurations. The separate charge configurations then become

$$\psi_\pi \left[\left(1d_{5/2}\right)^{n_1}_{\alpha_1 J_1} \left(2s_{1/2}\right)^{n_2}_{\alpha_2 J_2} \left(1d_{3/2}\right)^{n_3}_{\alpha_3 J_3} \right] J_\pi^{n_\pi} \tag{3.201}$$

and

$$\psi_\nu \left[\left(1d_{5/2}\right)^{n'_1}_{\alpha'_1 J'_1} \left(2s_{1/2}\right)^{n'_2}_{\alpha'_2 J'_2} \left(1d_{3/2}\right)^{n'_3}_{\alpha'_3 J'_3} \right] J_\nu^{n_\nu} \tag{3.202}$$

with $n_\pi = n_1 + n_2 + n_3$; $n_\nu = n'_1 + n'_2 + n'_3$ and $\alpha_i, J_i(\alpha'_i, J'_i)$ the proton (neutron) quantum numbers needed to specify the internal configurations in an unique way.

As an example, in ^{28}Si, the 12-particle state with $M = 0$ and $T_z = 0$ has a dimension 93,710 in the m-scheme and, in the (J, T) scheme, the $J = 3$, $T = 1$ 12 particle basis has a dimension 6,706. We also illustrate the case of ^{28}Si, in particular, for the binding energy of the $(1d_{5/2})^{12}(2s_{1/2})^0(1d_{3/2})^0$ closed sub-shell configuration, in Fig. 3.62. In increasing the number of particles excited from the $1d_{5/2}$ into the $2s_{1/2}$ and $1d_{3/2}$ orbitals (excitation order 1 to 12), the full configuration space rapidly 'explodes'. Only after all possible partitions of the 12

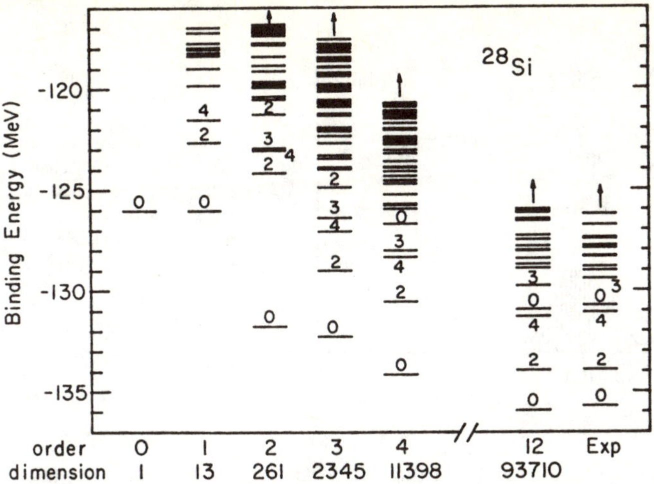

Fig. 3.62. Binding energy in the nucleus ^{28}Si as a function of the number n of nucleons excited out of the $1d_{5/2}$ orbital (with $0 \leq n \leq 12$)

particles have been considered is good agreement with the experimental spectrum achieved. At the same time, one notices the rather slow convergence of the energy spectrum towards its final form.

Similar types of shell-model studies have been carried out, according to the methods outlined here, in other mass regions. The *sd*-shell model space calculations, performed in great detail by Brown and Wildenthal (Brown, Wildenthal, 1988), can serve as a benchmark study in that respect.

3.6 Testing Single-Particle Motion in an Average Field

After a detailed discussion of the methods used to treat nucleon single-particle motion in an average field, as well as of the correlations induced by many-body effects through the residual interactions, and extensive comparison between theoretical and experimental energy spectra, one still has to prove as well as possible the very existence of independent-particle motion which is at the basis of the nuclear shell model.

One-nucleon transfer reactions (pick-up or stripping) have shown that nucleons can indeed be taken out or be put into single-particle orbitals. The ideal probe, however, to test nucleonic motion inside the nucleus is the electromagnetic interaction. When electrons scatter off the atomic nucleus, the electrons act as an almost ideal probe with which to study how nucleons move within the nucleus. By changing the electron energy ($\hbar\omega$) and its momentum ($\hbar q$), various information ranging from nuclear collective surface vibrations down to details related to the motion of individual nucleons can be obtained. In Fig. 3.63 we illustrate these

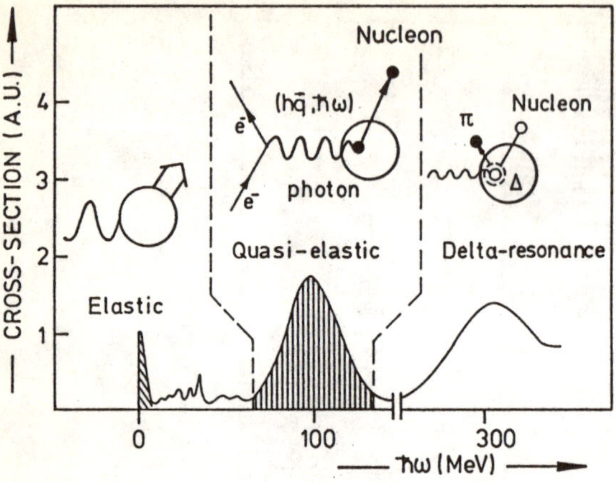

Fig. 3.63. Schematic representation of the cross-section for electron scattering off the atomic nucleus as a function of the energy transfer $\hbar\omega$ (in MeV). A number of energy regions are indicated, each characterized by some specific absorption mechanism, as illustrated (taken from De Vries et al., 1983)

various processes in a schematic way. At the point $\hbar\omega = 0$, elastic scattering takes place and the nucleus remains in its ground state. In the region $0 \leq \hbar\omega \leq 20\,\text{MeV}$, a large number of individual resonances in the nuclear many-body system start to appear. In the region $30\,\text{MeV} \leq \hbar\omega \leq 150\,\text{MeV}$, one-nucleon emission occurs as the dominant process. This is one of the most interesting regions since, by measuring simultaneously the energies and momenta of the outgoing nucleon and the scattered electron, information about the velocity distribution of the nucleon inside the nucleus prior to the interaction may be obtained.

A schematic representation of the above process is given in Fig. 3.64 where we adopt the most simple reaction mechanism: the detected nucleon is ejected in a one-step knock-out reaction in which all other nucleons remain as spectator particles. This approach is called the 'quasi-free' emission approximation. Momentum conservation at the two interaction vertices in Fig. 3.64, using the laboratory system with $|\boldsymbol{p}_A| = 0$, implies

$$\boldsymbol{p}_m = \boldsymbol{p}_a - \hbar\boldsymbol{q} = -\boldsymbol{p}_B\ . \tag{3.203}$$

The momentum $\boldsymbol{p}_m$ is customarily referred to as the 'missing' momentum and represents the momentum of the detected nucleon just before it was hit by the photon, characterizing the electromagnetic interaction, and thereby absorbing a momentum $\hbar\boldsymbol{q}$.

A nucleon moving inside a nucleus will be characterized by a velocity probability distribution (related to the Fourier transform of the coordinate wave function). Within the independent particle model, one can introduce a momentum distribution $\rho_a(\boldsymbol{p})$ corresponding to a given single-particle orbital, characterized by a (with $a \equiv \{n_a, l_a, j_a; \varepsilon_a\}$). This function $\rho_a(\boldsymbol{p})$ then describes the probability of finding the nucleon in the particular state a with momentum $\boldsymbol{p}$. A detailed derivation of this probability, or spectral function, has been given by Ryckebusch (1988) with, as a result,

$$\rho_a(\partial) = \frac{1}{2\pi^2\hbar^3}\left[\int dr\, r^2 j_{l_a}\left(\frac{pr}{\hbar}\right)\varphi_{n_a,l_a,j_a}(r)\right]^2 v_a^2(2j_a+1)\ , \tag{3.204}$$

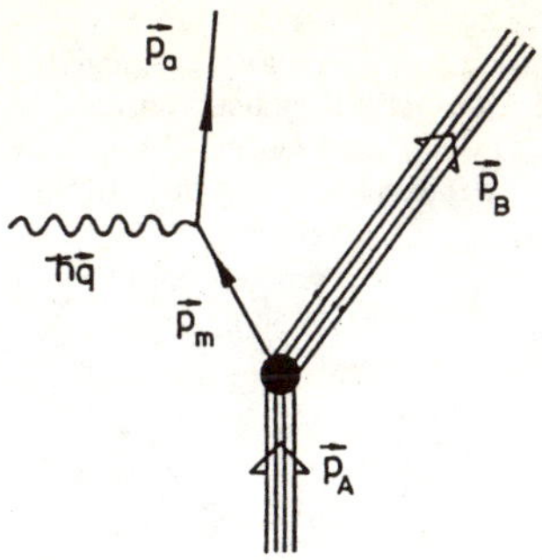

Fig. 3.64. One-photon exchange diagram for an electromagnetically induced one-nucleon emission process in the plane-wave impulse approximation. The various momenta ($\hbar q$, $\boldsymbol{p}_a$, $\boldsymbol{p}_m$, $\boldsymbol{p}_A$ and $\boldsymbol{p}_B$) are indicated

where $j_{l_a}(x)$ is the spherical Bessel function. So, in determining this momentum distribution $\rho_a(\boldsymbol{p})$, access to the nucleon motion inside the atomic nucleus is obtained through the Fourier-Bessel transformation. An example of a momentum distribution for the case of a $1p_{3/2}$ orbital in ^{16}O, together with the prediction in terms of the shell-model, is given in Fig. 3.65. We observe very good agreement with the original concept of a nucleon moving inside an orbital characterized by the $1p_{3/2}$ quantum numbers.

The electron accelerator in Amsterdam (the NIKHEF-K Medium Energy Accelerator MEA) provides electrons with an energy of up to $\simeq$ 600 MeV. The electrons that are scattered off the nucleus can cause the atomic nucleus to eject certain particles or even clusters of particles. Magnetic spectrometers are now able to accurately determine the momentum of both the scattered electron and the ejected particle(s). As outlined above, the microscopic structure of the nucleon single-particle motion can be reconstructed.

A survey of the knock-out from various valence orbitals is presented in Fig. 3.66 concerning the spectroscopic strength (in % of the maximal value which amounts to $2j_a+1$ for a given orbital j_a). The quite important depletion of strength with respect to 100 % values, which would be the case for fully occupied orbitals in an extreme single-particle picture, is still not fully understood. Deviations could be due to (i) admixtures of low-lying collective excitations in the ground-state wave function and, (ii) the strongly repulsive short-range two-body (and eventually many-body) correlations that could scatter particles out of the occupied orbitals.

Therefore, the existence of almost fully occupied single-particle orbitals in an average, one-body field seems rather well proven, at least for those levels lying quite close to the Fermi level. If we now consider deeply-bound orbitals and try to remove a nucleon from one of these, one might expect a pattern more complex than a single one-hole state. The single-hole strength will become spread out over a number of more complex configurations, coupling to the original single-hole configuration through the residual interaction. This process is given pictorially in Fig. 3.67, where the high density of close-lying complicated configurations in the vicinity of a single-hole configuration is drawn within the dash-line region. The residual interaction can then cause the single-hole strength to be fragmented in an

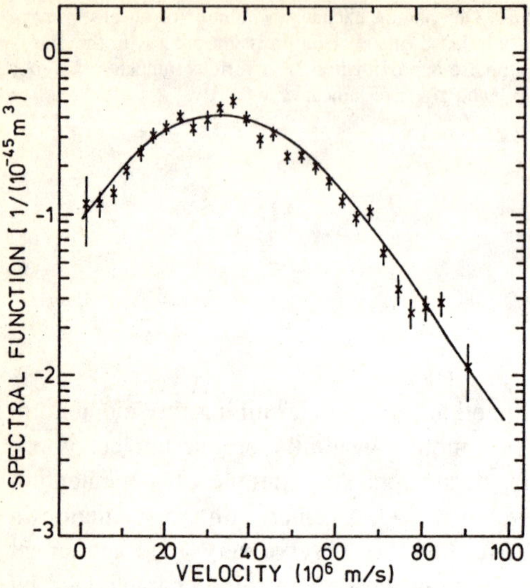

Fig. 3.65. Experimental spectral function for a $1p_{3/2}$ proton in ^{16}O, together with a theoretical prediction according to the nuclear shell model. The nucleon velocity is expressed in units of 10^6 m/s

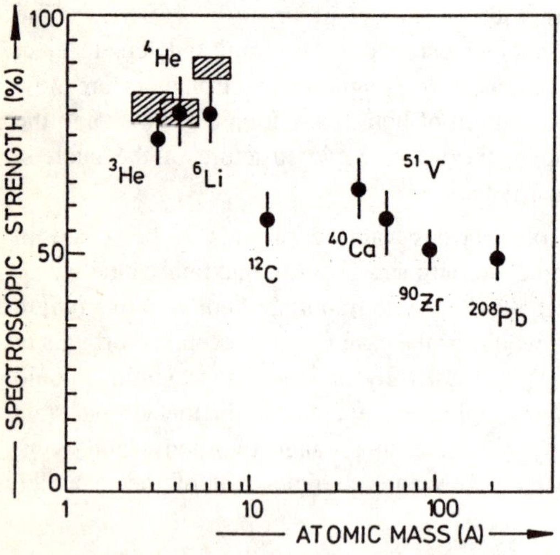

Fig. 3.66. Observed spectroscopic strength S (relative to the full shell-model value) for knock-out from the least bound orbital and as a function of the mass number A. The dashed regions indicate the range of theoretical values for $A \leq 6$ systems (taken from Dieperink et al., 1990)

important way. Thereby, the very idea of undisturbed pure independent-particle motion as a zero-order picture rapidly loses credence, since the total strength becomes spread out over very many states and over a large energy region. Detailed studies, making use of Green function methods, have been carried out and are discussed by Van Neck et al. (Van Neck, 1991).

To conclude this chapter on the shell model, various tests relating to how nucleons move inside the nucleus have been discussed. Using in particular the quasi-free one-step nucleon knock-out as an experimental probe, important infor-

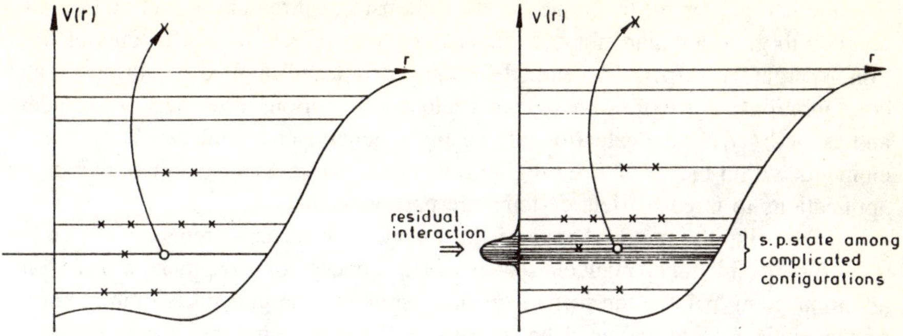

Fig. 3.67a, b. Pictorial representation of the way in which a nucleon is removed from a deeply-bound orbital in a potential well. On the left-hand side (**a**) a single-particle knock-out process is depicted, whereas in (**b**) interactions within a region of more complex configurations are indicated

mation on the particular nucleonic motion results. The large number of data on e.g. $(\gamma, n), (\gamma, p)(e, e'p), \ldots$ reactions all point towards a basic shell-model structure at least for these orbitals that are close to the Fermi level.

3.7 Summary

In this Chapter, having discussed the experimental evidence showing the need to consider independent-particle motion for nucleons moving in the average field built from the interactions of a given nucleon within the nucleus with all other nucleons, we described the basic structure of the nuclear shell model.

The radial equation for a harmonic oscillator average field is studied with the well-known energy spectrum. The importance of the spin-orbit interaction, which modifies the original spectrum so that it is in good agreement with the data, is discussed in detail. A number of illustrative examples of nuclei having one nucleon outside a closed shell (or one nucleon missing from a closed shell) are shown. A simple approach to Hartree-Fock theory, which underpins the microscopic structure of independent-particle motion, is given.

The next step is to consider two-particle states: the construction of appropriate two-body shell-model wave functions and a discussion of the possible forms of the residual two-body interaction is given. Starting from these basic ingredients, two-body matrix elements are evaluated in detail for central forces and, in particular, for the zero-range δ interaction. Detailed comparison with experimentally observed two-particle spectra is also given. In most nuclei, the two-particle states are not restricted to a single configuration. In most cases, all possible configurations resulting from partitioning the two particles over the available single-particle orbitals have to be constructed and the corresponding eigenvalue equation solved in this basis. Configuration mixing results and concepts such as model space, mode interaction, effective charges, ... are discussed.

Going one step further, one needs to construct three-particle configurations coupled to good angular momentum. The concept of fermion coefficients of fractional parentage (cfp) is developed in some detail and applied to various $(j)^n J$ configurations. An extension to n-particle configurations now becomes evident and is outlined. The evaluation of the most general one- and two-body matrix elements within these $(j)^n J$ configurations is discussed. Some detailed, numerical applications to three-particle systems are presented.

We develop the major characteristics of proton-neutron systems as well as the properties of identical nucleon systems. The concept of isospin is worked out, attention being paid to the formalism of isospin and the general isospin structure of the nuclear Hamiltonian. The isospin multiplet structure with corresponding energies is discussed. Subsequently, two- and many-particle isospin wave functions are built. The similarities and differences between the proton-neutron charge formalism and the isospin formalism are pointed out. Application to the two-nucleon(nn, pp, np) system and isospin triplets is made. Finally, the two-body matrix elements including isospin are derived and this theoretical framework is applied to the nucleus ^{42}Sc. A number of very special features of the proton-neutron force are pointed out: the parabolic rule and particle-particle properties are presented.

In the last two sections, a short excursion into the possibilities of present-day state-of-the-art large-scale shell-model calculations, with application to the sd-shell, as well as to the experimental evidence for the existence of single-particle motion in actual nuclei, is made. Here, the significance of electromagnetic interactions that probe the nucleon in its motion inside the atomic nucleus as an important proof of single-particle independent motion is clearly pointed out. Near the Fermi level, in particular, independent particle motion is shown to present a large body of evidence. For deep-lying hole states, however, important spreading over a large interval with many complex states becomes manifest.

Problems

3.1 The three-dimensional harmonic oscillator can also be solved by separating the eigenvalue equation for the energy in cartesian coordinates (x, y, z). Show that in this case $\varphi(x, y, z)$ becomes a product of three one-dimensional oscillator wave functions (product of Hermite functions). Show that a relation exists with the solutions of the eigenvalue equation in spherical (r, θ, φ) coordinates (see (3.16)).

3.2 Derive the Hartree-Fock equatins from the variational condition, expressed by (3.43).

3.3 Derive the antisymmetrized and normalized two-body matrix elements $\langle j_a j_b; JM|V|j_c j_d; JM\rangle_{\text{nas}}$ where V is a central two-body force $V(r)$ using the Moshinsky transformation brackets.

3.4 Study the three-level model, where two levels are degenerate at the energy ε and only interact via the intermediate of a third level at energy ε' having a strength V (see figure). Study the energy eigenvalues and wave functions as a function of $\Delta\varepsilon/V$ (with $\Delta\varepsilon = |\varepsilon' - \varepsilon|$).

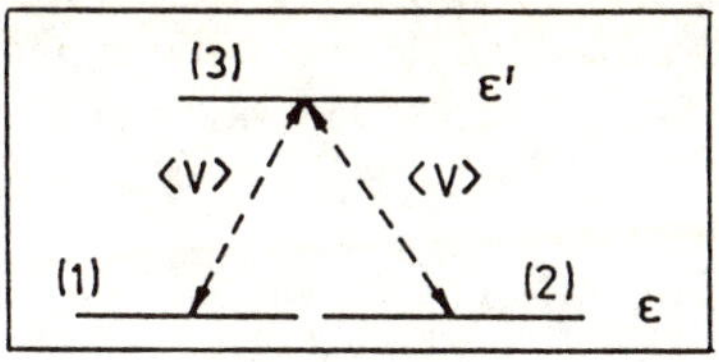

3.5 Show that perturbation theory summed to *all* orders in calculating the energy of the level (1) with unperturbed energy H_{11}, by including the interaction with a second level at $H_{22}(H_{22} > H_{11})$ with strength H_{12} gives the result

$$\lambda_- = \frac{H_{11} + H_{22}}{2} - \frac{1}{2}\left[(H_{11} - H_{22})^2 + 4H_{12}^2\right]^{1/2} .$$

3.6 Construct the three-particle wave functions for a $(j)^3 J = j$ configuration with $j \leq 7/2$. We give the normalized cfp coefficients

$$\left[j^2(J_1)jJ|\}j^3J\right] = \frac{\left[\delta_{J_1J_0} + 2\hat{J}_0\hat{J}_1\begin{Bmatrix} j & j & J_1 \\ J & j & J_0 \end{Bmatrix}\right]}{\left[3 + 6(2_{j_0} + 1)\begin{Bmatrix} j & j & J_0 \\ J & j & J_0 \end{Bmatrix}\right]^{1/2}} .$$

and

$$\begin{Bmatrix} j & j & 0 \\ j' & j' & J \end{Bmatrix} = (-1)^{j+j'+J}\left(\hat{j}\hat{j}'\right)^{-1} .$$

3.7 Show, by explicit calculation, that the $(1_{g_{9/2}})^3 J = 9/2$ configuration has *two* independent states with $J = j$, depending on the original angular momentum of the two-particle configuration (J_0) one starts from.

3.8 a) Show that the interaction energy in an n-particle state $|j^n\alpha JM\rangle$ with general $n(n \geq 2)$ can be expressed as a function of the two-body antisymmetrized matrix elements.

b) Show that in the calculation of a three-particle spectrum, starting from a two-body spectrum, one only needs the relative matrix elements

$$A_{J_1} = \langle j^2; J_1|V|j^2; J_1\rangle - \langle j^2; J_1 = 0|V|j^2; J_1 = 0\rangle .$$

*3.9 Given are n degenerate levels $j_1, j_2 \ldots, j_n(\varepsilon_{j_1} = \varepsilon_{j_2} = \ldots = \varepsilon_{j_n})$. Determine the energy spectrum for two particles in the configuration space with $J^\pi = 0^+$ if the interaction matrix elements

$$\langle (j_i)^2; J = 0|V|(j_k)^2; J = 0\rangle = -\frac{G}{2}\hat{j}_i\hat{j}_k \quad (G > 0) .$$

Determine also the wave function corresponding with the lowest $J^\pi = 0^+$ eigenstate.

3.10 In a nucleus, the two valence nucleons move in the $1_{g_{9/2}}$ and $2_{p_{1/2}}$ orbitals. The energy spectrum of 0^+ states looks like shown in the figure (a). If only the $2_{p_{1/2}}$ orbital would be considered, only one $J^\pi = 0^+$ state shown in the figure (b) results. Determine the relative energy difference between the $1_{g_{9/2}}$ and $2_{g_{1/2}}$ single-particle energies $\Delta\varepsilon$. Determine also the wave functions for the 0^+ ground state in case b.

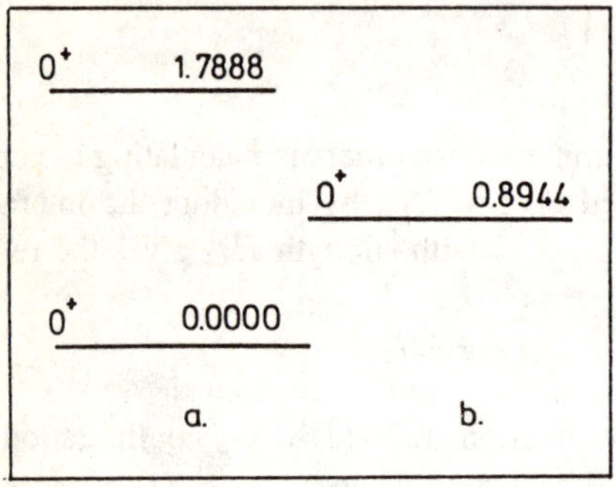

Given are the two-body matrix elements

$$\langle (j_a)^2; J=0|V_\delta|(j_c)^2; J=0\rangle = \frac{F^0}{2}(2j_a+1)(2j_c+1)(-1)^{j_a+j_c+l_a+l_c}$$
$$\times \begin{pmatrix} j_a & j_a & 0 \\ 1/2 & -1/2 & 0 \end{pmatrix} \begin{pmatrix} j_c & j_c & 0 \\ 1/2 & -1/2 & 0 \end{pmatrix} .$$

3.11 We give the two-particle spectrum in ${}^{148}_{66}\mathrm{Dy}_{82}$ with $Z = 64$, $N = 82$ as a closed-shell configuration. The relative spectrum is $E_{0^+} = 0$, $E_{2^+} = 1.677\,\mathrm{MeV}$, $E_{4^+} = 2.427\,\mathrm{MeV}$, $E_{6^+} = 2.731\,\mathrm{MeV}$, $E_{8^+} = 2.8323\,\mathrm{MeV}$ and $E_{10^+} = 2.9177\,\mathrm{MeV}$.
Calculate the three-particle spectrum in ${}^{149}_{67}\mathrm{Ho}_{82}$. Compare the calculated spectrum with the experimental situation $E_{11/2^-} = 0$, $E_{15/2^-} = 1.560\,\mathrm{MeV}$, $E_{19/2^-} = 2.287\,\mathrm{MeV}$, $E_{23/2^-} = 2.594\,\mathrm{MeV}$, $E_{27/2^-} = 2.738\,\mathrm{MeV}$. Discuss the agreement (or disagreement) between the data and the calculated spectrum. Enclosed: table of $(11/2)^3$ cfp coefficients.

3.12 The spectrum of ${}^{50}_{22}\mathrm{Ti}_{28}$ is given in the figure, with $J^\pi = 0^+, 2^+, 4^+$ and 6^+ spectrum.
a) The cfp $(7/2)^3$ coefficients are given in the table. Calculate the spectrum in ${}^{51}_{23}\mathrm{V}_{28}$ starting from the experimental spectrum in the figure.
b) In the ${}^{51}\mathrm{V}$ spectrum, a number of positive parity states ($J^\pi = 1/2^+, 3/2^+$) occur that cannot be described within a $(1f_{7/2})^3$ spectrum. Give some suggestions for the possible origin of these states.

3.13 Derive the explicit (T, T_z) dependence of the isoscalar, isovector and isotensor (rank 2) term energy $E(T, T_z)$ defined as $\langle T, T_z|H|T, T_z\rangle$.

Table P.2. The cfp coefficients $[(11/2)^2 J_1, 11/2; J|\}(11/2)^3 J]$ and this for the J = 3/2, 5/2, 7/2, 9/2, 11/2, 13/2, 15/2, 17/2, 19/2, 21/2, 23/2 and 27/2 states. For J = 9/2, 11/2 and 15/2, two-independent states can be constructed (two rows)

$$\left(\frac{11}{2}\right)^3$$

J = 3/2

V1	J1	V = 3
2	4	-0.476731
2	6	0.879C49

J = 5/2

V1	J1	V = 3
2	4	-0.77C977
2	6	-0.28426e
2	e	0.5699CC

J = 7/2

V1	J1	V = 3
2	2	-0.550482
2	4	0.250873
2	6	0.507519
2	e	0.613561

J = 9/2

V1	J1	V = 3	V = 3
2	2	-0.221937	-C.6S5312
2	4	0.355147	-0.4944e9
2	6	-0.7e0747	-C.C3C845
2	e	0.4609e5	-0.C4e8C0
2	1C	0.050416	C.575110

J = 11/2

V1	J1	V = 1	V = 3
0	C	-0.527C46	C.
2	2	C.2357C2	C.26C638
2	4	0.31622e	-C.67C224
2	6	0.3e0C5e	C.10C946
2	e	0.434614	C.61C164
2	1C	0.483C46	-C.316823

J = 13/2

V1	J1	V = 3
2	2	0.463eC8
2	4	-0.477255
2	6	-C.46729C
2	e	-0.085797
2	1C	C.575664

J = 15/2

V1	J1	V = 3	V = 3
2	2	-0.356966	-C.432121
2	4	0.29234C	-C.C4C212
2	6	-0.31145C	C.70C245
2	8	0.818537	-C.CC6146
2	1C	C.1418C5	C.566C47

J = 17/2

V1	J1	V = 3
2	4	C.463C9C
2	6	-C.578635
2	e	-0.372225
2	1C	C.55872e

J = 19/2

V1	J1	V = 3
2	4	-C.560449
2	6	C.21CC42
2	e	C.5C6C36
2	1C	C.621C53

J = 21/2

V1	J1	V = 3
2	6	C.383482
2	E	-C.794719
2	1C	0.470491

J = 23/2

V1	J1	V = 3
2	6	-C.545331
2	E	C.4C1C22
2	1C	0.736C6e

J = 27/2

V1	J1	V = 3
2	e	-C.486664
2	1C	C.873589

Table P.3. The cfp coefficients $[(7/2)^2(v_1 J_1), 7/2; J|\}(7/2)^3, vJ]$ and this for $J = 7/2$, 3/2, 5/2, 9/2, 11/2 and 15/2. The extra quantum number v, v_1 denotes the seniority quantum number

	v_1	0	2		
v	J \ J_1	0	2	4	6
1	$\frac{7}{2}$	$-\frac{1}{2}$	$\frac{\sqrt{5}}{6}$	$\frac{1}{2}$	$\frac{\sqrt{13}}{6}$
3	$\frac{3}{2}$		$\sqrt{\frac{3}{14}}$	$-\sqrt{\frac{11}{14}}$	
	$\frac{5}{2}$		$\frac{\sqrt{11}}{3\sqrt{2}}$	$\sqrt{\frac{2}{33}}$	$-\frac{\sqrt{65}}{3\sqrt{22}}$
	$\frac{9}{2}$		$\frac{\sqrt{13}}{3\sqrt{14}}$	$-\frac{5\sqrt{2}}{\sqrt{77}}$	$\frac{7}{3\sqrt{22}}$
	$\frac{11}{2}$		$-\frac{\sqrt{5}}{3\sqrt{2}}$	$\sqrt{\frac{13}{66}}$	$\frac{2\sqrt{13}}{3\sqrt{11}}$
	$\frac{15}{2}$			$-\sqrt{\frac{5}{22}}$	$\sqrt{\frac{17}{22}}$

3.14 Show, by acting with the isospin lowering operator $T_+ = \sum t_+(i)$ on the wave function corresponding to the nucleon distributions of Fig. 3.56, that these wave functions have a good isospin $T = T_z$ (maximal isospin).

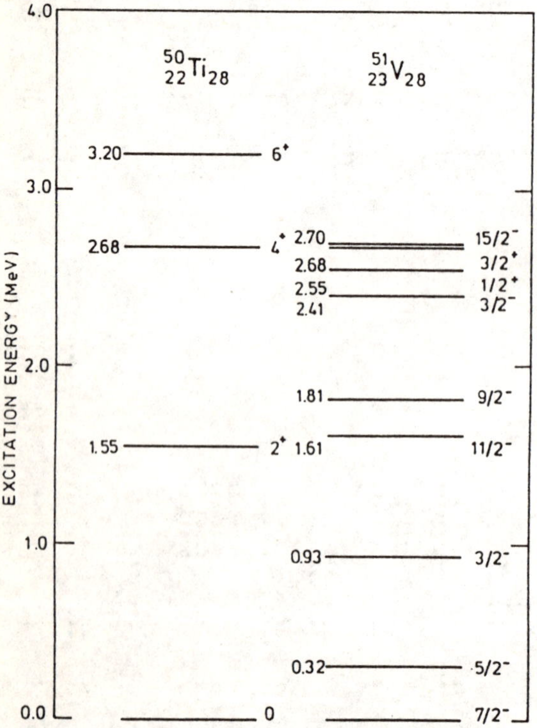

3.15 Construct the isospin wave functions for a two-nucleon system ($n-n$, $n-p, p-p$):

a) Calculate the expectation value for the above wave function of the interaction

$$V(1,2) = -V_0 \tfrac{1}{2}\,(1 + \boldsymbol{\tau}_1 \cdot \boldsymbol{\tau}_2)\ .$$

b) Determine the energy contribution for the above operator in the $(1f_{7/2})^2$ configurations in ^{42}Ca, ^{42}Sc and ^{42}Ti.

3.16 Show that the $k = 2$ multipole contribution in the evaluation of two-body matrix elements exhibits a quadratic dependence on the spin quantity $x = J(J+1)$.

*3.17 Prove that for very high multipole orders ($k \to \infty$), the specific contribution to a diagonal two-body matrix element is non-vanishing only for the $J = 0$ value.

3.18 Discuss possible experimental facets that all point towards the existence of a genuine shell structure in the atomic nucleus. Point out the importance of a spin-orbit term $\zeta(r)\boldsymbol{l}.\boldsymbol{s}$, and evaluate its matrix elements in a single-particle basis $|n,l,j,m\rangle$ with $j = l \pm 1/2$.

3.19 Construct, for the nucleus $^{26}_{12}$Mg having 10 valence particles distributed over the $1d_{5/2}$ (0.0 MeV), the $2s_{1/2}$ (0.87 MeV) and the $1d_{3/2}$ (5.08 MeV) levels, the unperturbed energy spectrum for the various possible partitions of these 10 particles. How many independent two-body matrix elements are needed to carry out a full sd-shell-model calculation?

*3.20 Show that by equating the mean-square radius, as obtained from the harmonic oscillator model for a nucleus with A nucleons, and from a homogeneous sphere of nuclear matter, the energy quantum $\hbar\omega$ of the oscillator becomes equal to $41.A^{-1/3}$.

3.21 Show that for a two-particle configuration with $j_1 = j_2 = j$

a) the two-particle wave function $\psi(J = 2j, M = 2j)$ is symmetric under the interchange of particles 1 and 2,

b) the two-particle wave function $\psi(J = 2j-1, M = 2j-1)$ is antisymmetric under the interchange of particles 1 and 2. Prove that the latter condition remains valid for all possible M values of $\psi(J = 2j-1, M)$.

3.22 In a two-level mixing model, described by the energy matrix

$$\begin{pmatrix} E_1^0 & V \\ V & E_2^0 \end{pmatrix}$$

where we denote by $R \equiv \frac{|\Delta E_{\text{unp.}}|}{V} = \frac{|E_2^0 - E_1^0|}{V}$ and $\Delta E_s = |E_1 - E_1^0| = |E_2 - E_2^0.|$ one obtaines the result that

$$\left|\frac{\Delta E_s}{\Delta E_{\text{unp.}}}\right| = \frac{1}{2}\left[\sqrt{1 + \frac{4}{R^2}} - 1\right]$$

3.23 For two particles, interacting via a short-range two-body interaction, and where the 'classical' angle between the individual angular momenta is denoted by θ, show that the interaction matrix elements $\Delta E(j^2; J)$ are proportional to $\tan(\theta/2)$.

*3.24 Consider the three-particle isospin wavefunction $\zeta^{1/2}_{+1/2}(1)\zeta^{1/2}_{+1/2}(2)\zeta^{1/2}_{+1/2}(3) = \zeta$ which is characterized by the total isospin of $T = 3/2$, $T_z = +3/2$.
a) show that the state obtained by acting with the isospin lowering operator $T_- = \sum_{i=1}^{3} t_-(i); T_-\zeta$ has isospin $T = 3/2$, $T_z = +1/2$,
b) construct the state orthogonal to $T_-\zeta$ which has $T = 1/2$, $T_z = +1/2$,
c) show that the latter state is neither fully symmetric nor fully antisymmetric for permutations of any two nucleons.

3.25 Show that the state constructed in the m-scheme $\Phi(jm = j, jm = j - 1, \ldots, jm = -j)$ in which all m values from $m = j$ until $m = -j$ occur is also an eigenstate of the total angular momentum operator $\hat{J}^2$ and of its J_z component, corresponding to the eigenvalues $J = 0$, $M = 0$, respectively.

*3.26 Show that the lowest 0^+ state, obtained by diagonalizing a surface delta interaction (SDI) in the two-particle model space with the $1d_{5/2}$, $2s_{1/2}$ and $1d_{3/2}$ orbitals (using the single-particle energies as shown in Fig. 3.37), is a coherent combination of the three separate two-particle configurations. Compare the binding energy, relative to the unperturbed energy of the two-particle configuration $(1d_{5/2})^2$. Compare the above binding energy gain with the result obtained in second-order perturbation theory using the same model space and interaction strength. Give an explanation for the difference between the results obtained using diagonalization and sccond-order perturbation theory.

4. Electromagnetic Properties in the Shell Model

4.1 General

Having studied the nuclear wave functions obtained from the nuclear secular equation, one has a first test of a good description of the nuclear Hamiltonian (average field, residual two-body interactions, ...). The study of nuclear decay rates via gamma decay is a much better test, however, of the nuclear wave functions obtained. Moreover, nuclear gamma-decay properties are some of the best indicators of nuclear spin and parity (J^π) of excited states. In this chapter we shall concentrate on the evaluation of transition rates and static moments. We do not, however, discuss a derivation of the electric and magnetic multipole operators themselves and refer the reader to (Brussaard, Glaudemans 1977) for an efficient introduction.

Transition rates per unit time are given by

$$T(L) = 8\pi ce^2/\hbar c(L+1)/\left(L[(2L+1)!!]^2\right)k^{2L+1}B(L)\,, \tag{4.1}$$

where $k = \omega/c \ll 1/R$ with R the nuclear radius, which means that we evaluate transition rates in the long wavelength limit (the excitation energy of $E_\gamma < 3\,\mathrm{MeV}$). Here, L describes the multipolarity and $B(L)$ the reduced transition probability which actually carries the nuclear structure information.

If we start from an initial state $|\alpha_i;\, J_iM_i\rangle$ going to a final state $|\alpha_f;\, J_fM_f\rangle$ with the operator $O(LM)$, then the reduced transition probability means summing for a given initial state (J_iM_i) over all final states (J_fM_f) and intermediate values of the transition operator (L, M), i.e.,

$$B(J_i \rightarrow J_f; L) = \sum_{M,M_f} |\langle\alpha_f;\, J_fM_f|O(LM)|\alpha_i;\, J_iM_i\rangle|^2\,, \tag{4.2}$$

and, when we apply the Wigner-Eckart theorem, this reduces to

$$B(J_i \rightarrow J_f; L) = \frac{1}{2J_i+1}|\langle\alpha_fJ_f\|O(L)\|\alpha_iJ_i\rangle|^2\,, \tag{4.3}$$

when no particular polarization of the initial orientation (J_iM_i) is implied. If the population of the initial substates M_i is dependent on M_i and given by certain occupation numbers $p(M_i)$, then (4.3) has to be modified.

The above arguments hold for *pure* states. If on the other hand, the initial and final state are expanded in a basis such that

$$
\begin{aligned}
|\alpha_i; J_i M_i\rangle &= \sum_k a_k(J_i)|k_i; J_i M_i\rangle \ , \\
|\alpha_f; J_f M_f\rangle &= \sum_l b_l(J_f)|l_f; J_f M_f\rangle \ ,
\end{aligned}
\tag{4.4}
$$

then the reduced transition probability in (4.3) is generalized into the expression

$$
B(J_i \to J_f; L) = \frac{1}{2J_i + 1}\left|\sum_{k,l} a_k(J_i) b_l(J_f)\langle l; J_f\|O(L)\|k; J_i\rangle\right|^2 . \tag{4.5}
$$

In the latter, interference effects can result. In some cases (collective transitions), all partial contributions in the sum over (k, l) have the same phase, resulting in enhanced transition rates.

4.2 Electric and Magnetic Multipole Operators

In the long wavelength approximation, the electric multipole operators are defined for a continuous distribution of charge $\varrho(\boldsymbol{r})$ as (Brussaard, Glaudemans 1977)

$$
O(\text{el.}, LM) = \frac{1}{e}\int \varrho(\boldsymbol{r}) r^L Y_L^M(\theta, \varphi) d\boldsymbol{r} \ . \tag{4.6}
$$

Similarly, the magnetic multipole operator which results from the convection current $\boldsymbol{j}(\boldsymbol{r})$ and the magnetization current $\boldsymbol{m}(\boldsymbol{r})$, can be written as

$$
\begin{aligned}
O(\text{mag.}, LM) = 1/(L+1) 1/\text{ec} \int &\boldsymbol{\nabla}\left[r^L Y_L^M(\theta, \varphi)\right] \\
&\times \{\boldsymbol{r} \times [\boldsymbol{j}(\boldsymbol{r}) + c\boldsymbol{\nabla} \times \boldsymbol{m}(\boldsymbol{r})]\} d\boldsymbol{r} \ .
\end{aligned}
\tag{4.7}
$$

For a system of point charges with given values $\tilde{e}_i$ moving with momenta $\boldsymbol{p}_i$ and intrinsic spin $\boldsymbol{s}_i$, the charge and current densities can be written as (Brussaard, Glaudemans 1977)

$$
\begin{aligned}
\varrho(\boldsymbol{r}) &= \sum_i \tilde{e}_i \delta(\boldsymbol{r} - \boldsymbol{r}_i) \ , \\
\boldsymbol{j}(\boldsymbol{r}) &= \sum_i \tilde{e}_i / m \boldsymbol{p}_i \delta(\boldsymbol{r} - \boldsymbol{r}_i) \ , \\
\boldsymbol{m}(\boldsymbol{r}) &= \sum_i \tilde{\mu}_i \boldsymbol{s}_i \delta(\boldsymbol{r} - \boldsymbol{r}_i) \ .
\end{aligned}
\tag{4.8}
$$

This situation conforms to a nucleus consisting of A point nucleons (Z protons, N neutrons) characterized by their charge $\tilde{e}_i$ and magnetization moment $\tilde{\mu}_i$. Here, already, we anticipate effective charges and moments that can differ from the *free* charges $(e, 0)$ for proton and neutron and (μ_p, μ_n) the free gyromagnetic moments.

Using the general expression for the reduced transition probability in (4.3) together with the electric and magnetic multipole operators where a system of point particles is considered, we obtain for the electric and magnetic reduced transition probabilities the following expressions.

i) Electric Transitions.

$$O(\text{el.}, LM) = \sum_i (\tilde{e}_i/e) r_i^L Y_L^M(\theta_i, \varphi_i) , \tag{4.9}$$

and

$$B(\text{el.}; J_i \to J_f; L) = \frac{1}{2J_i + 1} \times \left| \langle J_f \| \sum_i (\tilde{e}_i/e) r_i^L \boldsymbol{Y}_L(\theta_i, \varphi_i) \| J_i \rangle \right|^2 , \tag{4.10}$$

where we sum over all A nucleons in the nucleus.

ii) Magnetic Transitions. In Appendix G, we give a derivation where the magnetic multipole operator from (4.7) can be rewritten in a more convenient form. For the system of point particles, this becomes

$$O(\text{mag.}, LM) = \sum_i \boldsymbol{\nabla}_i \left(r_i^L Y_L^M(\theta_i, \varphi_i) \right) \cdot \frac{1}{e} \times \left[\frac{1}{(L+1)} \frac{\tilde{e}_i \hbar}{mc} \boldsymbol{l}_i + \tilde{\mu}_i \boldsymbol{s}_i \right] . \tag{4.11}$$

We use the relation

$$\boldsymbol{\nabla} \left[r^L Y_L^M(\theta, \varphi) \right] \cdot \boldsymbol{V} = [L(2L+1)]^{1/2} r^{L-1} \left[\boldsymbol{Y}_{L-1} \otimes \boldsymbol{V} \right]_M^{(L)} , \tag{4.12}$$

where $\boldsymbol{V}$ is a general vector operator. Further on, we shall measure the nuclear moment $\tilde{\mu}_i$ in units of nuclear magnetons (n.m.) such that $\tilde{\mu}_i = (e\hbar/2mc) g_s(i)$. We also introduce the total angular momentum operator $\boldsymbol{j} = \boldsymbol{l} + \boldsymbol{\sigma}/2 = \boldsymbol{l} + \boldsymbol{s}$.

Now the reduced magnetic transition probability becomes

$$B(\text{mag.}, J_i \to J_f; L) = \frac{1}{2J_i + 1} \Big| \langle J_f \| \sum_i \frac{\hbar}{mc} r_i^{L-1} \times \left\{ \frac{\tilde{e}_i}{e} \frac{1}{L+1} \cdot \left[\boldsymbol{Y}_{L-1} \otimes \boldsymbol{j}_i \right]^{(L)} + \frac{1}{2} \left(g_s(i) - \frac{\tilde{e}_i}{e} \cdot \frac{1}{L+1} \right) \times \left[\boldsymbol{Y}_{L-1} \otimes \boldsymbol{s}_i \right]^{(L)} \right\} (L(2L+1))^{1/2} \| J_i \rangle \Big|^2 . \tag{4.13}$$

4.3 Single-particle Estimates and Examples

If now the wave functions $|\alpha_i;\ J_iM_i\rangle$ and $|\alpha_f;\ J_fM_f\rangle$ are the single-particle wave functions constructed in Sect. 3.1, the reduced transition probabilities (4.10, 13) become the single-particle estimates. This means also that the sum will reduce to a single term: the remaining extra nucleon.

i) Electric Transitions. Starting from the expression

$$B_{\text{s.p.}}(\text{el.}, j_i \to j_f; L) = \frac{1}{2j_i+1}\left|\langle n_f l_f j_f \| r^L \boldsymbol{Y}_L \| n_i l_i j_i\rangle\right|^2 \left(\frac{\tilde{e}}{e}\right)^2 , \tag{4.14}$$

and the value of the $\boldsymbol{Y}_L$ reduced matrix element as evaluated in Appendix F, the above expression reduces to the closed form

$$\begin{aligned} B_{\text{s.p.}}(\text{el.}, j_i \to j_f; L) &= (2j_f+1)(2l_f+1)(2l_i+1)(2L+1)/4\pi \\ &\times \left(\frac{\tilde{e}}{e}\right)^2 \cdot \begin{pmatrix} l_f & L & l_i \\ 0 & 0 & 0 \end{pmatrix}^2 \begin{Bmatrix} l_f & j_f & \frac{1}{2} \\ j_i & l_i & L \end{Bmatrix} \langle r^L\rangle^2 , \end{aligned} \tag{4.15}$$

or, even more compact,

$$\begin{aligned} B_{\text{s.p.}}(\text{el.}, j_i \to j_f; L) &= (2j_f+1)(2L+1)/4\pi(\tilde{e}/e)^2 \\ &\times \begin{pmatrix} j_i & j_f & L \\ \frac{1}{2} & -\frac{1}{2} & 0 \end{pmatrix}^2 \langle r^L\rangle^2 . \end{aligned} \tag{4.16}$$

The radial integrals have been defined as

$$\langle r^L\rangle = \int u_{n_f l_f}(r) r^L u_{n_i l_i}(r) dr . \tag{4.17}$$

ii) Magnetic Transitions. Similarly, (4.13) reduces to the closed form

$$\begin{aligned} B_{\text{s.p.}}(\text{mag.}, j_i \to j_f; L) &= (\hbar/mc)^2 \langle r^{L-1}\rangle^2 \langle l_f \| \boldsymbol{Y}_{L-1} \| l_i\rangle^2 \\ &\times (2j_f+1)(2L+1)\frac{2}{L}\Bigg[(-1)^{j_f+l_f+1/2}\frac{L}{L+1}\hat{j}_i\sqrt{j_i(j_i+1)} \\ &\times \begin{Bmatrix} l_f & j_f & \frac{1}{2} \\ j_i & l_i & L-1 \end{Bmatrix} \begin{Bmatrix} j_f & L & j_i \\ 1 & j_i & L-1 \end{Bmatrix} \frac{\tilde{e}}{e} + \sqrt{\frac{3}{2}} \\ &\times \left(g_s L - \frac{\tilde{e}}{e}\frac{L}{L+1}\right) \begin{Bmatrix} l_f & \frac{1}{2} & j_f \\ l_i & \frac{1}{2} & j_i \\ L-1 & 1 & L \end{Bmatrix}\Bigg]^2 . \end{aligned} \tag{4.18}$$

For free nucleons we have to use in these single-particle estimates,

$$\tilde{e}_p = e , \quad \tilde{e}_n = 0 ; \quad g_s(p) = 5.58 , \quad g_s(n) = -3.82 .$$

Table 4.1. The conversion factors used to relate the reduced EL and ML transition probabilities (expressed in (fm)^{2L} and $(\hbar/2\text{mc})^2(\text{fm})^{2L-2}$, respectively) to the total transition probabilities (expressed in units s^{-1}). Here E_γ is expressed in units of MeV

$T(E1)$	$= 1.59 \times 10^{15}\ (E_\gamma)^3 \cdot B(E1)$
$T(E2)$	$= 1.22 \times 10^{9}\ (E_\gamma)^5 \cdot B(E2)$
$T(E3)$	$= 5.67 \times 10^{2}\ (E_\gamma)^7 \cdot B(E3)$
$T(E4)$	$= 1.69 \times 10^{-4}\ (E_\gamma)^9 \cdot B(E4)$
$T(M1)$	$= 1.76 \times 10^{13}\ (E_\gamma)^3 \cdot B(M1)$
$T(M2)$	$= 1.35 \times 10^{7}\ (E_\gamma)^5 \cdot B(M2)$
$T(M3)$	$= 6.28 \times 10^{0}\ (E_\gamma)^7 \cdot B(M3)$
$T(M4)$	$= 1.87 \times 10^{-6}\ (E_\gamma)^9 \cdot B(M4)$

Equations (4.16, 18) are single-particle estimates of the best kind as the angular momenta j_i, j_f still appear in the final expressions. Later we shall try to define estimates for the single-particle transition rates, values that will only depend on the multipolarity L and no longer on j_i, j_f. Thus, these (Weisskopf) estimates can be used as an easy measure of transition rates (Segré 1977).

First, we give in Table 4.1 the conversion factors between the reduced and total transition rates. We express $T(EL)$ and $T(ML)$ in s^{-1}, E_γ in MeV, $B(EL)$ in units $(fm)^{2L}$ and $B(ML)$ in units $(\hbar/2mc)^2 \cdot (fm)^{2L-2}$.

We now come to the Weisskopf estimate and point out the different restrictions that have to be imposed on (4.16, 18) in order to remove the j_i and j_f dependence in them.

i) We use constant radial wave functions when calculating the radial integrals (Fig. 4.1)

$$\begin{aligned} R(r) &= C \quad \text{for} \quad 0 < r < R\,, \\ &= 0 \quad \text{for} \quad r \geq R\,, \end{aligned} \tag{4.19}$$

and from the normalization condition, we find

$$C^2 \int_0^R r^2\,dr = 1 \quad \text{or} \quad C = \left(\frac{3}{R^3}\right)^{1/2} . \tag{4.20}$$

Now the radial integrals become

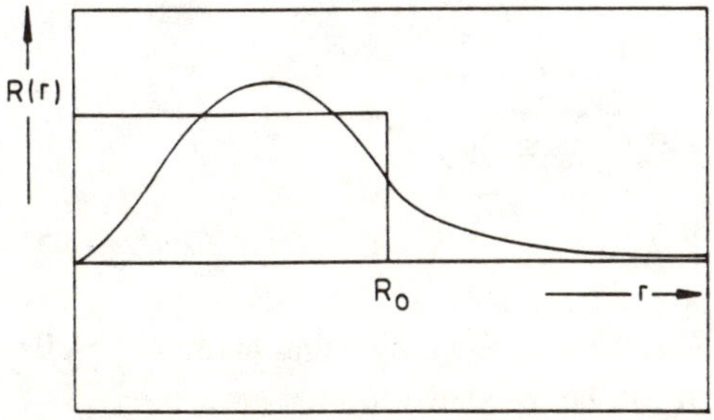

Fig. 4.1. Representation of the approximations made on the radial wave function $R_{nl}(r)$ (constant value over the nuclear interior), relative to a more realistic radial dependence, when deriving the Weisskopf estimate

$$\langle r^L \rangle = \int_0^R r^{L+2} \left(\frac{3}{R^3}\right) dr = \frac{3}{(L+3)} R^L \,. \tag{4.21}$$

ii) Next we calculate the values for $B_{\text{s.p.}}(\text{el})$ and $B_{\text{s.p.}}(\text{mag})$ for the transitions $j_i = L + \frac{1}{2}$, $j_f = \frac{1}{2}$. Under these restrictions, the single-particle transition rates simplify significantly and we get

$$B_{\text{s.p.}}(\text{el.}, L) = \frac{2(2L+1)}{4\pi} \begin{pmatrix} L+\frac{1}{2} & \frac{1}{2} & L \\ \frac{1}{2} & -\frac{1}{2} & 0 \end{pmatrix}^2 \left(\frac{3}{L+3}\right)^2 R^{2L} \,, \tag{4.22}$$

or,

$$W_E(L) = B_{\text{s.p.}}(\text{el.}, L) = \frac{(1.2)^{2L}}{4\pi} \cdot \left(\frac{3}{L+3}\right)^2 \cdot A^{2L/3} (fm)^{2L} \,. \tag{4.23}$$

For the magnetic transition rate, we note that the Wigner $6j$-symbol

$$\begin{Bmatrix} l_f & j_f & \frac{1}{2} \\ j_i & l_i & L-1 \end{Bmatrix} = 0 \,,$$

for the case that $j_i = L + \frac{1}{2}$ and $j_f = \frac{1}{2}$. For that case, we have $j_f + j_i = L + 1$ and $|j_i - j_f| = L$. Thus only one term contributes to the magnetic transition probability. Moreover, we use the restrictions (de-Shalit, Talmi 1963, Brussaard, Glaudemans 1977)

$$\tilde{e} = e \,, \qquad \left[L\left(g_s - \frac{\tilde{e}}{e} \cdot \frac{1}{L+1}\right)\right]^2 = \frac{5}{2} \,.$$

Then, the Weisskopf magnetic estimate reduces to

$$\begin{aligned} W_M(L) = B_{\text{s.p.}}(\text{mag.}, L) &= \frac{10}{\pi} (1.2)^{2L-2} \left(\frac{3}{L+3}\right)^2 A^{(2L-2)/3} \\ &\times (\hbar/2mc)^2 (fm)^{2L-2} \,. \end{aligned} \tag{4.24}$$

As an example we discuss the case of ^{17}O for the $1/2^+ \to 5/2^+$ $E2$ transition.

In (4.23, 24), all information on j_i and j_f is left out. When evaluating the rate for the $1/2^+ \to 5/2^+$ $E2$ transition which is mainly a $2s_{1/2} \to 1d_{5/2}$ single-particle transition (Fig. 4.2) in more detail, we use the more detailed expression (4.16) and get

$$\begin{aligned} B\left(E2; 1/2^+ \to 5/2^+\right) &= \frac{5.6}{4\pi} \left(\frac{\tilde{e}}{e}\right)^2 \begin{pmatrix} \frac{1}{2} & \frac{5}{2} & 2 \\ \frac{1}{2} & -\frac{1}{2} & 0 \end{pmatrix}^2 \langle r^2 \rangle^2 \,, \\ &= \frac{3}{4\pi} \left(\frac{\tilde{e}}{e}\right)^2 \langle r^2 \rangle^2 \,. \end{aligned} \tag{4.25}$$

Using a better estimate than the constant value of $R(r)$, by using harmonic oscillator wave functions, the radial integral can be evaluated to be

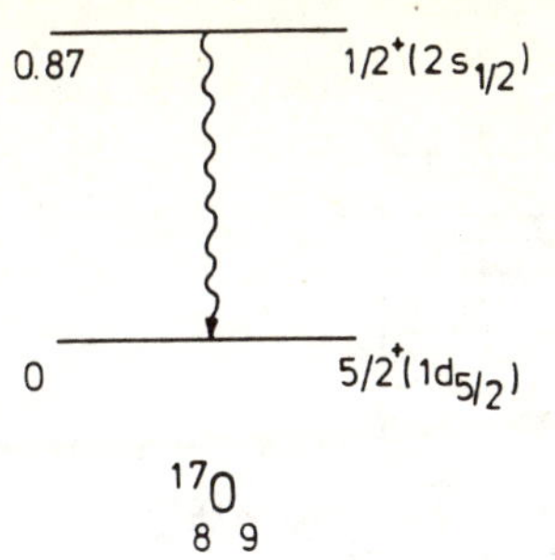

Fig. 4.2. The $E2$ gamma transition in ^{17}O, deexciting the $1/2^+$ level (mainly $2s_{1/2}$) into the ground state $5/2^+$ level (mainly $1d_{5/2}$) in a single-neutron approximation

$$\langle r^2 \rangle = \int_0^\infty r^2 u_{2s_{1/2}}(r) u_{1d_{5/2}}(r) dr \simeq 12\,\text{fm}^2 \,. \tag{4.26}$$

Combining the results from (4.25, 26), the $E2$ transition rate finally becomes

$$B(E2; 1/2^+ \to 5/2^+) \simeq 34.4\,\text{fm}^4\,(\tilde{e}/e)^2 \,. \tag{4.27}$$

The experimental result is 6.3 fm^4, so that an effective charge for the single-neutron transition becomes $\tilde{e} \cong 0.43\,$e, a value which is quite different from the free neutron charge e_n (free) = 0! We would like to briefly discuss the concept of an effective charge in the nucleus and the difference from the free nucleon charges (electric and magnetic charges).

In considering the nucleus as an A-nucleon system, if we solved the corresponding Schrödinger equation, an A-nucleon wave function $\psi(1, 2, , \ldots, A; JM)$ would result. The electromagnetic operators would be the sum of A one-body operators since each of the A nucleons can induce the transition. The related transition matrix elements would then read

$$\langle \psi_f(1, 2, \ldots, A; J_f M_f) | \sum_{i=1}^{A} O(LM; i) | \psi_i(1, 2, \ldots, A; J_i M_i) \rangle \,. \tag{4.28}$$

Working in a small model space, however, we wish to reproduce the matrix elements of (4.28) but now with model wave functions only using a restricted number of nucleons (only one neutron in the case of ^{17}O outside the closed ^{16}O core) and the transition operator acting in the model space only. For a reduced n-particle model space we have the matrix element

$$\langle \psi_f^M(1, 2, \ldots, n; J_f M_f) | \sum_{i=1}^{n} O^{\text{eff}}(LM; i) | \psi_i^M(1, 2, \ldots, n; J_i M_i) \rangle \,. \tag{4.29}$$

By equating the two matrix elements (4.28, 29) where in (4.28) the free nucleon charges $\tilde{e}_p = e$, $\tilde{e}_n = 0, \ldots$ occur, an implicit equation for the effective charges in the model operator $O^{\text{eff}}(LM; i)$ results once the model wave functions ψ^M and the full A nucleon wave functions ψ have been determined. This process is illustrated in Fig. 4.3. For the extreme case of ^{17}O, as discussed above, we go from a 17-nucleon problem (wave function and operators) to a 1-neutron problem. Therefore, the degrees of freedom of the 17-nucleon system have to be "absorbed"

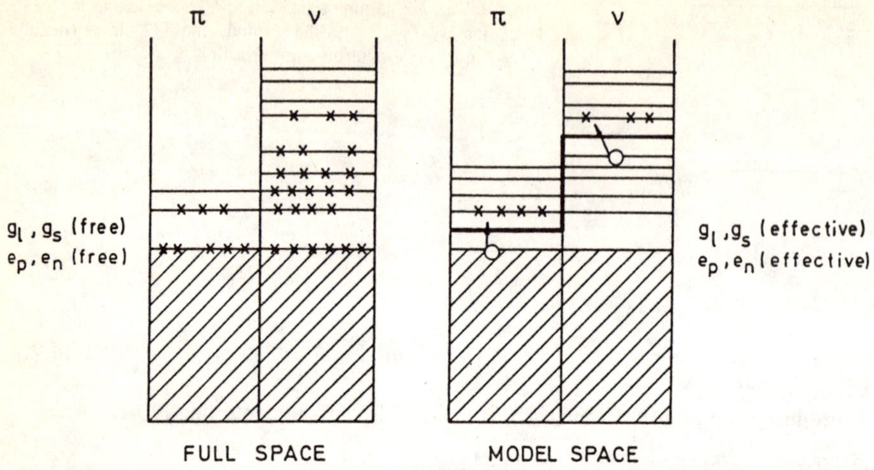

Fig. 4.3. Illustration of the concept of effective charge and effective gyromagnetic factors (e_p, e_n, g_l, g_s) depending on the model space used. On the left-hand side, besides a core, many valence protons and neutrons determine the nuclear low-lying properties with charges and gyromagnetic factors almost the free nucleon values (if no core would be present). On the right-hand side, the same physical situation is given but now, relative to a new reference state. In this case, only a few particle-hole excitations determine the low-lying properties. As a draw-back, effective charges and effective gyromagnetic factors have to be used, values that can differ much from the free values

in some way into the electromagnetic properties of the valence neutron. This concept defines model operators and effective "charges" (electric and magnetic) that are very much dependent on the model space. When the model space extends towards the full space, we have to go over to the free "charges" again. Thus, the charge for the $E2$ transition in ^{17}O for the $1/2^+ \to 5/2^+$ transition of $\tilde{e}_n \cong 0.43\,e$ reflects the neglect of the ^{16}O core with 8 protons and 8 neutrons. We show this in Fig. 4.4. Experimental results for electric quadrupole moments, $E2$ transitions and for magnetic dipole moments now have to be used in deducing the neutron (proton) effective charges and magnetic moments.

One observes an important "state" (nlj) dependence of the effective "charges", even for all cases where only a single-nucleon model is used to explain the experimental electromagnetic properties (Tables 4.2 and 4.3).

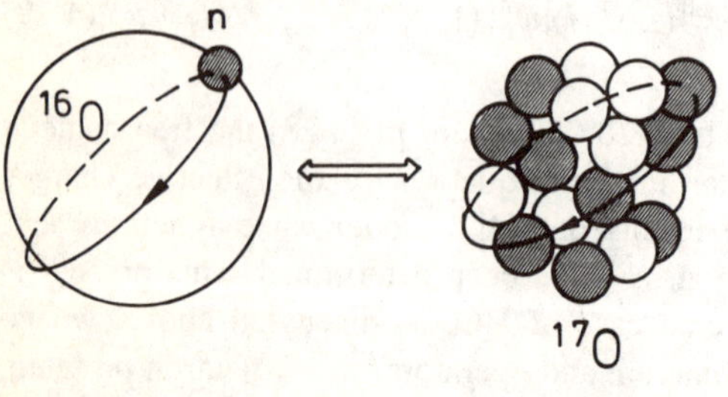

Fig. 4.4. Single-particle picture of properties in ^{17}O, described by a single neutron outside a ^{16}O core, and a 17 particle (8 protons + 9 neutrons) picture of ^{17}O in order to describe the properties of ^{17}O. The last neutron does not have a very specific position in the latter picture

Table 4.2. Effective, electric charges as deduced from known electric quadrupole moments (a) and $E2$-transition probabilities (b). In both cases, we concentrate on nuclear configurations that approach single-particle (or single-hole) configurations as close as possible. Therefore, we select doubly-closed shell nuclei (± 1 nucleon). Besides the data and the single-particle value, the effective charge (in units e) is given. For detailed references on the data, see (Bohr, Mottelson 1969), from which the table has been taken

(a) Quadrupole moments

Nucleus	lj	$Q_{obs}(10^{-24}\,cm^2)$	$Q_{sp}(10^{-24}\,cm^2)$	e_{eff}/e
$^{17}_{8}O$	$d_{5/2}$	-0.026	-0.066	0.40
$^{39}_{19}K$	$d^{-1}_{3/2}$	0.09	-0.052	1.8
$^{209}_{83}Bi$	$h_{9/2}$	-0.4	-0.26	1.6

(b) $E2$-transition probabilities

Nucleus	$(lj)_i$	$(lj)_f$	$B(E2)_{obs}(e^2 fm^4)$	$B(E2)_{sp}(e^2\,fm^4)$	e_{eff}/e
$^{15}_{7}N$	$p^{-1}_{3/2}$	$p^{-1}_{1/2}$	7.4	4.6	1.3
$^{17}_{8}O$	$s_{1/2}$	$d_{5/2}$	6.3	35	0.42
$^{17}_{9}F$	$s_{1/2}$	$d_{5/2}$	64	43	1.2
$^{41}_{20}Ca$	$p_{3/2}$	$f_{7/2}$	66	40	1.3
$^{41}_{21}Sc$	$p_{3/2}$	$f_{7/2}$	110	40	1.7
$^{207}_{82}Pb$	$f^{-1}_{5/2}$	$p^{-1}_{1/2}$	70	81	0.9
$^{207}_{82}Pb$	$p^{-1}_{3/2}$	$p^{-1}_{1/2}$	80	110	0.85
$^{209}_{82}Pb$	$s_{1/2}$	$d_{5/2}$	150	866	0.42
$^{209}_{83}Bi$	$f_{7/2}$	$h_{9/2}$	40 ± 20	2.3	4 ± 1.5

In discussing electromagnetic transitions, the motion of a proton or neutron is associated with a recoil of the rest of the nucleus since the center of mass remains at rest. This effect is of particular importance in the study of electric dipole ($E1$) transitions. Here, a one-particle effective charge results, expressed by the quantities

$$e_p = (1 - Z/A) \cdot e\,, \quad e_n = -(Z/A) \cdot e\,.$$

A detailed discussion of the derivation of these effective charges can be found in Eisenberg, Greiner 1970, Ring, Schuck 1980.

Table 4.3. Effective, gyromagnetic factors, illustrated by a comparison of the observed and single-particle (using free g_s and g_l factors) magnetic dipole moments. For detailed references on the data, see (Bohr, Mottelson 1969), from which the table has been taken

Nucleus	lj	μ_{obs}	μ_{sp}
H	$s_{1/2}^{-1}$	2.98	2.79
^{3}He	$s_{1/2}^{-1}$	−2.13	−1.91
^{15}N	$p_{1/2}^{-1}$	−0.28	−0.26
^{15}O	$p_{1/2}^{-1}$	0.72	0.64
^{17}O	$d_{5/2}$	−1.89	−1.91
^{17}F	$d_{5/2}$	4.72	4.79
^{39}K	$d_{3/2}^{-1}$	0.39	0.12
^{41}Ca	$f_{7/2}$	−1.59	−1.91
^{55}Co	$f_{7/2}^{-1}$	4.3 ± 0.3	5.79
^{207}Pb	$p_{1/2}^{-1}$	0.59	0.64
^{297}Pb	$f_{5/2}^{-1}$	0.65 ± 0.05	1.37
^{209}Bi	$h_{9/2}$	4.08	2.62

4.4 Electromagnetic Transitions in Two-particle Systems

In discussing two-particle systems, we use the general tensor reduction expressions discussed in Chap. 2 and Sect. 3.3. Using the $E2$ transition operator in shorthand notation

$$\boldsymbol{F}^{(2)} \equiv \sum_{k=1}^{2} \boldsymbol{f}^{(2)}(k) , \tag{4.30}$$

with $\boldsymbol{f}^{(2)}(k)$, the $E2$ operator of (4.9), we can derive the two-particle matrix element

$$\langle j^2; J \| \boldsymbol{F}^{(2)} \| j^2; J' \rangle = 2(-1)^{2j+2+J'} \hat{J} \hat{J}' \begin{Bmatrix} j & J & j \\ J' & j & 2 \end{Bmatrix} \langle j \| \boldsymbol{f}^{(2)} \| j \rangle . \tag{4.31}$$

This more general expression, in the situation of a $2^+ \rightarrow 0^+$ $E2$ transition becomes

$$\begin{aligned} \langle j^2; J = 0 \| \boldsymbol{F}^{(2)} \| j^2; J' = 2 \rangle &= 2(-1)^{2j} \sqrt{5} \begin{Bmatrix} j & 0 & j \\ 2 & j & 2 \end{Bmatrix} \langle j \| \boldsymbol{f}^{(2)} \| j \rangle \\ &= 2(-1)^{2j} \sqrt{5} (-1)^{2j} \frac{1}{\sqrt{5}} \frac{1}{\sqrt{2j+1}} \langle j \| \boldsymbol{f}^{(2)} \| j \rangle , \\ &= \frac{2}{\sqrt{2j+1}} \langle j \| \boldsymbol{f}^{(2)} \| j \rangle . \end{aligned} \tag{4.32}$$

We have now determined the $E2$ reduced matrix element for single-particle states to be

$$\langle j \| \boldsymbol{f}^{(2)} \| j \rangle = \langle r^2 \rangle (2j+1) \sqrt{\frac{5}{4\pi}} \frac{\tilde{e}}{e} \begin{pmatrix} j & j & 2 \\ \frac{1}{2} & -\frac{1}{2} & 0 \end{pmatrix} (-1)^{j-1/2} , \tag{4.33}$$

which becomes, when the explicit form of the Wigner $3j$-symbol is inserted,

$$\langle j \| \boldsymbol{f}^{(2)} \| j \rangle = \langle r^2 \rangle (2j+1) \sqrt{\frac{5}{4\pi}} \frac{\tilde{e}}{e} \frac{3/4 - j(j+1)}{(j(j+1)(2j-1)(2j+1)(2j+3))^{1/2}} . \tag{4.34}$$

In this way, the reduced $E2$ transition probability for the $2^+ \to 0^+$ transition reads

$$B\big(E2; j^2(2^+) \to j^2(0^+)\big) = \tfrac{1}{5} |\langle j^2; J = 0^+ \| \boldsymbol{F}^{(2)} \| j^2; J' = 2^+ \rangle|^2 ,$$

or

$$B\big(E2; j^2(2^+) \to j^2(0^+)\big) = \left(\frac{\tilde{e}}{e}\right)^2 \frac{\langle r^2 \rangle^2}{\pi} \frac{[3/4 - j(j+1)]^2}{j(2j-1)(2j+3)(j+1)} . \tag{4.35}$$

If we take the limiting value for $j \to \infty$, this expression "loses" its specific j dependence and gives

$$B\big(E2; 0^+ \to 2^+\big)_{j\to\infty} = \frac{5}{4\pi} \cdot \langle r^2 \rangle^2 \left(\frac{\tilde{e}}{e}\right)^2 . \tag{4.36}$$

If we introduce the assumption of constant radial wave functions when evaluating the radial integral in (4.36) as calculated before, we finally get

$$B\big(E2; 0^+ \to 2^+\big)_{\text{s.p.}} = 0.30\, \mathrm{A}^{4/3}\, \mathrm{fm}^4 . \tag{4.37}$$

This estimate is the extension of the Weisskopf estimate for single-particle transitions and can be used in even-even nuclei to "measure" $E2$ $0^+ \to 2^+$ transitions. The above estimate actually corresponds to (4.23) where we put $L = 2$ and take the statistical factor of $5 \equiv (2l_f + 1)$ into account.

Applying the above calculation of (4.35) to the case of ^{18}O where an $E2$ transition deexcites the 1.98 MeV 2^+ level, using $nlj = 1d_{5/2}$, we obtain

$$B\big(E2; 2^+ \to 0^+\big) = \frac{\langle r^2 \rangle^2}{\pi} \left(\frac{\tilde{e}}{e}\right)^2 \frac{8}{35} , \tag{4.38}$$

and using the radial integral

$$\langle r^2 \rangle \equiv \int_0^\infty u^2_{1d_{5/2}}(r) r^2 dr \simeq 12\, \mathrm{fm}^2 , \tag{4.39}$$

a theoretical value of $B(E2; 2^+ \to 0^+) = 10.5\, (\tilde{e}/e)^2\, \mathrm{fm}^4$, results. If we consider that the effective charge for the neutron does not change very much in going from

^{17}O to ^{18}O, for $\tilde{e}_n = 0.43\,$e, a value of $B(E2; 2^+ \to 0^+) = 1.9\,\text{fm}^4$ results. Comparing with the experimental value $B(E2; 2^+ \to 0^+)_{\text{exp}} = 7.05\,\text{fm}^4$, a discrepancy of a factor of 3 appears.

This can be interpreted in two ways:

i) By using a single $(1d_{5/2})^2$ pure configuration, a neutron effective charge of $\tilde{e}_n = 0.83\,$e is needed to reproduce the data. The variation when going from a one-particle to a two-particle model space is completely within the spirit of the concept of model or effective charges as discussed in Sect. 4.3.
ii) We use the effective charge $\tilde{e}_n = 0.43\,$e but take into account that $(1d_{5/2})^2$ is too strong a restriction in the model space for describing the 0^+ and 2^+ low-lying states. It was discussed in Sect. 3.2.4 that the $1d_{5/2}$ and $2s_{1/2}$ configurations will contribute to the 0^+ and 2^+ wave functions and we should evaluate the $B(E2; 2^+ \to 0^+)$ transitions for configuration mixed wave functions.

Thus for configuration mixing, in calculating the matrix element $\langle J_f \| \boldsymbol{F}^{(L)} \| J_i \rangle$, the initial and final states are described as linear combinations of basis states with the same J_f and J_i values.

Since we can write

$$|J_i M_i\rangle = \sum_{j,j'} c_{jj'}(J_i)|jj'; J_i M_i\rangle \,, \tag{4.40}$$

$$|J_f M_f\rangle = \sum_{j'',j'''} c_{j''j'''}(J_f)|j''j'''; J_f M_f\rangle \,, \tag{4.41}$$

the transition matrix element becomes

$$\sum_{\substack{j,j' \\ j'',j'''}} c_{jj'}(J_i) c_{j''j'''}(J_f) \langle j''j'''; J_f \| \boldsymbol{F}^{(L)} \| jj', J_i\rangle \,. \tag{4.42}$$

Because of the many configurations that can contribute, depending on the c-coefficients, coherence in the different contributions could result in transition rates enhanced relative to the Weisskopf estimate (collective $E2$, $E3$ transitions often occur in nuclei with many valence protons and neutrons outside closed shells).

Making use of the reduction formulas in calculating matrix elements (Chap. 2) and considering the antisymmetric character of the two-nucleon wave functions

$$|jj'; J_i M_i\rangle = \left[2(1+\delta_{jj'})\right]^{-1/2} \left[|jj'; J_i M_i\rangle - (-1)^{j+j'-J_i}|j'j; J_i M_i\rangle\right] \,, \tag{4.43}$$

the transition one-body matrix element is evaluated as

$$\begin{aligned}
&\langle j''j'''; J_f \| \boldsymbol{F}^{(L)} \| jj'; J_i\rangle \\
&= \left[(1+\delta_{j''j'''})(1+\delta_{jj'})\right]^{-1/2} \cdot \hat{J}_f \, \hat{J}_i \\
&\times \left\{ (-1)^{j''+j'''+J_i+L} \begin{Bmatrix} j'' & J_f & j''' \\ J_i & j & L \end{Bmatrix} \langle j'' \| \boldsymbol{f}^{(L)} \| j\rangle \delta_{j'j'''} \right.
\end{aligned}$$

$$+(-1)^{j'+j''+J_f+L}\begin{Bmatrix} j''' & J_f & j' \\ J_i & j' & L \end{Bmatrix}\langle j'''\|\boldsymbol{f}^{(L)}\|j'\rangle\delta_{jj''}$$
$$-(-1)^{J_i+J_f+L}\begin{Bmatrix} j''' & J_f & j'' \\ J_i & j & L \end{Bmatrix}\langle j'''\|\boldsymbol{f}^{(L)}\|j\rangle\delta_{j'j''}$$
$$+(-1)^{j'+j''+L}\begin{Bmatrix} j'' & J_f & j''' \\ J_i & j' & L \end{Bmatrix}\langle j''\|\boldsymbol{f}^{(L)}\|j'\rangle\delta_{jj'''}\,. \qquad (4.44)$$

This expression simplifies considerably for the case where $j = j' = j'' = j'''$, $J_f = J$ and $J_i = J'$. Only transitions

$$j^2 \to j^2\,,\quad j^2 \to jj'\,,\quad j'^2 \to jj'\,,\quad jj' \to jj'$$

are possible. Transitions $j^2 \to j'^2$ cannot occur (for $j \neq j'$) since there is only a one-body operator acting, indicating that only the quantum numbers of one nucleon can change and the other nucleon acts as a spectator [the Kronecker delta in (4.44) expresses this mathematically]. In Fig. 4.5 a Goldstone diagram illustrates the possible transitions in a two-particle configuration through this one-body character (shown by the external line $---\!\times$).

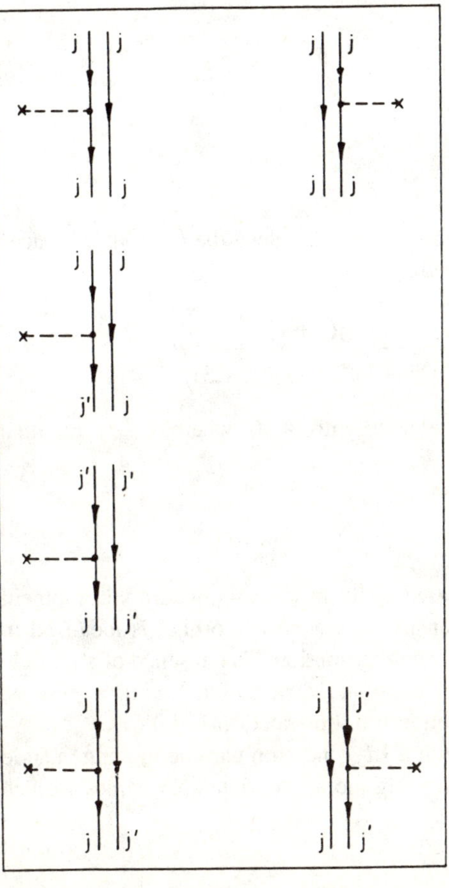

Fig. 4.5. Feynman-Goldstone diagrammatic way to describe the electromagnetic transitions, described by (4.43). Possible transitions are the $j^2 \to j^2$, $j^2 \to jj'$, $j'^2 \to jj'$ and $jj' \to jj'$ types. The fermion states are depicted by a downward line. The transition one-body operator which affects the properties of a single nucleon at most, is depicted by the dashed line and the vertex ($\times\!--$)

4.5 Quadrupole Moments

As a general definition of the quadrupole moment, we use the expectation value of $(3z^2 - r^2)$, within the angular momentum state $|JM\rangle$ with the maximal projection $J = M$. Since $(3z^2 - r^2)$ is proportional to the $E2$ operator with M component $M = 0$, the quadrupole moment is written as

$$Q \equiv \langle J; M = J | \sqrt{\frac{16\pi}{5}} \sum_i \frac{\tilde{e}_i}{e} r_i Y_2^0(\theta_i, \varphi_i) | J; M = J \rangle \,, \tag{4.45}$$

or, using the Wigner-Eckart theorem, as

$$Q = \begin{pmatrix} J & 2 & J \\ -J & 0 & J \end{pmatrix} \sqrt{\frac{16\pi}{5}} \langle J \| \sum_i \frac{\tilde{e}_i}{e} r_i^2 \boldsymbol{Y}_2(\theta_i, \varphi_i) \| J \rangle \,.$$

Filling out the Wigner $3j$-symbol explicitly, one obtains the resulting expression

$$Q = \left(\frac{J(2J-1)}{(2J+1)(2J+3)(J+1)} \right)^{1/2} \sqrt{\frac{16\pi}{5}} \times \langle J \| \sum_i \frac{\tilde{e}_i}{e} r_i^2 \boldsymbol{Y}_2(\theta_i, \varphi_i) \| J \rangle \,. \tag{4.46}$$

4.5.1 Single-particle Quadrupole Moment

For the case of a pure single-particle configuration, we need the $\langle j \| \boldsymbol{Y}_2 \| j \rangle$ reduced matrix element, which becomes (Appendix F)

$$\langle j \| \boldsymbol{Y}_2 \| j \rangle = (2j+1) \sqrt{\frac{5}{4\pi}} \frac{3/4 - j(j+1)}{\left(j(j+1)(2j-1)(2j+1)(2j+3) \right)^{1/2}} \,. \tag{4.47}$$

Now, putting the above reduced matrix element into (4.46) where $J \to j$, the final result simplifies very much to

$$Q_{\text{s.p.}}(j) = -\frac{(2j-1)}{(2j+2)} \frac{\tilde{e}}{e} \langle r^2 \rangle \,. \tag{4.48}$$

So, for a particle moving outside a closed shell, a negative quadrupole moment appears. This is because the particle density for a $j, m = j$ orbital is localized in the equatorial plane, giving rise to an oblate distribution. The absence of a particle (a hole) (Chap. 5) then corresponds to a prolate distribution giving a positive quadrupole moment. These types of motion are illustrated in Fig. 4.6.

As an example we take again ^{17}O with a $1d_{5/2}$ neutron particle moving outside the ^{16}O core in the equatorial plane. Inserting the appropriate spin values we get

$$Q_{\text{s.p.}}(1d_{5/2}) = -\frac{4}{7} \frac{\tilde{e}}{e} \langle r^2 \rangle \,. \tag{4.49}$$

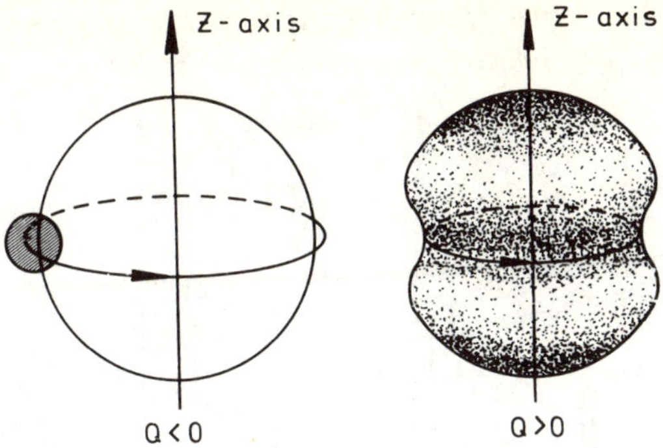

Fig. 4.6. Single-particle quadrupole moment for a nucleon moving in the equatorial plane inducing an oblate mass distribution, relative to the z-axis ($Q < 0$). When a nucleon is missing in the equatorial plane, an "effective" hole distribution gives rise to a prolate mass distribution relative to the z-axis ($Q > 0$)

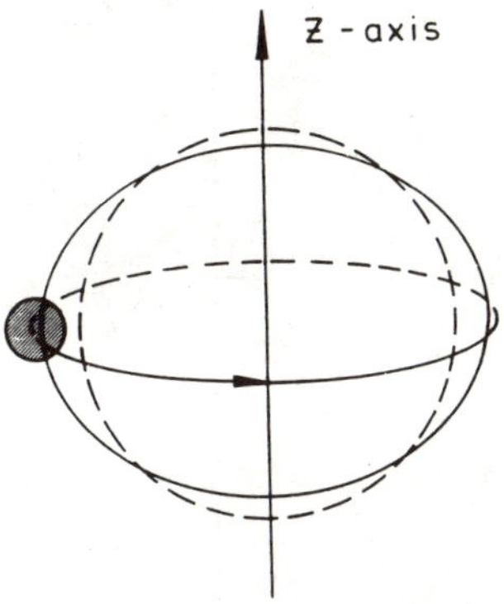

Fig. 4.7. The dynamic effect of the equatorial nucleon, polarizing the core around which the nucleon evolves, is illustrated. Thereby, an effective larger quadrupole moment shows up since the one-nucleon and core quadrupole moments add up and reinforce each other (coherence)

Using the value of the radial integral (4.39) and the effective neutron charge $\tilde{e}_n = 0.43\,\mathrm{e}$, a value of $Q_{\text{s.p.}}(1d_{5/2}) = -2.9\,\mathrm{fm}^2$ results, a value that is in very good agreement with the experimental value of $Q_{\text{exp}}(5/2^+) = -2.6\,\mathrm{fm}^2$. Thus the $E2$ transition in ^{17}O gives a charge consistent with the single-particle $E2$-moment.

Looking back to Figs. 4.6, 7, it is tempting to assume that the odd particle will "polarize" the core because of the nucleon-nucleon interaction and its short range, demanding an extra induced oblate core re-shaping. This polarization can be taken as the source of an extra or effective polarization charge that, for ^{17}O becomes $\tilde{e}_n = 0.43\,\mathrm{e}$. In some regions (Fig. 4.8), very large extra charges (large quadrupole moments) do appear (^{176}Lu, ^{167}Er) and indicate cases where the core is extremely polarizable and coherence in the odd-nucleon motion is present. Figure 4.8 shows the change of sign in Q for particle versus hole configurations. The closed shell configurations 8, 20, 28, 50, 82, 126 are indicated by arrows.

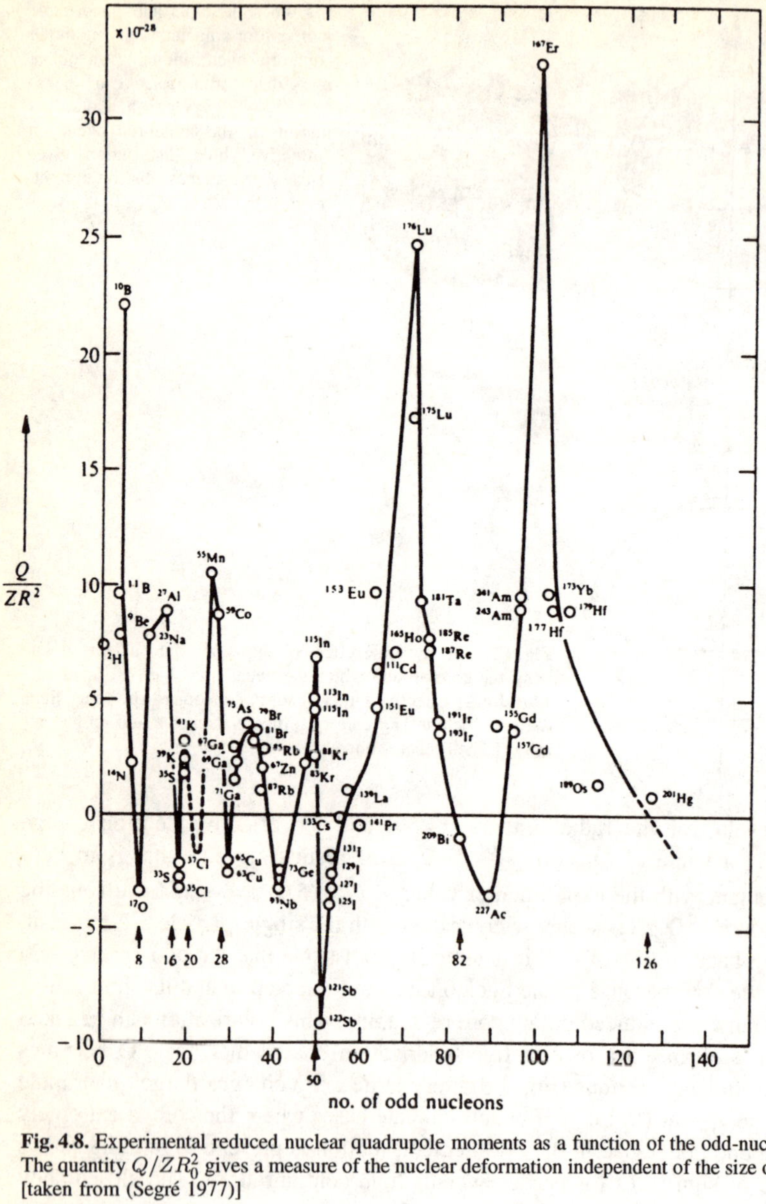

Fig. 4.8. Experimental reduced nuclear quadrupole moments as a function of the odd-nucleon number. The quantity Q/ZR_0^2 gives a measure of the nuclear deformation independent of the size of the nucleus [taken from (Segré 1977)]

4.5.2 Two-particle Quadrupole Moment

i) For $J^\pi = 0^+$ states, a trivial quadrupole moment $Q = 0$ results.
ii) For $J^\pi = 2^+$ states, by using the general expression (4.46) one obtains

$$Q(2^+) = \sqrt{\frac{2}{35}}\sqrt{\frac{16\pi}{5}}\langle j^2; J=2\|\sum_{i=1}^{2}\frac{\tilde{e}_i}{e}r_i^2\boldsymbol{Y}_2(\theta_i,\varphi_i)\|j^2; J=2\rangle\,. \tag{4.50}$$

Here again, we use the reduction formulas of Chap. 2 to get

$$\langle j^2; J=2\|\boldsymbol{F}^{(2)}\|j^2; J=2\rangle = 10(-1)^{2j}\begin{Bmatrix} j & 2 & j \\ 2 & j & 2\end{Bmatrix}\langle j\|\boldsymbol{f}^{(2)}\|j\rangle\,. \tag{4.51}$$

Combining this result with the single-particle quadrupole moment, one gets

$$Q(j^2;2^+) = -4\sqrt{\frac{10}{7}}\frac{\tilde{e}}{e}\langle r^2\rangle\begin{Bmatrix} j & 2 & j \\ 2 & j & 2\end{Bmatrix} \times (2j+1)\frac{3/4 - j(j+1)}{\left(j(j+1)(2j-1)(2j+1)(2j+3)\right)^{1/2}}\,. \tag{4.52}$$

For the example of the 2^+ level in ^{18}O, when considering this state as a $(1d_{5/2})^2$ pure configuration, we can evaluate the necessary quantities:

$$\begin{Bmatrix} \frac{5}{2} & 2 & \frac{5}{2} \\ 2 & \frac{5}{2} & 2\end{Bmatrix} = \frac{1}{7}\cdot\frac{1}{\sqrt{6}}\,,$$

and

$$Q(2^+) = 3.9\,(\tilde{e}/e)\mathrm{fm}^2\,.$$

Remember the positive sign of $Q(2^+)$ for a 2-particle configuration, in contrast to the one-particle case with a negative sign. These results can be put into a more general relationship between the one-particle and two-particle quadrupole moment:

$$Q(2^+) = -10\sqrt{\frac{2}{35}}\begin{Bmatrix} j & 2 & j \\ 2 & j & 2\end{Bmatrix} \times\left(\frac{(2j+1)(2j+3)(j+1)}{j(2j-1)}\right)^{1/2} Q_{\mathrm{s.p.}}(j)\,. \tag{4.53}$$

(The proof of this relation is an exercise in the problem set.)

Here, the $\{\substack{\cdots\\ \cdots}\}$ Wigner $6j$-symbol has a positive sign and thus $Q(2^+)$ and $Q_{\mathrm{s.p.}}(j)$ have opposite signs.

Extending this analysis to the high-spin $|j^2, JM\rangle$ states, one observes that the sign of the quadrupole moment has the sign of the coefficient $\left(-\begin{Bmatrix} j & j & 2 \\ J & J & j\end{Bmatrix}\right)$, which is positive for small J-values but changes sign at the aligned $J = 2j-1$ configuration. This changing relation, as well as the particular results for $Q(j)$ and $Q(j^2, 2^+)$, are illustrated in Fig. 4.9. For a single-particle situation, as is clear from the upper-left part, an oblate distribution with negative sign results. If now two

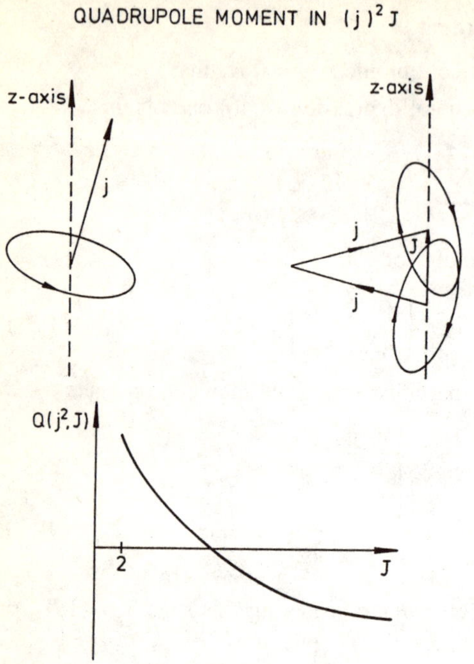

Fig. 4.9. Semi-classical illustration of the quadrupole moment for (i) a single-particle distribution, (ii) a configuration resulting from coupling two nucleon angular momenta to a small total angular momentum $J = 2$ state. In case (i) an oblate, in case (ii) a prolate distribution results, relative to the quantization z-axis. In the lower part, the variation in $Q(j^2, J)$ as a function of J starting from the $J = 2$ configuration is given

relatively large momenta j add up to the spin $J = 2$ values, and the quantization axis coincides with the J-axis, the two-particle density distribution shows a prolate distribution relative to the $\boldsymbol{J} = \mathbf{2}$ axis. For the aligned $J = 2j - 1$ state, the single-particle distributions align and become close to the upper-left part, again with a negative sign. This changing value of $Q(j^2, J)$ is illustrated in the lower part of Fig. 4.9.

4.5.3 Quadrupole Moments for n-Particle States

The quadrupole moment of a general $|j^n; JM\rangle$ configuration can be determined using the methods discussed in Sect. 3.3.2. The trans-lead region is an interesting one since in many nuclei, rather pure proton $(1h_{9/2})^n$ configurations are present. Using the angular momentum algebra techniques discussed earlier, the $Q(j^n, J)$ values can be related to the $Q(j)$ values. For the proton $1h_{9/2}$ particle configuration, a value of $-43(1)efm^2$ could be deduced (Neyens et al., 1993). For Po, At and Rn nuclei, the quadrupole moments are in quite good agreement with the calculated values (see Table 4.4). This is no longer the case for ^{214}Ra: theoretically, we expect a quadrupole moment with the same absolute values as for ^{212}Rn, but with the opposite sign if a pure linear filling of the $1h_{9/2}$ orbital describes the actual situation (see also Fig. 4.10 at the point with $x^2 = 0$). A much smaller value than the pure $Q\big((1h_{9/2})^6; 8^+\big)$ value is obtained experimentally. Configuration mixing with nearby shell-model configurations may provide an explanation. Using wave function of the form

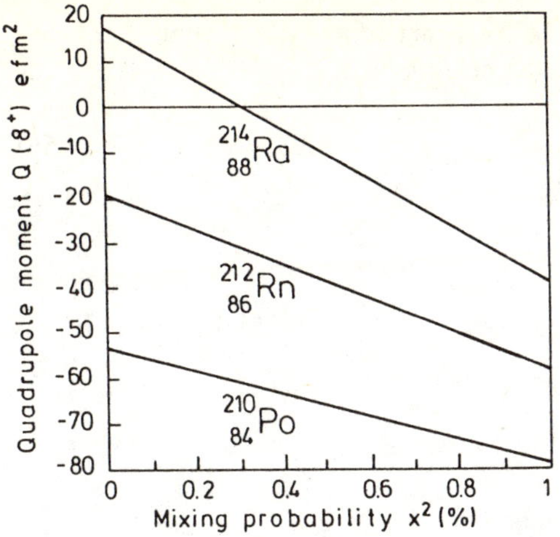

Fig. 4.10. Influence of configuration mixing (using (4.54)) on the quadrupole moment of $N = 126$ isotones in the Pb region. The major configuration is $(1h_{9/2})^n$ with n the number of valence protons outside of the $Z = 82$ closed shell

$$|8^+\rangle = \sqrt{1-x^2}\left|\left(1h_{9/2}\right)^n ; 8^+\right\rangle + x\left|\left(1h_{9/2}\right)^{n-1}\left(2f_{7/2}\right); 8^+\right\rangle \qquad (4.54)$$

together with the value calculated by Sagawa and Arima (1988), $Q(2f_{7/2}) = -39.32 efm^2$, a mixing amplitude of $x = 0.55$ almost reduces the quadrupole moment for ^{214}Ra to zero.

This discussion points out the powerful techniques for obtaining quadrupole moments when applying the reduction rules from n to two-or one-particle configurations. Similar methods could also be applied to the study of the particle-number dependence of other moments.

Table 4.4. Quadrupole moments extracted from experimental $B(E2)$ values and from angular momentum reduction into the single-particle quadrupole moment $Q_{\rm sp}(1h_{9/2}) = -43(1) efm^2$, assuming a pure $(1h_{9/2})^n$ configuration (taken from Neyens et al., 1993)

Isomer	Q (from $B(E2)$ values)	Q (from $Q_{\rm sp}(1h_{9/2})$)
^{210}Po, 8^+	56.8(3)	−57(1)
^{211}At, $21/2^-$	53(5)	−50(1)
^{212}Rn, 8^+	18(2)	−18.9(5)
^{213}Fr, $21/2^-$	7.2	0
^{214}Ra, 8^+	1.94(4)	+18.9(5)

4.6 Magnetic Dipole Moment

Here, according to convention, the magnetic dipole moment is defined as the expectation value of the dipole operator in the state with maximum M projection as

$$\mu \equiv \langle J; M = J | \sum_i g_l(i) l_{z,i} + \sum_i g_s(i) s_{z,i} | J; M = J \rangle \,. \qquad (4.55)$$

Here g_l and g_s are the orbital and spin gyromagnetic ratios. The operator in (4.55) can be reshaped so that the total single-particle momentum $j_{z,i}$ occurs. This gives

$$\mu \equiv \langle J; M = J | \sum_i \{ g_l(i) j_{z,i} + \left(g_s(i) - g_l(i) \right) s_{z,i} \} | J; M = j \rangle \,, \tag{4.56}$$

or, applying the Wigner-Eckart theorem

$$\mu = \frac{J}{\left(J(J+1)(2J+1) \right)^{1/2}} \langle J \| \sum_i \{ g_l(i) \boldsymbol{j}_i + \left(g_s(i) - g_l(i) \right) \boldsymbol{s}_i \} \| J \rangle \,. \tag{4.57}$$

4.6.1 Single-particle Moment: Schmidt Values

Here we use the reduced matrix elements

$$\langle j \| \boldsymbol{j} \| j \rangle = \left(j(j+1)(2j+1) \right)^{1/2} \,, \tag{4.58}$$

$$\langle j \| \boldsymbol{\sigma} \| j \rangle = \sqrt{3}(2j+1) \begin{Bmatrix} \frac{1}{2} & l & j \\ \frac{1}{2} & l & j \\ 1 & 0 & 1 \end{Bmatrix} \langle \tfrac{1}{2} \| \boldsymbol{\sigma} \| \tfrac{1}{2} \rangle (2l+1)^{1/2} \,. \tag{4.59}$$

The Wigner $9j$-symbol with zero angular momentum reduces to a Wigner $6j$-symbol that can still be written out in detail so that one gets (Brussaard, Glaudemans 1977, de Shalit, Talmi 1963)

$$\langle j \| \boldsymbol{\sigma} \| j \rangle = \left((2j+1)/j(j+1) \right)^{1/2} \left[j(j+1) + \tfrac{3}{4} - l(l+1) \right] \,. \tag{4.60}$$

Using the reduced matrix elements (4.58, 59), the single-particle dipole moments become finally

$$\begin{aligned} \mu \left(j = l + \tfrac{1}{2} \right) &= j g_l + \left(g_s - g_l \right)/2 \,, \\ \mu \left(j = l - \tfrac{1}{2} \right) &= j g_l - \left(g_s - g_l \right)/2 \cdot j/(j+1) \,. \end{aligned} \tag{4.61}$$

These results are the Schmidt values and are illustrated in Fig. 4.11, using $g_l(n) = 0$, $g_l\ (p) = 1$, $g_s(n) = -3.82$ n.m. and $g_s(p) = 5.58$ n.m. Thus, for the odd-proton, a steady, almost linear increase in μ for both the $j = l + \frac{1}{2}$ and $j = l - \frac{1}{2}$ orientations is present, in particular for large j values. This is not so for the neutron moments. Most of the experimental dipole moments fall in between the Schmidt lines, indicating again that the nucleons *in* the nucleus behave differently from *free* nucleons outside the nucleus. The effective moments or g-factors are model quantities as discussed in Sect. 4.3.

Fig. 4.11. The Schmidt single-particle magnetic moments for (**a**) proton single-particle states and (**b**) neutron single-particle states. Both the $j = l \pm \frac{1}{2}$ extreme lines are drawn [see (4.48)]. The data points, corresponding to effective g_s, g_l values fill up the domain in between the two Schmidt lines for a given charge state [taken from (Blin-Stoyle 1956)]

a)
b)

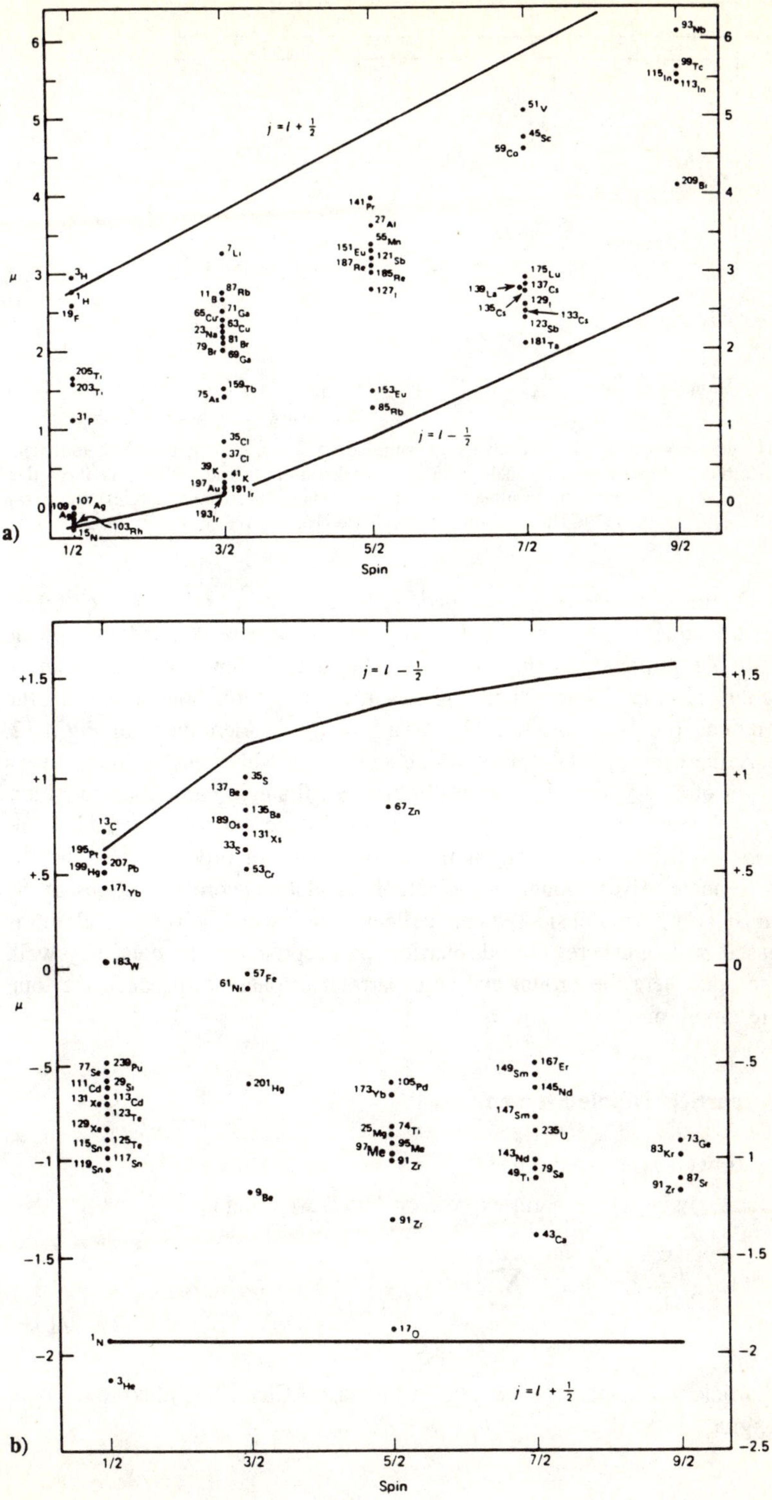

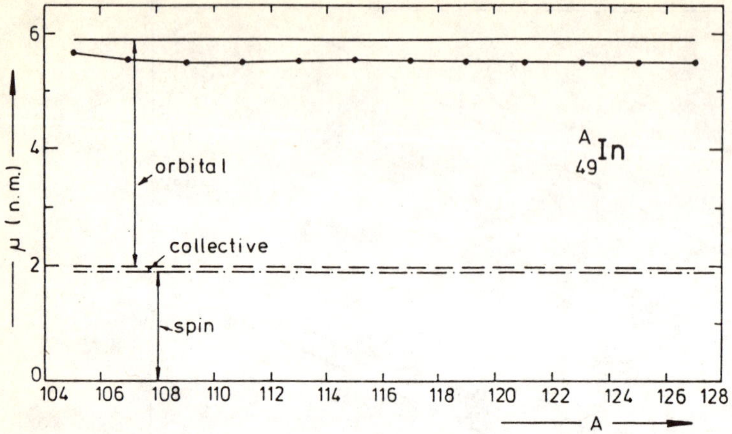

Fig. 4.12. The experimental dipole moments (μ_N) in the odd-A In nuclei with a single-hole moving in the $1g_{9/2}$ orbital below the $Z = 50$ shell closure [data taken from (Eberz 1987)]. The theoretical values [the separate contributions from orbital (g_l), spin (g_s) and collective admixtures] are shown (using $g_l = 1\mu_N$, $g_s^{\text{eff}} = 0.7 g_s^{\text{free}}$). The theoretical values have been derived in (Heyde 1978)

For ^{17}O, the example of $j = \frac{5}{2}$ ($1d_{5/2}$) gives a value of $\mu = -1.91$ n.m. (experimental value: -1.89 n.m.). We briefly discuss a more detailed case of the magnetic dipole moments for the odd-mass $_{49}$In nuclei. Even though the neutron number changes over a large shell (50, 82) the odd-proton hole moves in the same orbit, i.e., the $1g_{9/2}$ orbital. This is a $j = l + \frac{1}{2}$ orientation. In Fig. 4.12, we compare the experimental values which stay remarkably constant over a very large interval of $105 \leq A \leq 127$, with the theoretical single-particle calculations. We use $g_s = 1$ n.m and $g_s^{\text{eff}} = 0.7\, g_s^{\text{free}}$ to account for the effect of other nucleons that are *not* explicitly considered in the model space in order to describe the odd-mass In nuclei. Even though in the actual calculations core excitations of the even-even Sn core were considered (the collective part which is very small and is not discussed in detail, here) the odd-particle part reproduces the data very well. We also see that here the orbital and spin part act *in* phase to produce the total single-hole dipole moment.

4.6.2 Two-particle Dipole Moment

i) For $J = 0$, we trivially have $\mu = 0$,

ii) for $J^\pi = 2^+$, using the general expression (4.57) we obtain

$$\mu(j^2; 2^+) = \sqrt{\frac{2}{15}} \langle j^2; J = 2 \| \sum_{i=1}^{2} \{ g_l(i) \boldsymbol{j}_i + \left(g_s(i) - g_l(i) \right) \boldsymbol{s}_i \} \| j^2; J = 2 \rangle \,. \tag{4.62}$$

The two-particle to one-particle reduction formulas of Chap. 2, applied to a tensor of rank 1, give

$$\langle j^2; J=2 \| \boldsymbol{F}^{(1)} \| j^2; J=2 \rangle = 10 \begin{Bmatrix} j & 2 & j \\ 2 & j & 1 \end{Bmatrix} \langle j \| \boldsymbol{f}^{(1)} \| j \rangle \,. \tag{4.63}$$

Combining the results of Sect. 4.6.1 with the outcome of (4.62, 63), the two-particle dipole moment reads

$$\mu\left(j^2; J=2\right) = 2 \frac{\langle j \| \boldsymbol{f}^{(1)} \| j \rangle}{\left(j(j+1)(2j+1)\right)^{1/2}} = 2 g_j \,. \tag{4.64}$$

Here the moments are additive. For the $J^\pi = 2^+$ state the two nucleons each contribute their own g_j factor and yield the value $2g_j$.

4.6.3 Collective Admixtures

In a number of situations describing odd-A nuclei, where the odd-particle motion could be influenced by admixtures in the wave function coming from collective configurations (mainly from the low-lying collective quadrupole or octupole vibrations in the adjacent even-even core nuclei) with, as a result,

$$|\widetilde{jm}\rangle = \sqrt{1-\alpha^2}\, |jm\rangle + \alpha\, |2^+ \otimes j'; jm\rangle \ , \tag{4.65}$$

the dipole moment can still be calculated in closed form with the result

$$\mu(\tilde{j}) = \left(1-\alpha^2\right) \mu(j) + \alpha^2 \left\{ \frac{1}{4} \frac{1}{(j+1)} \left[6 + j(j+1) - j'(j'+1)\right] \mu\left(2^+\right) \right.$$
$$\left. + \frac{1}{2} \frac{1}{j'(j+1)} \left[j'(j'+1) + j(j+1) - 6\right] \mu(j') \right\} . \tag{4.66}$$

Here, the mixing amplitude α will be determined by the precise form of the way in which the odd particle couples to the collective core vibrations. This lies somewhat outside the field of discussion. A short and concise treatment is given by Brussaard an Glaudemans (1977) and, in addition, a computer code for particle-core coupling is described in Chap. 9.

Application of the above method for the Tl nuclei was carried out by Sagawa and Arima (1986) for the $1/2^+$ $3s_{1/2}^{-1}$ proton-hole configuration. The Sb and In ($Z = 51$ and $Z = 49$, respectively) nuclei form an ideal region for application of the above ideas. In Fig. 4.13, the collective, second-order magnetic dipole moment correction (the second term in (4.66)) is given as a function of the particle-core coupling strength ξ (which is proportional to α) for the $9/2_1^+$ (mainly the $1g_{9/2}^{-1}$ hole orbital) state in the odd-A In nuclei. Similar calculations could well be carried out for single-closed shell ± 1 nucleon configurations (i.e. the Sb, In, Tl, Bi, ...).

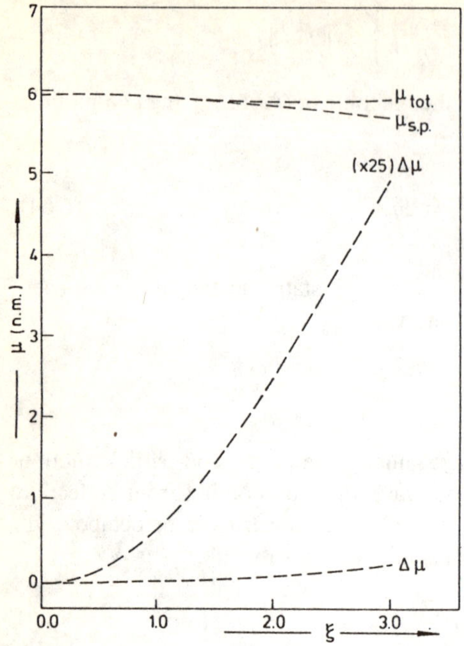

Fig. 4.13. The collective dipole moment admixture ($\Delta\mu$) as a function of the particle-core coupling strength ξ (or the admixture for the $|2^+ \otimes j'; jm\rangle$ configuration) for the $9/2_1^+$ state in the odd-A In nuclei. This configuration is mainly a $1g_{9/2}^{-1}$ hole configuration

4.7 Additivity Rules for Static Moments

Magnetic dipole and electric quadrupole moments have been measured in many odd-odd nuclei near closed shells (odd-odd In, odd-odd Sb, odd-odd Tl, ... nuclei). For such nuclei, starting again from a rather simple configuration for both the odd-proton and the odd-neutron cases, the known moments and rather general "additivity" rules can be derived by using simple angular momentum recoupling techniques (de-Shalit, Talmi 1963), and have been used in determining the composed moments.

If we call the eigenstate in the odd-proton nucleus $|J_p\rangle$ with $\mu(J_p)$, $Q(J_p)$, ... the corresponding moments and $|J_n\rangle$ the eigenstate for the odd-neutron nucleus with $\mu(J_n)$, $Q(J_n)$, ... the corresponding moments, under the assumption of weak coupling in obtaining the eigenstate $|J\rangle = |J_p \otimes J_n; J\rangle$ in the odd-odd nucleus, one obtains the expressions

$$\mu(J) = \frac{J}{2}\left[\frac{\mu(J_p)}{J_p} + \frac{\mu(J_n)}{J_n} + \left(\frac{\mu(J_p)}{J_p} - \frac{\mu(J_n)}{J_n}\right) \times \frac{J_p(J_p+1) - J_n(J_n+1)}{J(J+1)}\right], \tag{4.67}$$

$$Q(J) = \begin{pmatrix} J & 2 & J \\ -J & 0 & J \end{pmatrix} (-1)^{J_p+J_n+J} \cdot (2J+1)$$
$$\times \left[\begin{Bmatrix} J_p & J & J_n \\ J & J_p & 2 \end{Bmatrix} \frac{Q(J_p)}{\begin{pmatrix} J_p & 2 & J_p \\ -J_p & 0 & J_p \end{pmatrix}} + \begin{Bmatrix} J_n & J & J_p \\ J & J_n & 2 \end{Bmatrix} \frac{Q(J_n)}{\begin{pmatrix} J_n & 2 & J_n \\ -J_n & 0 & J_n \end{pmatrix}} \right] . \tag{4.68}$$

These equations also apply to the coupling between identical nucleons in going from the one-particle to the two-particle nucleus. Applying (4.67) to a system of two identical nucleons, then $\mu(J_p) = \mu(J_n) = \mu(j)$, $J_p = J_n = j$ and (4.67) reduces to (4.64) since

$$\mu(J) = J g_j \ , \tag{4.69}$$

an expression which applies for more general J, than just the value $J = 2$ that was taken in deriving (4.64). Analogously, for the addition of quadrupole moments with $Q(J_p) = Q(J_n) = Q_{\text{s.p.}}(j)$ and $J_p = J_n = j$, (4.68) reduces to

$$Q(J) = \frac{\begin{pmatrix} J & 2 & J \\ -J & 0 & J \end{pmatrix}}{\begin{pmatrix} j & 2 & j \\ -j & 0 & j \end{pmatrix}} (-1)^{J+1} \begin{Bmatrix} j & J & j \\ J & j & 2 \end{Bmatrix} Q_{\text{s.p.}}(j) \ , \tag{4.70}$$

an expression which, for $J = 2$, reduces to (4.52). Here again, (4.70) is more general since it applies to any J values.

These methods have been tested with remarkable success in the odd-odd In mass region by *Eberz* et al. (Eberz et al. 1987) for many magnetic dipole and electric quadrupole moments. (See Eberz et al. 1987 for a detailed discussion of the possible configurations.) In such an approach, rather complicated proton and neutron states are combined under the assumption that these states are not modified very much when coupling to form the final state in the odd-odd nucleus. In odd-odd nuclei near closed shells, when rather few configurations with the same J_p and J_n value are present in the odd-proton and odd-neutron nuclei, respectively, the additivity rules are expected to work well. The application to odd-odd In nuclei in the interval $104 \leq A \leq 126$ has been carried out for both the $\mu(J)$ and $Q(J)$ values and has considerable success in accounting for the data, as is illustrated in Fig. 4.14 and 4.15. These figures are taken from (Eberz et al. 1987).

For the odd-neutron case, where the neutron number N varies over a rather large interval $51 \leq N \leq 79$, a less unambiguous situation results. If we have the odd-proton nucleus eigenstates $|J_p^{(i)}\rangle$ and the odd-neutron eigenstates $|J_n^{(j)}\rangle$, where we can have different (J_p, i) values and (J_n, j) values in a single nucleus, weak-coupling may no longer hold, i.e., one obtains wave functions (Heyde 1989)

$$|J^{\text{odd-odd}}\rangle = |J_p^{(1)} \otimes J_n^{(1)}; J\rangle + \Sigma a \left(J_p^{(i)}, J_n^{(j)}; J\right) |J_p^{(i)} \otimes J_n^{(j)}; J\rangle \ . \tag{4.71}$$

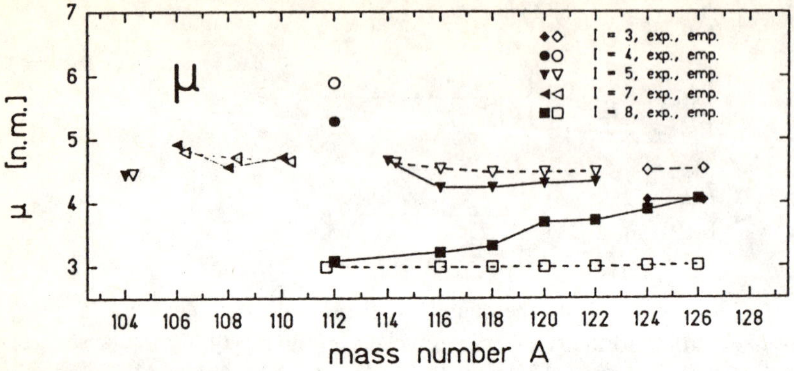

Fig. 4.14. Illustration of the magnetic dipole addition rules (4.67) in the case of the odd-odd In nuclei. The data are taken from (Eberz 1987). The full symbols represent the experimental values, the open symbols the results according to the addition rules. The moments needed to apply the addition rules are taken from the adjacent odd-proton (In) and odd-neutron (Sn) nuclei, when available. The symbols denote ◇, $J = 3$; ○, $J = 4$; ▽, $J = 5$; ◁, $J = 7$; □, $J = 8$ [taken from (Eberz 1987)]

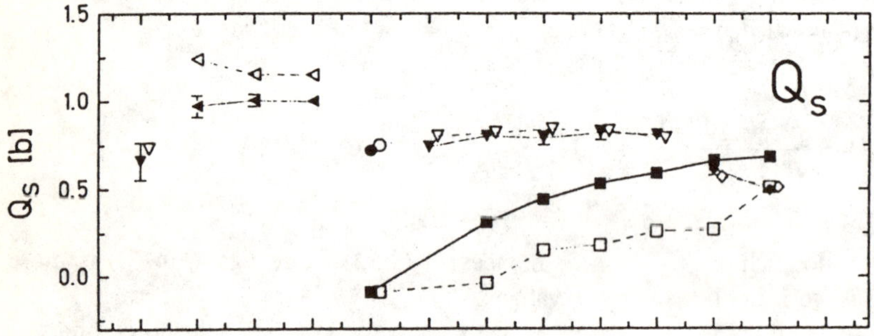

Fig. 4.15. See caption to Fig. 4.14, but now for the electric quadrupole moments and using (4.68)

In this case, extra components from configuration mixing in the final nucleus result. These terms give rise to extra "polarization" terms with respect to the original zero-order term. A good qualitative estimate of configuration mixing is obtained by studying the number of final states J in the odd-odd nucleus. If there is only a single J state over an interval of $\cong$ 1 MeV, weak-coupling will most probably be a good approximation. If, on the other hand, many J levels result at a small energy separation, chances for large configuration mixing are more likely to occur.

In using the additivity method, it is of the utmost importance to use the odd-mass moments as close as possible to the "unperturbed" odd-mass nuclei that are used to carry out the coupling in obtaining the final odd-odd nucleus. In the odd-odd In nuclei, in particular for the $\left(1g_{9/2}^{-1}(\pi)\,1h_{11/2}(\nu)\right)8^-$ configurations, some problems occur when comparing the measured moments and the "additivity" moments (Fig. 4.16). If one considers a pure $1g_{9/2}^{-1}$ proton-hole configuration and a $1h_{11/2}$ neutron one-quasiparticle configuration (and linear filling of the $1h_{11/2}$

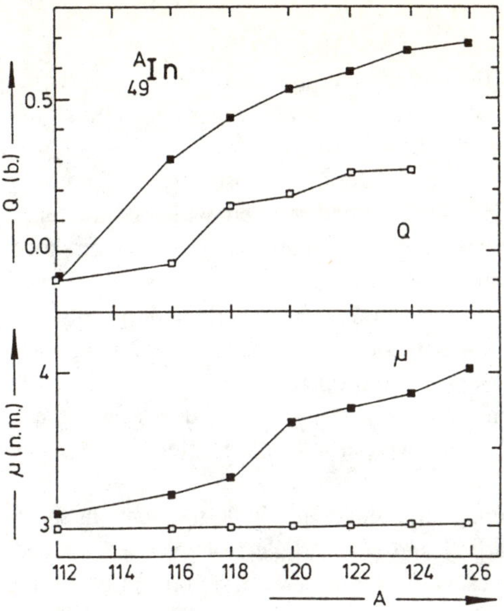

Fig. 4.16. Variation of $\mu(8_1^-)$ and $Q(8_1^-)$ in the odd-odd In nuclei ($112 \leq A \leq 126$). The experimental data points (■) are taken from (Eberz 1987). The additivity moments (□) [see (4.67) and (4.68)] are obtained, using the discussion in (Eberz 1987)

orbital with n valence neutrons), then the dipole, respectively, quadrupole moment would vary as (de-Shalit, Talmi 1963)

$$\mu(8^-) = a\mu\left(1g_{9/2}^{-1}\right) + b\mu\left(1h_{11/2}\right) , \tag{4.72}$$

$$Q(8^-) = a'Q\left(1g_{9/2}^{-1}\right) + b'(12 - 2n)/10\left(1h_{11/2}\right) , \tag{4.73}$$

which means a constant value for $\mu(8^-)$ and a linear increase in $Q(8^-)$ with n, the number of neutrons filling the $1h_{11/2}$ orbital. In the more specific case of more orbitals being filled at the same time, some modifications to this simple dependence on particle number can be expected (e.g. starting the filling of the $3s_{1/2}$, $2d_{3/2}$ orbitals before $N = 76$ and early filling of the $1h_{11/2}$ orbital before $N = 64$). This will result due to the pair correlations and the resulting pair distribution of neutrons over the five neutron single-particle states (Chap. 7).

We conclude this chapter, which is an intensive application of the rules and methods of Racah-algebra to spherical tensor operators that describe the electromagnetic properties and observables in the nucleus. We studied some of the most often used one- and two-particle transition rates and moments, concentrating on the $M1$ and $E2$ operator. The methods, however, are general and can easily be extended to other operators, starting from the general form of the electric and magnetic multipole operators discussed in Sect. 4.2. It is also possible to describe other observables: Beta-decay processes can be discussed analogously to electromagnetic decay processes. One can also address nucleon transfer reactions. In all these cases, one always has to first construct the appropriate spherical tensor operators when evaluating the nucleus matrix elements.

4.8 Summary

Having derived the basic properties of the nuclear shell model in Chap. 3, with emphasis on energies and wave functions, better tests of these nuclear wave functions can be obtained via the study of electromagnetic observables.

The multipole operators, describing the emission (or absorption) of gamma radiation for electric and magnetic transitions are discussed with, at the same time, the derivation of the elementary single-particle matrix elements (Weisskopf estimates). These matrix elements, which form the main estimates in discussing electromagnetic decay, are applied to simple nuclei with one-nucleon properties. The concepts of effective electric and magnetic "charges" are outlined and applied to ^{17}O, comparing the one-body and many-body properties.

Subsequently, general electromagnetic properties in two-particle systems are discussed. The relations between one- and two-particle matrix elements are derived.

Finally, we discuss electric quadrupole and magnetic dipole moments in some more detail. In each case, the single-particle and two-particle systems are described thoroughly. Extensive examples are presented (Schmidt values, application to the In nuclei, configuration mixing in the Po, Rn and Ra nuclei, ...). We also discuss general addition theorems for combining the moments from simple systems into more comples configurations.

Problems

4.1 Derive a general relation between the one-particle and the two-particle ($J^\pi = 2^+$) electric quadrupole moment.

4.2 Study the sign of the two-particle electric quadrupole moment for a $(j)^2J$ configuration as a function of J.

4.3 Derive the magnetic dipole moment for a two-particle configuration $\mu(j^2, J)$ as a function of the one-particle magnetic moment $\mu(j)$. Study the dependence on J (magnitude and sign).

4.4 Derive the additivity rules (the equivalent of (4.67) and (4.68)) for general ML and EL moments.

4.5 Derive, the n-dependence for the electric quadrupole and magnetic dipole moment in a $(j)^nJ$ configuration.

4.6 Derive the explicit value for $\mu(\tilde{j})$ and $Q(\tilde{j})$ when the single-particle configuration contains a collective admixture of the type $|2^+ \otimes j'; jm\rangle$ with amplitude α (Hint: use the explicit, analytic expression of the Wigner $6j$-symbols as given by Biedenharn and Rose (1953)).

4.7 Show that the $E2$ matrix element connecting a 2^+ state (which appears as a coherent superposition of $(j)^2 2^+$ configuration) to the 0^+ ground-state (also described by a coherent superposition of the same $(j)^2 0^+$ configurations) can become largely enhanced with respect to the single-particle $E2$ matrix element values.

4.8 Discuss a possible reason for having to use effective and 'quenched' gyromagnetic g-factors in order to correctly describe magnetic dipole moments near closed shell configurations.

*4.9 In many nuclei with two-valence particles (or holes) outside a closed-shell configuration (nuclei such as Cd, Te, Hg, Po, ...), the energy difference between high-spin $(j^2)J$ and $(j)^2J'$ configurations with $|J - J'| = 2$ can become rather small and of the order of $\approx 100\,\text{keV}$.
i) Derive the $B(E2; (j)^2 8^+ \rightarrow 6^+)$ value and use harmonic oscillator wave functions to evaluate the radial integral of $\langle r^2 \rangle$,
ii) deduce the actual $T_{1/2}(8^+)$ value and point out the possibility of obtaining an isomerism (i.e. a level with a lifetime in the microsecond region).

*4.10 Derive a closed expression for the linear energy weighted electric dipole sum rule

$$\sum_f (E_f - E_i) \left| \langle a_f | \boldsymbol{r} | a_i \rangle \right|^2 ,$$

where it holds that $\boldsymbol{H}|a_b\rangle = E_b|a_b\rangle$, $\boldsymbol{H}|a_i\rangle = E_i|a_i\rangle$. Here, $\boldsymbol{H}$ denotes the Hamiltonian for a particle moving in a Coulomb potential i.e. $\boldsymbol{H} = \boldsymbol{p}^2/2m - Ze^2/|\boldsymbol{r}| + V(|\boldsymbol{r}|)$.

5. Second Quantization

5.1 Creation and Annihilation Operators

In Chaps. 3, 4, we formulated the nuclear shell model using the coordinate representation. This means that when considering identical particles in the nucleus, the Pauli principle implies the explicit construction of antisymmetrized wave functions in the exchange of all coordinates of any two of the A particles. If we call, as before, the single-particle wave functions $\varphi_\alpha(\boldsymbol{r}_i, \boldsymbol{\sigma}_i, \boldsymbol{\tau}_i)$, denoted by $\varphi_\alpha(i)$ with $\alpha \equiv n_a, l_a, j_a, m_a, t_{z_a}$, wave functions that constitute a complete orthonormal set, then the A-particle antisymmetrized wave function reads

$$\Psi_{\alpha_1,\alpha_2,\dots\alpha_A}(1, 2, \dots, A) = \frac{1}{\sqrt{A!}} \begin{bmatrix} \varphi_{\alpha_1}(1) & \varphi_{\alpha_1}(2) & \dots & \varphi_{\alpha_1}(A) \\ \varphi_{\alpha_2}(1) & \varphi_{\alpha_2}(2) & \dots & \varphi_{\alpha_2}(A) \\ \vdots & & & \\ \varphi_{\alpha_A}(1) & \varphi_{\alpha_A}(2) & \dots & \varphi_{\alpha_A}(A) \end{bmatrix} . \quad (5.1)$$

In such a wave function there is superfluous information: the important point is that the A single-particle orbitals α_1, α_2, ... α_A are occupied by a nucleon. The Pauli principle gives no extra information: it simply includes the antisymmetrization meaning that any particle with coordinates 1 ($\boldsymbol{r}_1$, $\boldsymbol{\sigma}_1$, $\boldsymbol{t}_1$), 2, ... A could be in any single-particle state.

Thus, we shall try to go to a formalism that gives the minimum of information, i.e., which orbitals are occupied [also called the occupation number representation (Brussaard, Glaudemans 1977)] but has the Pauli principle built in, making use of the properties of the determinant wave function in (5.1).

We point out that the name "second quantization" formalism can be misleading since we are not introducing quantum field theory nor the subsequent quantization of standard quantum mechanics. The term "occupation number representation" is more precise since one simply uses an alternative formulation of the usual quantum mechanical description given in Chaps. 3, 4. However, it turns out to be a very useful way of handling an interacting many-body system such as the atomic nucleus.

We denote a vacuum state with no particles by $|\rangle$. By acting with an operator $a^+_{\alpha_i}$ on this vacuum state, a particle in the quantum state α_i is created and is denoted by

$$|\alpha_i\rangle = a^+_{\alpha_i}|\rangle . \quad (5.2)$$

Thus we put a one-to-one correspondence between the single-particle wave functions $\varphi_{\alpha_i}(1)$ and the states $|\alpha_i\rangle$

$$|\alpha_i\rangle \leftrightarrow \varphi_{\alpha_i}(\boldsymbol{r}_1, \boldsymbol{\sigma}_1, \boldsymbol{\tau}_1) . \tag{5.3}$$

The A-particle state can then be obtained by the successive action of A such operators, acting in a given order. The A particle state is made by the action of an operator on the $A-1$ particle state as

$$a^+_{\alpha_A}|\alpha_1, \alpha_2, \ldots, \alpha_{A-1}\rangle = |\alpha_1, \alpha_2, \ldots, \alpha_A\rangle , \tag{5.4}$$

or, working this out

$$|\alpha_1, \alpha_2, \ldots, \alpha_A\rangle = a^+_{\alpha_A} a^+_{\alpha_{A-1}} \ldots a^+_{\alpha_1}|\rangle . \tag{5.5}$$

We now need a one-to-one correspondence between the states (5.1) and (5.5), or

$$|\alpha_1, \alpha_2, \ldots, \alpha_A\rangle \leftrightarrow \Psi_{\alpha_1,\alpha_2,\ldots,\alpha_A}(1, 2, \ldots, A) . \tag{5.6}$$

The properties of the determinant wave function (5.1) indicate that it is multiplied by a factor of -1 when interchanging any two rows. This leads to the following properties of the operators $a^+_{\alpha_i}$, i.e.,

$$\Psi_{\alpha_1,\alpha_2,\ldots,\alpha_A}(1, 2, \ldots, A) = -\Psi_{\alpha_2,\alpha_1,\ldots,\alpha_A}(1, 2, \ldots, A) , \tag{5.7}$$

or

$$|\alpha_1, \alpha_2, \ldots, \alpha_A\rangle = -|\alpha_2, \alpha_1, \ldots, \alpha_A\rangle . \tag{5.8}$$

From the definition of the many-particle states given by (5.5), this gives

$$a^+_{\alpha_A} \cdots\cdots \left(a^+_{\alpha_2} a^+_{\alpha_1} + a^+_{\alpha_1} a^+_{\alpha_2}\right)|\rangle = 0 , \tag{5.9}$$

for any A-particle state and any interchange of two rows. Thus we get

$$a^+_{\alpha_i} a^+_{\alpha_j} + a^+_{\alpha_j} a^+_{\alpha_i} = \left\{a^+_{\alpha_i}, a^+_{\alpha_j}\right\} = 0 , \tag{5.10}$$

for any α_i, α_j, and $\{A, B\}$ is the anticommutator of the two operators A, B, i.e., $AB + BA$.

Equation (5.10) expresses the Pauli principle within the occupation or creation-operator formalism. The fact that two particles cannot move in the same orbital is easily expressed by (5.10) putting $\alpha_i \equiv \alpha_j$.

Using Hermitian conjugation, we obtain

$$\langle\alpha_i| = \langle\,|a_{\alpha_i} = \left(a^+_{\alpha_i}|\rangle\right)^+ = (|\alpha_i\rangle)^+ , \tag{5.11}$$

with a_{α_i} the Hermitian conjugate operator to $a^+_{\alpha_i}$. We can easily rewrite the above A-particle kets via Hermitian conjugation and obtain the results

$$\begin{aligned}\langle \alpha_1, \alpha_2, \ldots, \alpha_A| &= \left(|\alpha_1, \alpha_2, \ldots, \alpha_A\rangle\right)^+ \\ &= \langle\, | \left(a^+_{\alpha_A} \ldots a^+_{\alpha_2} a^+_{\alpha_1}\right)^+ \\ &= \langle\, | a_{\alpha_1} a_{\alpha_2} \ldots a_{\alpha_A} \,.\end{aligned} \tag{5.12}$$

Thus, the operators a_{α_i} acting to the left act again like creation operators. We also get the Hermitian conjugate of the commutation relation (5.10) as

$$\left\{a_{\alpha_i}, a_{\alpha_j}\right\} = 0\,. \tag{5.13}$$

We now want to know what the meaning of the a_{α_i} operators, acting to the right is.

In order to obtain the result of the operator a_λ acting to the right on a general state $|\alpha, \beta, \gamma, \ldots\rangle$, we calculate the components of the vector

$$a_\lambda|\alpha, \beta, \gamma, \ldots\rangle\,, \tag{5.14}$$

in the basis formed by all 1, 2, …, A, … particle states $|\alpha', \beta', \gamma', \ldots\rangle$ for all $\alpha', \beta', \gamma', \ldots$.

We know that

$$\begin{aligned}\langle \alpha', \beta', \gamma', \ldots |a_\lambda|\alpha, \beta, \gamma, \ldots\rangle &= \langle \alpha', \beta', \gamma', \ldots, \lambda|\alpha, \beta, \gamma, \ldots\rangle \\ = 0 \quad &\text{if} \quad \{\alpha', \beta', \gamma', \ldots, \lambda\} \neq \{\alpha, \beta, \gamma, \ldots\} \\ = \pm 1 \quad &\text{if} \quad \{\alpha', \beta', \gamma', \ldots, \lambda\} = \{\alpha, \beta, \gamma, \ldots\}\,.\end{aligned} \tag{5.15}$$

i) The case $\lambda \notin \{\alpha, \beta, \gamma, \ldots, \omega\}$. It follows that for the full set of states $\{\alpha', \beta', \gamma', \ldots, \omega'\}$ all matrix elements $\langle \alpha', \beta', \gamma', \ldots, \omega'|a_\lambda|a, \beta, \gamma, \ldots, \omega\rangle$ vanish. Since all components of the vector $a_\lambda|a, \beta, \gamma, \ldots, \omega\rangle$ vanish, the state should have zero length and we note

$$a_\lambda|\alpha, \beta, \gamma, \ldots, \omega\rangle = 0 \quad \text{for} \quad \lambda \notin \{\alpha, \beta, \gamma, \ldots, \omega\}\,. \tag{5.16}$$

ii) On the other hand, we have

$$\begin{aligned}&\langle \alpha, \beta, \gamma, \ldots, \omega|a_\lambda|\alpha, \beta, \gamma, \ldots, \omega, \lambda\rangle \\ &\quad = \langle \alpha, \beta, \gamma, \ldots, \omega, \lambda|\alpha, \beta, \gamma, \ldots, \omega, \lambda\rangle = +1\,,\end{aligned}$$

and we obtain the result

$$a_\lambda|\alpha, \beta, \gamma, \ldots, \omega, \lambda\rangle = |\alpha, \beta, \gamma, \ldots, \omega\rangle \quad \text{for} \quad \lambda \notin \{\alpha, \beta, \gamma, \ldots, \omega\}\,. \tag{5.17}$$

Thus the operator a_λ acting to the right on a given state acts as an annihilation operator: if λ is unoccupied we get zero, if λ is occupied we can get the result ± 1 depending on the order of the operators. We give some examples:

$$\begin{aligned}&a_\lambda|\,\rangle = 0\,, \\ &a_\lambda|\lambda\rangle = 1 \quad \text{for all} \quad \lambda\,, \\ &a^+_\beta|\beta\rangle = 0\,, \\ &a^+_\beta|\alpha\rangle = |\alpha\beta\rangle\,, \\ &\alpha_\beta|\alpha\beta\rangle = |\alpha\rangle\,.\end{aligned} \tag{5.18}$$

Having obtained the anticommutation relation for two creation or two annihilation operators, we also want to derive the anticommutation relation for a creation and an annihilation operator. We therefore use the following method: We let the operators $a^+_{\alpha_A} a_{\alpha_{A-1}}$ and $a_{\alpha_{A-1}} a^+_{\alpha_A}$ act on a single state $|\alpha_1, \alpha_2, \ldots, \alpha_{A-2}, \alpha_{A-1}\rangle$. We find

$$\begin{aligned} a^+_{\alpha_A} a_{\alpha_{A-1}} |\alpha_1, \alpha_2, \ldots, \alpha_{A-1}\rangle &= a^+_{\alpha_A} |\alpha_1, \alpha_2, \ldots, \alpha_{A-2}\rangle \\ &= |\alpha_1, \alpha_2, \ldots, \alpha_{A-2}, \alpha_A\rangle \,. \end{aligned} \tag{5.19}$$

Carrying out the same action but in the reverse order, we obtain

$$\begin{aligned} & a_{\alpha_{A-1}} a^+_{\alpha_A} |\alpha_1, \alpha_2, \ldots, \alpha_{A-2}, \alpha_{A-1}\rangle \\ &\quad = a_{\alpha_{A-1}} |\alpha_1, \alpha_2, \ldots, \alpha_{A-2}, \alpha_{A-1}, \alpha_A\rangle \\ &\quad = -a_{\alpha_{A-1}} |\alpha_1, \alpha_2, \ldots, \alpha_{A-2}, \alpha_A, \alpha_{A-1}\rangle \\ &\quad = -|\alpha_1, \alpha_2, \ldots, \alpha_{A-2}, \alpha_A\rangle \end{aligned} \tag{5.20}$$

Adding (5.19) and (5.20) we obtain

$$\left(a^+_{\alpha_A} a_{\alpha_{A-1}} + a_{\alpha_{A-1}} a^+_{\alpha_A}\right) |\alpha_1, \alpha_2, \ldots, \alpha_{A-2}, \alpha_{A-1}\rangle = 0 \,, \tag{5.21}$$

for any state $| \ldots\rangle$, giving that

$$\{a^+_{\alpha_i}, a_{\alpha_j}\} = 0 \quad \text{for} \quad \alpha_i \neq \alpha_j \,. \tag{5.22}$$

If $\alpha_i = \alpha_j$, the above arguments lead to

$$\begin{aligned} &\text{i)} \quad a^+_{\alpha_i} a_{\alpha_i} = 1 \quad \text{and} \quad a_{\alpha_i} a^+_{\alpha_i} = 0 \,, \quad \text{or} \\ &\text{ii)} \quad a^+_{\alpha_i} a_{\alpha_i} = 0 \quad \text{and} \quad a_{\alpha_i} a^+_{\alpha_i} = 1 \,, \end{aligned} \tag{5.23}$$

if the state α_i is occupied or unoccupied, respectively. In conclusion, we can state the basic content of the occupation number representation or the representation using creation and annihilation operators as given by the anticommutation relations

$$\begin{aligned} &\left\{a^+_{\alpha_i}, a^+_{\alpha_j}\right\} = 0 \,, \\ &\left\{a_{\alpha_i}, a_{\alpha_j}\right\} = 0 \,, \\ &\left\{a^+_{\alpha_i}, a_{\alpha_j}\right\} = \delta_{\alpha_i, \alpha_j} \,, \end{aligned} \tag{5.24}$$

for any α_i, α_j.

We work out a few examples:

i) $\langle \,|a_\alpha a^+_\beta|\, \rangle = \langle \,|\, \rangle \delta_{\alpha\beta} - \langle \,|a^+_\beta a_\alpha|\, \rangle = \delta_{\alpha\beta}$.

ii) $\langle \,|a_\alpha a_\beta a^+_\gamma a^+_\delta|\, \rangle = \delta_{\alpha\delta}\delta_{\beta\gamma} - \delta_{\alpha\gamma}\delta_{\beta\delta}$.

iii) We call $N_\alpha \equiv a^+_\alpha a_\alpha$ the number operator since N_α acting on an occupied orbital α gives 1, and the result 0 when acting on an unoccupied state α. Then

$$\mathcal{N} \equiv \sum_\alpha N_\alpha = \sum_\alpha a^+_\alpha a_\alpha \tag{5.25}$$

is the total number operator.

5.2 Operators in Second Quantization

Having indicated that there exists a one-to-one correspondence between the Slater determinant wave functions (5.1) and the states obtained via creation operators acting on a vacuum state, we would also like to find out the way in which the one- and two-body operators can be represented in second quantization (Brussaard, Glaudemans 1977). We first study one-body operators.

A general, one-body operator for an A-nucleon system reads

$$O^{(1)} = \sum_{k=1}^{A} O(\boldsymbol{r}_k, \boldsymbol{\sigma}_k, \boldsymbol{\tau}_k) \,. \tag{5.26}$$

(Below we use the notation $\boldsymbol{r}_k \equiv \boldsymbol{r}_k, \boldsymbol{\sigma}_k, \boldsymbol{\tau}_k, \ldots$.)

The one-body matrix element for single-particle states $\varphi_\alpha(\boldsymbol{r})$ and $\varphi_\beta(\boldsymbol{r})$ becomes

$$\langle\alpha|O|\beta\rangle = \int \varphi_\alpha^*(\boldsymbol{r}) O(\boldsymbol{r}) \varphi_\beta(\boldsymbol{r})\, d\boldsymbol{r} \,. \tag{5.27}$$

We shall demonstrate that the equivalent form to $O^{(1)}$ in second quantization reads

$$\boldsymbol{O} = \sum_{\alpha,\beta} \langle\alpha|O|\beta\rangle a_\alpha^+ a_\beta \,. \tag{5.28}$$

We show the correctness of (5.28) by evaluating the one body matrix elements

$$\begin{aligned} &\langle\alpha_1, \alpha_2, \ldots, \alpha_A|\boldsymbol{O}|\alpha_1, \alpha_2, \ldots, \alpha_A\rangle \\ &\quad = \sum_{\alpha,\beta} \langle\alpha|O|\beta\rangle \langle\, | a_{\alpha_1} a_{\alpha_2} \ldots a_{\alpha_A} a_\alpha^+ a_\beta a_{\alpha_A}^+ \ldots a_{\alpha_2}^+ a_{\alpha_1}^+ |\,\rangle \,. \end{aligned} \tag{5.29}$$

For each case $\alpha = \beta = \alpha_i$ ($i = 1, 2, \ldots, A$), the value of the matrix element containing the creation and annihilation operators is +1. Thus the matrix element (5.29) becomes

$$\langle\alpha_1, \alpha_2, \ldots, \alpha_A|\boldsymbol{O}|\alpha_1, \alpha_2, \ldots, \alpha_A\rangle = \sum_{\alpha_i} \langle\alpha_i|O|\alpha_i\rangle \,. \tag{5.30}$$

Also, non-diagonal matrix elements different from zero can result if from the A particle quantum numbers, $A - 1$ are identical in the bra and ket state, or,

$$\langle\alpha_1, \alpha_2, \ldots, \alpha_A|\boldsymbol{O}|\gamma_1, \alpha_2, \ldots, \alpha_A\rangle = \langle\alpha_1|O|\gamma_1\rangle \;(\alpha_1 \neq \gamma_1) \,. \tag{5.31}$$

One can illustrate the effect of one-body operators in a diagram (Fig. 5.1) where the operators in the initial and final states are indicated by lines going up and the one-body operator acts as an internal probe changing (eventually) the quantum numbers of not more than one particle at a time. Thus we can illustrate (5.30, 31) schematically.

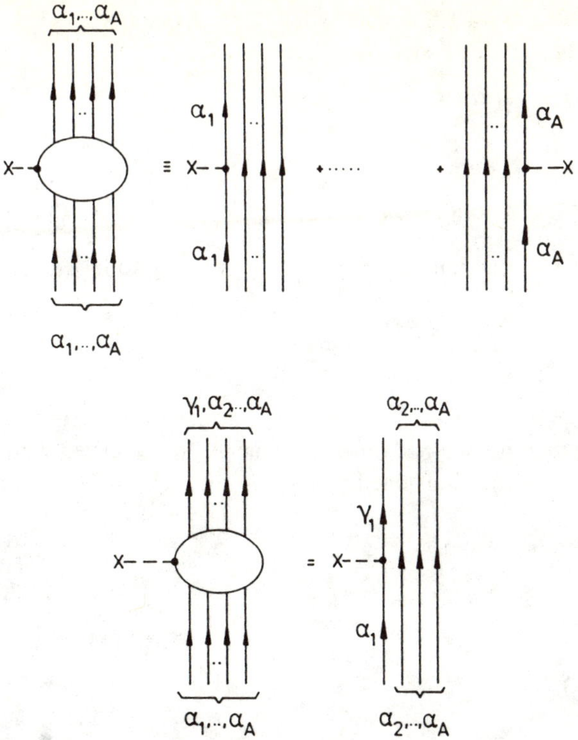

Fig. Fig. 5.1. Illustration of the one-body operator matrix elements taken between the A-particle states $|\alpha_1, \alpha_2, \ldots, \alpha_A\rangle$ as obtained in (5.30) and (5.31). We use a Feynman-Goldstone diagrammatic way to represent the above processes. (*i*) Diagonal elements (*upper part*) are given by the sum over A one-body processes acting on each separate line $\alpha_1, \alpha_2, \ldots$ up to α_A. The interaction zone of the one-body operator with the A-particle state is indicated by the "blob" ($\times --$). (*ii*) Non-diagonal elements (*lower part*) where only a single one-body scattering process remains ($\alpha_1 \rightarrow \gamma_1$). Orthogonality takes the $(A-1)$-particle state away

For two-body operators we have in coordinate space

$$O^{(2)} = \sum_{i<j=1}^{A} O(\boldsymbol{r}_i, \boldsymbol{r}_j) \qquad (\boldsymbol{r}_i \equiv \boldsymbol{r}_i, \boldsymbol{\sigma}_i, \boldsymbol{\tau}_i) \,, \tag{5.32}$$

and the two-body matrix elements

$$\langle\alpha\beta|O|\gamma\delta\rangle_{\text{nas}} = \int \varphi^*_\alpha(\boldsymbol{r}_1)\varphi^*_\beta(\boldsymbol{r}_2)O(\boldsymbol{r}_1, \boldsymbol{r}_2)(1 - P_{12})\varphi_\gamma(\boldsymbol{r}_1)\varphi_\delta(\boldsymbol{r}_2)d\boldsymbol{r}_1 d\boldsymbol{r}_2 \,, \tag{5.33}$$

with P_{12} an operator that exchanges the coordinates of particles 1 and 2. In second quantization, the corresponding two-body operator becomes

$$\boldsymbol{O} = \frac{1}{4} \sum_{\alpha,\beta,\gamma,\delta} \langle\alpha\beta|O|\gamma\delta\rangle_{\text{nas}} a^+_\alpha a^+_\beta a_\delta a_\gamma \,. \tag{5.34}$$

We verify, by a direct calculation, that (5.34) gives the correct two-body matrix element when used within an A-particle state. We thus calculate

$$\begin{aligned}&\langle \alpha_1, \alpha_2, \ldots, \alpha_A | \boldsymbol{O} | \alpha_1, \alpha_2, \ldots, \alpha_A \rangle \\ &= \frac{1}{4} \sum_{\alpha,\beta,\gamma,\delta} \langle \alpha\beta | O | \gamma\delta \rangle_{\text{nas}} \\ &\times \langle \, | a_{\alpha_1} a_{\alpha_2} \ldots a_{\alpha_A} a^+_\alpha a^+_\beta a_\delta a_\gamma a^+_{\alpha_A} \ldots a^+_{\alpha_2} a^+_{\alpha_1} | \, \rangle \, .\end{aligned} \tag{5.35}$$

In the evaluation, in order to get non-vanishing results, we have to take

$$\begin{aligned}\gamma = \alpha_i \, , \quad \delta = \alpha_j \quad &\text{and} \quad \alpha = \alpha_i \, , \quad \beta = \alpha_j \\ &\text{or} \quad \alpha = \alpha_j \, , \quad \beta = \alpha_i \, .\end{aligned}$$

We can similarly change to $\gamma = \alpha_j$ and $\delta = \alpha_i$, with again the two cases for α and β. When using the fact that the two-body matrix elements are evaluated with antisymmetrized two-body wave functions one has

$$\langle \alpha_i \alpha_j | O | \alpha_i \alpha_j \rangle_{\text{nas}} = -\langle \alpha_j \alpha_i | O | \alpha_i \alpha_j \rangle_{\text{nas}} = \ldots \, . \tag{5.36}$$

Thus in total, the matrix element becomes

$$\begin{aligned}&\langle \alpha_1, \alpha_2, \ldots, \alpha_A | \boldsymbol{O} | \alpha_1, \alpha_2, \ldots, \alpha_A \rangle \\ &= \sum_{\alpha_i, \alpha_j} \langle \alpha_i \alpha_j | O | \alpha_i \alpha_j \rangle_{\text{nas}} \, .\end{aligned} \tag{5.37}$$

Similarly, we can obtain non-diagonal matrix elements

$$\begin{aligned}&\langle \alpha_1, \alpha_2, \ldots, \alpha_A | \boldsymbol{O} | \gamma_1, \alpha_2, \ldots, \alpha_A \rangle \\ &= \sum_{\alpha_j} \langle \alpha_1 \alpha_j | O | \gamma_1 \alpha_j \rangle_{\text{nas}} \qquad (\alpha_1 \neq \gamma_1) \, , \\ &\langle \alpha_1, \alpha_2, \ldots, \alpha_A | \boldsymbol{O} | \gamma_1, \gamma_2, \ldots, \alpha_A \rangle \\ &= \langle \alpha_1 \alpha_2 | O | \gamma_1 \gamma_2 \rangle_{\text{nas}} \qquad (\alpha_1, \alpha_2 \neq \gamma_1, \gamma_2) \, .\end{aligned} \tag{5.38}$$

Here, too, by using the Goldstone diagrams one can show how a two-body operator connects two incoming and two outgoing lines, leaving all other $A-2$ nucleon lines unaffected (Fig. 5.2).

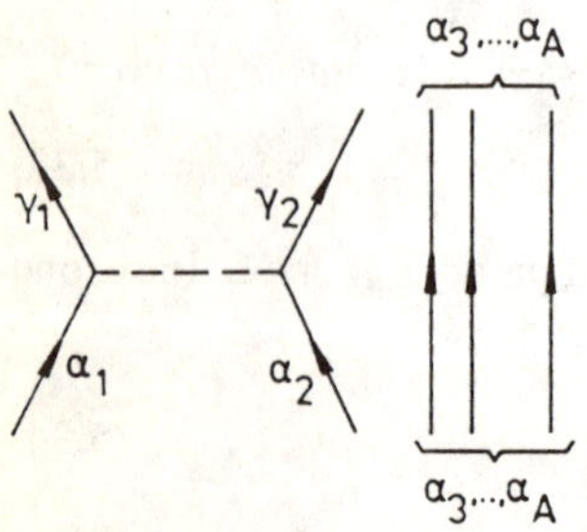

Fig. 5.2. Illustration of the two-body matrix element taken between the A-particle state $|\alpha_1, \alpha_2, \ldots, \alpha_A\rangle$ as obtained from (5.38). Because we represent a non-diagonal element, only a single process (with the two-body operator represented by the diagram $> - - <$) scattering the state α_1, α_2 into the states γ_1, γ_2 remains. The $A-2$ nucleon state remains unaffected by the two-body operator

Having a two-body operator in coordinate space, it is always possible to give the second quantized form corresponding to it by evaluating all two-body matrix elements. It is also possible by using second quantization, to define an interaction immediately in second quantization. Such an interaction is, e.g., the pairing interaction which will be discussed in more detail in Chap. 7. For the pairing interaction, there is no easy way to get a corresponding expression in coordinate space. We shall comment on this point here.

A constant pairing interaction (Rowe 1970) in second quantization reads

$$V = -G \sum_{\substack{m,m' \\ (>0)}} (-1)^{j+m}(-1)^{j+m'} a^+_{j,m} a^+_{j,-m} a_{j,-m'} a_{j,m'} , \tag{5.39}$$

and indicates that a nucleon pair is annihilated in the states $(j, m')(j, -m')$ (a 0^+ coupled pair) and scattered via the two-body interaction V into any other substate $(j, m)(j, -m)$ with constant strength G independent of the m-value. [In Chap. 7, we shall discuss the particular relevance of the phase factors appearing in (5.39).] This means fully isotropic scattering in the space of magnetic substates m. One can easily show that the interaction (5.39) only affects the $J^\pi = 0^+$ state and leaves all other $J^\pi = 2^+, 4^+, \ldots, (2j-1)^+$ states degenerate at zero energy. Thus the matrix giving the two-body matrix elements where column and row label the magnetic substate quantum number has constant elements for the pairing interaction

$$\begin{array}{c|cccc} m \backslash m' & j & j-1 & j-2 & \cdots \\ \hline j & 1 & -1 & 1 & \cdots \\ j-1 & -1 & 1 & -1 & \cdots \\ j-2 & 1 & -1 & 1 & \cdots \\ \vdots & \vdots & \vdots & \vdots & \vdots \\ & & & & \cdots \end{array} \; (-G) \, . \tag{5.40}$$

If we construct a similar matrix to (5.40) for a δ-interaction, scattering is almost isotropic but with a slight preference for diagonal over non-diagonal scattering. Speaking intuitively, the δ-interaction is the ultimate short-range interaction. We see that the pairing interaction has an even "shorter" range by inspecting the scattering matrix in the space of the different magnetic substate quantum numbers. For very low multipole forces, e.g.,

$$V(1,2) = r_1^2 r_2^2 P_2(\cos\theta_{12}) , \tag{5.41}$$

one can show (see problem set) that the scattering matrix equivalent to (5.40) is almost diagonal for large j-values.

In Fig. 5.3, we compare the δ- and pairing force spectrum for a $(j)^2 J$ configuration. We also compare in Fig. 5.4 the scattering in m-space for the pairing force compared to the low multipole (quadrupole) force.

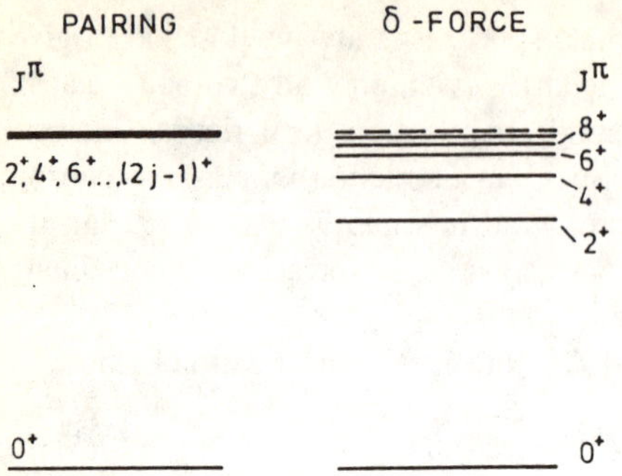

Fig. 5.3. Comparison of the two-particle spectra $(j)^2 J$ for a pairing force [defined in second quantization in (5.39)] and a δ-force

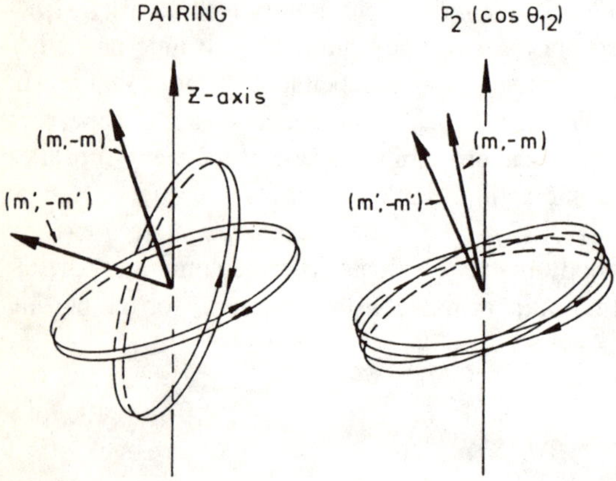

Fig. 5.4. Comparison of the scattering in the space of magnetic substates (in m-space) [the matrix of (5.40)] for a pure pairing force and a low-multipole force [e.g. a P_2 $(\cos\theta_{12})$ force]. In case (*i*), scattering is fully isotropic i.e. the probability for scattering from a state $(m, -m)$ to any state $(m', -m')$ is equal, independent of m and m'. In case (*ii*), scattering is predominantly in the "forward direction", which means that the final pair $(m', -m')$ obtained from $(m, -m)$ is such that $m' \cong m$

5.3 Angular Momentum Coupling in Second Quantization

In the above discussion, we showed how to approximate the many-particle states in describing the motion of A nucleons in the nuclear average field. As was shown, however, one needs coupled angular momentum states. The same methods as in Chap. 4 can also be used here. Because of the anticommutation properties of the creation and annihilation operators, the antisymmetry aspects that were demanded by the Pauli principle in coordinate space are automatically included. For two particles, one can construct the state

$$|j_a j_b; JM\rangle = N_{ab} \sum_{m_a, m_b} \langle j_a m_a, j_b m_b | JM\rangle a^+_{j_b m_b} a^+_{j_a m_a} |\rangle \,. \tag{5.42}$$

The condition for normalization then determines N_{ab}. Performing this calculation in detail, one has

$$\begin{aligned} 1 &= \langle j_a j_b; JM | j_a j_b; JM \rangle \\ &= N_{ab}^2 \sum_{m_a, m_b, m'_a, m'_b} \langle j_a m_a, j_b m_b | JM \rangle \langle j_a m'_a, j_b m'_b | JM \rangle \\ &\quad \times \langle \, | a_{j_a m'_a} a_{j_b m'_b} a^+_{j_b m_b} a^+_{j_a m_a} | \rangle \,. \end{aligned} \tag{5.43}$$

Using orthogonality properties of the Clebsch-Gordan coefficients and the fact that the matrix element $\langle | \ldots | \rangle$ in (5.43) becomes

$$\left(\delta_{m_a m'_a} \delta_{m_b m'_b} - \delta_{ab} \delta_{m_a m'_b} \delta_{m_b m'_a} \right) ,$$

one gets

$$\begin{aligned} 1 &= N_{ab}^2 \left(1 - \delta_{ab} (-1)^{j_a + j_b - J} \right) \\ &= N_{ab}^2 (1 + \delta_{ab}) \qquad (J\text{: even}) \,. \end{aligned} \tag{5.44}$$

Thus, the normalized state for two identical particles becomes

$$| j_a j_b; JM \rangle = \left(1 + \delta_{ab} \right)^{-1/2} \sum_{m_a, m_b} \langle j_a m_a, j_b m_b | JM \rangle a^+_{j_b m_b} a^+_{j_a m_a} | \rangle \,, \tag{5.45}$$

or, using the notation of tensor coupling for the two creation operators, one can also write

$$| j_a j_b; JM \rangle = - \left(1 + \delta_{ab} \right)^{-1/2} \left[\boldsymbol{a}^+_{j_a} \otimes \boldsymbol{a}^+_{j_b} \right]^{(J)}_M | \rangle \,. \tag{5.46}$$

If we include the possibility of non-identical nucleons by including the isospin quantum numbers in the particle state, one has

$$\begin{aligned} | j_a j_b; JM, TM_T \rangle &= \left(1 + \delta_{ab} \right)^{-1/2} \sum_{m_a, m_b, t_{z_a}, t_{z_b}} \langle j_a m_a, j_b m_b | JM \rangle \\ &\quad \times \langle \tfrac{1}{2} t_{z_a}, \tfrac{1}{2} t_{z_b} | TM_T \rangle a^+_{j_b m_b, 1/2 t_{z_b}} a^+_{j_a m_a, 1/2 t_{z_a}} | \rangle \,, \end{aligned} \tag{5.47}$$

or, in coupled notation

$$| j_a j_b; JM, TM_T \rangle = - \left(1 + \delta_{ab} \right)^{-1/2} \left[\boldsymbol{a}^+_{j_a, 1/2} \otimes \boldsymbol{a}^+_{j_b, 1/2} \right]^{(J,T)}_{M, M_T} | \rangle \,, \tag{5.48}$$

We can now use the above method to easily construct coupled (and antisymmetrized) 3, 4, … n particle states, by coupling the creation operators and calculating the overlap and norms of the different possible states. For three-particle states, one can construct the states

$$| (j_a j_b) J_1 j_c; JM \rangle \,, \tag{5.49}$$

as a notation for

$$\begin{aligned} | (j_a j_b) J_1 j_c; JM \rangle &= \mathcal{N} \sum_{m_a, m_b, M_1, m_c} \langle j_a m_a, j_b m_b | J_1 M_1 \rangle \langle J_1 M_1, j_c m_c | JM \rangle \\ &\quad \times a^+_{j_c m_c} a^+_{j_b m_b} a^+_{j_a m_a} | \rangle \,. \end{aligned} \tag{5.50}$$

In calculating the norms of the states (5.49, 50) and the overlaps, the non-vanishing states will give the cfp coefficients for the three particle states that were discussed in Sect. 3.3 (see problem set).

Not only the basis states but also the one- and two-body operators can be represented in an angular momentum coupled form. We show this for the two-body residual interaction that is written as

$$H_{\text{res}} = \frac{1}{4} \sum_{j_a, \ldots, j_d; m_a, \ldots, m_d} \langle j_a m_a, j_b m_b | V | j_c m_c, j_d m_d \rangle \times a^+_{j_a m_a} a^+_{j_b m_b} a_{j_d m_d} a_{j_c m_c} \,. \tag{5.51}$$

Now, by using the orthogonality of the Clebsch-Gordan coefficients, we can carry out the angular momentum coupling in *both* the matrix element and the operators, and using the definition $\tilde{a}_{j,m} \equiv (-1)^{j+m} a_{j,-m}$ (Sect. 5.4), one gets for the residual interaction Hamiltonian

$$H_{\text{res}} = -\frac{1}{4} \sum_{j_a, \ldots, j_d} \langle j_a j_b; JM | V | j_c j_d; JM \rangle \left((1+\delta_{ab})(1+\delta_{cd}) \right)^{1/2} \hat{J} \times \left[\left[\boldsymbol{a}^+_{j_a} \otimes \boldsymbol{a}^+_{j_b} \right]^{(J)} \otimes \left[\tilde{\boldsymbol{a}}_{j_c} \otimes \tilde{\boldsymbol{a}}_{j_d} \right]^{(J)} \right]^{(0)}_0 , \tag{5.52}$$

where, indeed, a scalar operator (coupling to tensor rank 0) results. Extension to include isospin is straightforward.

5.4 Hole Operators in Second Quantization

If we consider a fully occupied single j-shell (with $2j + 1$ particles, since all magnetic substates $m = -j, \ldots, m = +j$ are occupied) a closed shell has total angular momentum $J = 0$ and $M = 0$. If we now consider the excitation of one particle out of the occupied j-shell into a j'-orbital, a particle-hole excitation is formed with respect to the fully occupied j-shell (Fig. 5.5). There now exists a method of using specific particle and hole creation operators. This allows using a more economic method than considering a $[(j)^{2j}\ J = j, j']\ 2j + 1$ particle configuration (de-Shalit, Talmi 1963, Rowe 1970). We now define the closed j-shell state as

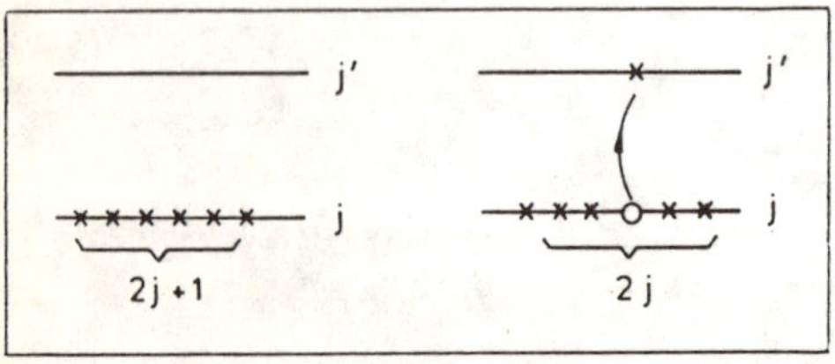

Fig. 5.5. Schematic representation of a full shell ($2j + 1$ particles) and a hole state (identical to a $2j$ particle state) in the orbital j (here, a particle needs to become excited into another configuration j' *if* conservation of particle number is fulfilled)

$$|0\rangle \equiv |j^{2j+1}; J = 0, M = 0\rangle \ , \tag{5.53}$$

and

$$|(jm)^{-1}\rangle \propto |j^{2j}; J = j, M = -m\rangle \ , \tag{5.54}$$

indicating that taking a particle out of the magnetic substate m, a $2j$-particle state remains with magnetic quantum number $M = -m$. We of course need to have a better definition of the hole state.

We know that the particle creation operator acting on a vacuum state $|\,\rangle$, $a^+_{j,m}|\,\rangle$ transforms as a spherical tensor of rank j such that

$$\boldsymbol{R} a^+_{j,m}|\,\rangle = \sum_{m'} D^{(j)}_{m',m}(R) a^+_{j,m'}|\,\rangle \ . \tag{5.55}$$

One can easily prove that a full shell such as the state $|0\rangle$ with

$$|0\rangle \equiv a^+_{j,j} a^+_{j,j-1} \cdots a^+_{j,-j}|\,\rangle \ , \tag{5.56}$$

is invariant under rotation ($J = 0$, $M = 0$ state) and thus

$$\boldsymbol{R}|0\rangle = |0\rangle \ . \tag{5.57}$$

One can now ask for the transformation properties of the state $a_{j,m}|0\rangle$. This leads to a more tedious calculation (see problem set) but one finally obtains

$$\boldsymbol{R} a_{j,m}|0\rangle = \sum_{m'} (-1)^{m-m'} D^{(j)}_{-m',-m}(R) a_{j,m'}|0\rangle \ , \tag{5.58}$$

or

$$\boldsymbol{R}(-1)^{j+m} a_{j,-m}|0\rangle = \sum_{m'} D^{(j)}_{m',m}(R)(-1)^{j+m'} a_{j,-m'}|0\rangle \ . \tag{5.59}$$

Thus, the operator $(-1)^{j+m}\, a_{j,-m}$ transforms as the creation operator $a^+_{j,m}$ (Brown 1964).

We then define as the hole state

$$|j^{-1}, m\rangle \equiv (-1)^{j+m} a_{j,-m}|0\rangle \ , \tag{5.60}$$

or, in second quantization notation for the hole creation operator

$$\tilde{a}_{j,m} \equiv (-1)^{j+m} a_{j,-m} \ . \tag{5.61}$$

Thus, $\tilde{a}_{j,m}$ and $a^+_{j,m}$ both transform as spherical tensors of rank j under rotation and can thus be coupled in the same way as the angular momentum eigenvectors or spherical tensor operators using Clebsch-Gordan coefficients. One can easily show that creating a hole and again creating a particle in the same state (j, m) leads to the closed shell. Formally, one has

$$|0\rangle = \sum_m \langle jm, j-m|00\rangle a^+_{j,m}\tilde{a}_{j,-m}|0\rangle$$
$$= \sum_m \frac{1}{\sqrt{2j+1}} a^+_{j,m} a_{j,m}|0\rangle \ . \tag{5.62}$$

A general particle-hole state $|j_a j_b^{-1}; JM\rangle$ then becomes, in second quantization,

$$|j_a j_b^{-1}; JM\rangle = \sum_{m_a, m_b} \langle j_a m_a, j_b m_b|JM\rangle \tilde{a}_{j_b, m_b} a^+_{j_a, m_a}|0\rangle$$
$$= -\left[\boldsymbol{a}^+_{j_a} \otimes \tilde{\boldsymbol{a}}_{j_b}\right]^{(J)}_M |0\rangle \ . \tag{5.63}$$

In the above way one can indeed economize the notation for low-lying excitations in doubly-closed shell nuclei. We take the example of ^{16}O where, in a Hartree-Fock approach, the ground state is described by a 16-particle state

$$|0\rangle \equiv |^{16}O; 0^+_{\mathrm{gs}}\rangle$$
$$= \left(a^+_{1p_{1/2,m=-1/2}} \cdots a^+_{1s_{1/2,m=-1/2}} a^+_{1s_{1/2,m=+1/2}}\right)_\pi |\,\rangle_\pi$$
$$\times \left(a^+_{1p_{1/2,m=-1/2}} \cdots a^+_{1s_{1/2,m=-1/2}} a^+_{1s_{1/2,m=+1/2}}\right)_\nu |\,\rangle_\nu \ . \tag{5.64}$$

A particle-hole excited state with a particle in the orbital $1d_{5/2,m}$ and a hole in the $1p_{1/2,+1/2}$ shell model state can then be written, for a proton $p-h$ excitation,

$$\left(a^+_{1d_{5/2,m}} \tilde{a}_{1p_{1/2,+1/2}}\right)_\pi |0\rangle_\pi \ , \tag{5.65}$$

and we do not need to write the 16 operators again (Fig. 5.6).

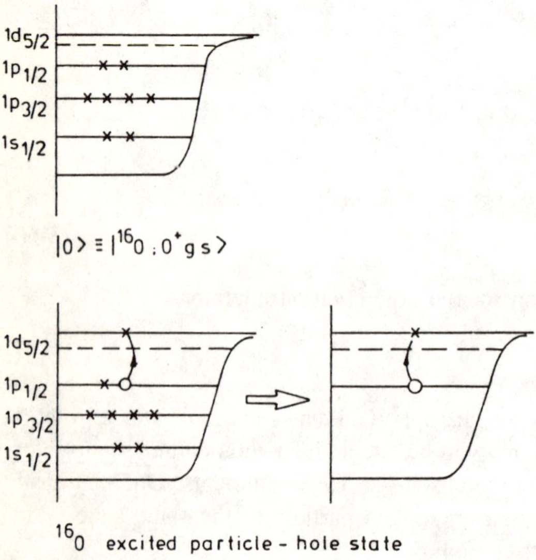

Fig. 5.6. Representation of the ^{16}O ground-state distribution of 8 nucleons (protons and neutrons) over the lowest $1s_{1/2}$, $1p_{3/2}$, $1p_{1/2}$ orbitals (5.64). The state is indicated by $|0\rangle$. In the lower part (*left-hand side*), an excited 16 particle state is shown relative to the vacuum state of nucleons. The same state can be represented (*right-hand side*) in a much simpler way as a $1p-1h$ $1p^{-1}_{1/2}\,1d_{5/2}$ configuration, relative to the new vacuum state $|0\rangle$

Including isospin in the formalism we have in exactly the same way the operator

$$\tilde{a}_{j,m;1/2,t_z} \equiv (-1)^{j+m+1/2+t_z} a_{j,-m;1/2,-t_z} , \tag{5.66}$$

and the operator in (5.66) acting on the closed shell produces a hole state

$$\tilde{a}_{j,m;1/2,t_z}|0\rangle = |j^{-1}, m; 1/2^{-1}, t_z\rangle . \tag{5.67}$$

If we now consider particle-hole excitations that can be both a proton $p-h$ and a neutron $p-h$ excitation, we can form specific linear combinations by vector coupling in isospin.

Using the definition of (5.66, 67), a proton $p-h$ state becomes

$$\begin{aligned} |ph^{-1}\rangle_\pi &= \frac{1}{\sqrt{2}}\{-|ph^{-1};T=1\rangle + |ph^{-1};T=0\rangle\} , \\ |ph^{-1}\rangle_\nu &= \frac{1}{\sqrt{2}}\{|ph^{-1};T=1\rangle + |ph^{-1};T=0\rangle\} , \end{aligned} \tag{5.68}$$

and the symmetric and antisymmetric combinations of (5.68) give

$$\begin{aligned} |ph^{-1};T=0\rangle &= \frac{1}{\sqrt{2}}\{|ph^{-1}\rangle_\pi + |ph^{-1}\rangle_\nu\} , \\ |ph^{-1};T=1\rangle &= \frac{1}{\sqrt{2}}\{|ph^{-1}\rangle_\nu - |ph^{-1}\rangle_\pi\} . \end{aligned} \tag{5.69}$$

This is in agreement with the generalized Pauli principle: a symmetric spatial-spin wave function in (5.69) is a $T = 0$ (antisymmetric in charge space) state and vice-versa. We shall see in Chap. 6 when studying low-lying excited states in doubly-closed shell nuclei that the lowest states are the $T = 0$ states and the $T = 1$ levels are the higher-lying ones.

5.5 Normal Ordering, Contraction, Wick's Theorem

In the foregoing discussion we used the notation $|\,\rangle$ for the real vacuum state and $|0\rangle$ for a closed-shell reference state. In the latter we can form particle-hole excitations (Fig. 5.7).

Let us call $h_1, h_2, h_3, \ldots$ the quantum numbers of the occupied states in $|0\rangle$ and $p_1, p_2, p_3, \ldots$ the quantum numbers of the unoccupied states relative to the reference state $|0\rangle$. We then define new operators such that

$$\begin{aligned} \xi_p^+ &\equiv a_p^+ , \\ \xi_p &\equiv a_p , \\ \xi_h^+ &\equiv \tilde{a}_h , \\ \xi_h &\equiv \tilde{a}_h^+ . \end{aligned} \tag{5.70}$$

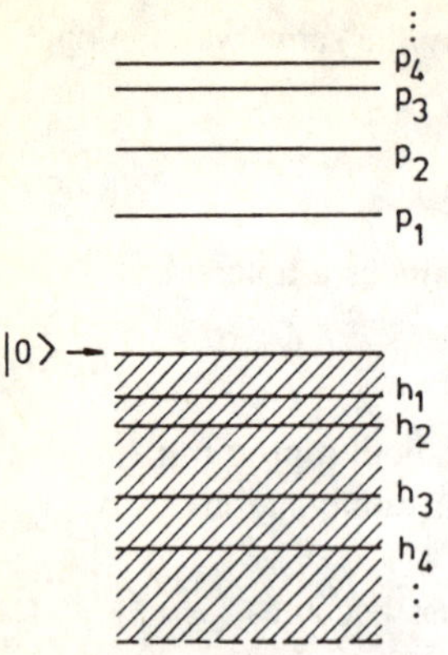

Fig. 5.7. Separation of particle configurations $(p_1, p_2, \ldots, p_i, \ldots)$ above the Fermi level and hole configurations $(h_1, h_2, \ldots, h_j, \ldots)$ below the Fermi level. The reference state $|0\rangle$ corresponds to the filling up to a sharp Fermi level with A nucleons

The operator ξ_α^+ (with α either p or h) now creates a particle (for unoccupied states) and a hole (for occupied states), and, in all cases, we have

$$\xi_\alpha|0\rangle = 0\,, \qquad \text{(for all } \alpha\text{)} \tag{5.71}$$

$$\langle 0|\xi_\alpha^+ = 0\,. \tag{5.72}$$

The anticommutation relations now remain for the new operators ξ_α^+ or

$$\begin{aligned} \{\xi_\alpha^+, \xi_\beta^+\} &= 0\,, \\ \{\xi_\alpha, \xi_\beta\} &= 0\,, \\ \{\xi_\alpha^+, \xi_\beta\} &= \delta_{\alpha\beta}\,. \end{aligned} \tag{5.73}$$

i) Normal Product (Fetter, Walecka 1971). A product of operators is in "normal" order relative to a given reference state $|0\rangle$ when *all* creation operators are at the left of the annihilation operators. In bringing a product of operators to the normal order, we often have to carry out a number of permutations. One does not, however, consider possible Kronecker δ-symbols when interchanging appropriate creation and annihilation operators. The sign becomes ± 1 according to the nature of the permutation needed to bring the original operator product in normal ordered form. As a notation one uses $N(ABC\ldots)$ or $:ABC\ldots:$.

As an example:

$$\begin{aligned} N\left(a_{h_1}^+ a_{p_1}^+ a_{p_2} a_{h_2}\right) &= N\left(\tilde{\xi}_{h_1} \xi_{p_1}^+ \xi_{p_2} \tilde{\xi}_{h_2}^+\right) \\ &= -\xi_{p_1}^+ \tilde{\xi}_{h_2}^+ \tilde{\xi}_{h_1} \xi_{p_2} \\ &= -a_{p_1}^+ a_{h_2} a_{h_1}^+ a_{p_2}\,, \end{aligned} \tag{5.74}$$

One clearly sees that the result depends on the choice of the reference state.

An interesting outcome is that

$$\langle 0|N(ABC\ldots)|0\rangle = 0\,. \tag{5.75}$$

ii) Contraction (Fetter, Walecka 1971). The contraction of two operators is the expectation value of the operator product with respect to the reference state. The result is a pure number and we use the notation AB. Thus, one has

$$\overline{a^+_\alpha a_\beta} \equiv \langle 0|a^+_\alpha a_\beta|0\rangle \ , \tag{5.76}$$

$$\overline{a^+_\alpha a^+_\beta a_\gamma} a_\delta \equiv -\langle 0|a^+_\alpha a_\gamma|0\rangle a^+_\beta a_\delta \ , \tag{5.77}$$

and the permutation factor is needed, ± 1 for even or odd permutation, in order to bring the contracted operators together. One can also have contractions within a normal product of operators, i.e.,

$$N\left(\overline{a^+_\alpha a^+_\beta a_\gamma} a_\delta\right) = -\langle 0|a^+_\alpha a_\gamma|0\rangle N\left(a^+_\beta a_\delta\right) \ , \tag{5.78}$$

and

$$N\left(\overline{a^+_\alpha a^+_\beta a^+_\gamma a_\delta a_\varepsilon} a_\zeta\right) = -\langle 0|a^+_\alpha a_\delta|0\rangle \langle 0|a^+_\beta a_\varepsilon|0\rangle N\left(a^+_\gamma a_\zeta\right) . \tag{5.79}$$

The contractions between two creation or two annihilation operators $a^+_\alpha a^+_\beta$ and $a_\alpha a_\beta$ become zero for the sharp reference state as defined via Fig. 5.7. One could, however, think of even more generalized reference states where the above contractions become non-vanishing. Such a reference state, denoted by $|\tilde{0}\rangle$, will imply a distribution in the occupation over the various orbitals and, thus, a precise difference between 'particle' and 'hole' excitation is replaced by a 'quasi-particle' excitation. The operators, generalizing th ξ^+_α and ξ_α operators, such that $\xi_\alpha|\tilde{0}\rangle = 0$ remains fulfilled, become the 'quasi-particle' operators defined by the relations

$$\begin{aligned} \xi^+_\alpha &= u_a a^+_\alpha + v_a \tilde{a}_\alpha \\ \xi_\alpha &= u_a a_\alpha + v_a \tilde{a}^+_\alpha \ . \end{aligned} \tag{5.80}$$

A more detailed discussion of the interpretation of the amplitudes u_a, v_a, and how to optimize these quantities for a given nucleus, is given in Chap. 7.

We are now in a position to formulate Wick's theorem.

iii) Wick's Theorem (Fetter, Walecka 1971). A product of a number of operators can be written as the sum of all contracted normal ordered products of the operators (considering partially and fully contracted terms, as well as the uncontracted ones).

Thus we have

$$\begin{aligned} A_1 A_2 A_3 \ldots A_N = {} & N(A_1, A_2, A_3 \ldots A_N) \\ & + \sum_{\alpha<\beta} N\left(A_1 A_2 \ldots \overline{A_\alpha \ldots A_\beta} \ldots A_N\right) \\ & + \sum_{\substack{\alpha<\beta \\ \gamma<\delta}} N\left(A_1 A_2 \ldots \overline{A_\alpha \ldots A_\gamma \ldots A_\beta \ldots A_\delta} \ldots A_N\right) \\ & + \ldots \\ & + \text{fully contracted terms.} \end{aligned} \tag{5.81}$$

As an example, we give

$$A_\alpha A_\beta = N\left(A_\alpha A_\beta\right) + \overline{A_\alpha A_\beta}\ . \tag{5.82}$$

We do not give a proof but refer the reader to (March et al. 1967).

Relative to the reference state $|0\rangle$ we also have

$$\overline{a^+_\alpha a^+_\beta} = \overline{a_\alpha a_\beta} = 0\ . \tag{5.83}$$

5.6 Application to the Hartree-Fock Formalism

We start from the one- and two-body A-particle Hamiltonian describing the kinetic energy and two-body interactions of A fermions in a nucleus as

$$H = \sum_{\alpha,\gamma} \langle\alpha|T|\gamma\rangle a^+_\alpha a_\gamma + \tfrac{1}{4} \sum_{\alpha,\beta,\gamma,\delta} \langle\alpha\beta|V|\gamma\delta\rangle_{\text{nas}} a^+_\alpha a^+_\beta a_\delta a_\gamma\ . \tag{5.84}$$

Applying Wick's theorem to the Hamiltonian one has,

$$\begin{aligned} H = &\sum_{\alpha,\gamma} \langle\alpha|T|\gamma\rangle \left\{ N\left(a^+_\alpha a_\gamma\right) + \overline{a^+_\alpha a_\gamma} \right\} \\ &+ \tfrac{1}{4} \sum_{\alpha,\beta,\gamma,\delta} \langle\alpha\beta|V|\gamma\delta\rangle_{\text{nas}} \left\{ N\left(a^+_\alpha a^+_\beta a_\delta a_\gamma\right) \right. \\ &+ N\left(a^+_\alpha a_\gamma\right) \overline{a^+_\beta a_\delta} - N\left(a^+_\alpha a_\delta\right) \overline{a^+_\beta a_\gamma} + \ldots \\ &\left. + \overline{a^+_\alpha a_\gamma}\, \overline{a^+_\beta a_\delta} - \overline{a^+_\alpha a_\delta}\, \overline{a^+_\beta a_\gamma} \right\}\ . \end{aligned} \tag{5.85}$$

The contractions disappear, except in those cases $a^+_\alpha a_\gamma$ where $\alpha = \gamma$ and α, γ describe occupied states. If, moreover, we keep track of the antisymmetry of the two-body matrix elements, i.e., $\langle\alpha\beta|V|\gamma\delta\rangle_{\text{nas}} = -\langle\alpha\beta|V|\delta\gamma\rangle_{\text{nas}} = -\langle\beta\alpha|V|\gamma\delta\rangle_{\text{nas}} = \langle\beta\alpha|V|\delta\gamma\rangle_{\text{nas}}$, and calling $h, h', h'', \ldots$ the quantum number of occupied states in the reference state $|0\rangle$, one can rewrite the Hamiltonian as

$$\begin{aligned} H = &\sum_{h} \langle h|T|h\rangle + \tfrac{1}{2} \sum_{h,h'} \langle hh'|V|hh'\rangle_{\text{nas}} \\ &+ \sum_{\alpha,\gamma} \left\{ \langle\alpha|T|\gamma\rangle + \sum_{h} \langle\alpha h|V|\gamma h\rangle_{\text{nas}} \right\} N\left(a^+_\alpha a_\gamma\right) \\ &+ \tfrac{1}{4} \sum_{\alpha,\beta,\gamma,\delta} \langle\alpha\beta|V|\gamma\delta\rangle_{\text{nas}} N\left(a^+_\alpha a^+_\beta a_\delta a_\gamma\right)\ . \end{aligned} \tag{5.86}$$

If we now take the expectation value of H relative to the reference state $|0\rangle$ of occupied orbitals $h, h', h'', \ldots$, the normal products make no contribution and one has the energy E_0

$$E_0 \equiv \langle 0|H|0\rangle = \sum_h \langle h|T|h\rangle + \tfrac{1}{2}\sum_{h,h'} \langle hh'|V|hh'\rangle_{\text{nas}} \, . \tag{5.87}$$

For an appropriate choice of the basis states $|\alpha\rangle, \ldots$, the single-particle Hamiltonian can be diagonalized and one obtains (we consider the single-particle energy ε_a to be independent of the magnetic quantum number m_a)

$$\langle\alpha|T|\gamma\rangle + \sum_h \langle\alpha h|V|\gamma h\rangle_{\text{nas}} = \varepsilon_a \delta_{\alpha\gamma} \, , \tag{5.88}$$

and

$$\varepsilon_a = \langle\alpha|T+U|\alpha\rangle \, , \tag{5.89}$$

with the one-body average field defined as

$$\langle\alpha|U|\alpha\rangle \equiv \sum_h \langle\alpha h|V|\alpha h\rangle_{\text{nas}} \, . \tag{5.90}$$

It is interesting to note that the energy of the reference state E_0 is not equal to the sum of the occupied single-particle energies i.e. $E_0 \neq \sum_h \varepsilon_h$ but, correctly, one obtains the relation

$$E_0 = \sum_h \varepsilon_h - 1/2. \sum_{h,h'} \langle h,h'|V|h,h'\rangle_{\text{nas}} \, . \tag{5.91}$$

From the self-consistent definition of the single-particle energy ε_h in Eq. (5.89), the two-body interactions with all other nucleons are taken into account. So, the summation $\sum_h \varepsilon_h$ involves a double-counting of the two-body interactions.

In this way, an average field is defined in terms of the two-body interaction according to (5.90). One can finally rewrite the Hamiltonian as

$$H = E_0 + \sum_\alpha \varepsilon_a N\left(a^+_\alpha a_\alpha\right) + \tfrac{1}{4}\sum_{\alpha,\beta,\gamma,\delta} \langle\alpha\beta|V|\gamma\delta\rangle_{\text{nas}} N\left(a^+_\alpha a^+_\beta a_\delta a_\gamma\right) , \tag{5.92}$$

and the total Hamiltonian has thus been separated into a core term E_0, the energy of the reference state $|0\rangle$, the single-particle energy contributions given by the ε_a, and the residual interaction given by the normal ordered product $N\left(a^+_\alpha a^+_\beta a_\delta a_\gamma\right)$. In spherically symmetric nuclei where the single-particle energy ε_α is independent of the orientation of the angular momentum j_a, we denote the single-particle energy as ε_a $\left(a = n_a, l_a, j_a \text{ and } \alpha \equiv \{a, m_a\}\right)$.

This application shows the power of Wick's theorem in separating the Hamiltonian into different terms. The Hamiltonian (5.92) can now serve as a starting point in describing the elementary excitation modes in (i) closed-shell nuclei and (ii) nuclei with a number of valence nucleons in open shells. We shall now study these two cases in some detail in Chaps. 6 and 7, respectively. There, the methods of second quantization will be fully used. We shall also study both exactly solvable, approximate models as well as realistic cases using modern, effective interactions. In Chap. 8 we shall discuss state-of-the-art shell-model calculations in a way similar to that in Chaps. 6, 7 but using only the Skyrme interaction.

5.7 Summary

Even though most of the formalism for the shell model has been presented using nucleon coordinates, explictly incorporating the Pauli principle through the construction of antisymmetrized many-particle wave functions, a more economical 'representation' can be used to study the atomic nucleus and its structure.

The second quantization formalism or the occupation number formalism allows for a concise formulation of the many-nucleon properties (wave functions, observables, operators,...) in terms of creation- and annihilation operators. The basic anticommutation properties $\{a_\alpha^+, a_\beta^+\} = \{a_\alpha, a_\beta\} = 0$ and $\{a_\alpha^+, a_\beta\} = \delta_{\alpha\beta}$ are derived.

Both one- and two-body operator expressions using second quantization are derived and this method is used to discuss certain operators, such as the pairing force operator, in second quantization form directly.

Angular momentum coupling for the many-body states in second quantized form is presented. An illustration of the method by evaluating the norm of two- and three-particle state vectors is given. A second quantized and angular momentum coupled form of the two-body residual interaction is also discussed.

The concept of "particle" and "hole" operators, relative to a reference vacuum state which then corresponds to a filling of orbitals with a sharp Fermi distribution, is presented. The importance of constructing hole operators with a correct spherical tensor character is stressed. Some applications to the doubly-closed shell nucleus ^{16}O are given.

Finally, a number of concepts relative to evaluating general expectation values of operators in second quantization relative to many-body states are presented. Normal ordering, contraction and Wick's theorem are given, but without detailed derivations. The practical use of the above concepts is illustrated by rewriting a general one- and two-body Hamiltonian such that a constant, reference (Hartree-Fock closed shell) energy, a one-body Hamiltonian and two-body residual interactions become clearly separated. This Hamiltonian will be the starting point for a study of (i) elementary modes of excitation inclosed-shell nuclei (Chapt. 6) and, (ii) pairing correlations in open-shell nuclei (Chap. 7).

Problems

5.1 Calculate the energy spectrum for the two-particle configurations $|j^2; JM\rangle$ (for $J = 0, 2, \ldots, 2j-1$) starting from the pairing interaction (5.39). Discuss the results and compare these with the energy spectrum resulting from a δ-interaction.

*5.2 Derive the matrix (5.40) for a δ-force interaction and also for a $P_2(\cos\theta_{12})$ force. Show that for the latter force in the limit of large $j(j \to \infty)$, forward scattering predominantly results which is equivalent to the dominance of $m' = m$ scattering.

5.3 Derive the angular momentum coupled form of the two-body interaction of (5.52) in second quantization.

5.4 Determine the particle-hole two-body matrix elements

$$\langle ph^{-1}; JM|H_{\text{res}}|ph^{-1}; JM\rangle$$

in terms of the particle-particle two-body matrix elements.

5.5 Show that the quadrupole moment for a hole state and for a particle state have opposite sign but the same absolute value (neglecting any core contribution). Show that the dipole moment remains identical for a particle or a hole configuration.

5.6 Decompose the uncoupled proton and neutron particle-hole excitations $|ph^{-1}\rangle_\pi$ and $|ph^{-1}\rangle_\nu$ in the isospin coupled basis with $T = 1$ and $T = 0$ states.
Hint: make a use of the relation $\tilde{a}_{1/2,t_z} = (-1)^{1/2+t_z} a_{1/2,-t_z}$, and the values of the particular Clebsch-Gordan coefficients $\langle 1/2\,1/2, 1/2 - 1/2|00\rangle = \langle 1/2\,1/2, 1/2 - 1/2|10\rangle = 1/\sqrt{2}$, $\langle 1/2 - 1/2, 1/2\,1/2|00\rangle = -1/\sqrt{2}$, $\langle 1/2 - 1/2; 1/2\,1/2|10\rangle = 1/\sqrt{2}$.

5.7 Prove that $E_0 \neq \sum \varepsilon_h$ but equal to $\sum \varepsilon_h - 1/2. \sum \langle hh'|V|hh'\rangle_{\text{nas}}$.

*5.8 Prove that the $n \to n-1$ cfp coefficients are related to the reduced matrix elements according to the expression:

$$\left[j^{n-1}(\alpha_1 J_1)\, jJ|\}j^n \alpha J\right] = \frac{\langle j^n, \alpha J\|a_j^+\|j^{n-1}, \alpha_1 J_1\rangle}{\sqrt{n}\sqrt{2J+1}}$$

*5.9 Derive the norm of the three-particle configuration $|j^3; JM\rangle$ using the occupation number representation. Discuss the relation with the $3 \to 2\, cfp$ coefficients as derived earlier in Sect. 3.3.

5.10 Prove that the particle number operator commutes with the nuclear one- plus-two-body Hamiltonian i.e. $[H, \hat{N}] = 0$, indicating that particle number is a conserved quantity.

5.11 Compare the particle-hole proton-neutron $|j_p j_n^{-1}; JM\rangle$ and the particle-particle proton-neutron $|j_p j_n; JM\rangle$ spectrum using a zero-range two-body proton-neutron interaction. How are the results modified if the force is taken to be a pure quadrupole-quadrupole force?

5.12 The annihilation and creation operators for a nucleon moving in infinite nuclear matter are characterized by their momentum $\boldsymbol{k}$ and intrinsic spin $m(\pm 1/2)$ and denoted by $a_{\boldsymbol{k},m_s}$ and $a^+_{\boldsymbol{k},m_s}$. They obey the standard fermion anti-commutation relations.
In the theory of superconductivity in nuclear matter (BCS theory), one can define correlated pairs via the operators

$$c_{\boldsymbol{k}} \equiv a_{-\boldsymbol{k},-1/2} a_{\boldsymbol{k},+1/2} \quad \text{and} \quad c^+_{\boldsymbol{k}} = a_{\boldsymbol{k},+1/2} a^+_{-\boldsymbol{k},-1/2} \,.$$

Show that the latter operators obey the commutation relations $[c_{\boldsymbol{k}}, c_{\boldsymbol{k}'}] = [c^+_{\boldsymbol{k}}, c^+_{\boldsymbol{k}'}] = 0$. Determine the commutator $[c_{\boldsymbol{k}}, c^+_{\boldsymbol{k}'}]$.

5.13 Show that when 'creating' a hole (annihilation of a particle) in a $|j^{2j+1}; J = 0\, M = 0\rangle$ configuration, the operator which can be used in angular momentum coupling as a spherical tensor operator with rank j and projection m is given by $\tilde{a}_{j,m} = (-1)^{j\pm m} a_{j,-m}$.

6. Elementary Modes of Excitation: Particle-Hole Excitations at Closed Shells

6.1 General

Having derived in the final part of Chap. 5 the second quantized expression for the nuclear many-body Hamiltonian (containing one-body and two-body operators), we shall now study how elementary excitations of the doubly-closed shells, via particle-hole excitations, can give rise to the low-lying excited states that have been observed in these nuclei. In this chapter we shall describe the basic methods of treating the residual interactions, induced by the Hamiltonian of (5.92), in doubly-closed shell nuclei. In addition to approximation methods that highlight the salient features of these elementary excitation modes (exactly solvable models), we shall also discuss a realistic application to the case of ^{16}O as a doubly-closed shell nucleus.

We start from the Hamiltonian of (5.92),

$$H = E_0 + \sum_{\alpha} \varepsilon_a N\left(a_\alpha^+ a_\alpha\right) + \frac{1}{4} \sum_{\alpha,\beta,\gamma,\delta} \langle \alpha\beta|V|\gamma\delta\rangle N\left(a_\alpha^+ a_\beta^+ a_\delta \alpha_\gamma\right) , \tag{6.1}$$

where, for the remaining discussion, we shall leave out the reference energy E_0 corresponding to the static Hartree-Fock energy of the doubly-closed shell nucleus. As a basis for treating the residual interaction [the third term of (6.1)], we can construct the zero-order wave function of the diagonal part of H. For a doubly-closed shell nucleus, this basis consists of all $1p-1h$, $2p-2h$, $3p-3h$, ..., $np-nh$ configurations (Fig. 6.1) which we generally call the zero-order eigenstates ψ_β.

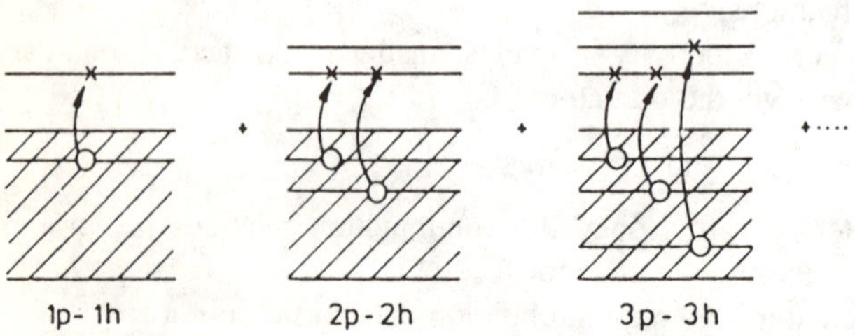

Fig. 6.1. Hierarchy of excitations, relative to a closed-shell nucleus, that form the basis for expanding the actual wave functions $\Psi_\alpha(J^\pi)$. The states are divided according to the number of $1p-1h$ components: $1p-1h$, $2p-2h$, $3p-3h$, ... (both for protons and neutrons). In this way, a classification according to unperturbed energy is made at the same time that can serve as a guide in truncating the model space

Because of the normal ordering part $N(a_\alpha^+ a_\alpha)$, the eigenvalue for a $1p-1h$ state is $\varepsilon_p - \varepsilon_h$, for a $2p-2h$ state $\varepsilon_p + \varepsilon_p{}' - (\varepsilon_h + \varepsilon_h{}')$, etc. We then obtain the secular equation for determining the total wave functions of H by expanding in the basis ψ_β as

$$\Psi_\alpha = \sum_\beta c_\beta^{(\alpha)} \psi_\beta \,. \tag{6.2}$$

We should note that here $\{\alpha, \beta, \ldots\}$ are labels for identifying the zero-order $np{-}nh$ wave functions completely. Thus for a $1p{-}1h$ state the label $\beta \equiv (p, h, J^\pi)$ or $\beta \equiv (p, h, J^\pi, T)$. Using the wave function (6.2), this secular equation becomes (Chap. 3)

$$\begin{aligned} &\sum_\beta c_\beta^{(\alpha)} \langle\gamma|H|\beta\rangle = E_\alpha c_\gamma^{(\alpha)} \qquad \text{or} \\ &\sum_\beta \left\{ c_\beta^{(\alpha)} \langle\gamma|H_0|\beta\rangle + c_\beta^{(\alpha)} \langle\gamma|H_{\text{res}}|\beta\rangle \right\} = E_\alpha c_\gamma^{(\alpha)} \,. \end{aligned} \tag{6.3}$$

Here, the part including H_0 is diagonal and becomes $E_0^{(\beta)} \cdot \delta_{\gamma\beta}$, with $E_0^{(\beta)}$ the unperturbed energy of the related $np-nh$ basis states. The eigenvalue equation can still be written as

$$\sum_\beta \left\{ \left(E_0^{(\beta)} - E_\alpha \right) \delta_{\gamma\beta} + \langle\gamma|H_{\text{res}}|\beta\rangle \right\} c_\beta^{(\alpha)} = 0 \,. \tag{6.4}$$

In principle, the dimension is infinite. Truncation effects can now effectively reduce the dimensions to a reasonable number by, e.g., restricting to $1p-1h$, $2p-2h$ configurations in light nuclei (^{16}O, ^{40}Ca, ...) and taking up to $1p-1h$ configurations only in the heavy nuclei (^{208}Pb) where the oscillator shells are much larger.

The matrix equation (determinantal equation) related to (6.4) becomes, more explicitly,

$$\begin{bmatrix} \left(E_0^{(1)} - E\right) + \langle 1|H_{\text{res}}|1\rangle & \ldots & \langle n|H_{\text{res}}|1\rangle \\ \langle 1|H_{\text{res}}|2\rangle & \ldots & \langle n|H_{\text{res}}|2\rangle \\ \vdots & \ldots & \vdots \\ \langle 1|H_{\text{res}}|n\rangle & \ldots & \left(E_0^{(n)} - E\right) + \langle n|H_{\text{res}}|n\rangle \end{bmatrix} = 0 \,. \tag{6.5}$$

Two major approximations for treating the doubly-closed shell nuclei are (Rowe 1970, Ring, Schuck 1980): i) the Tamm-Dancoff approximation (TDA); ii) the Random-Phase approximation (RPA); and we shall discuss both methods in some detail.

6.2 The TDA Approximation

If, as a ground state $|g\rangle$, we use the reference vacuum state $|0\rangle$ obtained by filling all Hartree-Fock orbitals for a given doubly-closed shell nucleus with the A nucleons present, then low-lying excited states will be obtained starting from $1p-1h$ configurations relative to this reference or vacuum state.

Thus, we start from the definitions

$$|g\rangle \equiv |0\rangle\ , \tag{6.6}$$

and

$$|mi^{-1}\rangle \equiv a^{+}_{m}a_{i}|0\rangle\ , \tag{6.7}$$

as the $1p-1h$ configurations. (We use the uncoupled basis. Angular momentum coupling makes all expressions somewhat more complicated when evaluating the necessary matrix elements, but all essential elements remain.) When carrying out angular momentum coupling, the particle-hole states

$$|mi^{-1}; JM\rangle \equiv \left[\boldsymbol{a}^{+}_{m} \otimes \tilde{\boldsymbol{a}}_{i}\right]^{(J)}_{M}|0\rangle\ , \tag{6.8}$$

are more conveniently defined by using the creation and annihilation operators $\boldsymbol{a}^{+}_{m}$ and $\tilde{\boldsymbol{a}}_{i}$, respectively. In the uncoupled representation, the basis (6.7) is more convenient since the extra phase factors $(-1)^{j_i+m_i}$ do not appear to make expressions more cumbersome.

The secular equation in the $1p-1h$ space becomes

$$H\overline{|\hbar\omega\rangle} = \hbar\omega\overline{|\hbar\omega\rangle}\ , \tag{6.9}$$

with the eigenstate $\overline{|\hbar\omega\rangle}$ at the energy $\hbar\omega$, expressed as

$$\overline{|\hbar\omega\rangle} = \sum_{m,i} X_{m,i}|mi^{-1}\rangle\ . \tag{6.10}$$

More explicitly, (6.9) leads to the equation

$$\left(\varepsilon_{mi} - \hbar\omega\right)X_{m,i} + \sum_{n,j}\langle mi^{-1}|H_{\text{res}}|nj^{-1}\rangle X_{n,j} = 0\ , \tag{6.11}$$

and

$$H_0|mi^{-1}\rangle = \varepsilon_{mi}|mi^{-1}\rangle \equiv \left(\varepsilon_m - \varepsilon_i\right)|mi^{-1}\rangle\ , \tag{6.12}$$

gives the diagonal $1p-1h$ energy denoted in the following as ε_{mi}.

We can evaluate the interaction matrix element in (6.11) by using the explicit form of the residual interaction H_{res} as

$$\begin{aligned}\langle mi^{-1}|H_{\text{res}}|nj^{-1}\rangle &= \tfrac{1}{4}\sum_{\alpha,\beta,\gamma,\delta}\langle g|a^{+}_{i}a_{m}N\left(a^{+}_{\alpha}a^{+}_{\beta}a_{\delta}a_{\gamma}\right)a^{+}_{n}a_{j}|g\rangle\langle\alpha\beta|V|\gamma\delta\rangle_{\text{nas}}\\ &= \langle mj|V|in\rangle_{\text{nas}}\ .\end{aligned} \tag{6.13}$$

Thus the secular equation for the $1p-1h$ subspace reads

$$\left(\varepsilon_{mi}-\hbar\omega\right)X_{m,i}+\sum_{n,j}\langle mj|V|in\rangle_{\text{nas}}X_{n,j}=0\,. \tag{6.14}$$

In coordinate space $(\boldsymbol{r},\boldsymbol{\sigma})$ (or even $(\boldsymbol{r},\boldsymbol{\sigma},\boldsymbol{\tau})$ when including the charge character in the isospin formalism), the matrix element $\langle mj|V|in\rangle_{\text{nas}}$ is the sum of a direct and an exchange term and can be expressed as

$$\begin{aligned}&\int\varphi_m^*(1)\varphi_j^*(2)V(1,2)\varphi_i(1)\varphi_n(2)d\mathbf{1}d\mathbf{2}\\&\quad-\int\varphi_m^*(1)\varphi_j^*(2)V(1,2)\varphi_n(1)\varphi_i(2)d\mathbf{1}d\mathbf{2}\,,\end{aligned} \tag{6.15}$$

where $\mathbf{1}$, is a notation for $\boldsymbol{r}_1$, $\boldsymbol{\sigma}_1$ $(\boldsymbol{\tau}_1)$. In diagrammatic form, the two terms of (6.15) can be given as the direct and exchange term, using the methods of Sect. 5.2. In the direct term, the given hole j^{-1} and particle state n combine to form the ground state which, through the residual interaction, is broken up again into a particle-hole state mi^{-1}. In the exchange part, the particle states m,n scatter off the hole state (i^{-1},j^{-1}) (Fig. 6.2).

Now in the actual calculations in TDA, the electromagnetic transition strength to the low-lying collective excitations is in almost all cases underestimated. This finds its origin in the asymmetry which is built into the approximation between the ground state $|g\rangle$ and the excited states $|\overline{\hbar\omega}\rangle$ since the latter contain $1p-1h$ correlations whereas $|g\rangle$ is a static reference state without any particle-hole correlations. We shall later learn how to remedy this asymmetry within the RPA.

In Sect. 6.4 we will study the specific applications of a TDA calculation in the case of ^{16}O and thus, at present, will not discuss realistic cases. A general outcome is that because of the short-range attractive character of the nucleon-nucleon interaction and since the direct and exchange matrix elements are almost equal in magnitude, the particle-hole interaction is attractive in light nuclei for the $T=0$ states and repulsive in the $T=1$ states (Rowe 1970).

This becomes clear when we consider the following. In light doubly-closed shell nuclei, proton $1p-1h$ and neutron $1p-1h$ excitations are almost degenerate in excitation energy, i.e.,

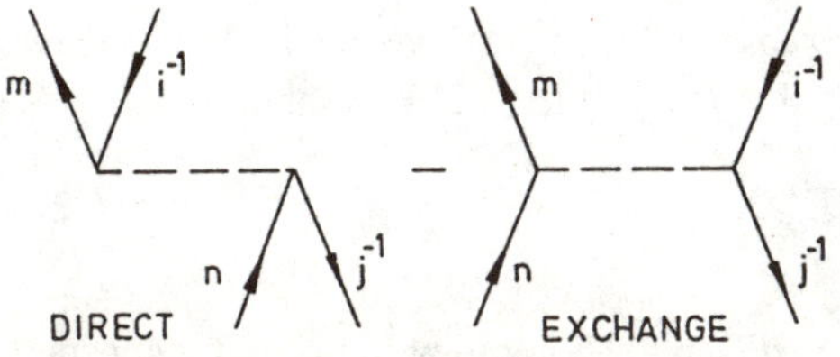

Fig. 6.2. The particle-hole matrix element $\langle mi^{-1}|H_{\text{res}}|nj^{-1}\rangle$ (6.13), separated into the direct and exchange contribution. In the direct term, a $1p-1h$ is annihilated by H_{res} and created again in another configuration (the $1p-1h$ states can even change in charge character through the action of H_{res}). In the exchange term, the particle and hole states are scattered off each other via the interaction: this is possible for equal charge character for mi^{-1} and nj^{-1} only

$$\left(\varepsilon_p - \varepsilon_h\right)_\pi \simeq \left(\varepsilon_p - \varepsilon_h\right)_\nu . \tag{6.16}$$

Having constructed the particle-hole $T = 0$ and $T = 1$ combinations in Sect. 5.4 (5.69), one has

$$|ph^{-1}\rangle \equiv \frac{1}{\sqrt{2}}\left\{|ph^{-1}\rangle_\nu \pm |ph^{-1}\rangle_\pi\right\} . \tag{6.17}$$

Assuming an attractive particle-particle interaction one obtains for the respective matrix elements

$${}_\pi\langle ph^{-1}|H_{\text{res}}|ph^{-1}\rangle_\pi = D - E ,$$

and

$${}_\pi\langle ph^{-1}|H_{\text{res}}|ph^{-1}\rangle_\nu = D , \tag{6.18}$$

where D and E stand for the direct and exchange terms, respectively. The final particle-hole interaction matrix elements for the states (6.17) read

$$\begin{aligned}\langle ph^{-1}|H_{\text{res}}|ph^{-1}\rangle_{T=0} &= 2D - E ,\\ \langle ph^{-1}|H_{\text{res}}|ph^{-1}\rangle_{T=1} &= -E .\end{aligned} \tag{6.19}$$

Since for a short-range attractive force $D \cong E$, the particle-hole interaction becomes attractive in the $T = 0$ channel and repulsive in the $T = 1$ channel ($D \cong E < 0$). These results will help us to construct and study an exactly solvable model, when we put in the above ingredients on the particle-particle and particle-hole interaction matrix elements.

Thus we discuss here the solvable model as proposed by *Brown* and *Bolsterli* (Brown, Bolsterli 1959). From (6.11), we can write that

$$\left(\hbar\omega - \varepsilon_{mi}\right)X_{m,i} = \sum_{n,j}\langle mi^{-1}|H_{\text{res}}|nj^{-1}\rangle X_{n,j} . \tag{6.20}$$

If we now assume a separable force that for the diagonal particle-hole interaction matrix elements agrees with the results of (6.19), we can write

$$\langle mi^{-1}|H_{\text{res}}|nj^{-1}\rangle = -\chi D_{m,i}D^*_{n,j} . \tag{6.21}$$

(With $\chi > 0$ we get the attractive $T = 0$ case and for $\chi < 0$ we obtain the repulsive $T = 1$ case in a simple way.) Bringing this result into (6.20) we obtain

$$\left(\hbar\omega - \varepsilon_{mi}\right)X_{m,i} = -\chi D_{m,i}\sum_{n,j} D^*_{n,j}X_{n,j} . \tag{6.22}$$

In this sum, the variables (n, j) go over *all* $1p - 1h$ configurations and a constant value (called N/χ) will result. Thus we can solve (6.22) for the $X_{m,i}$ as

$$X_{m,i} = \frac{ND_{m,i}}{\left(\varepsilon_{mi} - \hbar\omega\right)} . \tag{6.23}$$

The coefficient N can be determined through the normalization condition for the wave function expanded in the $1p-1h$ basis since

$$\sum_{m,i} |X_{m,i}|^2 = 1 \,, \qquad \text{or}$$

$$N^{-2} = \sum_{m,i} \frac{|D_{m,i}|^2}{\left(\hbar\omega - \varepsilon_{mi}\right)^2} \,. \tag{6.24}$$

A dispersion relation for the eigenvalues $\hbar\omega$ is then obtained by summing

$$\sum_{m,i} D^*_{m,i} X_{m,i} = -\sum_{m,i} \frac{\chi |D_{m,i}|^2}{\hbar\omega - \varepsilon_{mi}} \sum_{n,j} D^*_{n,j} X_{n,j} \,, \tag{6.25}$$

or,

$$\frac{1}{\chi} = \sum_{m,i} \frac{|D_{m,i}|^2}{-\hbar\omega + \varepsilon_{mi}} \,. \tag{6.26}$$

The roots $\hbar\omega$ of this equation are now the crossing points between the left hand side χ^{-1} and the right hand side which is an expression which goes to infinity at the values $\hbar\omega = \varepsilon_{mi}$ (the unperturbed $1p-1h$ energies in the system, Fig. 6.3).

If we take the degenerate case that all $1p-1h$ energies $\varepsilon_{mi} \equiv \varepsilon_m - \varepsilon_i$ are equal to ε, all eigenvalues remain degenerate at $\hbar\omega = \varepsilon$ except one:

$$\hbar\omega = \varepsilon - \sum_{m,i} \chi |D_{m,i}|^2 \,. \tag{6.27}$$

Thus, for $\chi > 0$ (attractive $p-h$ interaction), the one state that is affected in energy can become much lower in energy than the unperturbed $1p-1h$ energy (T = 0 states) whereas the case of $\chi < 0$ (repulsive $p-h$ interaction) is the one where

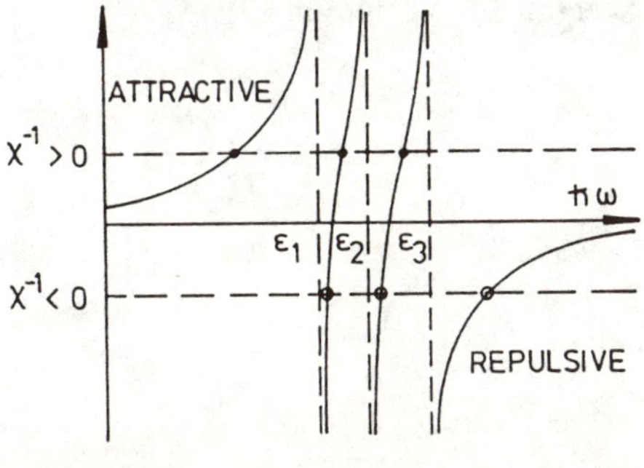

Fig. 6.3. The TDA secular equation (6.26) illustrated in a schematic way. The left-hand side of (6.26) ($1/\chi$) is a straight-line. The right-hand side represents the curve with a number of roots (where $\Sigma \ldots$ intersects the $1/\chi$ line). The unperturbed $1p-1h$ energies ε_{mi} are illustrated here as ε_1, ε_2, ε_3, ... The line ($\chi > 0$) gives the low-lying collective isoscalar state (•), the line ($\chi < 0$) the high-lying collective isovector state (o)

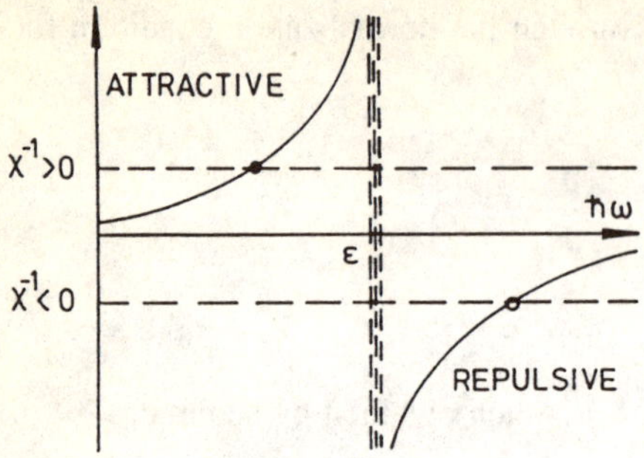

Fig. 6.4. Limit of Fig. 6.3 where all ε_{mi} become degenerate at a single point $\varepsilon_{mi} = \varepsilon$. The collective root is given (its energy $\hbar\omega$) by (6.27) and becomes attracted (•) or repelled (o) in excitation energy, starting from the ε value. All other roots remain degenerate at $\hbar\omega = \varepsilon$

the energy of the linear combination of $1p - 1h$ configurations gets pushed up in excitation considerably above the unperturbed value of ε (Fig. 6.4). In the $\chi^{-1} > 0$ (or $T = 0$) case, one can easily check from the wave functions that the proton and neutron particle-hole excitations move in phase whereas in the $\chi^{-1} < 0$ case ($T = 1$), proton-neutron particle motion is out of phase (Brown, Bolsterli 1959). It is the latter motion which goes up in energy in light nuclei even above the neutron emission threshold and so becomes a resonance. In particular, in light nuclei ^{16}O, ^{40}Ca, ..., the $J^{\pi} = 1^{-}$ $T = 1$ resonance or giant dipole resonance is obtained and can thus be described as a linear superposition of $1p - 1h$ configurations with all amplitudes $X_{m,i}$ having the same phase so that coherence in the electromagnetic decay or excitation probability results (Speth, Van der Woude 1981). In this way, a connection between a purely collective out- or in-phase motion of protons versus neutrons (Fig. 6.5) and a microscopic $1p-1h$ description could be obtained (Brink 1957). In Sect. 6.4, we discuss the realistic case for ^{16}O. In Fig. 6.5, the simple, schematic representation of the collective wave function (in a macroscopic and a microscopic way) is given as well as the concentration of $E1$ excitation strength (expressed in % of the total $1p - 1h$ dipole sum rule) for ^{16}O comparing both the unperturbed and interacting $1p - 1h$ 1^{-} spectra (Elliott, Flowers 1957).

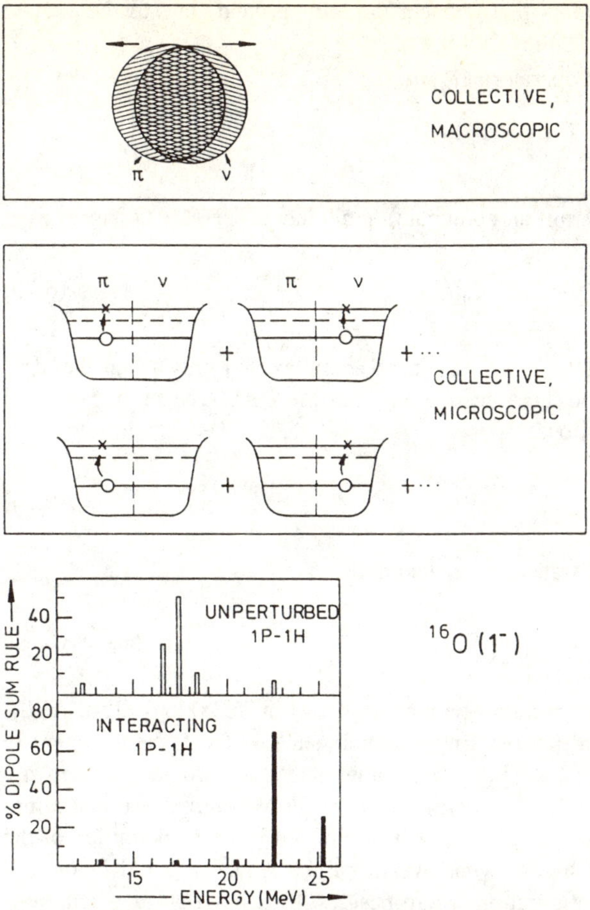

Fig. 6.5. Illustration of the macroscopic and microscopic point of view in describing the nuclear giant electric dipole collective state. In the collective model, the proton and neutron distributions undergo a translational, dynamic oscillatory motion (electric dipole). In the microscopic model, a linear coherent superposition of proton $1p-1h$ and neutron $1p-1h$ states makes up the collective state (repulsive part of the TDA dispersion relation corresponding to $\chi < 0$). Some $1p-1h$ configurations are given as an illustrative example. In the lowest part, the unperturbed $1p-1h$ spectrum and interacting $1p-1h$ spectrum in ^{16}O (for 1^- levels) is given. The height gives the % of the dipole sum rule (proportional to the $E1$ transition probability $|\langle 0\|\boldsymbol{D}(E1)\|1^-\rangle|^2$)

6.3 The RPA Approximation

We pointed out in the TDA approximation that a basic asymmetry between the ground state $|g\rangle$ and the excited $1p-1h$ $\overline{|\hbar\omega\rangle}$ states exists. In the RPA approximation, we use a ground state that is no longer described by the reference state $|0\rangle$ but is treated on an equal footing with the excited states, i.e., $p-h$ excitations or correlations are also present. We leave out a precise description of the new ground state but will define operators that connect the RPA ground state to the excited states (Rowe 1970, Ring, Schuck 1980).

We define the creation operators $Q^{+}_{\alpha}(m,i)$

$$Q^{+}_{\alpha}(m,i) \equiv X^{(\alpha)}_{m,i} a^{+}_{m} a_{i} - Y^{(\alpha)}_{m,i} a^{+}_{i} a_{m} , \tag{6.28}$$

and the corresponding (Hermitian conjugate) annihilation operator $Q_{\alpha}(m,i)$ as

$$Q_{\alpha}(m,i) \equiv X^{(\alpha)^{*}}_{m,i} a_{i} a^{+}_{m} - Y^{(\alpha)^{*}}_{m,i} a_{m} a^{+}_{i} . \tag{6.29}$$

The amplitudes $X^{(\alpha)}_{m,i}$ and $Y^{(\alpha)}_{m,i}$ are in general complex and only when the $Y^{(\alpha)}_{m,i}$ amplitudes disappear is the TDA recovered from the RPA. The ground state is now defined via the condition

$$Q_{\alpha}(m,i)|\tilde{0}\rangle = 0 , \tag{6.30}$$

for all $\alpha_{(m,i)}$. The excited states are created as

$$|\alpha\rangle \equiv Q^{+}_{\alpha}(m,i)|\tilde{0}\rangle . \tag{6.31}$$

From the definition of the creation operator $Q^{+}_{\alpha}(m,i)$ in (6.28) and $Q^{+}_{\alpha}(m,i)|\tilde{0}\rangle$, we extract the information that the single-particle state m has to be occupied in the state $|\tilde{0}\rangle$ in order for a_m to be able to annihilate that particle, but then a^{+}_{m} cannot create another particle. So, arguing in pure fermion terms, we come into conflict with the Pauli principle which is partly violated from the definition of the operators Q^{+}_{α} and Q_{α}. In a somewhat oversimplified way, we could picture an RPA ground state as a configuration where besides $|0\rangle$, $2p-2h$, ..., excitations outside $|0\rangle$ are present at the same time such that particles $(m, n, \ldots)$ above the Fermi level can be annihilated and particle states *below* the Fermi level $(i, j \ldots)$ can be created (Fig. 6.6).

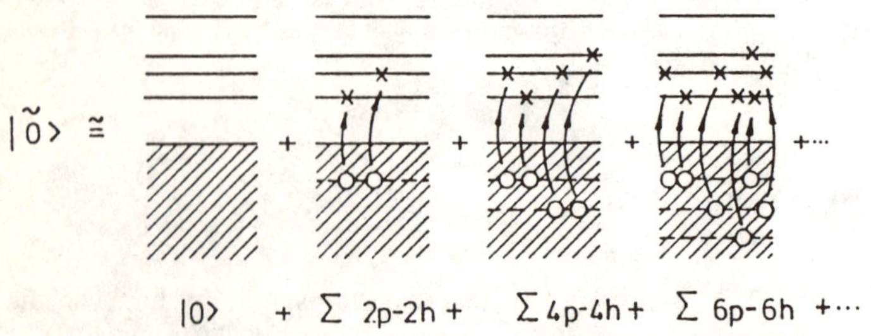

Fig. 6.6. A possible description of the RPA ground-state $|\tilde{0}\rangle$ which is defined through (6.30). In order to be able to create excited states via the operator $Q^{+}_{\alpha}(m,i)$ of (6.28), $2p-2h$, $4p-4h$, $6p-6h$, ... excitations relative to the system with a sharp distribution of nucleons up to the Fermi level have to be present

In RPA, one implies boson commutation relations for the Q^+_α and Q_α operators and the approximation is therefore also called the Quasi-Boson Approximation (QBA) (Lane 1964), where one enforces the commutator condition[1]

$$\left[Q_\alpha(m,i),\, Q^+_\beta(n,j)\right] = \delta_{\alpha\beta}\delta_{mn}\delta_{ij} \; . \tag{6.32}$$

The derivation of the RPA secular equations is somewhat more complicated than the derivation of the TDA secular equation. Imposing

$$H|\alpha\rangle = E_\alpha|\alpha\rangle \; , \tag{6.33}$$

one obtains the set of coupled equations for the $X^{(\alpha)}_{m,i}$ and $Y^{(\alpha)}_{m,i}$ amplitudes which form the RPA secular equations.

$$\begin{aligned} &\left(\varepsilon_{mi} - E_\alpha\right)X^{(\alpha)}_{m,i} + \sum_{n,j}\langle mj|V|in\rangle_{\text{nas}}X^{(\alpha)}_{n,j} \\ &\quad + \sum_{n,j}\langle mn|V|ij\rangle_{\text{nas}}Y^{(\alpha)}_{n,j} = 0 \; , \\ &\left(\varepsilon_{mi} + E_\alpha\right)Y^{(\alpha)}_{m,i} + \sum_{n,j}\langle in|V|mj\rangle_{\text{nas}}Y^{(\alpha)}_{n,j} \\ &\quad + \sum_{n,j}\langle ij|V|mn\rangle_{\text{nas}}X^{(\alpha)}_{n,j} = 0 \; , \end{aligned} \tag{6.34}$$

or, in matrix form

$$\begin{bmatrix} A & B \\ -B^* & -A^* \end{bmatrix} \begin{bmatrix} X \\ Y \end{bmatrix} = E_\alpha \begin{bmatrix} X \\ Y \end{bmatrix} . \tag{6.35}$$

Here we have defined the matrices A and B via their elements labeled by the particle-hole indices (m,i) as row and (n,j) as column indices

$$\begin{aligned} A_{mi,nj} &\equiv \varepsilon_{mi}\delta_{mi,nj} + \langle mj|V|in\rangle_{\text{nas}} \; , \\ B_{mi,nj} &\equiv \langle mn|V|ij\rangle_{\text{nas}} \; , \end{aligned} \tag{6.36}$$

or

$$\begin{aligned} A_{mi,nj} &\equiv \langle mi^{-1}|H|nj^{-1}\rangle \; , \\ B_{mi,nj} &\equiv \langle (mi^{-1})(nj^{-1})|H|\tilde{0}\rangle \; . \end{aligned} \tag{6.37}$$

Thus, the dimension of the RPA secular equation, given in (6.35), has become twice that of the TDA. For each solution formally written as $[X, Y]$ at an energy E_α, a solution $[Y^*, X^*]$ exists at the energy $-E_\alpha$. The latter values are on the "unphysical" branch as only positive energy eigenvalues can be taken to represent

[1] This commutator relation has to be considered as an expectation value relative to the RPA ground state.

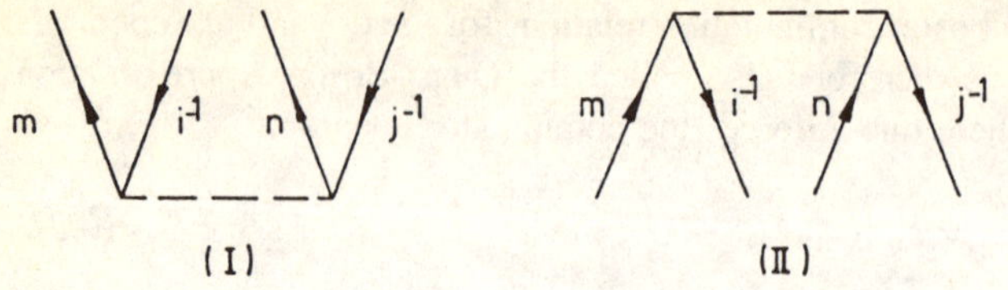

Fig. 6.7. Creation (I) [or annihilation (II)] of a $2p-2h$ configuration $(mi^{-1})(nj^{-1})$ out of the ground state. This process, described by the matrix elements $B_{mi,nj}$ is shown by the corresponding Feynman-Goldstone diagrams

the actual excited states in atomic nuclei. If we set the matrix $B \equiv 0$, the TDA secular equations result as a particular case. This means that those processes expressed by matrix elements of B, i.e., the creation of a $2p-2h$ configuration from the static ground state $|0\rangle$, can be obtained (Fig. 6.7).

In the same way as in TDA, a separable interaction can be used with the properties of attractive matrix-elements (in $T = 0$ channel) and repulsive matrix elements (in $T = 1$ channel) (Rowe 1970).

Again we take

$$\langle mi^{-1}|V|nj^{-1}\rangle = -\chi D_{m,i} D^*_{n,j} , \tag{6.38}$$

and obtain from (6.34)

$$\begin{aligned} X_{m,i} &= \frac{1}{\varepsilon_{mi} - \hbar\omega}\chi D_{m,i}\left[\sum_{n,j}\left(D^*_{n,j}X_{n,j} + D_{n,j}Y_{n,j}\right)\right] , \\ Y_{m,i} &= \frac{1}{\varepsilon_{mi} + \hbar\omega}\chi D^*_{m,i}\left[\sum_{n,j}\left(D^*_{n,j}X_{n,j} + D_{n,j}Y_{n,j}\right)\right] . \end{aligned} \tag{6.39}$$

The expression between square brackets [...] has a constant value and is denoted by N/χ. Thereby, we obtain the result

$$\begin{aligned} X_{m,i} &= \frac{ND_{m,i}}{\left(\varepsilon_{mi} - \hbar\omega\right)} , \\ Y_{m,i} &= \frac{ND^*_{m,i}}{\left(\varepsilon_{mi} + \hbar\omega\right)} . \end{aligned} \tag{6.40}$$

Multiplying (6.39) by $D^*_{m,i}$, (respectively $D_{m,i}$), and summing over all (n,j) particle-hole pairs, we obtain the dispersion equation for the energy eigenvalues $\hbar\omega$ as

$$\frac{1}{\chi} = 2\sum_{m,i}\frac{|D_{m,i}|^2\varepsilon_{mi}}{\varepsilon^2_{m,i} - (\hbar\omega)^2} . \tag{6.41}$$

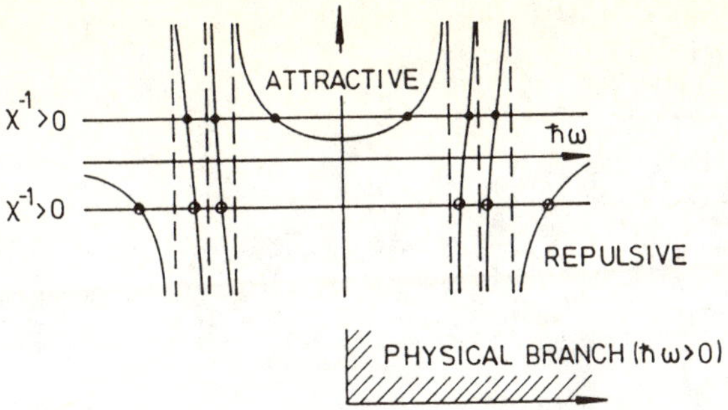

Fig. 6.8. The RPA eigenvalue equation (6.41) is shown in a schematic way. The left-hand side in (6.41) is equal to the TDA equation (see Fig. 6.3). The right-hand side is quadratic in $\hbar\omega$ and results in both positive and negative solutions for the energies $\hbar\omega$. The physical branch is the sector $\hbar\omega > 0$. Because of the doubling of solutions, possibilities exist that the lower, collective solution (•) becomes imaginary (no intersection with the $1/\chi$ line)

This equation resembles the TDA secular equation. However, we have a quadratic expression in $\hbar\omega$ in the denominator and also an extra factor of 2 in the numerator. The factor 2 indicates that for the same strength χ, the lowest (or highest) RPA eigenvalue for $\hbar\omega$ is quite a bit lower (or higher) than the corresponding TDA eigenvalue. We represent schematically the solutions to (6.41) in Fig. 6.8. For the degenerate cases where all $\varepsilon_{mi} = \varepsilon$, one gets the eigenvalue equation (6.41) into the form

$$(\hbar\omega)^2 = \varepsilon^2 - 2\varepsilon\chi \sum_{m,i} |D_{m,i}|^2 \,. \tag{6.42}$$

If we take the particular case that ($\chi > 0$)

$$\chi = \frac{\varepsilon}{\left(2 \sum\limits_{m,i} |D_{m,i}|^2\right)} , \tag{6.43}$$

one obtains the lowest eigenvalue (RPA) at $\hbar\omega = 0$. On the contrary, for the same strength χ of (6.43), the lowest TDA eigenvalue occurs at $\hbar\omega = \varepsilon/2$. This application again indicates that in the RPA, the lowest-lying collective in-phase motion of protons and neutrons is lower in energy. It can even become so low that for particular strong collective states (e.g., 3^- in ^{40}Ca) the lowest root becomes imaginary. In Fig. 6.9, we indicate besides the graphical solution to the RPA dispersion relation of (6.41), the particular case of a strength in the degenerate case given by (6.43).

We now define the quantity S (the "non-energy weighted" sum rule for the relevant one body operator expressed by $D_{m,i}$),

$$S = \sum_{m,i} |D_{m,i}|^2 \,. \tag{6.44}$$

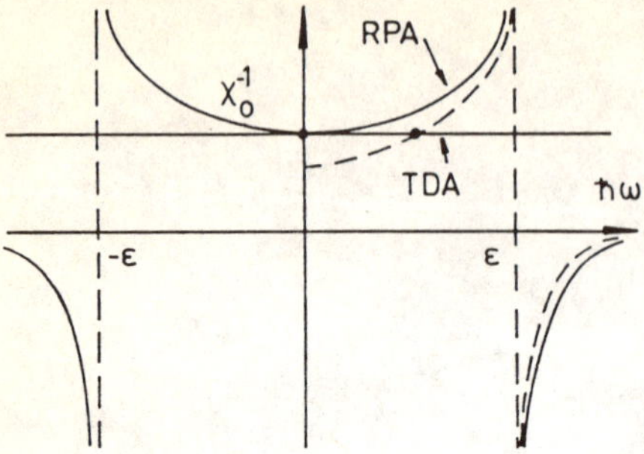

Fig. 6.9. Degenerate limit (all $\varepsilon_{mi} = \varepsilon$) of Fig. 6.8 for the RPA approximation. We illustrate the particular situation that the value of $\chi = \chi_0$ is given by (6.43) and the lowest (collective) RPA root equals $\hbar\omega = 0$! (*full line*). We draw on the same figure the TDA solution with the same value of χ. Here, the collective TDA root occurs at $\hbar\omega(\text{TDA}) = \varepsilon/2$. So, for a *given* strength χ, the RPA root becomes lower than the corresponding TDA root

We obtain the approximate wave function for the collective excitation

$$
\begin{aligned}
X_{m,i}^{(1)} &= \frac{D_{m,i}\left(\varepsilon + \hbar\omega^{(1)}\right)}{\left(4\hbar\omega^{(1)}\varepsilon S\right)^{1/2}} \simeq S^{-1/2} D_{m,i} \,, \\
Y_{m,i}^{(1)} &= \frac{D_{m,i}^{*}\left(\varepsilon - \hbar\omega^{(1)}\right)}{\left(4\hbar\omega^{(1)}\varepsilon S\right)^{1/2}} \simeq S^{-1/2} D_{m,i}^{*} \left(\frac{-\hbar\omega^{(1)} + \varepsilon}{2\varepsilon}\right) .
\end{aligned}
\tag{6.45}
$$

Here, onc observes that in most cases the amplitudes

$$
Y_{m,i}^{(1)} \ll X_{m,i}^{(1)} \,,
$$

because of the extra factor

$$
\frac{\left(\varepsilon - \hbar\omega^{(1)}\right)}{2\varepsilon} \ll 1 \,,
\tag{6.46}
$$

in actual cases.

The transition probability of this collective state to the RPA ground state then becomes generally

$$
\overline{\langle \hbar\omega^{(1)} | \boldsymbol{D} | \tilde{0} \rangle} = \sum_{m,i} \left[X_{m,i}^{*} D_{m,i} + Y_{m,i}^{*} D_{m,i}^{*} \right] = \frac{N^{*}}{\chi} \,.
\tag{6.47}
$$

For the degenerate case and its collective, low-lying (or high-lying) state, the transition matrix element (6.47) simplifies to

$$
\overline{\langle \hbar\omega^{(1)} | \boldsymbol{D} | \tilde{0} \rangle} = 2N^{*} \frac{\varepsilon S}{\left[\varepsilon^2 - \left(\hbar\omega^{(1)}\right)^2\right]} \,.
\tag{6.48}
$$

Using the expression for the normalization coefficient N and its relation to the solution, this gives the transition probability

$$\overline{|\langle\hbar\omega^{(1)}|\boldsymbol{D}|\tilde{0}\rangle|^2} = \frac{\varepsilon}{\hbar\omega^{(1)}}S\ . \tag{6.49}$$

Thus, the collectivity of this transition is expressed by the coherence of S in *all* particle-hole contributions with the same sign (which one also obtains in the TDA case), but now multiplied by an extra factor $\varepsilon/\hbar\omega^{(1)}$ which easily becomes 2.

This extra enhancement in the transition probability comes in particular from the fact that the RPA ground state $|\tilde{0}\rangle$ contains $p-h$ correlations and that an excited state $|\hbar\omega\rangle$ can decay into the ground state in different ways by using a one-body transition operator $\boldsymbol{D}$.

In Appendix H we shall discuss a slightly more elaborate two-group RPA model which, however, bears a resemblance to many actual situations in real nuclei.

Before discussing a detailed study of ^{16}O, we illustrate the results of a study of the 3^- collective isoscalar state in ^{208}Pb as was studied by *Gillet* et al. (Gillet et al. 1966). As a function of the dimension of the $1p-1h$ configuration space both the lowest TDA and RPA eigenvalues are illustrated in Fig. 6.10. One first observes that the RPA eigenvalue is always lower than the corresponding TDA eigenvalue but that at the same time, for a given strength of the residual interaction (in this case a Gaussian residual interaction has been used), the dimension of the 3^- configuration space strongly affects the particular excitation energy.

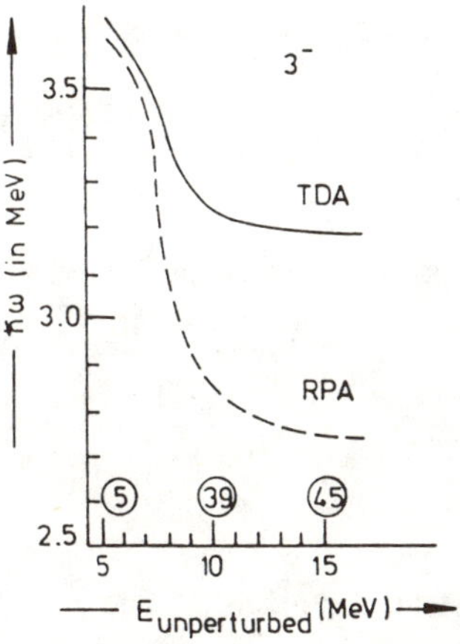

Fig. 6.10. The position of the octupole state in ^{208}Pb using TDA and RPA as a function of the dimension of the configuration space. On the abscissa, we give both the number and unperturbed energy of the $1p-1h$ configurations. A Gaussian interaction $V(r) = V_0\exp(-r^2/\mu^2)[W + MP(r) + BP_\sigma + HP_\sigma P(r)]$ was used with $V_0 = -40$ MeV, $\mu = 1.68$ fm and exchange admixture W, M, B and H as discussed in (Gillet 1966) [Fig. 4, adapted from (Gillet 1966)]

6.4 Application of the Study of $1p-1h$ Excitations: ^{16}O

For the doubly-closed shell nucleus ^{16}O, the low-lying excited states will be within the space of $1p-1h$ configurations where, due to the charge independence of the nucleon-nucleon interaction and the almost identical character of the nuclear average field, proton $1p-1h$ and neutron $1p-1h$ excitations are nearly degenerate in unperturbed energy (Fig. 6.11). It is the small mass difference between proton

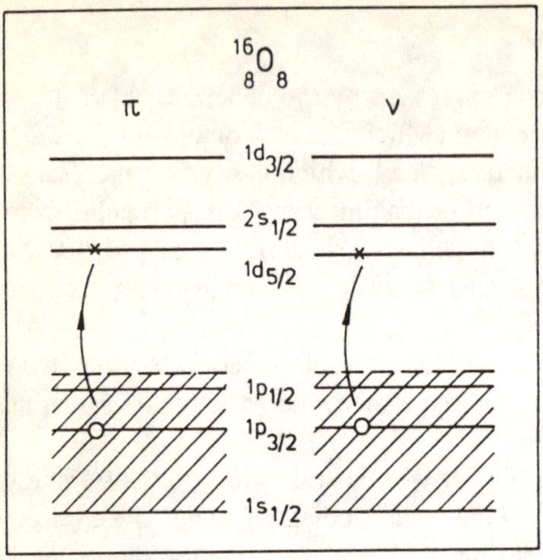

Fig. 6.11. The model space used to carry out the $1p-1h$ calculations discussed in Sect. 6.4. The full (sd) space for unoccupied and (sp) space for occupied configurations are considered in constructing the relevant energy matrices

and neutron and the Coulomb interaction that induces slight perturbations on the isospin symmetry of the picture. Although we could work in an isospin formalism we shall use the explicit difference between proton and neutron $1p-1h$ configurations and later check the isospin purity of the eigenstates. Since we know how to couple the proton $1p-1h$ and neutron $1p-1h$ configurations to definite isospin,

$$
\begin{aligned}
|ph^{-1}; J^\pi\rangle_\pi &= \frac{1}{\sqrt{2}}\left[-|ph^{-1}; J^\pi, T=1\rangle + |ph^{-1}; J^\pi, T=0\rangle\right] , \\
|ph^{-1}; J^\pi\rangle_\nu &= \frac{1}{\sqrt{2}}\left[|ph^{-1}; J^\pi, T=1\rangle + |ph^{-1}; J^\pi, T=0\rangle\right] ,
\end{aligned}
\tag{6.50}
$$

we can, in the linear combination of the wave functions expressed in the (π, ν) basis, substitute expressions (6.50) and obtain an expansion in the isospin J^π, T coupled basis. Formally, the wave function

$$
|i; J^\pi\rangle = \sum_{p,h,\varrho} c\big(ph^{-1}(\varrho); iJ\big)|ph^{-1}; J^\pi\rangle_\varrho , \tag{6.51}
$$

where i denotes the rank number labeling the eigenstates for given J^π and ϱ the charge quantum number ($\varrho \equiv \pi, \nu$) can be rewritten in the basis

$$
|i; J^\pi\rangle = \sum_{p,h,T} \bar{c}\big(ph^{-1}; iJ, T\big)|ph^{-1}; J^\pi, T\rangle , \tag{6.52}
$$

where now for each (p, h) combination we sum over $T = 0$ and $T = 1$ states. From the diagonalization (see later) it follows that low-lying states are mainly $T = 0$ in character and the high-lying ones are $T = 1$ states.

The Hartree-Fock field for ^{16}O was determined by using the $SkE4^*$ interaction (Chap. 8) with its proton and neutron single-particle orbitals, energies, with the

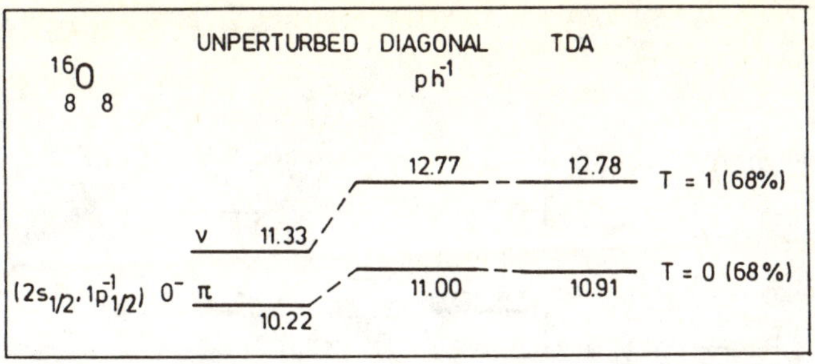

Fig. 6.12. Using the $SkE4^*$ force (see Chap. 8) in determining both the Hartree-Fock field (single-particle energies ε_{nlj}) and the residual interaction, we present: (i) the unperturbed proton and neutron $0^-(2s_{1/2}\,1p_{1/2}^{-1})$ configurations, (ii) the energies, when adding the diagonal ph^{-1} matrix elements (repulsive), (iii) the final TDA results after diagonalizing the energy matrix for $J^\pi = 0^-$. Isospin purity is also given in $T\,(\%)$

same interaction $SkE4^*$ being used as a residual interaction (Waroquier 1982, Waroquier et al. 1983b). We illustrate schematically that $\varepsilon_j(\pi) \neq \varepsilon_j(\nu)$ in Fig. 6.11. This is even more clear for the case of those $1p - 1h$ excitations that form the 0^- state. The $(2s_{1/2}, 1p_{1/2}^{-1})0^-$ configurations are given in Fig. 6.12 with a net difference in unperturbed energy of $\Delta\varepsilon = 1.11$ MeV. The diagonal $(ph^{-1}, 0^-)$ matrix elements are slightly repulsive and finally, the interaction in non-diagonal form is small. Thus, very pure proton ph^{-1} and neutron ph^{-1} states result. When expressed in the basis of (6.52), this gives very impure isospin wave functions (Fig. 6.12) with 68 % of the main component only.

Relating to the 1^- states within the $1\hbar\omega\,(\Delta N = 1)$ space, we can only form the following 1^- configurations where

$$
\begin{aligned}
&|2s_{1/2}(1p_{1/2})^{-1}; 1^-\rangle\,, \\
&|2s_{1/2}(1p_{3/2})^{-1}; 1^-\rangle\,, \\
&|1d_{3/2}(1p_{1/2})^{-1}; 1^-\rangle\,, \\
&|1d_{3/2}(1p_{3/2})^{-1}; 1^-\rangle\,, \\
&|1d_{5/2}(1p_{3/2})^{-1}; 1^-\rangle\,,
\end{aligned}
$$

where both proton and neutron $1p - 1h$ configurations can result, leading to a 10-dimensional model space only. Using now the methods discussed in Sects. 6.2, 3, one can set up the TDA and RPA eigenvalue equations. The results for $0^-, 1^-, 2^-$, 3^- and 4^- are given in Fig. 6.13, where we give in all cases:

i) the unperturbed energy with the charge character of the particular ph^{-1} excitation,
ii) the TDA diagonalization result (with isospin purity),
iii) the RPA diagonalization result,
iv) the experimental data.

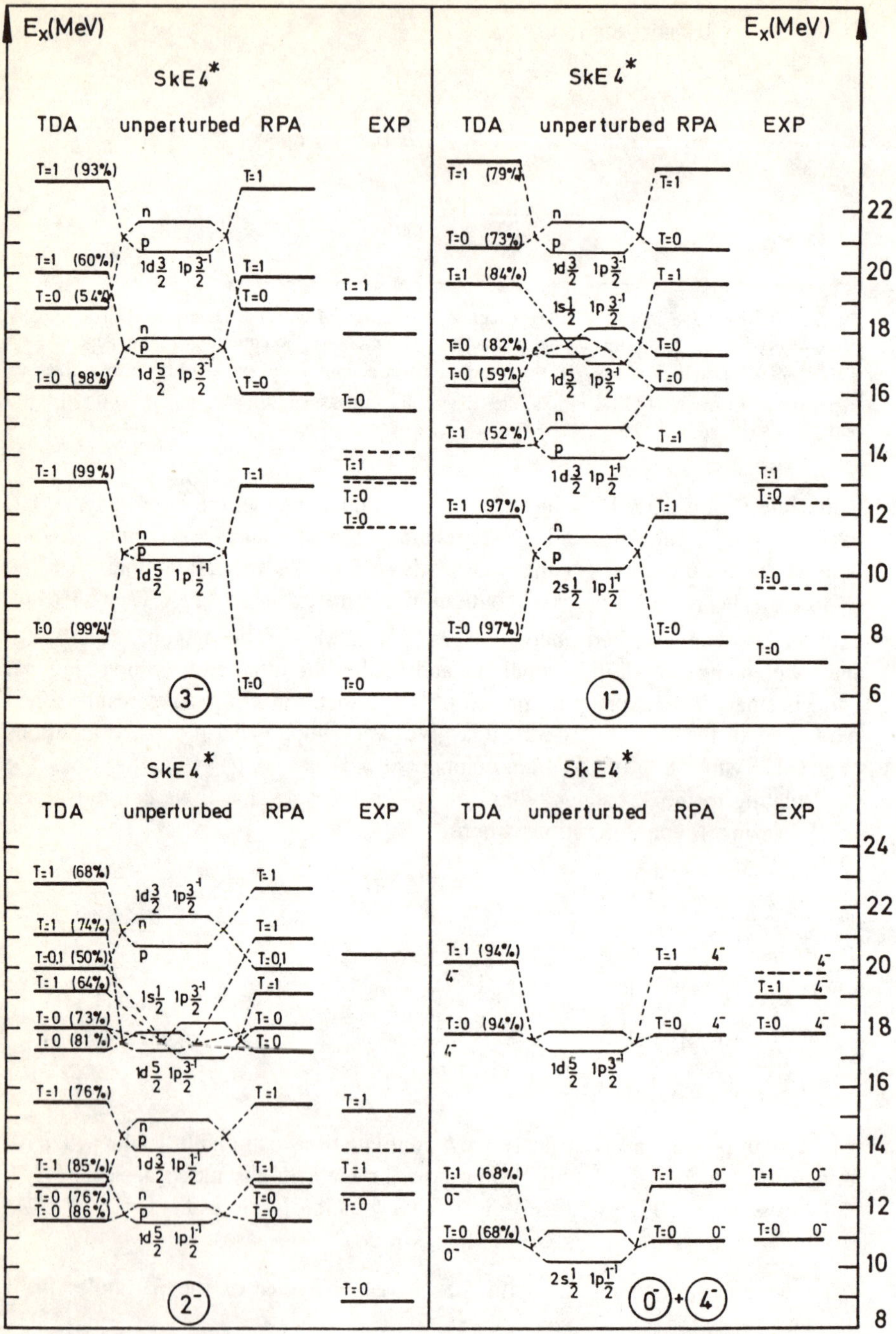

Fig. 6.13. Negative parity states in ^{16}O. Comparison between the TDA and RPA results using $SkE4^*$ and the data is given. The unperturbed structure is indicated (n: neutron, p: proton). The isospin purity (%) is also given. Experimental levels, drawn with a dashed line, have predominant $3p-3h$ character [taken from (Waroquier 1983b)]

Table 6.1. The wave functions for the 3^- $T = 0$ and $T = 1$ states in ^{16}O, obtained using TDA and the $SkE4^*$ interaction. The wave functions are expanded in both a charge basis, giving the $|ph^{-1}; J^\pi\rangle_\varrho$ ($\varrho \equiv p, n$) components and an isospin basis, giving the $|ph^{-1}; J^\pi, T\rangle$ components. The total isospin purity is also expressed in the last column

	p / h^{-1}	$1f_{7/2}$ / $1s_{1/2}^{-1}$	$1f_{5/2}$ / $1s_{1/2}^{-1}$	$1d_{5/2}$ / $1p_{3/2}^{-1}$	$1d_{3/2}$ / $1p_{3/2}^{-1}$	$1d_{5/2}$ / $1p_{1/2}^{-1}$	isospin purity
3^-	p	−0.01	+0.01	+0.15	−0.20	+0.72	
E_x = 7.88 MeV	n	−0.02	+0.01	+0.15	−0.18	+0.60	
	$T = 0$	−0.03	+0.02	+0.22	−0.27	+0.93	99 %
	$T = 1$	+0.00	−0.00	+0.00	−0.01	+0.08	1 %
3^-	p	−0.03	+0.00	+0.25	−0.05	−0.64	
E_x = 13.10 MeV	n	+0.03	+0.00	−0.13	−0.04	+0.71	
	$T = 0$	−0.00	+0.00	+0.09	−0.06	+0.05	1 %
	$T = 1$	−0.04	−0.00	+0.26	−0.01	−0.96	99 %

One observes, in particular for the lowest $1^-, 3^-$ levels, the very pure $T = 0$ isospin character. Since the unperturbed proton and neutron $1p-1h$ configurations are almost degenerate, the diagonalization induces a definite symmetry ($T = 0$ or $T = 1$) according to the lower- or higher-lying states. In the case of 3^- ($T = 0$) and 3^- ($T = 1$), we also give the wave functions in both the charge quantum number basis and the isospin quantum number basis (Table 6.1) to illustrate the above general discussion on isospin (Sect. 3.4) and the discussion at the beginning of this section. In Chap. 8, when discussing some state-of-the-art shell-model calculations, we shall discuss some more examples as well as outline the self-consistent methods used in determining both the average field and the residual interaction.

One could, of course, discuss other topics related to the TDA and, in particular, to the RPA method. One can show that the RPA equations correspond to the small amplitude limit of the time-dependent Hartree-Fock method (harmonic approximation) and thus the term quasi-boson approximation is related to the dynamics described by the RPA eigenvalue equations (Lane 1964, Ring, Schuck 1980, Rowe 1970). The RPA secular problem reduces to a non-Hermitian matrix problem which we will not discuss in detail since it is outside the scope of this book.

6.5 Summary

In the present chapter, we study the elementary excitation modes of doubly-closed shell nuclei. After a general formulation of the secular equation within a one-particle one-hole ($1p-1h$) basis, a number of approximations to this more general problem of studying the eigenstates are described.

The Tamm-Dancoff approximation (TDA) builds the $1p-1h$ excitations out of a sharp Fermi distribution of totally filled (hole) and fully empty (particle)

orbitals. Here, too, the secular equation is reformulated. Making use of a schematic model (Brown and Bolsterli model) and with the use of a separable particle-hole interaction describing the residual interactions, a simple dispersion relation is obtained relating the strength of the separable interaction to the unperturbed single-particle energies and the energy eigenvalues. The situation of a configuration space with fully degenerate $1p-1h$ excitations is described.

The Random-Phase approximation (RPA) is a more realistic approximation to a description of the particle-hole excitation in closed shell nuclei. Here, the operator creating the elementary excitation modes contains, besides the creation of a $1p-1h$ configuration, the annihilation part for the $1p-1h$ configuration. The ground state to which these new operators refer is a more general vacuum state than that used in the TDA approximation. It allows at the same time for a better description of collective phenomena in closed shell nuclei. Here, as in the TDA approach, a schematic model is discussed and the corresponding RPA dispersion relation derived. The salient features of the solutions to the secular equation i.e. doubling of the number of energy eigenvalues, critical interaction strength for real solutions, the observation of isoscalar and isovector solutions, are discussed in some detail. The transition strenth from a given state with angular momentum L to the ground state and its collective character is derived.

We present in this chapter a detailed, numerical application for the $1p-1h$ study of excitation in ^{16}O using an effective interaction consisting of zero-range two- and three-body force components (Skyrme-type force, see Chap. 8). The results for the angular momentum $0^-, \ldots, 4^-$ states are presented, and the wave functions are discussed within both the isospin basis and the proton-neutron charge basis. Some specific properties such as collectivity, influence of the dimension of the $1p-1h$ space on the energy eigenvalues, possible imaginary solutions to the lowest eigenvalues,... are pointed out.

Problems

6.1 Derive the transition probability to the ground state $|\langle 0|\boldsymbol{D}|\overline{\hbar\omega_1}\rangle|^2$ and $|\langle 0|\boldsymbol{D}|\overline{\hbar\omega_{\text{degenerate}}}\rangle|^2$ for the collective and non-collective states, obtained from the schematic TDA model. Show the coherence of the different $p-h$ components in the transition probability occurring in the collective state.

6.2 Derive the secular equation for the angular momentum coupled $p-h$ representation $|ph^{-1}; JM\rangle$ in the case of the TDA and RPA approximations.

6.3 Study the general properties of the RPA non-hermitian eigenvalue problem (normalization, orthogonality, metric,...).

6.4 Starting from the schematic TDA (or RPA) model, show that the excitation energy of the collective root is inversely proportional, for a given strength, to the model space dimension.

*6.5 Construct the full $1p-1h$ basis for ^{16}O, using the p-model space for the hole excitations and the sd-model space for the particle excitations. Construct the unperturbed energy spectrum using proton and neutron separation energies and the excitation energies in the respective one-particle (proton and neutron) and one-hole (proton and neutron) nuclei around ^{16}O (use the data from Bohr and Mottelson (1969)). What will be the effect of diagonalizing the residual interaction for the 1^- states (try a numerical calculation using a residual zero-range interaction and the available codes). In ^{16}O, a low-lying 0^+ state (at $\simeq 6$ MeV) has been observed. Suggest a possible origin for this 0^+ excitation.

6.6 Using a separable particle-hole interaction (discarding the exchange term), show that the particle-hole matrix element $\langle mi^{-1}|H|nj^{-1}\rangle$ separates as $-\chi D_{m,i}D^*_{n,j}$.

6.7 Show that the lowering in energy of the collective root in the TDA and RPA method (for a given interaction strength) is proportional to the dimension of the model space. Study the dependence in some detail by considering Fig. 6.10.

6.8 If the interaction between two particles is an attractive constant

$$V_{\alpha\beta\gamma\delta} = -V(\delta_{\alpha\gamma}\delta_{\beta\delta} - \delta_{\alpha\delta}\delta_{\beta\gamma}) ,$$

prove that the particle-hole interaction results in repulsive values.

6.9 Show that, for $\chi < 0$ (Eqs. (6.27) and (6.42)), the collective root corresponds to an antisymmetric state in the proton and neutron character for the particle-hole configurations (isovector state) and a symmetric state ($\chi > 0$) in the proton and neutron character (isoscalar state).

*6.10 Prove that, for charge-exchange excitations, one has the following model-independent sum rules:

(i) $$\sum_\nu \{|\langle\nu|T_-|gs\rangle|^2 - |\langle\nu|T_+|gs\rangle|^2\} = N - Z ,$$

(with $T_- = \sum_i t_-,i$ and $T_+ = \sum_i t_+,i$).

(ii) $$\sum \left\{ \left|\langle\nu|\sum_i \sigma_i t_-,i|gs\rangle\right|^2 - \left|\langle\nu|\sum_i \sigma_i t_+,i|gs\rangle\right|^2 \right\} = 3(N - Z) .$$

Hint: $\nu \equiv (ph^{-1})$ are states formed from the ground-state. Use closure relations in evaluating the squares of the matrix elements (i.e. $\sum_\nu |\langle\nu|O|gs\rangle|^2 = \langle gs|OO^+|gs\rangle$).

7. Pairing Correlations: Particle-Particle Excitations in Open-Shell Nuclei

7.1 Introduction

Until now we always considered a sharp Fermi level with a very particular distribution of level occupation probabilities. Up to the Fermi level, all levels are fully occupied, while above, all are unoccupied. This conforms with a Hartree-Fock static ground state with $1p - 1h$, $2p - 2h$, ... excitations as elementary modes of excitation (see Sect. 6.2 for a discussion on the TDA approximation). A short range force of sufficient strength is now able to scatter couples of particles across the sharp Fermi level. For nuclei at or near doubly-closed shell configurations, this leads to a "diffuse" ground state with $2p - 2h$, $4p - 4h$, ... correlations, much like the RPA ground state $|\tilde{0}\rangle$ as discussed in Sect. 6.3. When, however, many nucleons outside closed shells are present, this pair can scatter, under certain conditions, lead to a stable, smooth probability distribution for the occupation of the single-particle orbitals. So, the pairing correlations set in and modify the nuclear ground state distribution of nucleons in nuclei in a major way (see Fig. 7.1).

In Chap. 7, we shall discuss these modifications in detail, starting from the most simple, solvable model of a single j-shell with n particles interacting with the pairing interaction of Sect. 5.2 and going on to detailed pairing calculations.

Before attacking the new aspects brought into the nucleus by means of the pairing correlations, we recall briefly the major aspects that determine the nucleon

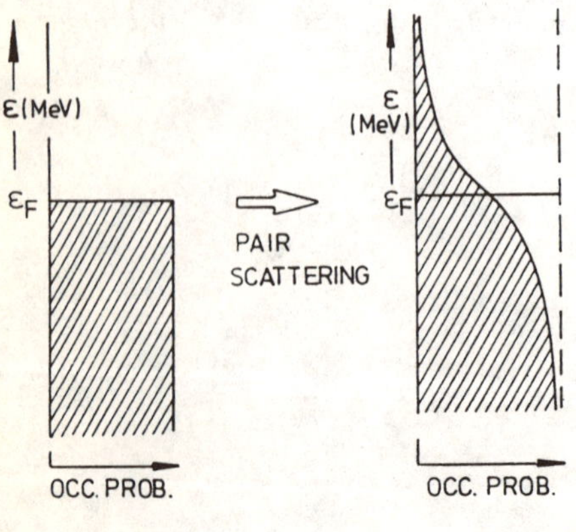

Fig. 7.1. Schematic distribution of nucleon pairs in one case where all nucleons occupy the low-lying orbitals pairwise up to the Fermi level (closed shell) and in another case, where a smeared out pair distribution occurs

motion in the nucleus (Rowe 1970). We have indicated how the Hartree-Fock theory has succeeded in separating the A-body nuclear Hamiltonian into

$$H = H_0 + H_{\text{res}} \,, \tag{7.1}$$

where H_0 becomes the Hartree-Fock Hamiltonian H_{HF} when going to the appropriate single-particle basis. The importance of this separation in (7.1) depends on the strength of $\langle H_{\text{res}} \rangle$ compared to $\langle H_{\text{HF}} \rangle$. In light nuclei like ^{16}O, ^{40}Ca,... the energy gap between major oscillator shells, $\hbar\omega_0$, is such that

$$\hbar\omega_0 \gg \langle H_{\text{res}} \rangle \,. \tag{7.2}$$

So, the Hartree-Fock ground state remains stable with respect to the residual interactions and independent particle-hole excitations out of this Hartree-Fock ground state are well defined (see Fig. 7.2).

For non-closed-shell nuclei ($^{A}_{50}$Sn with $A \cong 110 - 120$), the Hartree-Fock solution to the ground-state nucleon distribution can sometimes lead to a small gap so that

$$\hbar\omega_0 \gtrsim \langle H_{\text{res}} \rangle \,. \tag{7.3}$$

In those cases, instability against pair excitations or deformation of the nucleus could set in. We know that the two-body interaction $V(i,j)$ cannot completely be absorbed into the average Hartree-Fock field. We can ask exactly what part goes into the average field.

In the Hartree-Fock approximation (Sect. 3.1), we have shown that

$$U(\boldsymbol{r}_i) = \int V(\boldsymbol{r}_i, \boldsymbol{r}_j)\varrho(\boldsymbol{r}_j)d\boldsymbol{r}_j \,. \tag{7.4}$$

If we now make a multipole expansion of the two-body interaction as

$$V(\boldsymbol{r}_i, \boldsymbol{r}_j) = \sum_{\lambda,\mu} f^{\lambda}(r_i, r_j) \frac{4\pi}{2\lambda+1} Y_{\lambda}^{\mu}(\hat{\boldsymbol{r}}_i) Y_{\lambda}^{\mu^*}(\hat{\boldsymbol{r}}_j) \,, \tag{7.5}$$

and substitute this expression in (7.4) one obtains

$$U(\boldsymbol{r}_i) = \sum_{\lambda,\mu} U_{\lambda}^{\mu}(\boldsymbol{r}_i) \,, \tag{7.6}$$

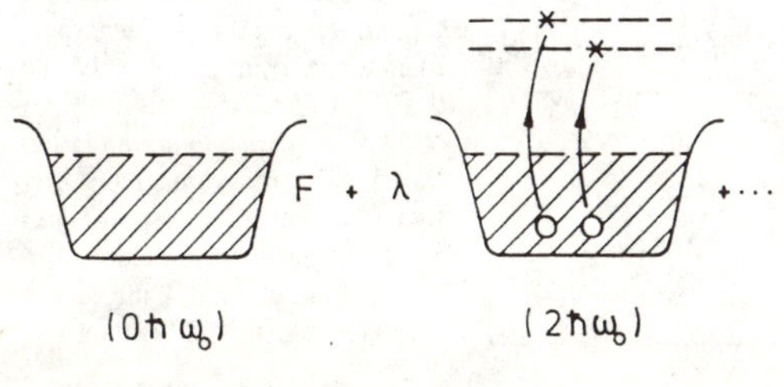

Fig. 7.2. The Hartree-Fock ground state for doubly-even nuclei is rather stable against $2p - 2h$ excitations ($\lambda \ll 1$) in those cases where $\sim\omega_0$ (separation between major oscillator shells) is large compared to the residual interaction energy $\langle H_{\text{res}} \rangle$

with

$$U_\lambda^\mu(\boldsymbol{r}_i) = Y_\lambda^\mu(\hat{\boldsymbol{r}}_i)\frac{4\pi}{2\lambda+1}\int f^\lambda(r_i, r_j)Y_\lambda^{\mu^*}(\hat{\boldsymbol{r}}_j)\varrho(\boldsymbol{r}_j)d\boldsymbol{r}_j\ ,$$
$$= Y_\lambda^\mu(\hat{\boldsymbol{r}}_i)a_\lambda^\mu(r_i)\ . \tag{7.7}$$

So, the average field also presents different multipoles according to the folding of the density with corresponding multipoles of the two-body residual interaction $V(\boldsymbol{r}_i, \boldsymbol{r}_j)$. For $\lambda = 0$, we recover the spherical, central static field component. The $\lambda = 1$ part relates to an overall shift of the centre-of-mass of the nucleus and does not relate to specific "internal" modes of the nuclear field. Higher multipoles, $\lambda = 2$, $\lambda = 3$, ... represent the quadrupole, octupole, ... deformation of the average field, respectively. The low multipoles of $V(\boldsymbol{r}_i, \boldsymbol{r}_j)$ ($\lambda = 0$, $\lambda = 2$, $\lambda = 3$) are called the field-producing components (Bohr, Mottelson 1975, Ring, Schuck 1980).

In contrast to the "long-range" components of $U(\boldsymbol{r}_i)$, short-range components of the interaction scatter nucleons out of their individual orbitals. Contributions to all λ are obtained from short-range forces such as the $\delta(\boldsymbol{r}_i - \boldsymbol{r}_j)$ zero-range force or the pairing interaction (defined in second quantization in Sect. 5.2). These forces show the tendency to correlate nucleons in zero-coupled pairs ($J^\pi = 0^+$) and thus restore the spherical symmetry in the nucleus. This correlation is well-known from the energy spectra in nuclei with just two nucleons outside a doubly-closed shell configuration. We show, in Fig. 7.3, the case of ${}^{210}_{82}\mathrm{Pb}_{128}$ with two neutrons outside

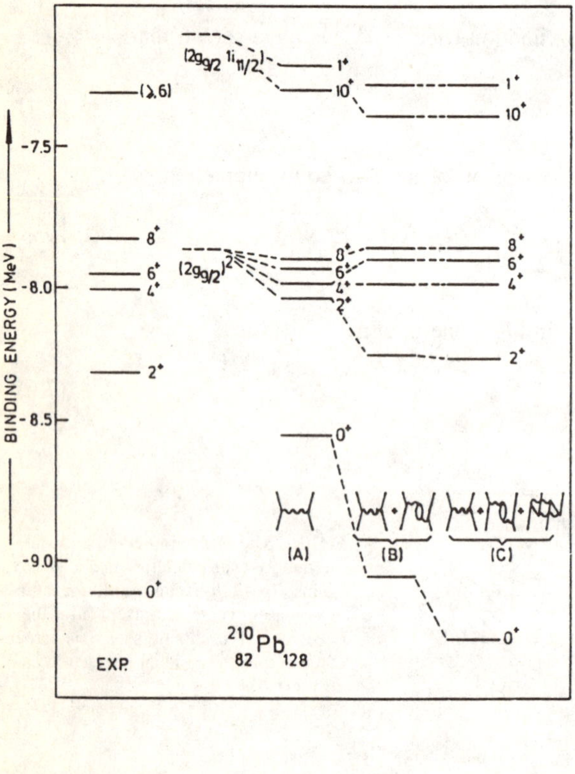

Fig. 7.3. Typical example of the nucleon-nucleon pairing properties in a heavy nucleus ${}^{210}_{82}\mathrm{Pb}$ with 2 valence neutrons outside the ${}^{208}\mathrm{Pb}$ doubly closed-shell core. On the left-hand side, the experimental spectrum is indicated. On the right-hand side various theoretical spectra are given, illustrated by the diagrams in columns (*A*), (*B*), and (*C*). Here, (*C*) represents the more realistic situation, (*A*) the "bare" interaction (see Herling, Kuo 1972)

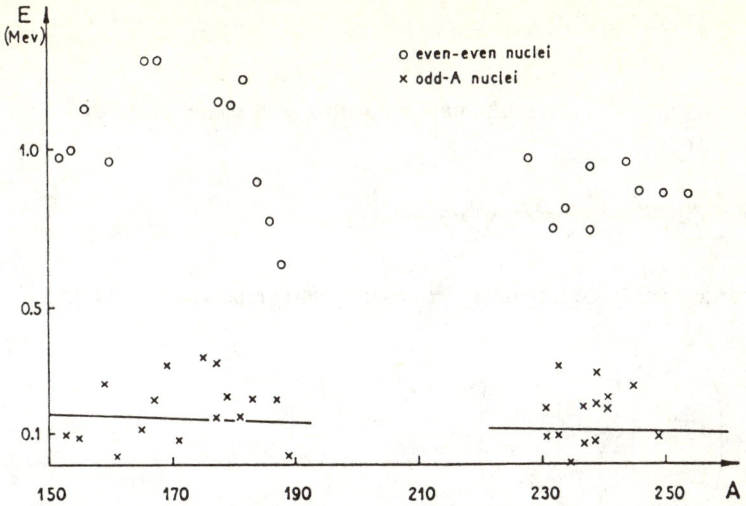

Fig. 7.4. Energies of the first excited *intrinsic* states in deformed nuclei as a function of the mass number. The solid line represents the average distance between intrinsic levels in the odd-mass nuclei. The figure contains data in the $150 < A < 190$ and $A > 228$ regions. We have not included the low-lying $K = 0$ states in even-even nuclei in the Ra, Th region since these states represent collective odd-parity oscillations (taken from (Bohr, Mottelson, Pines 1958))

the ^{208}Pb core, moving in the $2g_{9/2}$ orbital as the lowest orbital. A comparison with the theoretical two-particle spectra as determined by Herling and Kuo (Herling, Kuo 1972), using the Hamada-Johnston force, is carried out.

This particular pairing effect was recognized before and introduced by Racah in 1942 (Racah 1942a, 1942b), in the seniority scheme in atomic physics. Later, the 0^+ property of even-even nuclei throughout the whole nuclear mass region, was recognized by M.G. Mayer in 1950 (Mayer 1950), and applied to the nucleus by Flowers, Racah and Talmi in 1952 (Flowers 1952, Racah, Talmi 1952, Talmi 1952). Pairing was incorporated in a more elaborate way using the BCS theory of superconductivity as applied to even-even atomic nuclei (Bohr et al. 1958) from which we take their original illustration of the energy of the first excited state in deformed nuclei (intrinsic state) for both odd-mass and even-even nuclei (see Fig. 7.4).

We shall treat the salient features presented by not just one extra pair but by many pairs outside closed-shell configurations.

7.2 Pairing in a Degenerate Single j-Shell

Starting from the above pairing aspects, it is tempting to use an interaction that has this pairing property of acting only in $J^\pi = 0^+$ states. Knowing that a given two-body interaction, given in coordinate space can be written in a second-quantized form

$$V(1,2) \to \tfrac{1}{4} \sum_{\alpha,\beta,\gamma,\delta} \langle\alpha\beta|V|\gamma\delta\rangle_{\text{nas}} a^+_\alpha a^+_\beta a_\delta a_\gamma \,, \tag{7.8}$$

we can define an interaction Hamiltonian, immediately in second quantized form as (see Sect. 5.2)

$$H = -G \sum_{m,m'>0} a^+_{jm} a^+_{j-m} a_{j-m'} a_{jm'} (-1)^{2j+m+m'} \,. \tag{7.9}$$

In the two-particle subspace (m-states), H has the matrix representation (with $j+\frac{1}{2}$ rows and columns)

$$H = -G \begin{pmatrix} 1 & -1 & 1 & \dots \\ -1 & 1 & -1 & \dots \\ 1 & -1 & 1 & \dots \\ \vdots & \vdots & \vdots & \end{pmatrix} \,. \tag{7.10}$$

It can be shown that the state

$$\frac{1}{\sqrt{\Omega}} \begin{pmatrix} 1 \\ 1 \\ 1 \\ \cdot \\ \cdot \end{pmatrix} = \frac{1}{\sqrt{\Omega}} \sum_{m>0} (-1)^{j+m} a^+_{jm} a^+_{j-m} |0\rangle = |(j)^2; J=0\ M=0\rangle \,, \tag{7.11}$$

is the lowest energy eigenstate of (7.10). Here $|0\rangle$ denotes the closed-shell wave function and Ω is the shell degeneracy or $\Omega \equiv j+\frac{1}{2}$. The energy of the state (7.11) is $E_0 = -G\Omega$.

To solve now for the n-particle system the pairing interaction problem, we define the pair creation operator

$$S^+_j = \frac{1}{\sqrt{\Omega}} \sum_{m>0} (-1)^{j+m} a^+_{jm} a^+_{j-m} \,, \tag{7.12}$$

which creates the two-particle $J^\pi = 0^+$ state. Using this operator S^+_j, the Hamiltonian (7.9) can be rewritten as

$$H = -G\Omega S^+_j S_j \,. \tag{7.13}$$

The commutation relations are

$$\left[S_j, S^+_j\right] = 1 - \frac{n}{\Omega} \,, \tag{7.14}$$

where n is the number operator

$$n = \sum_m a^+_{jm} a_{jm} = \sum_{m>0} \left(a^+_{jm} a_{jm} + a^+_{j-m} a_{j-m} \right) . \tag{7.15}$$

A class of eigenstates of H [see (7.9)] is now given by

$$\left[H, S_j^+\right] = -GS_j^+(\Omega - \boldsymbol{n}) = -G(\Omega - \boldsymbol{n} + 2)S_j^+ \; . \tag{7.16}$$

Starting from zero valence particles, one gets

$$\begin{aligned} HS_j^+|0\rangle &= -G\Omega S_j^+|0\rangle \; , \\ H\left(S_j^+\right)^2|0\rangle &= -2G(\Omega - 1)\left(S_j^+\right)^2|0\rangle \; , \\ &\vdots \qquad\qquad \vdots \\ H\left(S_j^+\right)^{n/2}|0\rangle &= -\frac{G}{4}n(2\Omega - n + 2)\left(S_j^+\right)^{n/2}|0\rangle \; . \end{aligned} \tag{7.17}$$

Here all particles are coupled pairwise to $J^\pi = 0^+$. They are described as *seniority* $v = 0$ *states*, where the seniority quantum number v denotes the number of unpaired particles. In shorthand we can write

$$|n, v = 0\rangle \equiv \left(S_j^+\right)^{n/2}|0\rangle \; , \tag{7.18}$$

and

$$E_{v=0}(n) = -\frac{G}{4}n(2\Omega - n + 2) \; . \tag{7.19}$$

One can generalize this procedure in adding to the S_j^+ operator, the $\Omega - 1$ operators B_J^+ which create pairs coupled to angular momentum $J(J \neq 0)$ defined as

$$B_J^+ = \sum_{m>0} (-1)^{j+m} \langle jmj - m|J0\rangle a_{jm}^+ a_{j-m}^+ \; . \tag{7.20}$$

They obey the relation

$$\left[H, B_J^+\right]|0\rangle = 0 \; . \tag{7.21}$$

So we can construct a set of seniority $v = 2$ states

$$\begin{aligned} HB_J^+|0\rangle &\equiv H|n = 2,\, v = 2,\, J\rangle &&= 0 \; , \\ HS_j^+B_J^+|0\rangle &\equiv H|4,\, 2,\, J\rangle &&= -G(\Omega - 2)|4,\, 2,\, J\rangle \; , \\ &\vdots && \vdots \\ H\left(S_j^+\right)^{(n/2-1)}B_J^+|0\rangle &\equiv H|n,\, 2,\, J\rangle &&= -\frac{G}{4}(n - 2)(2\Omega - n)|n,\, 2,\, J\rangle \; . \end{aligned} \tag{7.22}$$

For $v > 2$ we cannot continue in this way since an overcomplete set of states $B_J^+B_J^+|0\rangle, \ldots$ arises. Still we can calculate the energy since a state with maximum seniority $v = n$ has energy zero or $H|n, v = n\rangle = 0$. We can now calculate successively

$$\begin{aligned}
H|n, n-2\rangle &= H\left(S_j^+\right)|n-2, n-2\rangle &&= -G(\Omega - n + 2)|n, n-2\rangle\,, \\
H|n, n-4\rangle &= H\left(S_j^+\right)^2|n-4, n-4\rangle &&= -G(2\Omega - 2n + 6)|n, n-4\rangle\,, \\
&\vdots && \vdots \\
H|n, v\rangle &= H\left(S_j^+\right)^{(n-v)/2}|v, v\rangle &&= -\frac{G}{4}(n-v)(2\Omega - n - v + 2)|n, v\rangle\,,
\end{aligned} \tag{7.23}$$

giving the general expression for the spectrum

$$E_v(n) - E_0(n) = \frac{G}{4}v(2\Omega - v + 2)\,. \tag{7.24}$$

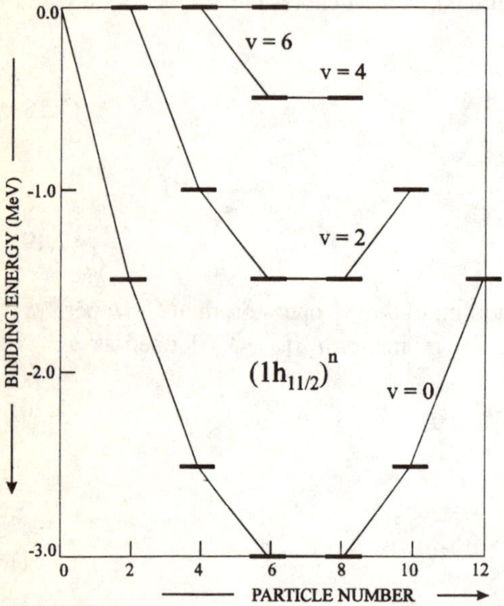

Fig. 7.5. The spectrum for a pure pairing force within the $1h_{11/2}$ orbital i.e. for the $(1h_{11/2})^n$ spectrum with seniority $v = 0$, $v = 2$, $v = 4$ and $v = 6$ as a function of particle number n. The pairing strength G was chosen as $G = 0.25$ MeV (see (7.24))

The spectrum relating to (7.24), is illustrated in Fig. 7.5 for the $1h_{11/2}$ shell where we steadily fill the orbital with 2, 4, 6, 8, 10, 12 particles. The pairing strength G was taken as $G = 0.25$ MeV. Here, one again observes that the $J^\pi = 0^+$ state is the lowest in all cases, next come the states with one "broken-pair" (see Sect. 7.6) (seniority $v = 2$), etc. It appears that the binding energy of the pairs is maximal when the orbital is half-filled. It is also remarkable to observe that the energy difference (7.24) is *independent* of n as shown in Fig. 7.5.

For a more realistic case: the $N = 50$ single-closed shell nuclei where the proton number represents 4 holes, 6 holes, 8 holes, 10 holes respectively in the $Z = 50$ proton core, the pairing properties are experimentally well established (see the 0^+, 2^+, 4^+, 6^+, 8^+ spectra) (see Fig. 7.6) (Sau et al. 1983).

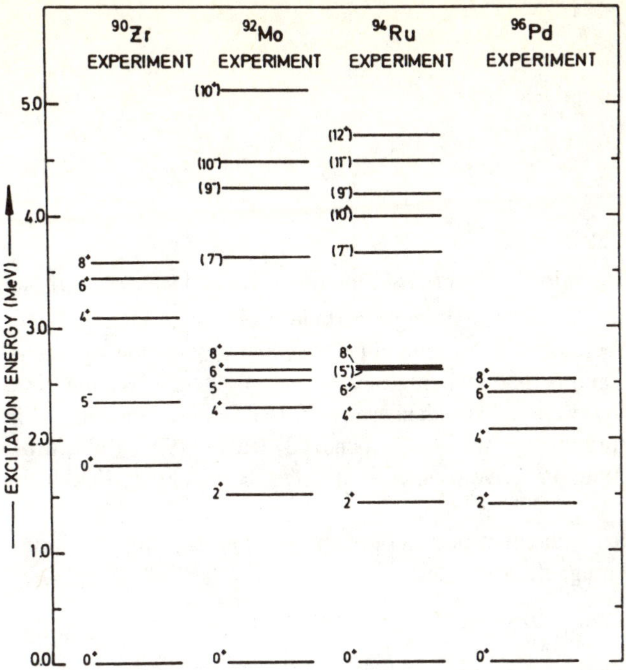

Fig. 7.6. The experimental spectra for the N = 50 single closed-shell nuclei ^{90}Zr, ^{92}Mo, ^{94}Ru, ^{96}Pd up to an excitation energy of $E_x \cong 5\,\text{MeV}$. From ^{92}Mo on, the seniority v = 2 spectrum of nucleons moving in the $1g_{9/2}$ orbital is clearly observed (taken from (Sau et al. 1983))

7.3 Pairing in Non-Degenerate Levels: Two-Particle Systems

If we now consider, first 2 (later n) particles in a number of non-degenerate orbitals much of the simplicity of the above exact solvable model is lost but the main results of a possible classification in the energy spectra according to seniority $v = 0$, $v = 2$, ... remain approximately realized in actual nuclei.

So, let us suppose that the different single-particle orbitals are such that $\Delta\varepsilon_a < G$ on the average. Moreover, for the short-range interactions discussed before, the $J^\pi = 0^+$ matrix elements are attractive and much larger than the other J^π matrix elements.

We shall, in order to make the discussion applicable to deformed nuclei too, work in the basis where only the magnetic number is denoted. So, the orbitals we use could be the Hartree-Fock orbitals in a deformed nucleus where the angular momentum is no longer a good quantum number, but only its projection on the symmetry axis for axially symmetric systems (Rowe 1970, Ring, Schuck 1980, Bohr, Mottelson 1969) is. It is then always possible to expand in a spherical basis (general case)

$$
\begin{aligned}
a_{\mu}^{+} &= \sum_{j,m} c_{j,m}^{\mu} a_{j,m}^{+} , \\
a_{\bar{\mu}}^{+} &= \sum_{j,m} c_{j,m}^{\mu} a_{j,\bar{m}}^{+} ,
\end{aligned}
\tag{7.25}
$$

where

$$a_{j,\bar{m}}^{+} \equiv (-1)^{j+m} a_{j,-m}^{+} , \tag{7.26}$$

and indicates for $a_{\bar{\mu}}^{+}$ the creation of a particle in the time-reversed state of μ. If one uses the particular choice of the orbital single-particle wave functions $i^l Y_l^{(m)}(\hat{\boldsymbol{r}})$, then the time-reversal operator or the rotation operator, $\mathcal{R}_y(\pi)$ (performing a rotation of π around the y-axis) acting on a single-particle state $|j,m\rangle$ gives the same result. So, the above phase choice (Biedenharn-Rose (BR) phase-convention) is quite often used in pairing theory (see also Appendix I) (Rowe 1970, Bohr, Mottelson 1969, Alder, Winther 1971, Waroquier et al. 1979, Brussaard 1970, Allaart 1971).

Using the above BR convention, one can obtain the *same* sign for all pairing matrix elements, in the magnetic substates

$$\langle \mu\bar{\mu}|V|\nu\bar{\nu}\rangle = -G \quad \text{with} \quad G > 0 . \tag{7.27}$$

So, pairing interactions are described by the Hamiltonian

$$H = \sum_{\nu} \varepsilon_{\nu} a_{\nu}^{+} a_{\nu} - G \sum_{\mu,\nu>0} a_{\mu}^{+} a_{\bar{\mu}}^{+} a_{\bar{\nu}} a_{\nu} , \tag{7.28}$$

where ε_{ν} are the single-particle energies. If we now define the eigenstate

$$|\alpha\rangle \equiv S_{\alpha}^{+}|0\rangle = \sum_{\nu>0} X_{\nu}^{(\alpha)} a_{\nu}^{+} a_{\bar{\nu}}^{+}|0\rangle , \tag{7.29}$$

then $X_{\nu}^{(\alpha)}$ and E_{α} are determined via the secular equation

$$H|\alpha\rangle = E_{\alpha}|\alpha\rangle , \tag{7.30}$$

or

$$\left(2\varepsilon_{\nu} - E_{\alpha}\right) X_{\nu}^{(\alpha)} = G \sum_{\mu>0} X_{\mu}^{(\alpha)} . \tag{7.31}$$

The formal solution therefore becomes

$$X_{\nu}^{(\alpha)} = N_{\alpha} \frac{G}{2\varepsilon_{\nu} - E_{\alpha}} , \tag{7.32}$$

with

$$N_{\alpha} = \sum_{\mu>0} X_{\mu}^{(\alpha)} . \tag{7.33}$$

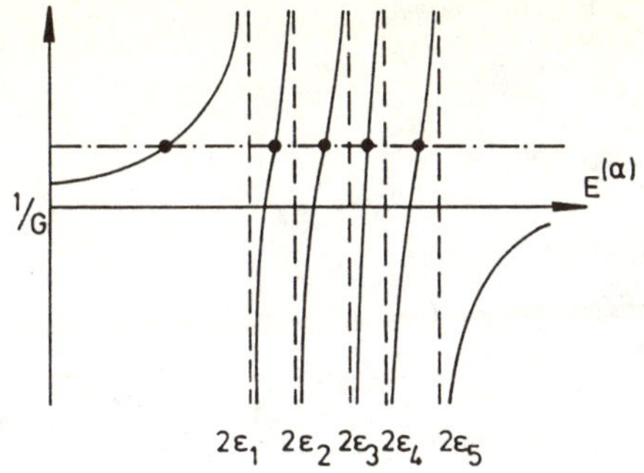

Fig. 7.7. The TDA secular (or dispersion) equation for the pairing properties of *two* particles in a number of non-degenerate orbitals at unperturbed energies $2\varepsilon_1$, $2\varepsilon_2$, $2\varepsilon_3, \ldots$ for the two-particle configurations (see (7.34)). The intersection between the horizontal line at $1/G$ with the right-hand side of (7.34) gives the eigenvalues (roots)

The normalization of the state $|\alpha\rangle$ then leads to the condition

$$\sum_{\nu} |X_{\nu}^{(\alpha)}|^2 = 1 ,$$

or,

$$\frac{1}{G} = \sum_{\nu>0} \frac{1}{2\varepsilon_{\nu} - E_{\alpha}} . \tag{7.34}$$

Since $G > 0$, we obtain a low-lying 0^+ state that comes down from the unperturbed energies $2\varepsilon_1$, $2\varepsilon_2$, ... to a low-energy. This coherent combination of 0^+ pairs is the new ground state that takes the short-range interaction optimally into account (Fig. 7.7).

One could carry through the above discussion within a spherical single-particle basis where the creation- and annihilation operators are characterized by the quantum numbers (j, m).

7.4 n Particles in Non-Degenerate Shells: BCS-Theory

The situation where many pairs move in a number of non-degenerate orbitals is most close to the actual nuclei one can observe e.g. the even-even Sn nuclei. Here, the proton number remains at $Z = 50$ but the neutron number varies between $50 \leq N \leq 82$, indicating that between 0 and 32 nucleons (0 to 16 pairs) are distributed over the available five neutron orbitals $2d_{5/2}$, $1g_{7/2}$, $1h_{11/2}$, $3s_{1/2}$, $2d_{3/2}$. Still, the first 2^+ level remains remarkably stable around an excitation energy of $E_x \cong 1.2$ MeV and needs to be explained as an important property of the nuclear many-body problem (Fig. 7.8).

Within the most general case of n particles (or $n/2 = N$ pairs), an exact treatment as in Sect. 7.2 and 7.3 is no longer possible and different approximation methods to obtain the nucleon pair distribution have been introduced: (BCS) Bardeen-Cooper-Schrieffer theory (Bardeen et al. 1957), (GS) generalized seniority (Talmi 1971), (BPA) broken-pair approximation (Allaart et al. 1988), ...

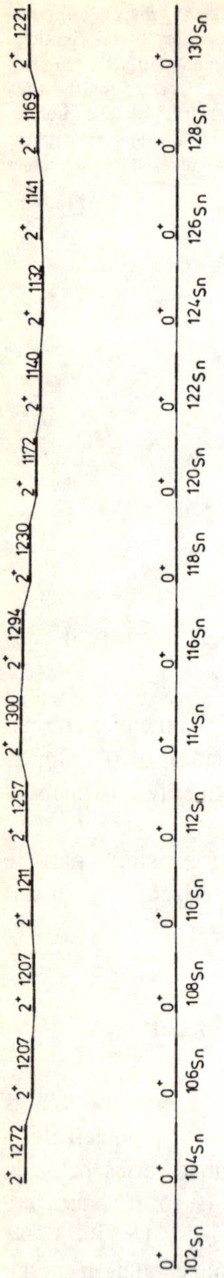

Fig. 7.8. The systematics of the first 2^+ level in the even-even Sn nuclei ($102 \le A \le 130$)

We shall later on discuss BPA in some detail since, with this approach, one most easily gains contact with the spherical shell-model that was studied in detail earlier, and also with the IBM (Interacting Boson model) of Sect. 7.7.

As a trial approach to the ground-state 0^+-wave function we use a product state of $n/2$ pair states as discussed in Sect. 7.3 and we obtain

$$|n\rangle \equiv \left(\sum_{\nu>0} c_\nu(n) a_\nu^+ a_{\bar{\nu}}^+ \right)^{n/2} |0\rangle . \tag{7.35}$$

Then, one has to determine the coefficients $c_\nu(n)$ (for each n separately) so as to study the variational problem

$$\delta\langle n|H|n\rangle = 0 . \tag{7.36}$$

The above wave function $|n\rangle$ is no longer an independent-particle wave function which makes the variational problem difficult to carry out in an analytic way.

A way out is to consider a more general wave function $|\tilde{0}\rangle$ which, projected on the space of $n/2$ pairs, gives back the ground state $|n\rangle$ of (7.35). This new trial wave function,

$$|\tilde{0}\rangle \equiv \prod_{\nu>0} \left(u_\nu + v_\nu a_\nu^+ a_{\bar{\nu}}^+ \right) |0\rangle , \tag{7.37}$$

is known as the BCS ground state, but has no constant particle number. At this stage, the coefficients u_ν and v_ν are still to be determined. Later, we shall show that it turns out that v_ν^2 is the occupation probability of the level ν.

Carrying out the projection via the operator P_n, one gets

$$P_n|\tilde{0}\rangle = \left(\left(\frac{n}{2} \right)! \right)^{-1} \left(\prod_{\mu>0} u_\mu \right) \left(\sum_{\nu>0} \frac{v_\nu}{u_\nu} a_\nu^+ a_{\bar{\nu}}^+ \right)^{n/2} |0\rangle . \tag{7.38}$$

[Prove this relation (7.38).]

The state $|\tilde{0}\rangle$ has the peculiar property that it is the vacuum state for a new kind of generalized fermion annihilation operator c_μ, such that

$$c_\mu|\tilde{0}\rangle = 0 \qquad \text{for all } \mu . \tag{7.39}$$

Here,

$$c_\mu \equiv u_\mu a_\mu - v_\mu a_{\bar{\mu}}^+ . \tag{7.40}$$

The BCS-transformation (canonical transformation) that transforms from the particle creation and annihilation operators to the new "quasi" particle operators is given by (Rowe 1970, Ring, Schuck 1980, Lane 1964, de-Shalit, Feshbach 1974)

$$\begin{aligned} c_\nu^+ &= u_\nu a_\nu^+ - v_\nu a_{\bar{\nu}} , \\ c_\nu &= u_\nu a_\nu - v_\nu a_{\bar{\nu}}^+ , \end{aligned} \tag{7.41}$$

and the inverse transformation

$$\begin{aligned} a_\nu^+ &= u_\nu c_\nu^+ + v_\nu c_{\bar{\nu}} , \\ a_\nu &= u_\nu c_\nu + v_\nu c_{\bar{\nu}}^+ . \end{aligned} \tag{7.42}$$

From (7.41) one observes, by inspection, that the new operators reduce to

$$\begin{aligned} c_\nu^+ &= a_\nu^+ \,, \quad c_\nu = a_\nu \quad (u_\nu = 1 \,,\ c_\nu = 0) \,, \\ c_\nu^+ &= a_{\bar\nu} \,, \quad c_\nu = a_{\bar\nu}^+ \quad (u_\nu = 0 \,,\ v_\nu = -1) \,, \end{aligned} \tag{7.43}$$

so that u_ν^2 and v_ν^2 can, although still in a loose way, be interpreted as occupation probabilities. Far above the Fermi level of occupied states the quasi-particle operators reduce to the regular particle operators. Far below the Fermi level the quasi-particle creation (annihilation) operators become creation (annihilation) operators of hole excitations. Near the Fermi level, some composite structure results.

From the constraint of anticommutation of the new operators $\{c_\mu^+, c_\nu\} = \delta_{\mu\nu}$, one derives

$$u_\nu^2 + v_\nu^2 = 1 \,. \tag{7.44}$$

We would like to stress again that the new vacuum state $|\tilde{0}\rangle$ does not contain a fixed particle number. In Fig. 7.9a and b, we respectively indicate the state $|\tilde{0}\rangle$, the one quasi-particle state $c_\nu^+|\tilde{0}\rangle$, as well as the population of the single-particle levels ε_ν depending on the pairing force strength G.

The quantities v_ν (or u_ν) are now determined by performing a constrained variational calculation. So, we minimize the "Hamiltonian"

$$\mathcal{H} = H - \lambda n \,, \tag{7.45}$$

$$\mathcal{H} = \sum_{\nu>0} (\varepsilon_\nu - \lambda)\left(a_\nu^+ a_\nu + a_{\bar\nu}^+ a_{\bar\nu}\right) - G \sum_{\mu,\nu>0} a_\mu^+ a_{\bar\mu}^+ a_{\bar\nu} a_\nu \,. \tag{7.46}$$

Here, λ is a Lagrange multiplier, and is chosen such that the average particle number agrees with the actual number of valence nucleons or

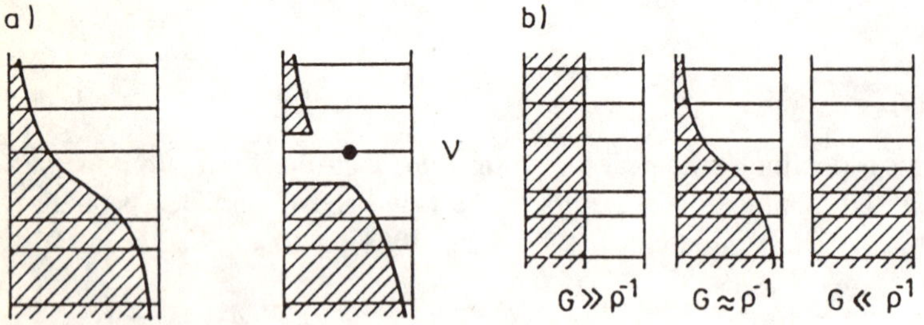

Fig. 7.9. (**a**) Representation, on a graph with the occupation probability distribution for the different orbitals, of the quasi-particle vacuum state $|\tilde{0}\rangle$ (*left-hand part*) and of a one-quasi particle (1qp) excitation in the orbital ν denoted by $c_\nu^+|\tilde{0}\rangle$. In such a case, the orbital ν is occupied by a single particle since one has annihilated in $|\tilde{0}\rangle$ the state ν for that part which was occupied and created a particle in ν for that part which was unoccupied. (**b**) Probability for the occupation of the orbitals $\nu(\varepsilon_\nu)$ for different ratios of the pairing strength G to the average distance (ϱ^{-1}) between single-particle levels. For $G \gg \varrho^{-1}$, an almost constant occupation for *all* levels ν is obtained. For $G \cong \varrho^{-1}$, a smeared out distribution and for $G \ll \varrho^{-1}$, a sharp distribution of full occupation up to a given level (Fermi-level) is obtained

$$\langle \tilde{0}|\boldsymbol{n}|\tilde{0}\rangle = n \ , \tag{7.47}$$

So, we get $\lambda = \lambda(n)$ via the condition

$$\frac{\partial}{\partial n}\langle \tilde{0}|\mathcal{H}|\tilde{0}\rangle = 0 \ , \tag{7.48}$$

or

$$\lambda = \frac{\partial}{\partial n}\langle \tilde{0}|H|\tilde{0}\rangle \ , \tag{7.49}$$

indicating that λ is the chemical potential of the A-nucleon system.

In order to determine the quantities u_ν, v_ν, we start by rewriting the Hamiltonian $\mathcal{H}$, using Wick's theorem, with respect to the new vacuum state $|\tilde{0}\rangle$ and transforming the creation and annihilation operators in (7.46), making use of the expressions (7.42).

In short, the expression turns out to have the structure

$$\mathcal{H} = U_0 + H_{11} + H_{20} + H_{02} + H_{\text{res}} \ , \tag{7.50}$$

where U_0 is the energy, corresponding to the reference state $|\tilde{0}\rangle$. Furthermore, the other terms have as general structure

$$\begin{aligned} &H_{11} \propto c^+c \ , \\ &H_{20} \propto c^+c^+ \ , \\ &H_{\text{res}} \propto c^+c^+c^+c^+ + c^+c^+c^+c + c^+c^+cc + \text{hc} \ . \end{aligned} \tag{7.51}$$

In more detail, the ground state energy reads

$$\begin{aligned} U_0 = &\sum_{\nu>0} (\varepsilon_\nu - \lambda)\ \{a_\nu^+ a_\nu + a_{\bar{\nu}}^+ a_{\bar{\nu}}\} \\ &- G \sum_{\mu,\nu>0} \{a_{\bar{\mu}}^+ a_{\bar{\nu}} a_\mu^+ a_\nu + a_\mu^+ a_{\bar{\mu}}^+ a_{\bar{\nu}} a_\nu\} \ . \end{aligned} \tag{7.52}$$

If we now evaluate the different contractions, where also the terms a^+a^+ give non-zero contributions, we obtain for the energy U_0

$$U_0 = \sum_{\nu>0}\left[2(\varepsilon_\nu - \lambda)v_\nu^2 - Gv_\nu^4\right] - G\left[\sum_{\nu>0} u_\nu v_\nu\right]^2 \ . \tag{7.53}$$

Since, in all other terms of the Hamiltonian, normal order products of operators appear, the variational problem of calculating $\langle \tilde{0}|\mathcal{H}|\tilde{0}\rangle$ reduces to evaluating the above expression U_0 and one obtains

$$\frac{\partial}{\partial v_\nu}\langle \tilde{0}|\mathcal{H}|\tilde{0}\rangle = \frac{\partial}{\partial v_\nu} U_0 = 0 \ . \tag{7.54}$$

This derivative leads to the condition

$$2(\varepsilon'_\nu - \lambda)u_\nu v_\nu = \Delta\left(u_\nu^2 - v_\nu^2\right) \ , \qquad \text{with} \tag{7.55}$$

$$\varepsilon'_\nu \equiv \varepsilon_\nu - Gv_\nu^2 ,$$
$$\Delta \equiv G \sum_{\mu>0} u_\mu v_\mu . \tag{7.56}$$

The new single-particle energy ε'_ν ($\equiv \varepsilon_\nu - Gv_\nu^2$) contains the self-energy correction of a particle in a given orbital ν interacting, via the constant pairing force, with an extra pair of nucleons. It describes the changing single-particle energy ε_ν as a function of n when starting from the constant Hartree-Fock energy ε_ν as determined for a doubly-closed shell nucleus. In many BCS calculations where a set of single particle energies is deduced from the data, the ε'_ν are used from the beginning.

From (7.55) one obtains the solutions

$$u_\nu^2 = \frac{1}{2}\left[1 + \frac{(\varepsilon'_\nu - \lambda)}{\left[(\varepsilon'_\nu - \lambda)^2 + \Delta^2\right]^{1/2}}\right] ,$$
$$v_\nu^2 = \frac{1}{2}\left[1 - \frac{(\varepsilon'_\nu - \lambda)}{\left[(\varepsilon'_\nu - \lambda)^2 + \Delta^2\right]^{1/2}}\right] . \tag{7.57}$$

One now needs two equations to determine the chemical potential λ and the quantity Δ. From (7.55) which is also called the "gap" equation, one can derive an equivalent form

$$\sum_{\nu>0}\left[\left(\varepsilon'_\nu - \lambda\right)^2 + \Delta^2\right]^{-1/2} = \frac{2}{G} . \tag{7.58}$$

The particle number condition then leads to

$$\sum_{\nu>0}\left[1 - \frac{\varepsilon'_\nu - \lambda}{\left[\left(\varepsilon'_\nu - \lambda\right)^2 + \Delta^2\right]^{1/2}}\right] = n . \tag{7.59}$$

In practice, one has now to solve (7.58) and (7.59) simultaneously for a given set of energies ε_ν, given n and pairing strength G for the unknowns λ, Δ. This highly non-linear set of two coupled equations can be solved using the iterative Newton-Raphson method in a rapid way. Once Δ, λ are known, all other quantities v_ν^2, u_ν^2, ε'_ν can be obtained. In Chap. 9 we enclose a code for solving these equations using a constant pairing force and a number of single-particle states.

The one-body part of the Hamiltonian H_{11}, together with the part $H_{20} + H_{02}$ can be obtained by considering the single contractions in (7.46)

$$\begin{aligned}
&\sum_{\nu>0}(\varepsilon_\nu - \lambda)\left\{N\left(a_\nu^+ a_\nu\right) + N\left(a_{\bar\nu}^+ a_{\bar\nu}\right)\right\} \\
&\quad - G \sum_{\mu,\nu>0}\left\{a_\mu^+ a_{\bar\mu}^+ N\left(a_{\bar\nu} a_\nu\right) + N\left(a_\mu^+ a_{\bar\mu}^+\right) a_{\bar\nu} a_\nu \right. \\
&\quad \left. + a_{\bar\mu}^+ a_{\bar\nu} N\left(a_\mu^+ a_\nu\right) + N\left(a_{\bar\mu}^+ a_{\bar\nu}\right) a_\mu^+ a_\nu\right\} .
\end{aligned} \tag{7.60}$$

Using the contractions

$$\begin{aligned} a_\alpha^+ a_\beta &= \delta_{\alpha\beta} v_\alpha^2 \\ a_\alpha^+ a_\beta^+ &= a_\beta a_\alpha = \delta_{\bar{\alpha}\,\bar{\beta}}\, u_\alpha v_\alpha = -\delta_{\bar{\alpha}\,\bar{\beta}}\, u_\beta v_\beta \ , \end{aligned} \tag{7.61}$$

the part corresponding to H_{11} reads

$$H_{11} = \sum_{\nu>0} E_\nu \left(c_\nu^+ c_\nu + c_{\bar{\nu}}^+ c_{\bar{\nu}} \right) . \tag{7.62}$$

Herein, we have used the definition

$$E_\nu = \left[\left(\varepsilon'_\nu - \lambda \right)^2 + \Delta^2 \right]^{1/2} . \tag{7.63}$$

If we now could show that $H_{02} + H_{20}$ is not very important, it could be shown that going to the new representation of quasi-particle operators, a set of new elementary excitation modes are defined that absorb a large part of the residual interaction (pairing part) into the new basis.

Evaluating $H_{20} + H_{02}$ from the remaining terms in (7.60), one obtains

$$H_{20} + H_{02} = \sum_{\nu>0} \left[2\left(\varepsilon'_\nu - \lambda \right) u_\nu v_\nu - \Delta \left(u_\nu^2 - v_\nu^2 \right) \right] \left(c_\nu^+ c_{\bar{\nu}}^+ + c_{\bar{\nu}} c_\nu \right) . \tag{7.64}$$

This term is identically zero, as the variational problem implies (7.55), and therefore the factor in square brackets in (7.64) vanishes. In retrospect, one can reinterpret the quasi-particle representation as that representation in which the two-particle scattering processes across the Fermi level become absorbed in defining a new basis and reference state. This condition [vanishing term in (7.64)] implies the minimum in energy of U_0.

Finally, the total Hamiltonian becomes (relative to the energy U_0)

$$\mathcal{H} = \sum_{\nu>0} E_\nu \left(c_\nu^+ c_\nu + c_{\bar{\nu}}^+ c_{\bar{\nu}} \right) + H_{40} + H_{31} + H_{22} + \text{hc} \ , \tag{7.65}$$

with the latter terms (H_{40}, H_{31}, H_{22}, ...) to be considered as the residual interaction H_{res} relative to the new zero-order basis.

With (7.65) we are back to a well-known problem of the nuclear shell-model with a number of valence nucleons in open shells. One can again consider the ground state $|\tilde{0}\rangle$, one-, two-, three-,... quasi-particle excitations relative to this ground state. In even-even nuclei one shall consider

$0 + 2\,\text{qp}$ excitations ,

and in odd-mass nuclei

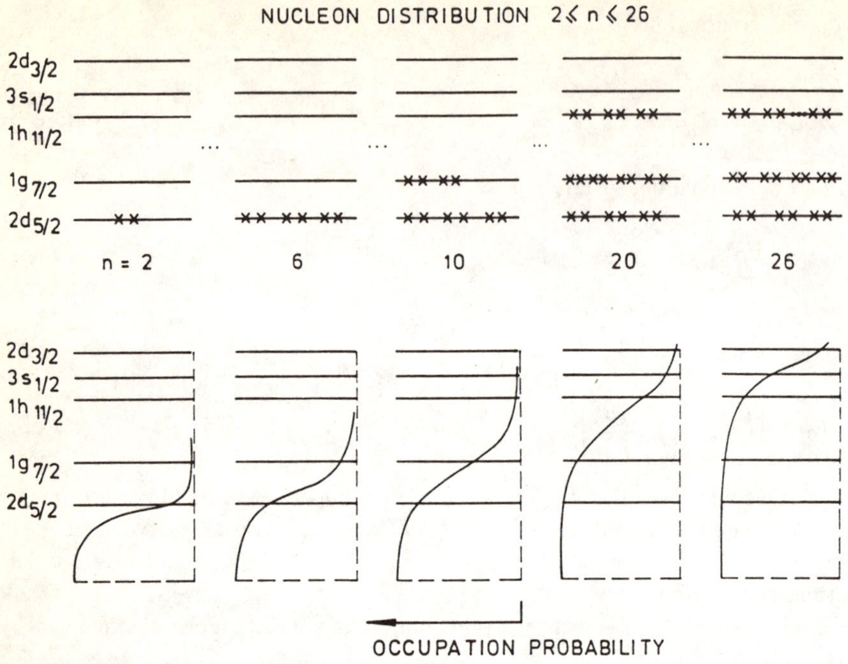

Fig. 7.10. Illustration of the distribution of a number of nucleons $n(2 \leq n \leq 26)$ over the five orbitals $2d_{5/2}$, $1g_{7/2}$, $1h_{11/2}$, $3s_{1/2}$ and $2d_{3/2}$. In the upper part, we depict one (for each n) of the many possible ways the n nucleons can be distributed over the five available orbitals. This number of possible distributions grows very fast with increasing n. In the lower part we give, for increasing n, the optimal pair distribution according to the BCS prescription (7.54) of minimal ground-state energy $\langle\tilde{0}|\mathcal{H}|\tilde{0}\rangle$

$$1 + 3\,\text{qp} \quad \text{excitations (Baranger 1960, 1961, Kuo et al 1966a, 1966b)}\,.$$

The above configurations, when pictured in a particle representation of *all* nucleons outside the closed shell, correspond to highly complicated nucleon distributions. So, the enormous advantage here is that with numerical calculations that are not more difficult than performing a two-particle shell-model calculation, one studies a large chain of even-even nuclei. In Fig. 7.10, we illustrate this for the even-even Sn nuclei where we consider a number of cases: in the upper part we consider 2, 6, 10, 20 and 26 neutrons to be distributed over the available five orbitals. Although we only draw the most trivial distribution we should in principle draw, as n increases, a large increasing set of distributions for describing the ground state wave function. In the lower part, the BCS pair distribution, adapted to each case by solving the gap equations, gives already within the 0 qp space only a very good description of the actual pair distribution. In considering also the excited states, the 2 qp excitations have to be taken into account, but even here this leads maximally (for $J^\pi = 2^+$) to the diagonalization of a 9 dimensional matrix.

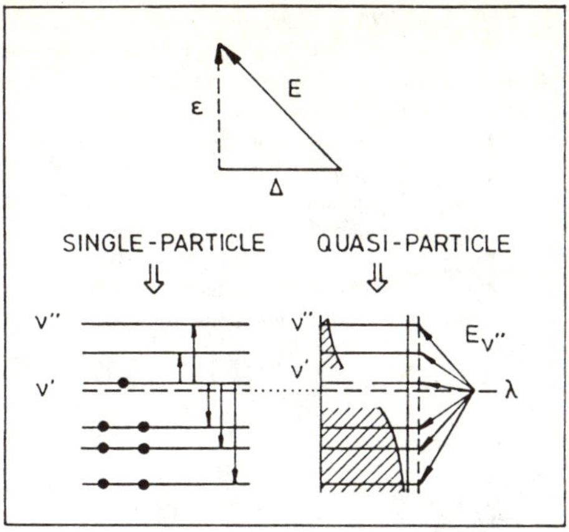

Fig. 7.11. The difference between single-particle excitations $(\varepsilon_{\nu''} - \varepsilon_{\nu'})$ and one-quasi particle excitations $E_{\nu''} - E_{\nu'}$, illustrated in a schematic way. The Fermi level λ is also given. The single-particle energy difference is measured by the distance between two given single-particle configurations. The one quasi-particle energy difference follows from a geometrical construction since $E_\nu = \sqrt{(\varepsilon_\nu - \lambda)^2 + \Delta^2}$ (see the insert in Fig. 7.11)

In odd nuclei, the one-quasi particle excitations are the corresponding low-lying excited states that correspond to single-particle (or single-hole) excitations at doubly-closed shell nuclei. Here, the effects of the pair distribution on the odd-particle comes about via the change of the single-particle energy ε_ν into the one quasi-particle energy E_ν. This becomes minimal Δ (for $\varepsilon_\nu = \lambda$).

In Fig. 7.11 we compare the distinctly different features between excitations in odd-mass nuclei: starting from the Fermi level, single-particle energy differences will always be larger than the corresponding quasi-particle energy differences (because of the dominant Δ factor). We also indicate the geometric way to interpret the one-quasi particle energy.

In even-even nuclei, the lowest excited states result at a minimal value of $E_x \cong 2\Delta$ (which in most cases is of the order of 2 MeV) (see Fig. 7.12).

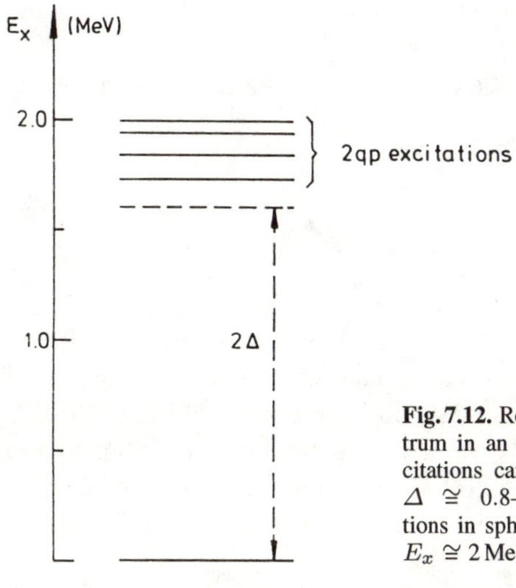

Fig. 7.12. Representation of the two quasi-particle (2qp) spectrum in an even-even nucleus. The lowest possible 2qp excitations cannot show up below 2Δ. For typical values of $\Delta \cong 0.8$–1.0 MeV, this indicates that most 2qp excitations in spherical, single closed-shell nuclei show up around $E_x \cong 2$ MeV

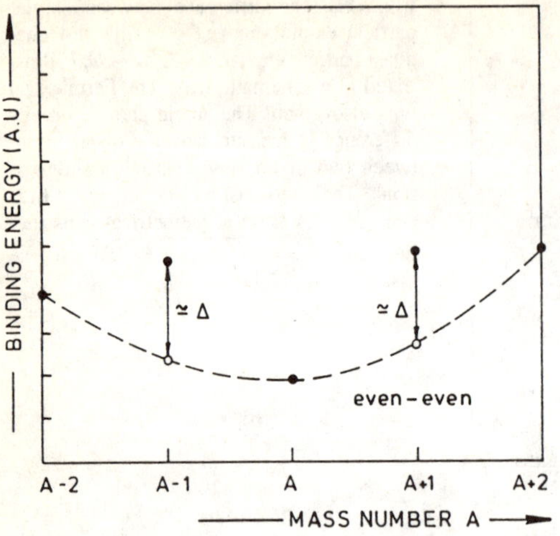

Fig. 7.13. Illustration of the relation between the pairing gap (Δ) or better, the lowest one-quasi particle energy, corresponding with the ground-state configuration E_{1qp} (ground-state) and the binding energy of a number of nuclei. In a linear approximation, one obtains the expression E_{1qp} (ground-state) $\simeq \Delta = 1/2(\mathrm{BE}(A+2) + \mathrm{BE}(A) - 2\mathrm{BE}(A+1))$ with BE(A) the binding energy in the nucleus with A nucleons

The energy in an even-even nucleus and the adjacent odd-A nuclei is, in lowest order described by U_0, $U_0 + E_\nu$, $U_0 + E_{\nu'}$ and so by studying the difference in binding energy in the ground states in adjacent nuclei, an empirical estimate of Δ (upper limit) can be obtained (Bohr, Mottelson 1969). This relation is expressed in Fig. 7.13.

Later on, in Sect. 7.5, we shall discuss some more realistic cases where conditions like constant pairing matrix elements are relaxed and thus, a better agreement with experiment can be realized.

First, we discuss briefly the effect of H_{res} in the quasi-particle scheme. The main effect is that the number of quasi-particles gets mixed because of the structure of the different terms. H_{40} has 4 quasi-particle creation operators, H_{31} three quasi-particle creation and one annihilation operator, etc., depicted diagrammatically in Fig. 7.14. The diagrams H_{40} (H_{04}), H_{31} (H_{13}) relate to RPA-type correlations whereas H_{22} is the only quasi-particle number conserving interaction. It has the same form as in the ordinary particle representation.

We give some attention to the existence of solutions to the BCS gap equations

$$\begin{aligned} \Delta &= G \sum_\nu u_\nu v_\nu \;, \\ n &= 2 \sum_{\nu>0} v_\nu^2 \;. \end{aligned} \tag{7.66}$$

These equations always contain the trivial solution $\Delta = 0$, $u_\nu = 1, 0$ and $v_\nu = 0, 1$, corresponding with a sharp Fermi distribution. Sometimes, a non-trivial solution $\Delta \neq 0$ exists. If G is too small, or $\Delta\varepsilon$ too large compared to G, no BCS solution outside the trivial one exists. If we take α as the highest, occupied and β as the lowest, unoccupied orbital, then, to fulfil the number equation of the gap equations, λ must occur in between ε_α and ε_β. If for such a value of λ, G is so small that

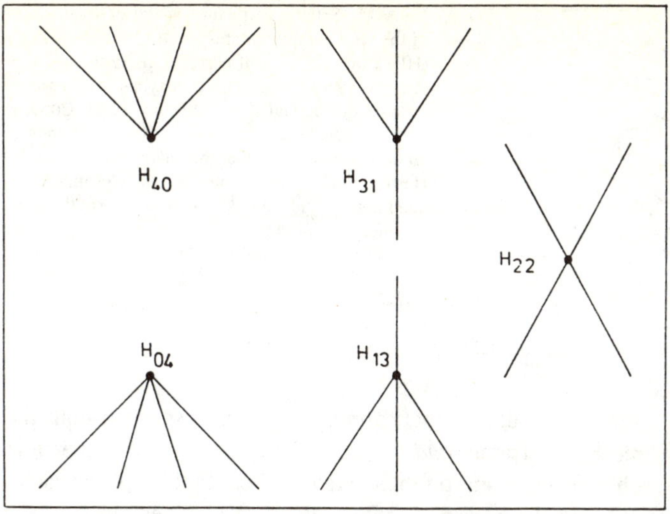

Fig. 7.14. Diagrammatic represention of the different contributions to the residual interaction in the quasi-particle representation (see (7.65)). The notation H_{nm} is such that n indicates (reading from below the interaction vertex to above the vertex) the number of outgoing lines and m the number of incoming lines or n indicates the number of qp creation operators and m the number of qp annihilation operators. The terms $H_{40}(H_{04})$ induce RPA-like correlations, $H_{31}(H_{13})$ the creation (annihilation) of 2qp states and H_{22} the scattering of 2qp states

$$\frac{G}{2}\sum_{\nu>0}|\varepsilon_\nu - \lambda|^{-1} < 1\,, \tag{7.67}$$

then clearly no solution for

$$\frac{G}{2}\sum_{\nu>0}|\varepsilon_\nu - \lambda|^{-1}\left[1+\frac{\Delta^2}{(\varepsilon_\nu-\lambda)^2}\right]^{-1/2} = 1\,, \tag{7.68}$$

exists. This is precisely the case for closed-shell nuclei where pairing is so weak that no scattering into the empty orbitals in the next major oscillator shell is possible. For nuclei with a partly filled, degenerate j-shell, a BCS solution ($\Delta \neq 0$) always exists no matter how small G becomes.

As examples we quote (Rowe 1970):

i) $2s$-$1d$ shell nuclei: deformed nuclei exist but a non-trivial BCS solution does not.
ii) single-closed shell Sn, $N = 82$ nuclei, ...: spherical solutions, BCS solutions exist.
iii) rare-earth nuclei: deformed solutions with a non-trivial BCS solution on top exist.

We finally point out that minimizing the energy U_0 to find a BCS solution does not immediately imply that we have obtained the lowest energy solution

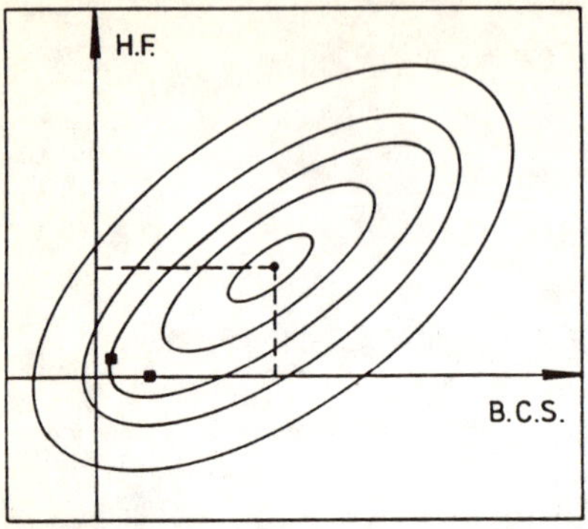

Fig. 7.15. Schematic picture of the optimalization of the total energy with respect to the Hartree-Fock (HF) and BCS type of correlations in the ground-state wave function. The two types are depicted on two orthogonal axes (HF and BCS). Contour lines of constant energy are shown. It is shown, albeit schematically, that the energy minima in the Hartree-Fock ground state and BCS ground state need not coincide and do not always form the lowest energy minimum

to H in an absolute sense. It could be that, minimizing with respect to both the structure of the Hartree-Fock orbitals *and* the pair distribution at the same time (HFB), one could reach a deeper point on the energy surface. Fully self-consistent HFB calculations have rarely been carried out. In the best of cases, an iterative set of HF + BCS minimizations are carried out, a method that can in most cases approach the absolute minimum in a good way (Ring, Schuck 1980) (see Fig. 7.15).

7.5 Applications of BCS

We shall now discuss a number of typical results that are obtained using the BCS method, using this time the more realistic set of gap equations, in which (7.56) has to be replaced by (we use spherical nuclei) (Kuo, Brown 1966, Kuo et al. 1966a, Heyde, Waroquier 1971, Waroquier, Heyde 1970, 71)

$$\Delta_a = (2j_a + 1)^{-1/2} \sum_c (2j_c + 1)^{1/2} u_c v_c \langle j_a^2; 0^+ | V | j_c^2; 0^+ \rangle \ . \tag{7.69}$$

We discuss (i) odd-even mass differences and one-quasi-particle energies, (ii) energy spectra in even-even and odd-mass nuclei, (iii) electromagnetic transition rates, (iv) spectroscopic factors and applications to neutron stars, respectively.

7.5.1 Odd-Even Mass Differences, $E_{1\,\mathrm{qp}}$

Because of the pairing correlations, the last odd nucleon is less bound in odd-mass nuclei compared to the even-even nuclei. A mass difference, in a linear approximation, can be defined via the relation (Bohr, Mottelson 1969)

$$E_{1\,\mathrm{qp}} = \tfrac{1}{2}\big(BE(n+1) + BE(n-1) - 2BE(n)\big) \ , \tag{7.70}$$

where n is the number of valence nucleons in the odd-mass nucleus and BE is the binding energy of the nucleus. If the ground state in the odd-mass nucleus would always be a pure one-quasi particle (1 qp) state, then one has

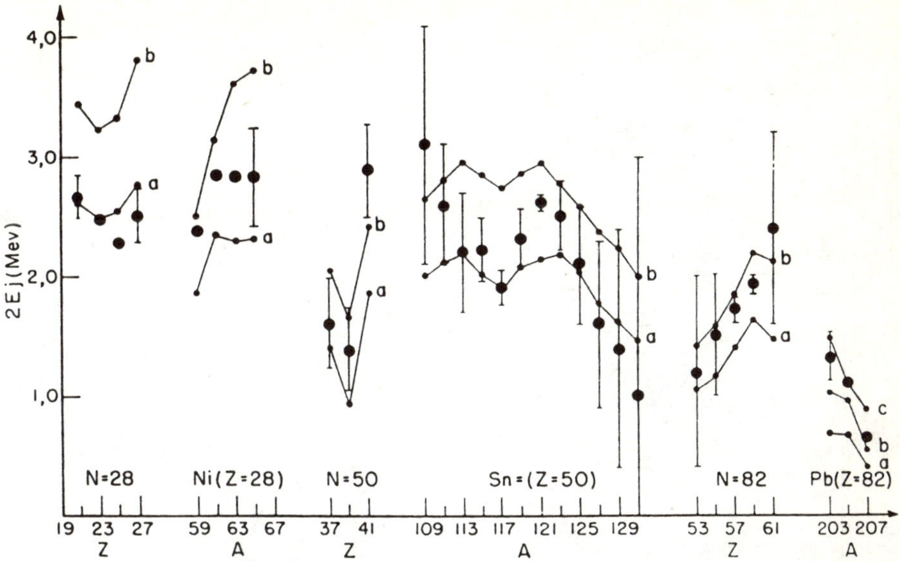

Fig. 7.16. Even-odd mass differences (see Fig. 7.13). The dots are the experimental differences BE($A+1$)+BE($A-1$) − 2BE(A). The theoretical curves are $2E_{1\text{qp}}(j)$, twice the energy of the lowest 1qp excitation for the odd-A nuclei. Curves (a) and (b) are for $G = 19/A$ and $23/A$. Curve (c), for the Pb nuclei only, corresponds with $G = 30/A$ (taken from (Kisslinger, Sorensen 1960))

$$E_{1\,\text{qp}}(\nu) = \left[(\varepsilon_\nu - \lambda)^2 + \Delta^2\right]^{1/2} \simeq \Delta\,, \tag{7.71}$$

if

$$\varepsilon_\nu \simeq \lambda\,.$$

We illustrate $2E_{1\,\text{qp}}$, as obtained from (7.70) throughout the nuclear mass region (Fig. 7.16). The data are compared with constant pairing strength calculations using $G = 23/A$ (line b), $G = 19/A$ (line a) and $G = 30/A$ (line c) for the Pb region only. This figure has been taken from (Kisslinger, Sorensen 1960) and illustrates that pairing is able to account in a very systematic way for the behaviour in odd-even mass differences.

We also illustrate, in some more detail, for the $N = 82$ nuclei, how the 1 qp energies change with increasing Z ($53 \le Z \le 63$) for given proton single-particle energies (Fig. 7.17) (see insert). The data and theoretical values are compared. For more details on the force used in the calculation, we refer to (Heyde, Waroquier 1971, Waroquier, Heyde 1970, 1971). The effect of filling particles and the subsequent variation of λ is very nicely illustrated in the interchange of $2d_{5/2}$ and $1g_{7/2}$ between $Z = 57$ and $Z = 59$ with, on top, a steady decrease of the $1h_{11/2}$, $3s_{1/2}$ and $2d_{3/2}$ 1 qp energies. A gap between the ($2d_{5/2}$, $1g_{7/2}$) and ($1h_{11/2}$, $3s_{1/2}$, $2d_{3/2}$) orbitals is evident.

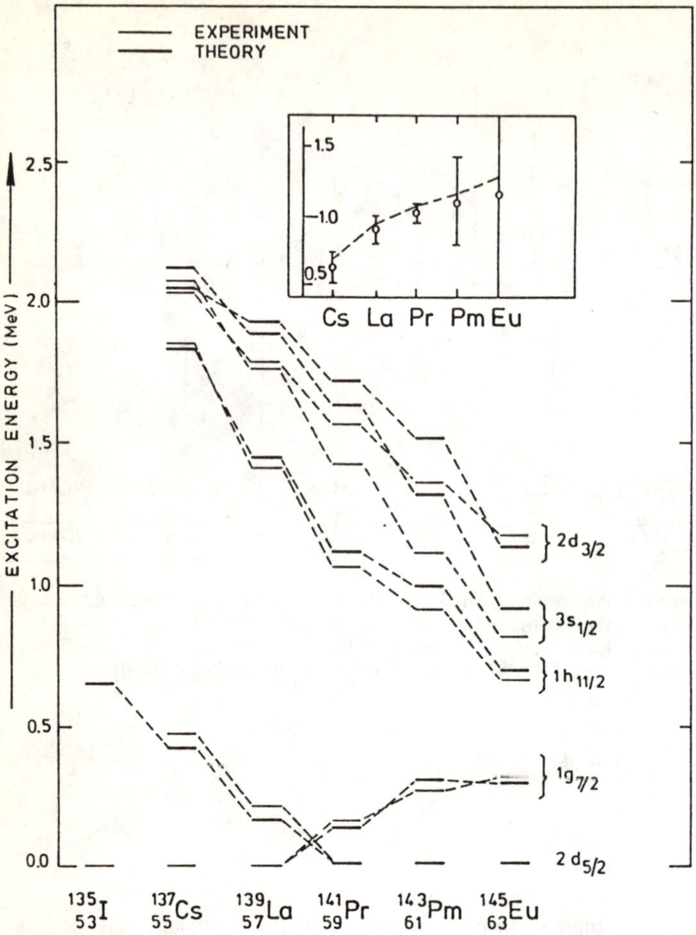

Fig. 7.17. Realistic calculation for the lowest $5/2^+$, $7/2^+$, $11/2^-$, $3/2^+$ and $1/2^+$ excitations in the $N = 82$ nuclei, that correspond to rather pure one-quasi particle configurations. Both the theoretical and experimental energies, as a function of Z, are given. In the insert, we give the lowest quasi-particle energy $E_a(A)$, as determined from the odd-even mass differences. We compare the data with the theoretical values

7.5.2 Energy Spectra

In even-even nuclei, as discussed before, one does not expect excited states to lower below the value of 2Δ in the unperturbed energy spectra. Since $\Delta \cong 1\,\text{MeV}$, this means that above $E_x > 2\,\text{MeV}$, the level density should rapidly increase, as is indeed the case for the $N = 82$ nuclei: ^{136}Xe, ^{138}Ba, ^{140}Ce (see Fig. 7.18). There are, however, some states that descend into the gap, in particular the $J^\pi = 2^+$ collective state, which occurs around $E_x \cong 1.2$-$1.4\,\text{MeV}$. Also, one 4^+ and 6^+ level decreases below the value 2Δ but it is indeed close to $E_x \cong 2$–$2.5\,\text{MeV}$ that many levels are observed and calculated in the $0 + 2\,\text{qp}$ calculation using a Gaussian residual interaction within the TDA approximation.

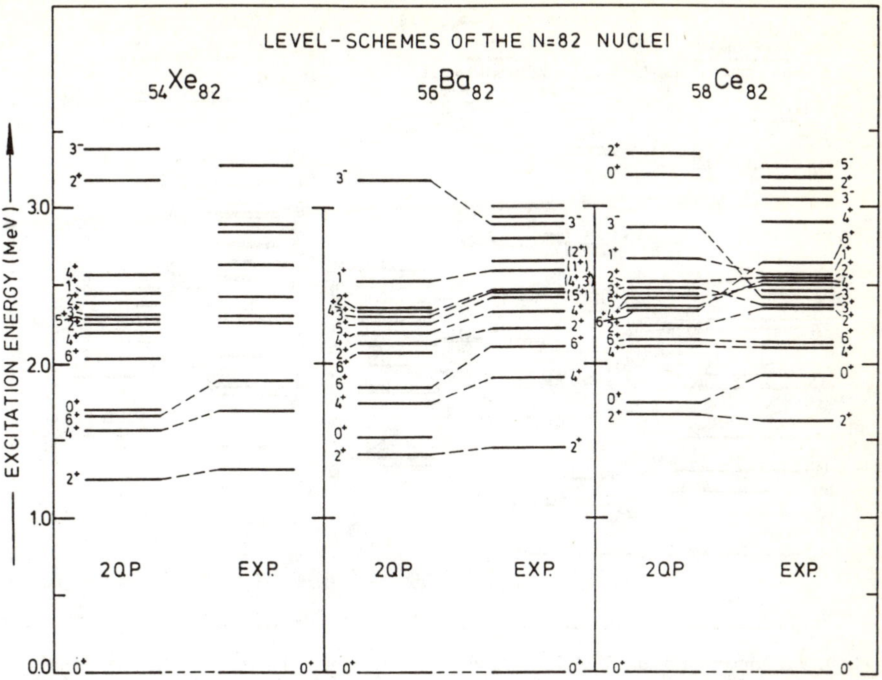

Fig. 7.18. Level schemes in the single closed-shell $N = 82$ nuclei ^{136}Xe, ^{138}Ba and ^{140}Ce. We compare the experimental spectra with the results of a 2qp calculation. In the calculations, a Gaussian radial shape was used with spin exchange, expressed by the spin projection operators P_S, P_T as $V(r) = -V_0 \exp[-\beta r^2](P_S + tP_T)$. In each case $t = +0.2$ was used and V_0 obtained via the inverse gap equation (IGE) method (see also (Waroquier, Heyde 1971))

In the odd-mass nuclei, on the other hand, no such gap shows up, as the lowest levels we encounter are the 1 qp excitations E_a, E_b, ... In Fig. 7.19, we illustrate the case of ^{137}Cs, where we compare the data (Holm, Borg 1967, Wildenthal 1969, Wildenthal et al. 1971), the results of a 1 + 3 qp calculation (Heyde, Waroquier 1971) and the exact shell-model calculation of Wildenthal, using a SDI-force (Wildenthal 1969, Wildenthal et al. 1971). The comparison between both calculations is striking but the 1 + 3 qp calculation needs much less numerically involved methods.

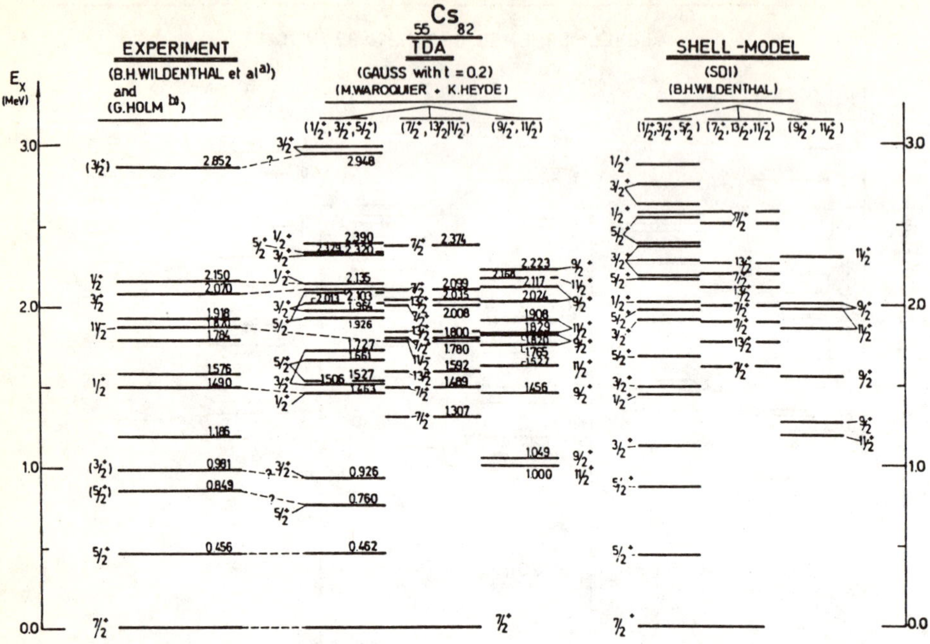

Fig. 7.19. Comparison of the TDA 1 + 3qp calculation, using a Gaussian interaction (see caption to Fig. 7.18) with spin exchange and the shell-model calculation of Wildenthal (Wildenthal 1969) using a SDI force with the data in ^{137}Cs. The data are from (Wildenthal et al. 1971) and (Holm, Borg 1967). Levels connected by dashed lines are well established states from one-nucleon transfer reactions (taken from (Heyde, Waroquier 1971))

7.5.3 Electromagnetic Transitions

The electromagnetic transition operators, that were discussed in detail in Chap. 4, and were one-body operators (Kuo, Brown 1966, Kuo et al. 1966a, Brussaard 1970, Allaart 1971, Waroquier et al. 1975)

$$\sum_{\alpha,\beta} \langle \alpha | 0_{\mu}^{(\lambda)} | \beta \rangle a_{\alpha}^{+} a_{\beta} \, , \tag{7.72}$$

can now be rewritten, using the quasi-particle transformation to the quasi-particle operators c_{α}^{+}, c_{β}. Using the Condon-Shortley phase convention for the single-particle orbital wave functions Y_l^m (also see Appendix I), one obtains for (7.72), the expression

$$\begin{aligned} &\sum_{a,b} (\hat{\lambda})^{-1} \langle a \| \mathbf{0}^{(\lambda)} \| b \rangle \tfrac{1}{2} \Big[\Big(u_a v_b + (-1)^{\lambda} u_b v_a \Big) \\ &\quad \times \Big(A^{+}(ab, \lambda\mu) + (-1)^{\mu} A(ab, \lambda - \mu) \Big] \\ &\quad + \sum_{a,b} (\hat{\lambda})^{-1} \langle a \| \mathbf{0}^{(\lambda)} \| b \rangle \Big(u_a u_b - (-1)^{\lambda} v_a v_b \Big) A^{0}(ab, \lambda\mu) \\ &\quad + \delta_{\lambda 0} \sum_{\alpha\beta} \langle \alpha | 0_{\mu}^{(\lambda)} | \beta \rangle v_a^2 \delta_{\alpha\beta} \, . \end{aligned} \tag{7.73}$$

Here,

$$A^+(ab,\lambda\mu) = \sum_{m_a,m_b} \langle j_a m_a, j_b m_b|\lambda\mu\rangle c^+_{j_a,m_a} c^+_{j_b,m_b} , \quad (7.74)$$

and

$$A(ab,\lambda\mu) \equiv \left(A^+(ab,\lambda\mu)\right)^+ .$$

We also define

$$A^0(ab,\lambda\mu) = \sum_{m_a,m_b} (-1)^{j_b-m_b} \langle j_a m_a, j_b - m_b|\lambda\mu\rangle c^+_{j_a,m_a} c_{j_b,m_b} .$$

The summation over greek variables means a summation over *all* quantum numbers i.e. $\alpha \equiv n_a, l_a, j_a, m_a$, whereas the summation over roman variables excludes the magnetic quantum number m_a i.e. $a \equiv n_a, l_a, j_a$.

Within the CS (Condon-Shortley) phase convention, one can find solutions to the BCS equations with all $v_a > 0$, and the phase $u_a = (-1)^{l_a}(1 - v_a^2)^{1/2}$. Using the BR convention, one can find solutions with both u_a, and v_a taken as positive quantities.

So, transitions 0 qp ⇔ 2 qp; 1 qp ⇔ 3 qp; 1 qp ⇔ 1 qp; 2 qp ⇔ 2 qp can occur. The corresponding reduction factors for the single-particle matrix elements, that show up in (7.73), are given in Table 7.1. One observes, that in particular $E\lambda(M\lambda)$

Table 7.1 The single-particle pairing reduction factors for various types of transitions and for the electric and magnetic multipole operators, according to (7.73). The table is such that it corresponds to the CS phase convention for defining the single-particle wave functions (i.e. one chooses $u_j v_j = (-1)^{l_j} |u_j v_j|$). The relation with the BR phase convention is discussed in the text

Operator	Type of transition	Pairing factor
Eλ	$1qp - 1qp$	$(u_i u_j - v_i v_j)$
Mλ	$1qp - 1qp$	$(u_i u_j + v_j v_i)$
Q	$1qp - 1qp$	$(u_i^2 - v_i^2)$
μ	$1qp - 1qp$	1
Eλ	$0qp - 2qp$	$(u_i v_j + u_j v_i)$
Mλ	$0qp - 2qp$	$(u_i v_j - u_j v_i)$

transitions without (with) a change in qp number can be much retarded (see the sign for destructive interference). Also the quadrupole moments Q can become heavily reduced (in the middle of a single j-shell) changing sign at the mid-shell configuration.

We now give a number of illustrations

i) The $B(E2; 6_1^+ \rightarrow 4_1^+)$ values in the $N = 82$ nuclei (Fig. 7.20). Since we have here a 2 qp → 2 qp transition, reduction factors $u_j^2 - v_j^2$ show up and are responsible for the very large modulation in $B(E2)$ values since we go through the $2d_{5/2}$ and $1g_{7/2}$ orbitals, respectively. A more detailed discussion is given in (Waroquier, Heyde 1974).

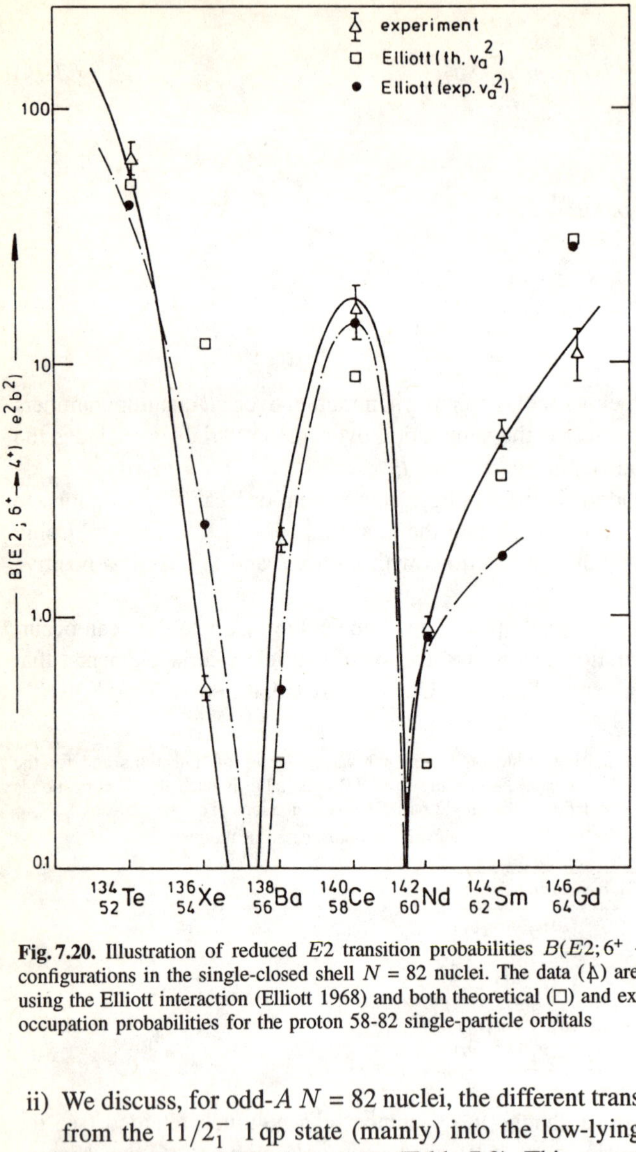

Fig. 7.20. Illustration of reduced $E2$ transition probabilities $B(E2; 6^+ \to 4^+)$ between mainly 2qp configurations in the single-closed shell $N = 82$ nuclei. The data (⍋) are compared with calculations using the Elliott interaction (Elliott 1968) and both theoretical (□) and experimental (•) values for the occupation probabilities for the proton 58-82 single-particle orbitals

ii) We discuss, for odd-A $N = 82$ nuclei, the different transitions that can deexcite from the $11/2_1^-$ 1 qp state (mainly) into the low-lying $7/2_1^+$ and $5/2_1^+$ levels that are also mainly 1 qp states (Table 7.2). This means that we have to study $M2$ and $E3$ transitions. We discuss the results for ^{137}Cs-^{145}Eu and indicate the very stable results. Here, it is also clear that the 1 qp ⇔ 3 qp components give a rather important contribution which, in all cases, corrects the 1 qp ⇔ 1 qp matrix elements towards the experimental values (Waroquier, Van den Berghe 1972).

Table 7.2. The different transition rates from the $J_i^\pi = 11/2_1^-$ state to the $7/2_1^+$ and $5/2_1^+$ states. We also indicate the half-life estimate for the $11/2_1^-$ level in the odd-proton $N = 82$ nuclei (taken from (Waroquier, Van den Berghe 1972))

			^{137}Cs	^{139}La	^{141}Pr	^{143}Pm	^{145}Eu
$11/2^- \to 7/2^+$	E_γ(exp) in MeV		1.870	1.420	0.965	0.688	0.383
	$B(M2)$ in $(eh/2Mc)^2 fm^2$	Single part. estimate	62.56	62.86	63.16	63.45	63.76
		$1QP-1QP$	33.70	25.28	18.64	18.63	13.22
		$1-3QP-1-3QP$	14.25	11.64	10.27	11.41	9.583
		Experiment				10.3 ± 0.7	
	$\frac{P_\gamma(M2;1h_{11/2}\to 1g_{7/2})}{P_\gamma(E3;1h_{11/2}\to 1g_{7/2})}$	Single part. estimate	3.47×10^2	5.95×10^2	1.28×10^3	2.49×10^3	7.79×10^3
		$1QP-1QP$	5.48×10^2	1.55×10^3	7.05×10^3	2.55×10^4	2.03×10^5
		$1-3QP-1-3QP$	3.42×10^2	1.39×10^3	1.41×10^4	7.34×10^4	1.43×10^6
		Experiment				very large	
$11/2^- \to 5/2^+$	$B(E3)$ in $e^2 fm^6$	Single part. estimate	1.526×10^4	1.548×10^4	1.570×10^4	1.593×10^4	1.615×10^4
		$1QP-1QP$	1.356×10^4	1.168×10^4	8.604×10^3	4.107×10^3	9.557×10^2
		$1-3QP-1-3QP$	1.266×10^4	1.065×10^4	6.450×10^3	3.542×10^3	8.292×10^2
		Experiment				1.098×10^4	
	E_γ(exp) in MeV		1.414	1.254	1.110	0.959	0.716
	Branching-ratio $\frac{P_\gamma(1h_{11/2}\to 1g_{7/2})}{P_\gamma(1h_{11/2}\to 2d_{5/2})}$	$1QP-1QP$	119	61	21	22	30
		$1-3QP-1-3QP$	54	31	15	15	25
		Experiment			9	4.7 ± 0.5	
	$T_{1/2}$ of the $1h_{11/2}$ level in ns	$1QP-1QP$	0.0658	0.345	3.13	17.0	431
		$1-3QP-1-3QP$	0.154	0.737	5.58	27.3	591
		Experiment		$0.5 \le T^{exp}_{1/2} 2$		26.0 ± 1.2	

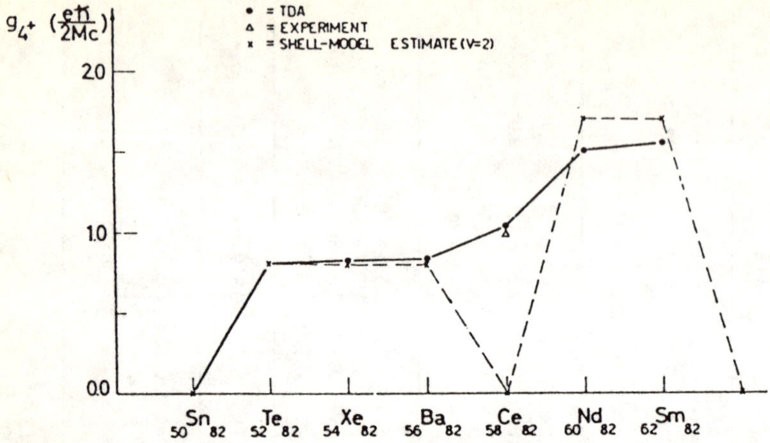

Fig. 7.21. The gyromagnetic g-factor for the $J_i^\pi = 4_1^+$ state in $N = 82$ nuclei, calculated using a proton 2qp TDA calculation. The dashed-dot line results from a pure, seniority $v = 2$ shell-model wave-function for the $J^\pi = 4^+$ level. In the case of a completely filled (or completely empty) single j-shell, only $J^\pi = 0^+$ states are formed in a pure shell-model picture. In this case $g(0^+)$ is given (taken from (Waroquier, Heyde 1971))

iii) Dipole moments $g(4^+)$ are presented for the $N = 82$ nuclei and compared to the pure shell model data, in the same way as we carried this out in (iii) (Fig. 7.21). The point in Ce is remarkable as is the agreement with the data. This clearly points out the importance of pairing correlations in obtaining rather good wave functions for these single closed shell nuclei (Waroquier, Heyde 1971).

iv) Quadrupole moments in even-even $N = 82$ nuclei are presented in Fig. 7.22. We compare, for the $Q(2_1^+)$ value with a simple-shell model estimate assuming a consecutive filling of the $2d_{5/2}$ and $1g_{7/2}$ orbitals. Already from Ba on, important deviations start to show up (Waroquier, Heyde 1971).

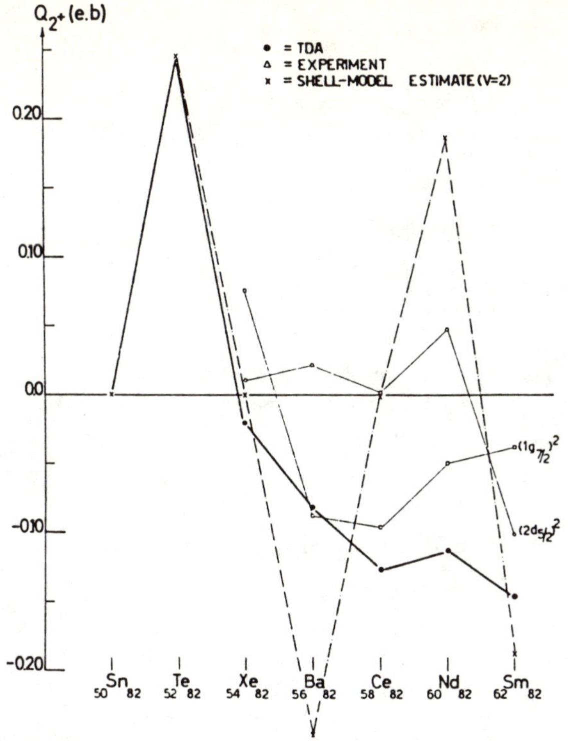

Fig. 7.22. The electric quadrupole moment for the $J_i^\pi = 2_1^+$ level in the $N = 82$ nuclei (*solid line*). The dashed line connects Q_{2^+} for a seniority $v = 2$ shell-model wave-function for the 2^+ state. For a fully occupied (or empty) shell, the same remark as in Fig. 7.21 holds. The separate contributions for $(1g_{7/2})^2_{2^+}$ and $(2d_{5/2})^2_{2^+}$ components to $Q(2_1^+)$ are also drawn (*light, solid line*) (taken from (Waroquier, Heyde 1971))

7.5.4 Spectroscopic Factors

Spectroscopic factors in one-nucleon transfer reactions are very important in that they provide information on the position and possible fragmentation of nuclear single-particle configurations (Brussaard, Glaudemans 1977, Kuo, Brown 1966, Kuo et al. 1966a).

If we describe the transfer of one-nucleon between two nuclei, the lighter nucleus being described by the wave function $\psi(J_0, M_0)$ and the heavier nucleus by $\psi(J, M)$, one defines the spectroscopic factor as

$$S(J; jJ_0) = |\langle \psi(J) \| a_j^+ \| \psi(J_0) \rangle|^2 \frac{1}{2J+1} , \tag{7.75}$$

or

$$S(J; jJ_0) = |\langle \psi(JM) | \Psi(jJ_0; JM) \rangle|^2 , \tag{7.76}$$

if $\Psi(jJ_0; JM)$ is the state obtained by coupling the odd nucleon j with the core J_0 to angular momentum J. If now the final state $\psi(J)$ would be precisely the odd particle j coupled to the core J_0, without fragmentation, the overlap integral would be unity and we should obtain a spectroscopic factor of 1.

Table 7.3 Single-nucleon transfer pairing reduction factors

Target	Final nucleus	Pairing factor
Z even, A even	$Z, A+1$	u_j^2
Z even, A even	$Z, A-1$	v_j^2
Z even, A odd	$Z, A+1$	v_j^2
Z even, A odd	$Z, A-1$	u_j^2

Suppose now that the lighter nucleus is an even-even nucleus, so $\psi(J_0, M_0) \equiv |\tilde{0}\rangle$ and that in the heavier nucleus we have a 1 qp configuration $\psi(J, M) \equiv c_{j,m}^{+}|\tilde{0}\rangle$, then the spectroscopic factor becomes

$$S(j; j0) = |\langle\tilde{0}|c_{jm}a_{jm}^{+}|\tilde{0}\rangle|^2 = u_j^2 \,. \tag{7.77}$$

We give the other combinations in Table 7.3, where all possible transitions are encountered.

We discuss now

i) occupation probabilities deduced from stripping reactions for nuclei throughout the nuclear mass region (Cr–Nd). We compare the data with the theoretical value (Table 7.4) of

Table 7.4 Comparison between the experimental and theoretical occupation factors from one-nucleon stripping to the 0^+ ground-state of spherical nuclei throughout the nuclear mass region. The theoretical value is $S_j = (2j+1)v_j^2 \cdot |C_{j,0}^{j}|^2$ where the latter coefficient gives the single-particle component in the odd-A nuclear wave function, describing the ground-state with angular momentum j (taken from (Sorensen, Lin 1966))

Target	J	l	S_{exp}	S_{theory}
^{53}Cr	3/2	1	0.91	0.97
^{57}Fe	1/2	1	0.072 ± 0.01	0.067
^{61}Ni	3/2	1	2.0 ± 0.3	1.7
^{67}Zn	5/2	3	2.2 ± 0.3	2.9
^{77}Se	1/2	1	0.68 ± 0.01	0.83
^{91}Zr	5/2	2	1.44 ± 0.21	1.6
^{95}Mo	5/2	2	2.48 ± 0.37	2.7
^{99,101}Ru	5/2	2	2.74 ± 0.4	3.3
^{105}Pd	5/2	2	1.74 ± 0.25	2.8
^{115}Sn	1/2	0	1.08 ± 0.16	1.03
^{117}Sn	1/2	0	1.4 ± 0.2	1.2
^{119}Sn	1/2	0	1.3 ± 0.2	1.4
^{125}Te	1/2	0	1.2 ± 0.3	0.99
^{135,137}Ba	3/2	2	2.4 ± 0.3	2.6
^{143,145}Nd	7/2	3	2.4 ± 1.2	1.4

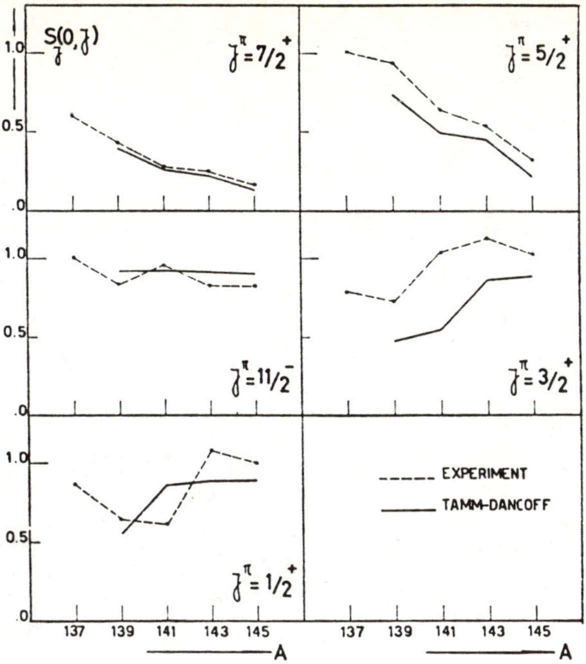

Fig. 7.23. Spectroscopic factors for (^{3}He,d) stripping to the strongest, excited $J^\pi = 1/2^+, 3/2^+, 5/2^+, 7/2^+$ and $11/2^-$ states in nuclei with a single closed-shell at $N = 82$ ($137 \leq A \leq 145$). The data (*dashed line*) and TDA 2qp calculation (*full lines*) are compared (see (7.77)) (taken from (Waroquier, Heyde 1970))

$$S_j = (2j+1)v_j^2 \left(C^j_{j,0}\right)^2 , \tag{7.78}$$

where $C^j_{j,0}$ indicates the single-particle component in the ground state j, as obtained from quasi-particle core coupling calculations (Sorensen, Lin 1966). This can give, in some cases, a large extra reduction over the 1 qp pure result of $(2j+1)v_j^2$.

ii) We compare detailed values for u_j^2 for the lowest $5/2^+$, $7/2^+$, $11/2^-$, $3/2^+$ and $1/2^+$ levels, as obtained from (^{3}He,d) reactions (Fig. 7.23). Comparison of the TDA 1 + 3 qp calculation with the data is presented and proves to be very good (Waroquier, Heyde 1970).

iii) The occupation probabilities for the $2d_{5/2}$, $1g_{7/2}$ orbitals when filling these orbitals with 4 to 13 particles (^{136}Xe to ^{145}Eu). We compare the extreme shell-model estimates (dashed line) with different calculations and the data (Fig. 7.24) (Heyde, Waroquier 1971).

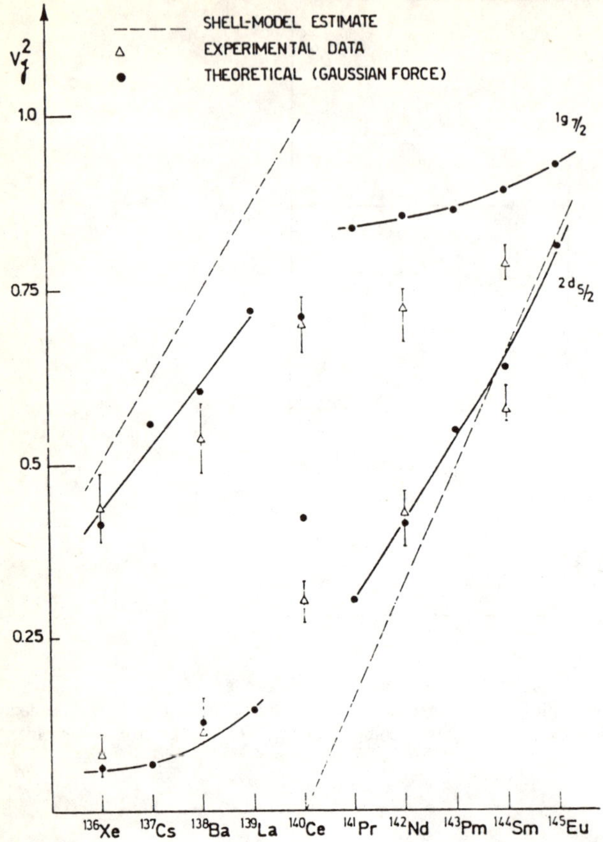

Fig. 7.24. The experimentally determined occupation probabilities for the $1g_{7/2}$ and $2d_{5/2}$ proton orbitals in the single-closed shell $N = 82$ nuclei (Wildenthal et al. 1971). A comparison with the theoretical values, obtained by solving the BCS equations, is carried out. Extreme (linear) filling of the $1g_{7/2}$ and $2d_{5/2}$ orbitals, in sequence, in a naive shell model is also presented

7.5.5 Superfluidity in Neutron Stars

The single-particle wave functions in infinite nuclear matter become plane wave solutions and are characterized by their momentum $\boldsymbol{k}$ and intrinsic spin $\chi_{m_s}^{1/2}$, or, more explicitly

$$\varphi_{\boldsymbol{k},m_s} = e^{i\boldsymbol{k}\cdot\boldsymbol{r}}\chi_{m_s}^{1/2} . \tag{7.79}$$

Pairs of mutually time-reversed states then become

$$S^+ = \sum_{\boldsymbol{k}} c_{\boldsymbol{k}}\varphi_{\boldsymbol{k},+1/2}\varphi_{\boldsymbol{k},-1/2} . \tag{7.80}$$

The expression (7.69) (or I.35) for the gap $\Delta_{\boldsymbol{k},m_s}$ in this situation becomes

$$\Delta_{\boldsymbol{k},m_s} = -1/2. \int \frac{d\boldsymbol{k}'}{(2\pi)^3} \sum_{m_s'} u_{\boldsymbol{k}',m_s'} v_{\boldsymbol{k}',m_s'} \times \langle \boldsymbol{k}, m_s; -\boldsymbol{k}, -m_s | V | \boldsymbol{k}', m_s'; -\boldsymbol{k}', -m_s' \rangle \ . \tag{7.81}$$

For an isotropic medium this simplifies into the expression

$$\Delta_k = -\frac{1}{2\pi^2} \int \boldsymbol{k}'^2 \, dk' \, u_{k',m_s'} v_{k',m_s'} V(k, k') \ , \tag{7.82}$$

with

$$V(k, k') = \langle \boldsymbol{k}, m_s; -\boldsymbol{k}, -m_s | V | \boldsymbol{k}', m_s'; -\boldsymbol{k}', m_s' \rangle \ . \tag{7.83}$$

It should be noted that one has

$$u_{k',m_s'} v_{k',m_s'} = 1/2.\Delta_{k'}/E_{k'} \ , \tag{7.84}$$

so that the gap equation (7.82) can be rewritten as

$$\Delta_k = -\frac{1}{2\pi^2} \int_0^\infty k'^2 \, dk' \, V(k, k') \Delta_{k'}/E_{k'} \ . \tag{7.85}$$

In order to understand this in more detail and carry out calculations for Δ_k, it is interesting to study first of all the integrand in (7.85). First of all, the factor uv in (7.84) is approximately equal to 1/2 for k' values near the Fermi momentum k_F and small for k' values far away from k_F. So, the largest contribution to Δ_k in (7.85) comes from a region near k_F. If $V(k, k')$ is a smooth function of k and k' near $k \simeq k' \simeq k_F$, one can conclude that

$$\Delta_k \propto k_F^2 \ . \tag{7.86}$$

Detailed and realistic calculations (Baldo et al., 1990) with the Paris (meson exchange) potential indeed exhibit the trend evident from (7.86), at least for low values of k_F. The quantity plotted in Fig. 7.25 is the value of the gap at the Fermi level $\Delta_{k=k_F}$. The nuclear matter (inside the nucleus) density corresponds to almost $k_F \cong 1.36\,\text{fm}^{-1}$. A direct conclusion is that superfluidity is mainly a property of the nuclear surface where the density corresponds to $k = 1.1\,\text{fm}^{-1}$. The calculated value of $\Delta_{k_F \approx 1.1\,\text{fm}^{-1}}$ agrees roughly with the gap parameter for the Sn nuclei.

The reason why the trend of (7.86) is not continued beyond a value of $k_F \gtrsim 1\,\text{fm}^{-1}$ is that $V(k, k')$ is not a constant, but becomes repulsive for large values of k and k' (see Fig. 7.26). This happens because of the short-range (high-k value) repulsive properties in the nucleon-nucleon interaction. For the Paris potential, the integral (7.85) has to be extended to at least $k' \simeq 50\,\text{fm}^{-1}$. This property of the nucleon-nucleon interaction is also reflected in the value of the gap parameter for $k \gg k_F$ (see Fig. 7.27).

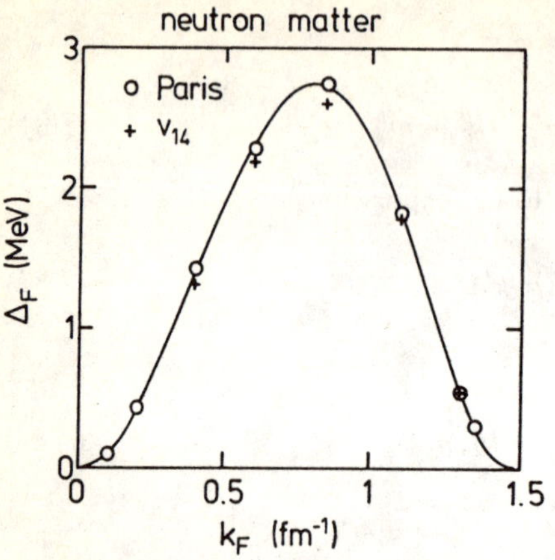

Fig. 7.25. Value of the gap at the Fermi surface Δ_{k_F} of neutron matter, as a function of the Fermi momentum, and for the Paris potential (open dots) and for the Argonne potential (crosses) (taken from Baldo et al., 1990)

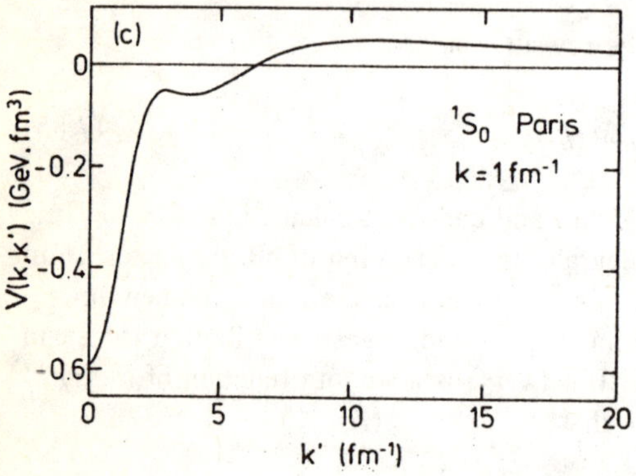

Fig. 7.26. Matrix element $V(k, k')$ of the Paris potential in the $^1S_0$ channel and this as a function of the momentum k' for the indicated value of $k = 1\,\mathrm{fm}^{-1}$ (taken from Baldo et al., 1990)

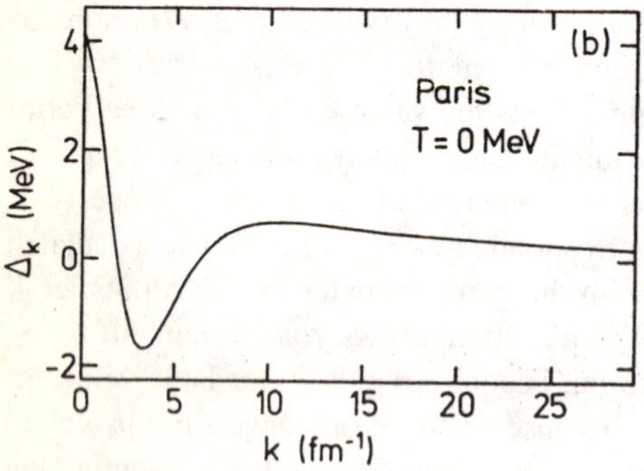

Fig. 7.27. The gap parameter Δ_{k_F} calculated for neutron matter with a Fermi momentum $k_F = 0.84\,\mathrm{fm}^{-1}$ using the Paris potential at temperature $T = 0$ (taken from Baldo et al., 1990)

7.6 Broken-Pair Model

7.6.1 Low-Seniority Approximation to the Shell-Model

Before considering the broken-pair model as such, we first consider for a system with n valence particles that are distributed over a number i of single-particle orbitals a slightly different approximation to the full shell model calculation. We call this the low-seniority shell-model (LSSM) (Racah 1942a, 1943, Racah, Talmi 1952, Talmi 1971) and recall some of the important ingredients of the model as compared to the general BCS method that was discussed in Sect. 7.4. We shall discuss the case with only one type of nucleons: extensions will be discussed later in this section.

We thereby start again from the Hamiltonian describing the interaction in an identical nucleon system

$$H = \sum_{\alpha} \varepsilon_a a_\alpha^+ a_\alpha + \tfrac{1}{4} \sum_{\alpha,\beta,\gamma,\delta} \langle \alpha\beta|V|\gamma\delta\rangle_{\mathrm{nas}} a_\alpha^+ a_\beta^+ a_\delta a_\gamma \; . \tag{7.87}$$

Considering n valence (identical) nucleons outside a doubly-closed core, the shell-model problem of constructing the n-particle basis states with fixed J^π quantum number, where nucleons are distributed over i orbitals, very soon (with n and i increasing) leads to prohibitively large energy matrices to be diagonalized. The BCS method, where a particular trial wave function was assumed (Sect. 7.4) is one approach but has the drawback of not having a correct particle number. The LSSM is a method that overcomes this point but still truncates the "model space" to tractable dimensions.

In the description of semi-magic nuclei, where one kind of active nucleons in the valence shells is considered, the pair scattering matrix elements $\langle (j_a)^2; 0^+|V|(j_c)^2; 0^+\rangle$ are by far the largest due to the short-range nature of the interaction (see Fig. 7.28). Therefore, the low-lying states in the energy spectrum of even-even nuclei are expected to be composed mainly of $J^\pi = 0^+$-coupled pairs. This concept is also supported by experimental evidence on ground-state spin of even-even semi-magic nuclei. They occur considerably lower than the $J^\pi \neq 0^+$ states. Therefore, it is meaningful to characterize the shell-model configurations of n particles distributed over the i orbitals by the number of particles combined in 0^+ pairs. For this purpose, the seniority quantum number was introduced (called v), denoting the number of unpaired particles. Thus $(n-v)/2$ angular momentum 0^+ – coupled pairs are present in a seniority v configuration with n particles. An orthonormal set of $v = 0$ basis states can be constructed using algebraic properties of quasi-spin operators (Kerman 1961, Kerman et al. 1961, Helmers 1961, Watanabe 1964, Arima, Kawasada 1964). These operators can be constructed from the S_j^+ (7.12), by using a slightly different normalization, i.e. one calls

$$S^+(j) = \sqrt{\Omega_j} S_j^+ \; . \tag{7.88}$$

The operator $S^+(j)$ is related to the pair creation operator

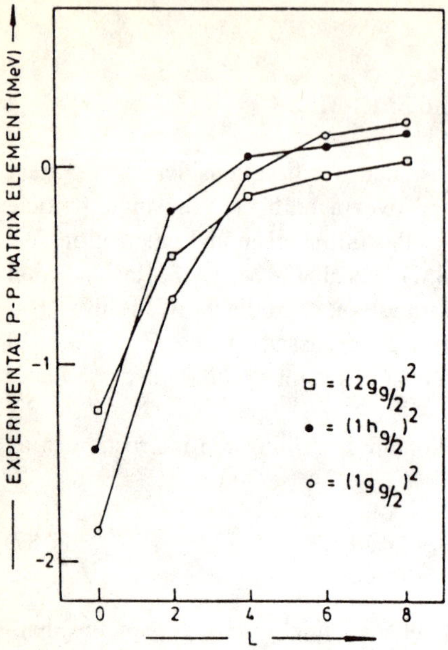

Fig. 7.28. Some typical two-body matrix elements for the $1g_{9/2}(\circ)$, $1h_{9/2}(\bullet)$ and $2g_{9/2}(\square)$ orbitals. The large preference of the formation of $J^\pi = 0^+$ and, to a smaller extent, of $J^\pi = 2^+$ pairs, is illustrated. All other (higher J^π values) remain almost degenerate at zero binding energy

$$A^+_{JM}(ab) \equiv \left[\boldsymbol{a}^+_a \otimes \boldsymbol{a}^+_b\right]^{(J)}_M , \tag{7.89}$$

via the condition

$$S^+(j) = \frac{1}{\sqrt{2}}\sqrt{\Omega_j}A^+_{00}\left(j^2\right) . \tag{7.90}$$

One so obtains for a system of n identical nucleons the shell-model $v = 0$ state

$$|\frac{n_1}{2}, \ldots , \frac{n_i}{2} ; v = 0\rangle = \prod_{k=1}^{i}\left[\frac{(\Omega_k - n_k/2)!}{(n_k/2)!\Omega_k!}\right]^{1/2} S^+(k)^{n_k/2}|0\rangle . \tag{7.91}$$

Here, the index k runs over *all* shells where some ordering is assumed. Furthermore Ω_k are the specific degeneracies $\Omega_k = j_k + 1/2$ and $n_k/2$ the number of pairs in the kth valence orbital. This satisfies

$$\begin{aligned} &\frac{n_k}{2} \le \Omega_k , \\ &\sum_{k=1}^{i}\frac{n_k}{2} = \frac{n}{2} . \end{aligned} \tag{7.92}$$

The total seniority v is now defined as the sum of the seniorities v_k for the particles in the subshells or

$$v = \sum_{k=1}^{i} v_k . \tag{7.93}$$

The general expression for a normalized state with $n_k = q_k + v_k$ particles in the shell k and with seniority v_k is

$$|n_k, v_k\rangle = N_{q_k,v_k}^{-1/2} \left[S^+(k)\right]^{q_k/2} |v_k, v_k\rangle \ , \tag{7.94}$$

where $|v_k, v_k\rangle$ is the normalized state with v_k particles in the shell k and seniority v_k. The normalization factor is

$$N_{q_k,v_k} = \frac{(q_k/2)!(\Omega_k - v_k)!}{(\Omega_k - v_k - q_k/2)!} \ , \tag{7.95}$$

if

$$q_k/2 \leq \Omega_k - v_k \ .$$

For $q_k > \Omega_k - v_k$, the state (7.94) vanishes. As an example one can write a normalized $v = 2$ basis state with unpaired nucleons in the shells a and b as

$$\begin{aligned} &|(ab)J; n_1'/2, \ldots, n_i'/2; \ldots \rangle v(ab) = 2\rangle \\ &\qquad = \left\{ (1 + \delta_{ab}) \frac{\Omega_a - n_a'/2}{\Omega_a} \frac{\Omega_b - n_b'/2 - \delta_{ab}}{\Omega_b - \delta_{ab}} \right\}^{1/2} \\ &\qquad \times A_{JM}^+(ab) | \frac{n_1'}{2}, \ldots, \frac{n_i'}{2}; v = 0\rangle \ , \end{aligned} \tag{7.96}$$

where now the relations

$$\sum_k \frac{n_k'}{2} = \frac{n}{2} - 1 \ , \tag{7.97}$$

and

$$\frac{n_k'}{2} \leq \Omega_k - \delta_{k,a} - \delta_{k,b} \ , \tag{7.98}$$

hold.

The relation (7.94) can now be used to calculate matrix elements of shell-model operators between low-seniority states.

The dimension of the $v = 0$ space is equal to the number of different ways in which $n/2$ pairs can be distributed over the valence subshells. This is typically of the order of a hundred for medium-weight and heavy semi-magic nuclei with five or six subshells. When more shells are included the number increases very rapidly. The size of the model space also increases considerably with increasing seniority. It is important to note that not all distributions of the pairs over the shells are equally important. One may adopt a special type of linear combinations, which can be written in the form of a state with $n/2$ identically distributed pairs. With this type of states one finds almost perfect overlap with the $v = 0$ shell-model ground state (Macfarlane 1966). This is the rationale of the broken-pair model.

7.6.2 Broken-Pair or Generalized-Seniority Scheme for Semi-Magic Nuclei

The ground state for a system of two identical valence nucleons is known to have $J^\pi = 0^+$. In the shell model it can be represented as

$$S^+|0\rangle = \sum_a \varphi_a S^+(a)|0\rangle \,, \tag{7.99}$$

where the coefficients φ_a characterize the distribution of the pair over the valence orbitals a. This distribution is produced by the short-range part of the shell-model residual interaction. The broken-pair scheme is now based on the observation that there is only one state (7.99) which has an energy much lower than all the other states, even though there may be several single-particle orbits that have almost degenerate single-particle energies. So one concludes that the interaction singles out one specific coherent pair structure (7.99), henceforth called S-pair, while other, orthogonal, superpositions of $J^\pi = 0^+$ configurations for two particles are as high in energy as the lowest $J^\pi \neq 0^+$ states. Because the two-body shell-model interaction favours the S-pair structure so strongly, one may now suppose that the ground state of nuclei with several valence nucleons is also predominantly composed of S-pairs only. This idea finds support in the observation that the lowest part of the spectrum for a semi-magic nucleus with several valence pairs is similar to the spectrum for one valence pair (see also Fig. 7.28). A possible interpretation is that in the lowest states of nuclei with several pairs all but one pair are S-pairs while the last pair has similar configurations as for the nucleus with a single valence pair. The number of nucleons that do not occur in S-pairs will be called generalized seniority v_g, in analogy with the seniority concept of the previous Sect. 7.6.1. One also speaks of *broken* pairs referring to particle pairs which are not S-pairs.

A shell-model basis which is labelled with this generalized-seniority quantum number v_g (or equivalently with the number of broken pairs) must be generated by explicit construction. For a system of n identical nucleons, this is done by adopting a step by step procedure, starting from the $v_g = 0$ or zero-broken-pair (0 bp) state, which is built from S-pairs only

$$|\psi_n(v_g = 0)\rangle = N_0 \left(S^+\right)^{n/2}|0\rangle \,. \tag{7.100}$$

Here N_0 is a non-trivial normalization factor. It has often been demonstrated that this multi-S-pair state is an excellent approximation to the low-seniority shell model LSSM or even to the exact shell-model ground state (Gambhir et al. 1973). This is true only if the coefficients φ_a in the S-pair operator (7.99) are determined by minimization of the multi-pair state (7.100). It turns out that these coefficients φ_a do not depend very much in practice on the pair number $n/2$. Unless stated otherwise, we shall mean by generalized seniority, the number of particles which do not occur in S-pairs and where the φ_a are calculated for each specific nucleus. Thus v_g is related to the number of broken S-pairs n_{bp} for a system with even particle number as

$$v_g = 2n_{bp} \, , \tag{7.101}$$

by definition.

Starting with this $v_g = 0$ (0 bp) state (7.100), one then constructs the $v_g = 2$ (1bp) states by replacing one S-pair creation operator in (7.100) by a two-particle creation operator (7.89), obtaining the 1bp configuration

$$A^+_{JM}(ab)\big(S^+\big)^{(n/2)-1}|0\rangle \, . \tag{7.102}$$

It is to be understood that for $J^\pi = 0^+$ one constructs linear combinations of such states which are orthogonal to the $v_g = 0$ state (7.100). This is similar to the seniority truncation scheme, with the difference that here the structure of $n/2 - 1$ pairs is frozen. This in fact is responsible for a large reduction of the dimensions of the broken-pair configuration space. As the structure of (7.102) is governed by the pair operator A^+_{JM} the number of $v_g \le 2$ (zero or one bp) states is equal to the number of two-particle shell-model configurations, irrespective of the total number of valence nucleons n under consideration. Similarly the $v_g = 4$ (2bp) states are constructed as linear combinations of

$$A^+_{J_1M_1}(ab)A^+_{J_2M_2}(cd)\big(S^+\big)^{(n/2)-2}|0\rangle \, , \tag{7.103}$$

orthogonalized to the previously obtained set of $v_g \le 2$ states. The dimension of the $v_g \le 4$ space is equal to that for the four-particle shell-model problem. By such a procedure one may continue to obtain a complete hierarchy of broken-pair basis states characterized by the quantum number v_g or equivalently the number of broken pairs according to the definition (7.101). One hopes that this classification of states provides a good truncation scheme, such that one may truncate the shell model space to include $v_g \le 2$ or $v_g \le 4$ only. Obviously the $v_g \le n$ space has the same dimension as the full shell-model space. The hope that one may apply a v_g truncation in calculations of low-lying states of semi-magic nuclei is mainly based on the empirical relation

$$E\big(v_g = 2\big) \ge E\big(v_g = 0\big) + v_g\bar{\Delta} \, , \tag{7.104}$$

where $\bar{\Delta}$ is half the energy gap between the ground states and the lowest (non-collective) excited states. One finds $\bar{\Delta} \cong 1.5\,\mathrm{MeV}$ for Ni isotopes and $N = 50$ isotones, $\bar{\Delta} \cong 1.2\,\mathrm{MeV}$ for Sn nuclei and for $N = 82$ isotones while $\bar{\Delta} \cong 0.9\,\mathrm{MeV}$ for the Pb isotopes. If the quantity $2\bar{\Delta}$ is interpreted as the energy that is required to break up each next pair, (7.104) may also be valid for $v_g > 2$. In that case v_g truncation corresponds to a rather well-defined energy truncation of the shell-model space. It should be emphasized, however, that (7.104) is only a rough estimate which may be violated for collective states which are pushed downwards in energy by the coherent action of multipole forces. It is also unclear whether (7.104) may indeed be applied for the case of large v_g. Nevertheless, a v_g truncation scheme seems the only reasonable way to treat the shell-model problem for semi-magic nuclei with many valence nucleons. It has been demonstrated (Allaart, Boeker 1971) that the $v_g \le 2$ model provides a very good tractable approximation of the

Table 7.5 Dimensions of model spaces with definite generalized seniority. The example of ^{112}Sn is shown with 12 valence particles in the orbitals $1g_{7/2}$, $2d_{5/2}$, $2d_{3/2}$, $3s_{1/2}$ and $1h_{11/2}$ (taken from (Allaart et al. 1988))

J^π	$v_g = 0$	$v_g = 2$	$v_g = 4$	$v_g = 6$	$v_g = 8$	total, all v_g
0^+	1	4	45	467	3318	56, 907
2^+	–	9	157	1967	15149	267, 720
4^+	–	7	190	2854	23372	426, 558
6^+	–	3	158	3006	26588	
8^+	–	2	133	2630	22668	
⋮						
16^+	–	–	3	255	4577	
22^-	–	–	–	2	184	
22^+	–	–	–	1	118	
31^+	–	–	–	–	–	1 $v_g = 12$
31^-	–	–	–	–	–	4

$v \leq 2$ shell model, which involves much larger dimensions of the model spaces. In Table 7.5 these dimensions are listed for a typical case. Indeed, $v_g \leq 4$ calculations have actually been performed and are discussed by (Allaart et al. 1988), while it has only recently been feasible to carry out $v \leq 4$ shell-model calculations for heavy semi-magic nuclei and then still with additional truncations (Scholten 1983).

A slightly different notation of broken-pair states that has been used quite often by Gambhir et al. (Gambhir et al. 1969, 1971, 1973) is

$$|\psi_n(v_g = 0)\rangle = \tau^+_{n/2}|0\rangle \,, \tag{7.105}$$

with

$$\tau^+_{n/2} \equiv \prod_k (u_k)^{\Omega_k} \frac{1}{(n/2)!} (S^+)^{n/2} \,. \tag{7.106}$$

We illustrate the above broken-pair model with the calculation of the $N = 50$ semi-magic nuclei ^{88}Sr and ^{90}Zr (Fig. 7.29). The calculations were carried out by Allaart et al. (Allaart et al. 1988), using a model space of 10 shells listed in Table 7.6 and using a Gaussian force.

$$V(|\boldsymbol{r}_1 - \boldsymbol{r}_2|) = V(P_{\mathrm{SE}} + tP_{\mathrm{TO}}) \exp\left\{-\left(\frac{1}{\mu}|\boldsymbol{r}_1 - \boldsymbol{r}_2|\right)^2\right\} \,. \tag{7.107}$$

The interaction parameters used were chosen as $V = -23\,\mathrm{MeV}$, $t = 0.6$ supplemented by a triplet-even force between protons and neutrons with strength $V_{\mathrm{TE}} = -20\,\mathrm{MeV}$ and a range parameter such that $\mu\nu = 0.90$ where ν is the harmonic oscillator parameter. The largest 1bp components are listed with a coding of $1 \equiv 2p_{1/2}$, $3 \equiv 2p_{3/2}$, $5 \equiv 1f_{5/2}$, $9 \equiv 1g_{9/2}$. The symbol "S'" means that no specific component is dominating the excited 0^+ state.

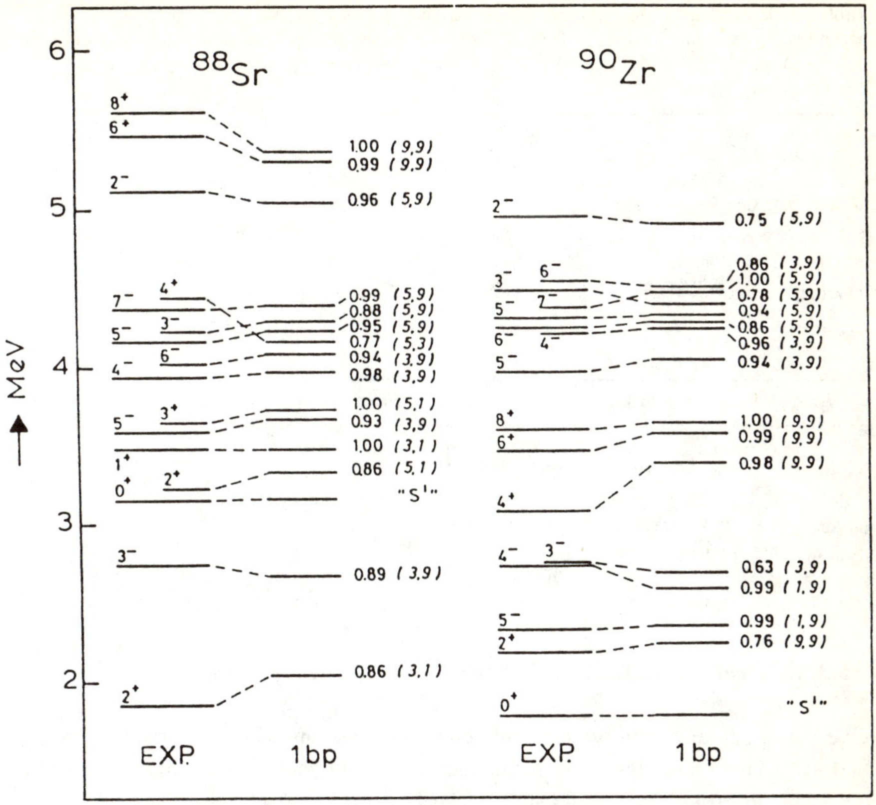

Fig. 7.29. One-broken-pair (1BPA) spectra for $^{88}_{38}Sr_{50}$ and $^{90}_{40}Zr_{50}$. The calculations were carried out in a model space of 10 subshells, listed in Table 7.6 and using a Gaussian force $V(r) = V_0(P_S + tP_T)\exp\{-r^2/\mu^2\}$ (sometimes supplemented with a triplet-even force (V_{TE}) between protons and neutrons). The parameters were $V_0 = -23$ MeV, $t = 0.6$, $V_{TE} = -20$ MeV and a range parameter $\mu\nu = 0.9$ with ν the harmonic oscillator range. The largest 1BP components are listed with the coding $1 \equiv 2p_{1/2}$; $3 \equiv 2p_{3/2}$; $5 \equiv 1f_{5/2}$; $9 \equiv 1g_{9/2}$. The symbol "S′" indicates an unspecified mixture of 0^+ pairs, orthogonal to the 0^+ ground state, taken from (Allaart et al. 1988)

Table 7.6 Single-particle energies and occupation probabilities for $N = 50$ isotones in a (broken-) pair model calculation. These were deduced from data on the adjacent odd-mass isotones with interaction parameters, given in the caption to Fig. 7.29. The neutron energies were the same for both nuclei (taken from (Allaart et al. 1988))

	nlj	$1f\frac{7}{2}$	$1f\frac{5}{2}$	$2p\frac{3}{2}$	$2p\frac{1}{2}$
^{88}Sr	ε_a(MeV)	−14.0	−11.2	−10.2	−8.8
protons	occ. %	98	95	85	23
^{90}Zr	ε_a(MeV)	−16.5	−14.0	−13.5	−11.5
protons	occ. %	98	97	94	66
neutrons	ε_a(MeV)	−18.5	−15.5	−12.5	−11.5
	occ. %	100	100	100	100

	$1g\frac{9}{2}$	$2d\frac{5}{2}$	$3s\frac{1}{2}$	$2d\frac{3}{2}$	$1g\frac{7}{2}$	$1h\frac{11}{2}$
^{88}Sr	−7.4	−3.5	−2.8	−2.0	−1.5	+0.3
protons	4.5	0.6	0.3	0.4	0.5	0.2
^{90}Zr	−10.5	−7.8	−7.0	−5.5	−4.5	−3.0
protons	10.5	1.3	0.5	0.6	0.5	0.3
neutrons	−11.1	−6.4	−5.3	−4.2	−3.7	−2.4
	100	0	0	0	0	0

7.6.3 Generalization to Both Valence Protons and Neutrons

One may generalize the broken-pair scheme to systems with both valence protons and neutrons in a straightforward manner by forming simply the product of proton and neutron broken-pair basis states. The zero-broken-pair state

$$|\psi_{p,n}\rangle = \tau^{+}_{p/2}\tau^{+}_{n/2}|0\rangle \ , \tag{7.108}$$

may now be obtained by minimizing its energy, including the coupling between the proton and neutron distributions due to the proton-neutron interaction Hamiltonian

$$H_{\pi\nu} = \sum_{\pi_1,\pi_2,\nu_1,\nu_2} \langle\pi_1\nu_1|V|\pi_2\nu_2\rangle a^{+}_{\pi_1}a^{+}_{\nu_1}a_{\nu_2}a_{\pi_2} \ . \tag{7.109}$$

We have used the symbols π for protons and ν for neutrons explicitly. The energy minimization condition for (7.108) results in a set of coupled equations for the S-pair structure coefficients φ_p and φ_n (Egmond, Allaart 1983). It should now be remarked, however, that the zero-broken-pair state (7.108) may not in general be a good approximation for the ground state of nuclei with both open proton and neutron shells, especially where the strong proton-neutron quadrupole force induces large admixtures of components with several broken pairs in the ground state. Empirically it is known that nuclei with several valence protons and neutrons are soft against quadrupole vibrations or even become deformed. Such a deformed state corresponds to a mixture of states with many broken pairs in the spherical representation. For such a case the model becomes less interesting. One should then work in a deformed representation. However, the generalization presented here may still be of interest for two reasons. *First*, one may hope that it is still applicable when either the proton or the neutron number of valence particles (or holes) is so small, typically one or two, that deformation effects are still negligible. We shall discuss a few applications for such cases. *Secondly*, this may be useful in the microscopic interpretation of the phenomenological Interacting Boson Model (Sect. 7.7), for one may assume that the S-pairs, together with a coherent pair structure with angular momentum $J = 2$, called D-pair, still play a prominent role in deformed nuclei. The broken-pair model with states built from these S- and D-pairs may then be useful in the microscopic analysis of the boson model parameters from a shell model microscopic viewpoint. We shall discuss this relation in Sect. 7.7.

Some applications when both proton and neutron broken pairs can be present are the following:

i) Studies for odd-mass nuclei with $v_g = 1$ and $v_g = 3$ can also be carried out. This is similar to a $v_g = 0+2$ calculation. For ^{89}Y, we compare such a $v_g \leq 3$ calculation (Fig. 7.30). We indicate the angular momentum and parity as $2J^{\pi}$. In the first calculation, only the proton $1f_{5/2}$, $2p_{3/2}$, $2p_{1/2}$ and $1g_{9/2}$ orbitals are included in the $v_g \leq 3$ model space. The spectrum called "core-excitation" includes all shells between magic numbers 20 and 82 for both protons and neutrons. A Gaussian effective interaction has been used (Allaart et al. 1988).

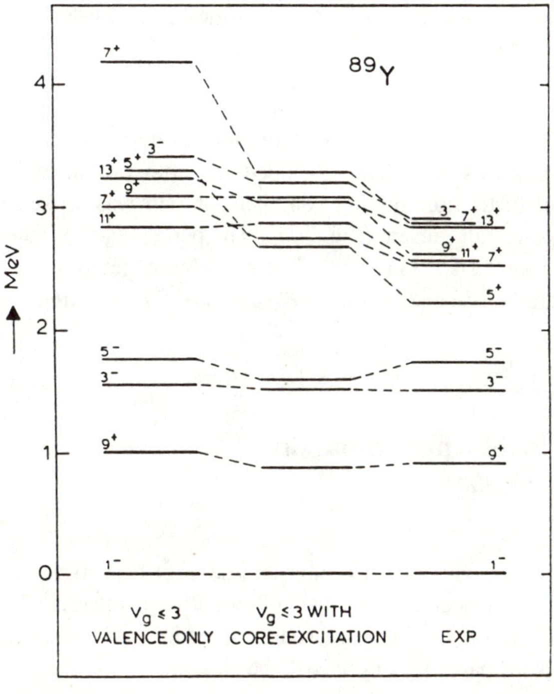

Fig. 7.30. Influence of core-excitations on the low-lying levels in $^{89}_{39}Y_{50}$ in a broken-pair ($v_g \leq 3$) model calculation. The numbers near the levels indicate twice the angular momenta. In the first calculation, the proton $1f_{5/2}$, $2p_{3/2}$, $2p_{1/2}$ and $1g_{9/2}$ shells are included in the $v_g \leq 3$ model space. The middle spectrum includes *all* shells between magic numbers 20 and 82 for *both* protons and neutrons. A Gaussian, effective interaction (see Fig. 7.29) was used. The data are from (Kocher 1975) (taken from (Allaart et al. 1988))

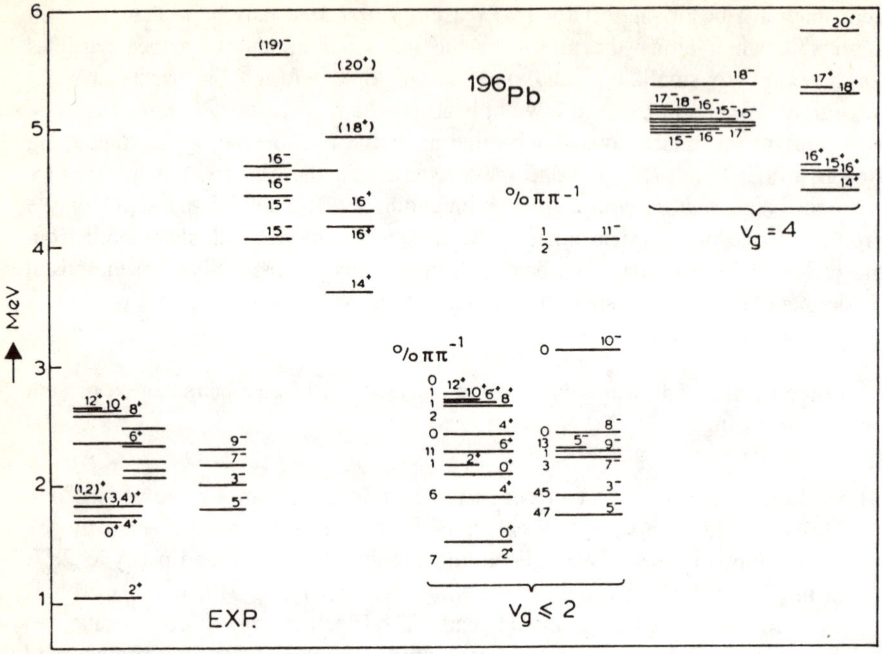

Fig. 7.31. Broken-pair calculation for ^{196}Pb. For angular momenta $J < 13$, as many as 20 subshells were included for both proton and neutron orbitals with, however, the restriction that only *one* neutron pair was broken or only one proton $1p - 1h$ pair was present. The percentages of these proton $1p - 1h$ configurations are listed. For $J > 13$, only valence neutrons in the $82 < N < 126$ shell were considered in a *two* broken-pair model ($v_g = 4$). The data are from (Van Ruyven et al. 1986) and (Roulet et al. 1977) (taken from (Allaart et al. 1988))

ii) For the even-even $Z = 82$ closed shell ^{196}Pb nucleus, for angular momenta $J < 13$, as many as 20 subshells have been included for both proton and neutron orbitals, but with the restriction that only one neutron pair was broken or only one proton particle-hole pair present (Fig. 7.31). The percentages of the proton $p - h$ configurations are listed. For $J > 13$, only valence neutrons in the $82 < N < 126$ shell were considered in a 2bp ($v_g = 4$) calculation (Allaart et al. 1988).

7.7 Interacting Boson-Model Approximation to the Nuclear Shell Model

Starting from the broken-pair model where both proton and neutron broken pairs can be present, it was alluded to in Sect. 7.6 that the proton-neutron interaction will almost inevitably mix the proton and neutron seniority strongly. In nuclei where such proton-neutron broken-pair excitations will be dominant (for nuclei with both *open* proton and neutron shells), one actually enters those regions of nuclei where the quadrupole degree of freedom starts to play a dominant role in determining nuclear structure. Collective quadrupole vibrational excitations and rotational motion and the transitional forms in between determine the low-lying excited states in those nuclei. Besides the geometric or shape variable models conceived by A. Bohr, B. Mottelson and J. Rainwater (Bohr 1951, 1952, 1976, Bohr, Mottelson 1953, 1955, 1969, 1975, Mottelson 1976, Rainwater 1950, 1976) (a topic which is outside the scope of the present book), an alternative description based on symmetries in an interacting boson model, a model where s ($L = 0$) and d ($L = 2$) bosons are considered, has been introduced by Arima and Iachello (Arima, Iachello 1975a, 1975b, 1976, 1978a, 1978b, Iachello, Arima 1988). Both models, although coming from different assumptions, give a good description of those nuclei where the quadrupole degree of freedom is dominant. This is exactly the region of open shell proton-neutron systems, where the broken-pair model, discussed above, is probably able to give some justification to the Interacting-Boson model (IBM). Both the Bohr-Mottelson model and the IBM have been discussed in great detail, in particular in recent years in a number of review articles and some books, and it is not our purpose to go again into these topics in much detail. Even though the original ideas of the Interacting Boson model (IBM) are deeply rooted within concepts of dynamical symmetries in nuclei, we shall not elaborate on that interesting aspect of the IBM.

We should like, however, to mention briefly how the concept of symmetries in physics has always been a guideline to unify seemingly different phenomena. So, let us quote some of the major steps using symmetry concepts in describing different aspects of the nuclear many-body system.

1932: The concept of isospin symmetry, describing the charge independence of the nuclear forces by means of the isospin concept with the SU(2) group as the underlying mathematical group (Heisenberg 1932). This is the simplest of all dynamical symmetries and expresses the invariance of the Hamiltonian against the exchange of all proton and neutron coordinates.

1936: Spin and isospin are combined by Wigner into the SU(4) supermultiplet scheme with SU(4) as the group structure (Wigner 1937). This concept has been extensively used in the description of light α-like nuclei ($A = 4 \times n$).

1948: The spherical symmetry of the nuclear mean field and the realization of its major importance for describing the nucleon motion in the nucleus has been put forward by Mayer (Mayer 1949), Haxel, Jensen and Suess (Haxel et al. 1949).

1958: Elliott remarked that in some cases, the average nuclear potential could be depicted by a deformed, harmonic oscillator containing the SU(3) dynamical symmetry (Elliott 1958, Elliott, Harvey 1963). This work opened a first possible connection between the macroscopic collective motion and its microscopic description.

1942: The nucleon residual interaction amongst identical nucleons is particularly strong in $J^\pi = 0^+$ and 2^+ coupled pair states. This "pairing" property is a corner stone in accounting for the nuclear structure of many spherical nuclei near closed shells in particular. Pairing is at the origin of seniority, itself related to the quasi-spin classification and group as used first by Racah in describing the properties of many-electron configurations in atomic physics (Racah 1943).

1952: The nuclear deformed field is a typical example of the concept of spontaneous symmetry breaking. The restoration of the rotational symmetry, present in the Hamiltonian, leads to the formation of nuclear rotating spectra. These properties were discussed earlier in a more phenomenological way by Bohr and Mottelson (Bohr 1951, 1952, Bohr, Mottelson 1953).

1974: The introduction of dynamical symmetries in order to describe nuclear collective motion starting from a many-boson system with only s ($L = 0$) and d ($L = 2$) bosons is introduced by Arima and Iachello (Arima, Iachello 1975a, 1975b, 1976, 1978a, 1979). The relation to the nuclear shell model and its underlying shell-structure has been studied extensively (Otsuka et al. 1978b). These boson models have given rise to a new momentum in nuclear physics research. These symmetries are depicted schematically in Fig. 7.32.

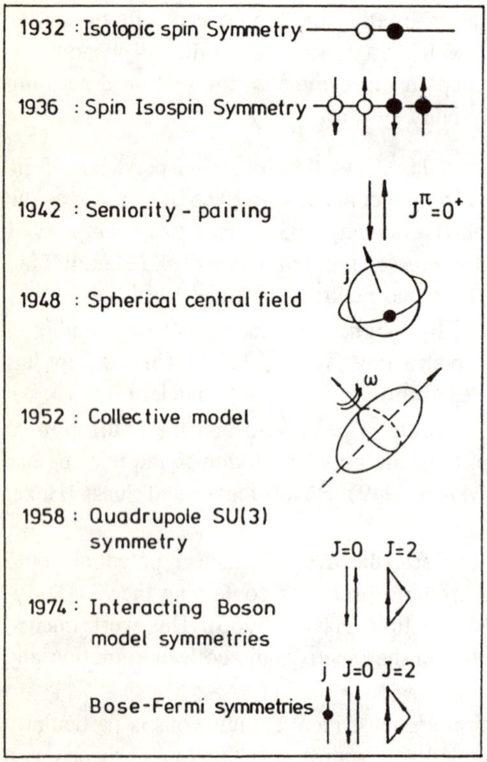

Fig. 7.32. Pictorial representation of the most important nuclear symmetries developed over the years

We should like, however, to point out in which way, starting from the shell-model techniques, described in this chapter (Sects. 7.1 to 7.6), contact can be made with the proton-neutron IBM model, called IBM-2. Thereby, we shall in particular point out in an almost schematic way, the different approximations and truncations needed in the exact shell-model calculation with *all* valence nucleons, in order to come within the IBM-2 model space.

First, however, we shall discuss a simpler model that allows for the onset of quadrupole motion in nuclei near vibrational regions of the mass table. Thereby, we shall point out the major importance of the quadrupole proton-neutron interaction in establishing smooth, quadrupole collective motion.

The nuclear mean field gives rise to a number of highly stable magic nucleon configurations so that nucleons outside these configurations can be treated as valence nucleons that mainly determine the low-lying nuclear collective degrees of freedom. *This process is a first act of truncation of the large shell-model space into a truncated valence shell-model space.* The residual nucleon-nucleon interaction in the $0\hbar\omega$ model space now selects mainly $J^\pi = 0^+$ and 2^+ coupled pairs, and therefore, will separate a highly coherent set of pairs for the low-lying configurations as a starting point when describing nuclear quadrupole collective motion (Fig. 7.28). As such, in more realistic cases with many valence nucleons outside closed shells, a BCS pair distribution (or 0^+ pair distribution with correct particle number) is a very good starting point in describing the distribution of nucleons over the available fermion orbitals (Sects. 7.6 and 7.7).

Starting from a second quantized form of the paired states, using (7.12)

$$S_j^+ = \frac{1}{\sqrt{\Omega}} \sum_{m>0} (-1)^{j+m} a_{jm}^+ a_{j-m}^+ , \tag{7.110}$$

[with $\Omega \equiv j + \frac{1}{2}$ the degeneracy and a_{jm}^+ a fermion creation operator in the orbit denoted by $(j,m) \equiv (n,l,j,m)$] and

$$D_j^+ = \sum_{m>0} (-1)^{j-m} \langle jm\, j-m | J0 \rangle a_{jm}^+ a_{j-m}^+ , \tag{7.111}$$

for a single j-orbit, the lowest 0^+ and 2^+ excitations for protons and neutrons separately can be described in the shell model as

$$\begin{aligned} &\left[\left(S_j^+\right)^{(n_\pi/2)-1}\left(D_j^+\right)\right]2^+|0\rangle ,\\ &\left[\left(S_j^+\right)^{(n_\nu/2)-1}\left(D_j^+\right)\right]2^+|0\rangle , \end{aligned} \tag{7.112}$$

$$\begin{aligned} &\left(S_j^+\right)^{(n_\pi/2)}|0\rangle ,\\ &\left(S_j^+\right)^{(n_\nu/2)}|0\rangle . \end{aligned} \tag{7.113}$$

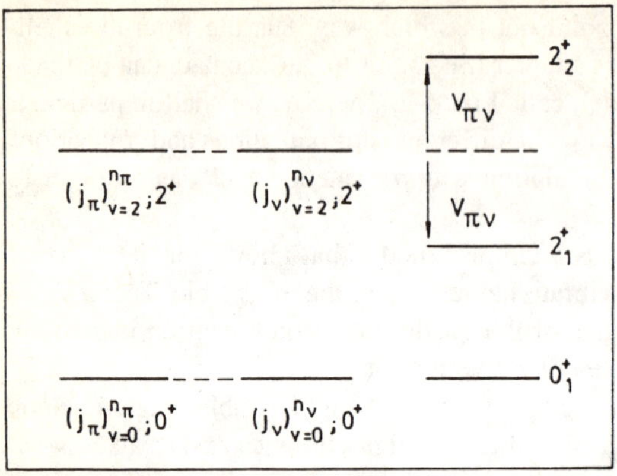

Fig. 7.33. The unperturbed seniority $v = 2$ proton and neutron shell-model configurations for situations with n_π (and n_ν) proton (neutron) particles in a single $j_\pi(j_\nu)$ shell, respectively (see (7.114)) as well as the two $J^\pi = 2^+$ levels obtained after diagonalizing the quadrupole proton-neutron force $-\kappa \boldsymbol{Q}_\pi \cdot \boldsymbol{Q}_\nu$ in the two-level model of (7.114) and (7.117) (taken from (Heyde, Sau 1986))

In the proton-neutron coupled basis, there are two independent 2^+ basis states i.e.

$$|2_\pi^+\rangle = \left[\left(S_j^+\right)^{(n_\pi/2)-1}\left(D_j^+\right)\right]2^+ \otimes \left(S_j^+\right)^{(n_\nu/2)}|0\rangle \,,$$
$$|2_\nu^+\rangle = \left[\left(S_j^+\right)^{(n_\nu/2)-1}\left(D_j^+\right)\right]2^+ \otimes \left(S_j^+\right)^{(n_\pi/2)}|0\rangle \,. \tag{7.114}$$

If the energy of the lowest 2^+ states in the proton and neutron space are nearly equal [which is often the case in experimental situations e.g. $E_x(2_1^+)$ in Sn and $E_x(2_1^+)$ in $N = 82$ nuclei] and taken as

$$\varepsilon^0_{2^+} = \varepsilon^0_{2^+_\pi} = \varepsilon^0_{2^+_\nu} \,, \tag{7.115}$$

then the eigenstates for the coupled proton-neutron system follow from diagonalizing in the 2×2 model space, the residual proton-neutron interaction (Fig. 7.33). Using a quadrupole proton-neutron interaction $-\kappa \boldsymbol{Q}_\pi \cdot \boldsymbol{Q}_\nu$ and using the Racah algebra reduction formulae (de-Shalit, Talmi 1963), one obtains for the matrix element

$$\begin{aligned}\langle V_{\pi\nu}\rangle &= \langle 2_\pi^+| - \kappa \boldsymbol{Q}_\pi \cdot \boldsymbol{Q}_\nu | 2_\nu^+\rangle \\ &= -\frac{\kappa}{5}\left(N_\pi\left(1-\frac{N_\pi}{\Omega_\pi}\right)N_\nu\left(1-\frac{N_\nu}{\Omega_\nu}\right)\right)^{1/2}\left(\frac{4}{(\Omega_\pi-1)(\Omega_\nu-1)}\right)^{1/2} \\ &\quad \times \langle j_\pi\|\boldsymbol{Q}_\pi\|j_\pi\rangle\langle j_\nu\|\boldsymbol{Q}_\nu\|j_\nu\rangle \quad \left(\text{with } N_\varrho \equiv \frac{n_\varrho}{2}\right) .\end{aligned} \tag{7.116}$$

The energy matrix becomes

$$\begin{bmatrix} \varepsilon^0_{2^+} & \langle V_{\pi\nu}\rangle \\ \langle V_{\pi\nu}\rangle & \varepsilon^0_{2^+} \end{bmatrix} , \tag{7.117}$$

with as lowest eigenvalue, the expression

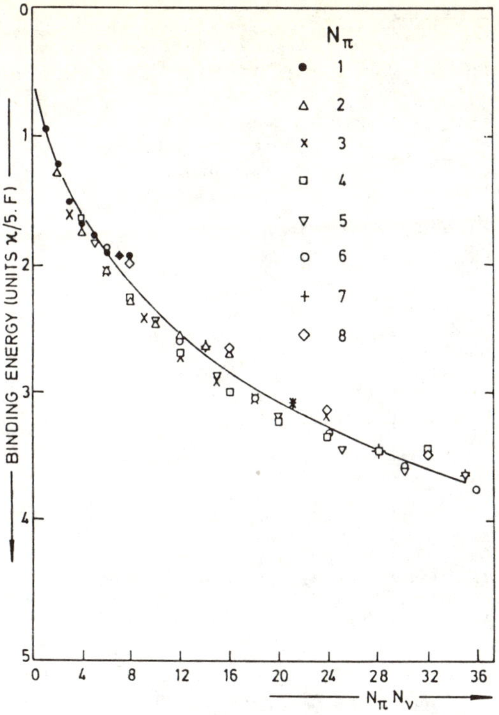

Fig. 7.34. The energy of the lowest $J_i^\pi = 2_1^+$ level (see (7.118)), in units of $(\kappa/5) \cdot F$ (where F is defined in (7.116)), plotted as a function of the product $N_\pi N_\nu$ for nuclei in the region $1 \leq N_\pi \leq 8$. Here, we denote with N_π (N_ν) the number of proton (neutron) pairs or $N_\varrho = n_\varrho/2(\varrho \equiv \pi, \nu)$ (taken from (Heyde, Sau 1986))

$$E\left(2_1^+\right) = \varepsilon_{2^+}^0 - \frac{\kappa}{5}\left(N_\pi N_\nu\left(1-\frac{N_\pi}{\Omega_\pi}\right)\left(1-\frac{N_\nu}{\Omega_\nu}\right)\right)^{1/2} F\,, \tag{7.118}$$

where

$$F = \langle j_\pi \| \boldsymbol{Q}_\pi \| j_\pi \rangle \langle j_\nu \| \boldsymbol{Q}_\nu \| j_\nu \rangle \left(\frac{4}{(\Omega_\pi - 1)(\Omega_\nu - 1)}\right)^{1/2} .$$

The corresponding wave function reads

$$|2_1^+\rangle = \frac{1}{\sqrt{2}}\left(|2_\pi^+\rangle + |2_\nu^+\rangle\right) . \tag{7.119}$$

So, the lowest 2^+ state is a symmetric linear combination of the unperturbed states and thus seniority becomes strongly mixed in the final 2^+ state. Also, the energy eigenvalue presents a specific $N_\pi N_\nu$ dependence shown in Figs. 7.34 and 7.35 which for small N_π/Ω_π, N_ν/Ω_ν is almost like $\sqrt{N_\pi N_\nu}$. Recently Casten et al. (Casten et al. 1981, Casten 1985a, 1985b, 1985c) have pointed out the very striking observation that experimental quantities such as $E_x(2_1^+)$, $E_x(4_1^+)/E_x(2_1^+)$ lie on a smooth curve when plotted as a function of the product $N_\pi N_\nu$. At least in the vibrational regime, the above simple shell-model calculation predicts such a behaviour for $E_x(2_1^+)$.

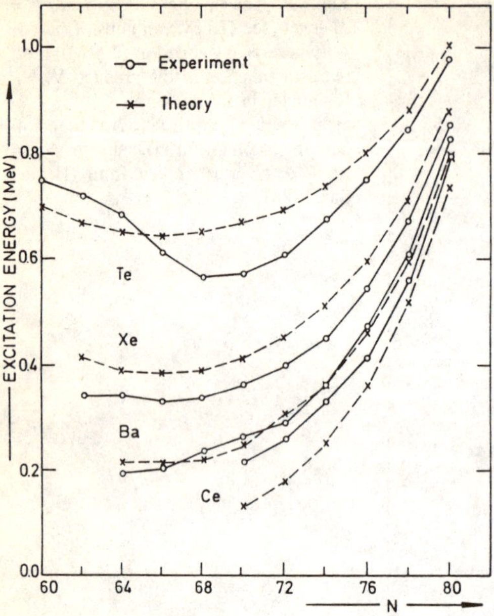

Fig. 7.35. A detailed fit, using (7.118), to the lowest $J_i^\pi = 2_1^+$ level in the even-even Te, Xe, Ba and Ce nuclei. Here, we use the degeneracies $\Omega_\pi = \Omega_\nu = 16$ and a constant value of $(\kappa/5)$. F = 0.365 MeV for *all* nuclei (taken from (Heyde, Sau 1986))

Since we should like to go beyond the vibrational region and also handle more realistic cases where particles move in many, non-degenerate shells, we have to use more general pair creation operators e.g. we use

$$
\begin{aligned}
S_\pi^+ &= \sum_{j_\pi} \alpha_{j_\pi} S^+(j_\pi) \,, \\
D_{\pi,\mu}^+ &= \sum_{j_\pi, j_{\pi'}} \beta_{j_\pi, j_{\pi'}} D_{j_\pi, j_{\pi'}, \mu}^+ \,, \\
G_{\pi,\mu}^+ &= \sum_{j_\pi, j_{\pi'}} \gamma_{j_\pi, j_{\pi'}} G_{j_\pi, j_{\pi'}, \mu}^+ \,,
\end{aligned}
\tag{7.120}
$$

with

$$
\begin{aligned}
S^+(j_\pi) &= \frac{\sqrt{\Omega_{j_\pi}}}{\sqrt{2}} \left(a_{j_\pi}^+ a_{j_\pi}^+\right)_0^{(0)} \,, \\
D_{j_\pi, j_{\pi'}, \mu}^+ &= \left(1 + \delta_{j_\pi j_{\pi'}}\right)^{-1/2} \left(a_{j_\pi}^+ a_{j_{\pi'}}^+\right)_\mu^{(2)} \,,
\end{aligned}
\tag{7.121}
$$

and where the sums go over all single-particle states. It should be noted that the S_π^+ operator used in (7.120) and within the generalized seniority scheme (see also Sect. 7.6) and in the further discussion, has a different normalization i.e. for a single j-shell $S_\pi^+ \equiv \sqrt{\Omega} S_j^+$. For degenerate orbitals, all α_j are equal $(\alpha_j = \sqrt{\frac{1}{2}(j + \frac{1}{2})})$ but for more realistic situations, deviations occur and the optimum pair distribution will have to be determined from a variational procedure for each nucleus (and in principle for each excited state). Using all possible angular momenta S, D, G, …, a huge shell-model basis still results.

Here, Our Next Step of Truncation Comes in. We truncate the huge space of pair states ($\cong 10^{14}$ down to 10^3 for e.g. ^{156}Sm) such that only the S and D pairs are considered (Otsuka et al. 1978a, 1978b). The argument of truncation here is again based on the fact that in studying low-lying collective quadrupole states we first of all have to take into account the most strongly bound pair states for the nuclear many-body system. The basis functions are now written as

$$\left\{\left[\left(S_\pi^+\right)^{N_{s\pi}}\left(D_\pi^+\right)^{N_{d\pi}}_{\gamma_\pi J_\pi}\right]\otimes\left[\left(S_\nu^+\right)^{N_{s\nu}}\left(D_\nu^+\right)^{N_{d\nu}}_{\gamma_\nu J_\nu}\right]\right\}^{(J)}_M |0\rangle\,, \tag{7.122}$$

with

$$\begin{aligned} N_{s_\pi}+N_{d_\pi} &= \frac{n_\pi}{2} = N_\pi\,,\\ N_{s_\nu}+N_{d_\nu} &= \frac{n_\nu}{2} = N_\nu\,, \end{aligned} \tag{7.123}$$

where $n_\pi(n_\nu)$ are the number of valence protons (neutrons) respectively.

As a technical point, we mention briefly that there are problems with the construction of orthonormal states when many D-pairs are present, e.g.

$$\begin{aligned} &\left(S_\varrho^+\right)^{N_s}\left(D_\varrho^+\right)^{N_d}_{\gamma JM}|0\rangle\,, \quad \text{and}\,,\\ &\left(S_\varrho^+\right)^{N_s+1}\left(\Gamma_\varrho^+\right)_{JM}|0\rangle\,,\\ &\quad\vdots \end{aligned} \tag{7.124}$$

where $(\Gamma_\varrho^+)_{JM}$ is a notation for operators which create $2(N_d-1)$ nucleons ($\varrho = \pi,\nu$) (e.g., the state $|D^2, J=0\rangle$ is not orthogonal to the state $|S^2, J=0\rangle$) (Rowe 1970). Thus, calculations using fermion pairs are still cumbersome. Although the fermion pair commutator relations resemble the boson ones, there are corrections that vanish only in the small n limit.

Here, the Third Approximation Comes in. We map the space built from S, D fermion pairs and the operators in the S, D space to a corresponding s, d-boson space (Fig. 7.36). Thus, the corresponding boson state of (7.122) reads

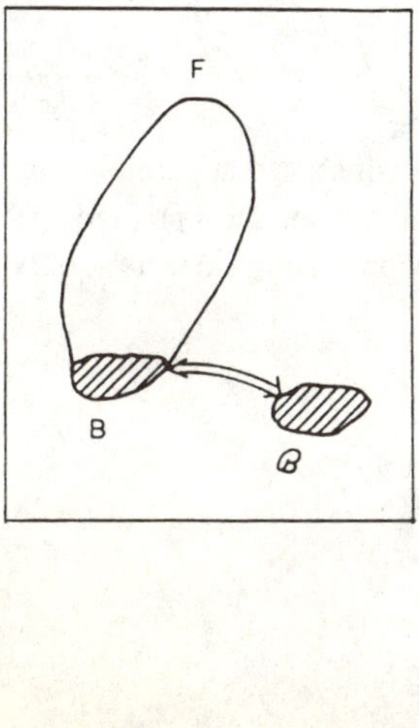

Fig. 7.36. A schematic representation of the mapping procedure, explained in (7.125, 126), in order to construct the boson Hamiltonian and boson transition operators. Here, F denotes the fermion complete space, B the S, D-pair fermion subspace and $\mathcal{B}$ the corresponding sd-boson space (taken from (Arima, Iachello 1984))

Table 7.7 Correspondence between lowest seniority SD fermion states and the corresponding sd-boson states (taken from (Arima, Iachello 1984))

Fermion space B		Boson space $\mathcal{B}$	
$n=0, v=0$	$\|0\rangle$	$N=0, n_d=0$	$\|0)$
$n=2, v=0$	$S^\dagger\|0\rangle$	$N=1, n_d=0$	$s^\dagger\|0)$
$v=2$	$D^\dagger\|0\rangle$	$n_d=1$	$d^\dagger\|0)$
$n=4, v=0$	$S^{\dagger 2}\|0\rangle$	$N=2, n_d=0$	$s^{\dagger 2}\|0)$
$v=2$	$S^\dagger D^\dagger\|0\rangle$	$n_d=1$	$s^\dagger d^\dagger\|0)$
$v=4$	$D^{\dagger 2}\|0\rangle$	$n_d=2$	$d^{\dagger 2}\|0)$
...	...	...	...

$$\left\{\left[\left(s_\pi^+\right)^{N_{s_\pi}}\left(d_\pi^+\right)^{N_{d_\pi}}_{\gamma_\pi J_\pi}\right]\otimes\left[\left(s_\nu^+\right)^{N_{s_\nu}}\left(d_\nu^+\right)^{N_{d_\nu}}_{\gamma_\nu J_\nu}\right]\right\}^{(J)}_M|\tilde{0}\rangle\,, \tag{7.125}$$

where now $|\tilde{0}\rangle$ is the boson vacuum state (see Table 7.7 for the fermion-boson correspondence for the lowest seniority states). The method used starts by equating matrix elements between corresponding fermion and boson states and thereby we obtain implicit equations for the boson operator

$$\langle\psi_F|H_F|\psi_F\rangle=\langle\psi_B|H_B|\psi_B\rangle\,. \tag{7.126}$$

Thereby, we ask that the lowest eigenvalues and transition matrix elements remain equal in the boson space as compared with the fermion space. Thus, we carry out a boson mapping, not a boson expansion. As a result one gets a number of implicit equations determining the boson-truncated Hamiltonian H_B as well as the boson truncated electromagnetic operator $\boldsymbol{O}_B$ as a function of the fermion quantities (single-particle energies, two-body fermion interaction matrix elements, fermion charges). This mapping, for a pairing-plus-quadrupole Hamiltonian, is now described.

Using (7.126), fermion matrix elements for low-seniority states can be calculated using Racah algebra. Since the starting Hamiltonian for discussing quadrupole collective motion consists of a pairing-plus-quadrupole part we discuss both terms separately (Otsuka et al. 1978b).

i) Pairing Term. This part of the Hamiltonian, for a pure pairing force between identical nucleons ϱ ($\varrho\equiv\pi,\nu$) can be written as

$$H_{\text{pair}}=-GS^+(j_\varrho)S(j_\varrho)\,. \tag{7.127}$$

Since H does not mix seniority, N_s and N_d will be conserved separately, and we only need the two-body matrix-elements for all states from the $(j)^2$ and $(j)^4$ configurations in order to determine, via (7.126), the boson Hamiltonian uniquely. After some calculations, one gets

$$\begin{aligned}H_B=&\varepsilon_s s^+s+\varepsilon_d d^+\cdot\tilde{d}+\frac{1}{2}\sum_L c_L\left(d^+d^+\right)^{(L)}\cdot(\tilde{d}\tilde{d})^{(L)}\\&+\tfrac{1}{2}u_0 s^+s^+ss+u_2\left(s^+d^+\right)^{(2)}\cdot(s\tilde{d})^{(2)}\,,\end{aligned} \tag{7.128}$$

with the coefficients given by

$$\begin{aligned}
\varepsilon_s &= \langle (j)^2\, J=0|\, H_{\text{pair}}\, |(j)^2\, J=0\rangle = -G\cdot\Omega\,,\\
\varepsilon_d &= \langle (j)^2\, J=2|\, H_{\text{pair}}\, |(j)^2\, J=2\rangle = 0\,,\\
c_L &= \langle (j)^4\, J|\, H_{\text{pair}}\, |(j)^4\, J\rangle - 2\,\varepsilon_d = 0\,,\\
u_0 &= \langle (j)^4\, J=0|\, H_{\text{pair}}\, |(j)^4\, J=0\rangle - 2\,\varepsilon_s = 2G\,,\\
u_2 &= \langle (j)^4\, J=2|\, H_{\text{pair}}\, |(j)^4\, J=2\rangle - \varepsilon_s - \varepsilon_d = 2G\,.
\end{aligned} \tag{7.129}$$

Thus, for a proton-neutron system and considering the pairing term only, a boson Hamiltonian

$$H_B = \varepsilon_{s_\nu} \boldsymbol{n}_{s_\nu} + \varepsilon_{d_\nu} \boldsymbol{n}_{d_\nu} + \varepsilon_{s_\pi} \boldsymbol{n}_{s_\pi} + \varepsilon_{d_\pi} \boldsymbol{n}_{d_\pi} + V_{\pi\pi} + V_{\nu\nu}\,, \tag{7.130}$$

($\boldsymbol{n}_{s_\varrho}$, $\boldsymbol{n}_{d_\varrho}$ are the number operators) is obtained which can be rewritten (since $\varepsilon_s < \varepsilon_d$) as

$$H_B = E_0 + \varepsilon_{d_\nu} \boldsymbol{n}_{d_\nu} + \varepsilon_{d_\pi} \boldsymbol{n}_{d_\pi} + V_{\pi\pi} + V_{\nu\nu}\,. \tag{7.131}$$

ii) Quadrupole Term. We map the fermion quadrupole operator, using the matrix element mapping method. Since the quadrupole operator has both seniority changing and non-changing parts, two different terms result in the boson quadrupole operator. For the seniority changing boson term one can write

$$Q'_B = \alpha_0 \left(d^+ s\right)^{(2)} + \sum_L \alpha_1^{(L)} \left(\left(d^+ d^+\right)^{(L)} \tilde{d}\right)^{(2)} s + \dots\,. \tag{7.132}$$

By mapping the corresponding fermion and boson matrix elements we get the equality

$$\langle s^{N-1} d \| \boldsymbol{Q}'_B \| s^N\rangle = \langle \left(S^+\right)^{N-1} D^+,\, J=2 \| \boldsymbol{Q}_F \| \left(S^+\right)^N J=0\rangle\,, \tag{7.133}$$

or

$$\alpha_0 = \sqrt{\frac{\Omega - N}{\Omega - 1}}\, \frac{1}{\sqrt{5}} \langle j^2; J=2 \| \boldsymbol{Q}_F \| j^2; J=2\rangle\,. \tag{7.134}$$

Similarly, the non-changing seniority part becomes

$$Q''_B = \beta_0 \left(d^+ \tilde{d}\right)^{(2)} + \sum_{LL'} \beta_1^{(L,L')} \left[\left(d^+ d^+\right)^{(L)} (\tilde{d}\tilde{d})^{(L')}\right]^{(2)} + \dots\,. \tag{7.135}$$

Mapping matrix elements gives

$$\langle s^{N-1} d \| \boldsymbol{Q}''_B \| s^{N-1} d\rangle = \langle \left(S^+\right)^{N-1} D^+; J=2 \| \boldsymbol{Q}_F \| \left(S^+\right)^{N-1} D^+; J=2\rangle\,, \tag{7.136}$$

or

$$\beta_0 = \frac{\Omega - 2N}{\Omega - 2}\, \frac{1}{\sqrt{5}} \langle j^2; J=2 \| \boldsymbol{Q}_F \| j^2; J=2\rangle\,. \tag{7.137}$$

The higher order coefficients $\alpha_1^{(L)}$, $\beta_1^{(L,L')}$ can, in principle, be determined using the same method as outlined here.

Combining all parts, the full image (in lowest order) of the fermion quadrupole operator becomes

$$Q_B^{(2)} = \kappa_\varrho\Big(\big(d^+s + s^+\tilde{d}\big)^{(2)} + \chi_\varrho\big(d^+d^+\big)^{(2)}\Big)\,, \tag{7.138}$$

with $\kappa_\varrho = \alpha_0$ and $\chi_\varrho = \beta_0/\alpha_0$.

A rather general IBM-2 Hamiltonian can now be written as

$$H_B = E_0 + \varepsilon_{d_\pi} n_{d_\pi} + \varepsilon_{d_\nu} n_{d_\nu} + \kappa \boldsymbol{Q}_\pi^{(2)} \cdot \boldsymbol{Q}_\nu^{(2)} + V_{\pi\pi} + V_{\nu\nu} + M_{\pi\nu}\,, \tag{7.139}$$

for which E_0, ε_{d_π}, ε_{d_ν}, κ, χ_π, χ_ν are related to the underlying shell-model structure that we now study. Here ε_{d_π} (ε_{d_ν}) is the proton (neutron) d-boson energy, n_{d_π} (n_{d_ν}) the proton (neutron) d-boson number operator, $\boldsymbol{Q}_\varrho^{(2)}$ the quadrupole boson operator consisting of two terms [see (7.138)] with χ_ϱ describing the relative strength and $V_{\pi\pi}$, $V_{\nu\nu}$, $M_{\pi\nu}$ the remaining boson residual interaction terms. In particular, the last term $M_{\pi\nu}$ describes the Majorana term which we do not discuss here. Using the above mapping, microscopic estimates for these quantities $\varepsilon_\varrho \equiv \varepsilon_{d_\varrho} - \varepsilon_{s_\varrho}$, κ, χ_ϱ are obtained (Scholten 1980) and are illustrated in Fig. 7.37.

We have determined $\varepsilon = \varepsilon_d - \varepsilon_s = G\Omega$ and this quantity has to be a constant value independent of the particle number (Fig. 7.37). For the quadrupole operator parameters, the following behaviour results:

i) the total strength $\kappa = \bar{\kappa}\kappa_\pi\kappa_\nu$ should be given by the function (Fig. 7.37)

$$\kappa = \kappa_0\sqrt{\big(\Omega_\nu - N_\nu\big)\big(\Omega_\pi - N_\pi\big)}\,. \tag{7.140}$$

(here $\bar{\kappa}$ is the strength of the original fermion quadrupole interaction).

ii) the relative value χ_ϱ then, is given by the expression

$$\chi_\varrho = \chi_\varrho^0\big(\Omega_\varrho - 2N_\varrho\big)/\sqrt{\Omega_\varrho - N_\varrho}\ (\varrho \equiv \pi, \nu)\,. \tag{7.141}$$

We show the dependence of ε, κ and χ_ϱ for a series of isotones in the 50–82 shell where all orbitals are considered as one large degenerate $j = 31/2$ shell ($\Omega = 16$) (Fig. 7.37) (Scholten 1980). More detailed calculations studying the parameter behaviour have been given by Duval, Barrett (for two j-shells) (Duval, Barrett 1981a, 1981b) and by Pittel et al. (for many non-degenerate shells) (Pittel et al. 1982) where the major trends in $\varepsilon, \kappa, \chi_\varrho$ are retained. A full calculation along the above lines, using a delta interaction and a quadrupole proton-neutron force was carried out by T. Otsuka (Otsuka et al. 1978a, 1978b). Results for the case $n_\pi = 6$ ($0 \le n_\nu \le 14$) are presented in Fig. 7.38a for a single j ($j = 31/2$) shell approximation together with a fit (using ε, κ, χ_ϱ, $V_{\pi\pi}$, $V_{\nu\nu}$ as parameters) for the Ba nuclei (Fig. 7.38b). Thus, it

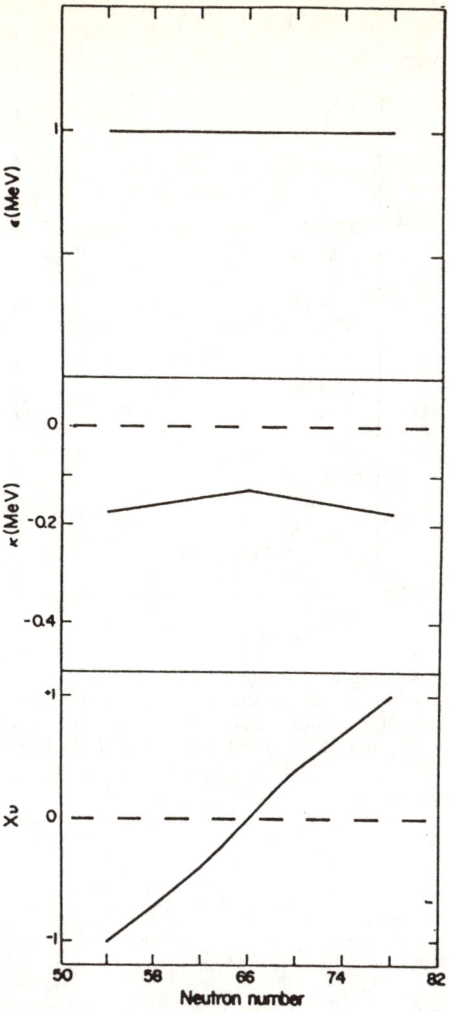

Fig. 7.37. The dependence of the important IBM parameters ε, κ, χ_ν on the neutron number n_ν as deduced using the shell-model theory with degenerate single-particle orbitals (taken from (Scholten 1980))

appears that many nuclear properties in medium-mass and heavy-mass regions are rather insensitive to details and only depend on gross features such as
– the existence of closed shells at 50, 82, 126
– the number of valence protons n_π and neutrons n_ν.

So, we have shown how, starting from general shell-model techniques and using the pairing and quadrupole force components, a rather interesting approximation to the shell-model could be determined. The basic approximations are the S, D pair truncation and subsequently, the boson (s, d) space mapping. A boson Hamiltonian is obtained which for the lowest-lying levels is determined via the shell-model S, D basis and shell-model interactions. This Hamiltonian is able to describe a large class of collective states in medium heavy and heavy nuclei. One

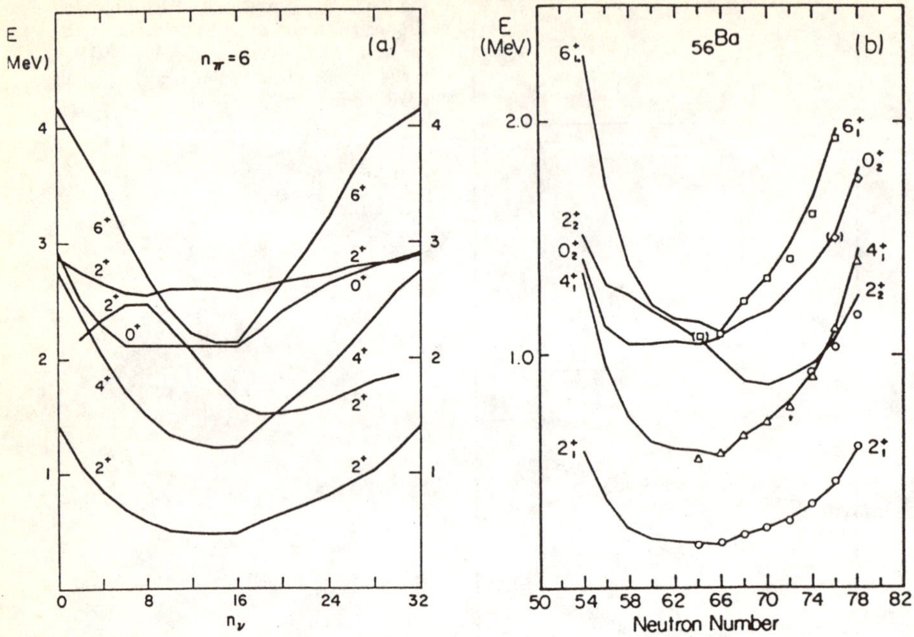

Fig. 7.38. (**a**) The energy spectrum of even-even nuclei, for a fixed proton number $n_\pi = 2N_\pi = 6$ and varying neutron number $0 \le n_\nu \le 32$ in the single-j shell approximation. We give the $J^\pi = 0^+$, 2^+, 4^+ and 6^+ levels (taken from (Otsuka et al. 1978a)). (**b**) Calculated energy spectra in the even-even Ba ($Z = 56$) nuclei. The circles (2^+), squares (6^+), triangles (4^+) and diamonds (0^+) denote the experimental data (taken from (Arima, Iachello 1984))

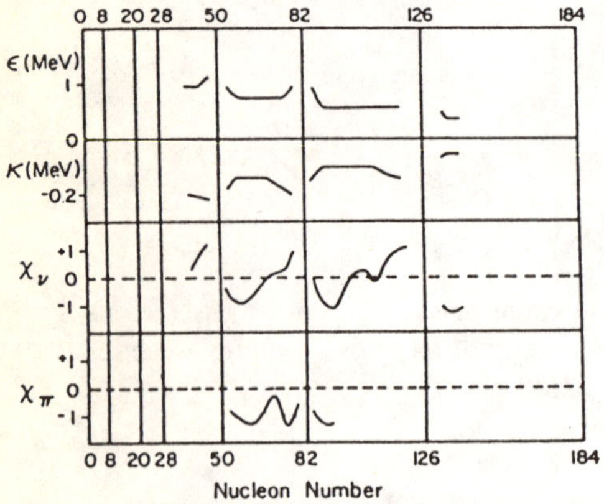

Fig. 7.39. Behavior of the IBM-2 parameters ε, κ and χ_ν as a function of N_π and of χ_π as a function of the nucleon number (taken from (Wood 1983) and (Wood 1987))

can now also, from here on, consider the parameters ε, κ, χ_ϱ, ..., as free parameters that are determined in order to describe nuclear excited states optimally (Wood 1983, 1987) (Fig. 7.39). In order to carry out such a program, the general Hamiltonian H_B needs to be diagonalized in the boson basis of (7.125). A general code has been written by T. Otsuka to perform this task (NPBOS) and I show

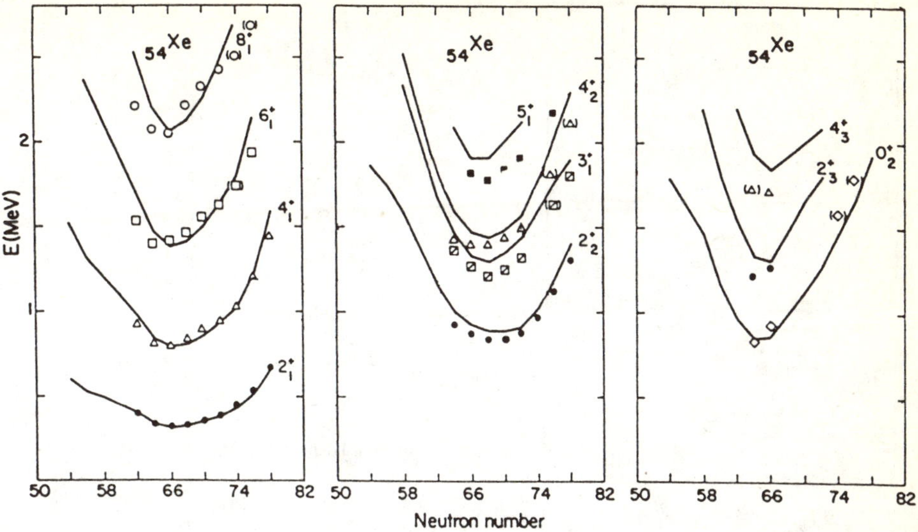

Fig. 7.40. Comparison between a typical IBM-2 calculated spectrum for even-even Xe nuclei and the data (taken from (Puddu et al. 1980))

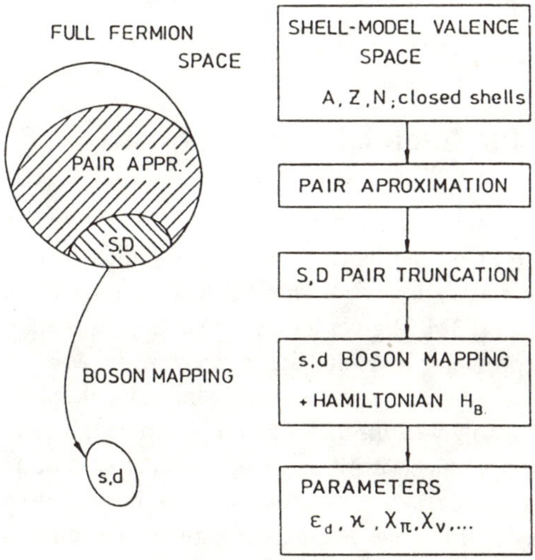

Fig. 7.41. Schematic representation of the approximations underlying the interacting boson model when starting from the complete shell-model valence space

some typical results (Fig. 7.40). The general strategy for carrying out an IBM-2 calculation is schematically depicted in Figs. 7.41 and 7.42.

So, from this small Sect. 7.7, with its relation to the general problem of treating a large number of valence protons and neutrons outside a doubly-close shell configuration, it should have become clear that, within a number of reasonable approximations, a nice link between the nuclear shell-model as such and a boson model description of low-lying collective quadrupole excitations can be made.

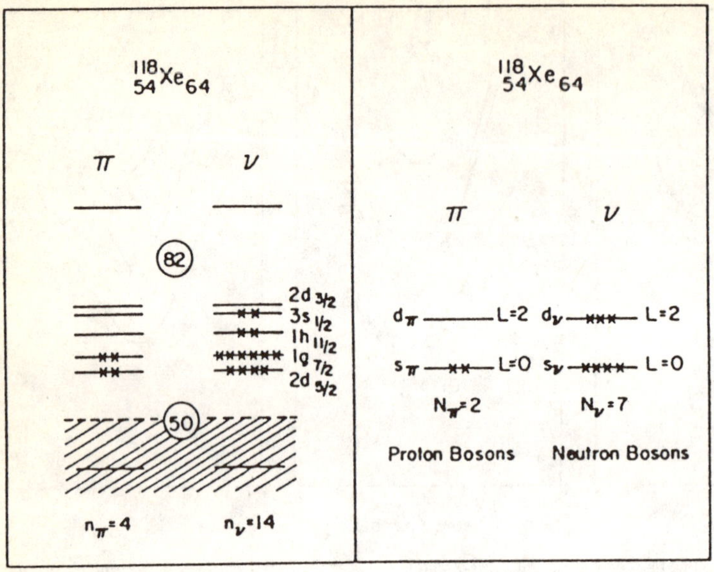

Fig. 7.42. Schematic outline of how to carry out an actual IBM-2 calculation for a nucleus like $^{118}_{54}Xe_{64}$: (i) First determine the nearest closed shell (50,50), (ii) Determine the number of bosons (valence particle (hole) number divided by two), (iii) Make an estimate of the major parameters $\varepsilon, \kappa, \chi_\pi$ and χ_ν, according to (7.140, 141) (taken from (Arima, Iachello 1984))

7.8 Symmetry Properties of the Interacting Boson Model

In the discussion in Sect. 7.7, we attempted to set up a shell-model description of nuclear collective quadrupole motion using the pairing (between identical nucleons) and quadrupole proton-neutron interactions. We showed that an interacting boson approximation to the shell model (IBM-2) gives a rather good approximation to the full shell-model many-body problem, and that the underlying parameters in it can be related to the nuclear shell-model interaction and shell-model space. We can now, however, adopt the philosophy of the IBM-2 as a phenomenological model defined immediately in the boson space itself and study collective motion starting from the interactions and possible symmetries within this boson space only. In such an approach, called the IBM-1, the distinction between proton and neutron variables is left out (and, thus, also the relation to the nuclear shell model). Here, we shall not discuss how to relate the IBM-2 and IBM-1 descriptions since the IBM-1 model simply forms a subspace of the IBM-2 where the charge variables occur in a symmetrized way (see Scholten (1980), Otsuka (1978), Arima and Iachello (1984)).

7.8.1 Symmetries

In many problems in physics, exact solutions can be obtained if the Hamiltonian has certain symmetries. Rotational invariance leads, in general, to the possibility of characterizing the angular eigenfunctions by the quantum number l, m, which are quantum numbers related to the rotation group $O(3)$ and $O(2)$ in three and two dimensions. The group structure of the boson model is, of course, more complex but can be discussed in a 6-dimensional space spanned by the s^+, d^+_μ ($\mu = -2, \ldots, +2$) bosons. The column vector

$$b^+_\mu = \begin{bmatrix} s^+ \\ d^+_2 \\ d^+_1 \\ \vdots \\ d^+_{-2} \end{bmatrix} , \tag{7.142}$$

then transforms according to the group $U(6)$ and the s and d bosons basis states

$$B = \underbrace{b^+_\mu b^+_\pi b^+_\kappa \ldots}\, |0\rangle , \tag{7.143}$$

therefore form the totally symmetric representations of the group $U(6)$, which is characterized by the number of bosons N. In the present section, where no distinction is made between proton and neutron bosons, we call N the total boson number. The 36 bilinear combinations $G_{\mu\nu} = b^+_\mu b_\nu$ ($\mu, \nu = 1, 2, \ldots, 6$) $b^+ \equiv (s^+, d^+)$ form a $U(6)$ Lie algebra, since these generators "close" under commutation. This is the case if, for the generators χ_a, one has the relation (see also Appendix I)

$$[\chi_a, \chi_b] = \sum_c C^c_{ab} \chi_c \, . \tag{7.144}$$

The general two-body Hamiltonian within this $U(6)$ Lie algebra consists of linear (single-boson energies) and quadratic (two-body interactions) terms in the generators $G_{\mu\nu}$

$$\left.\begin{aligned} H &= E_0 + \sum_{\mu\nu} \varepsilon_{\mu\nu} G_{\mu\nu} + \tfrac{1}{2} \sum_{\substack{\mu\nu \\ \kappa\sigma}} U_{\mu\nu\kappa\sigma} G_{\mu\nu} G_{\kappa\sigma} \\ H &= E_0 + H' , \end{aligned}\right\} \tag{7.145}$$

indicating that, in the absence of H', all possible states for a given N will be degenerate in energy. The interaction H' will then split the different possible states for given N and, in general, one needs a matrix diagonalization.

There are now situations where the Hamiltonian H can be rewritten exactly as a sum of Casimir (invariant) operators of a complete chain of subgroups of G, $G \supset G' \supset G'' \ldots$

$$H = \alpha C(G) + \alpha' C(G') + \alpha'' C(G'') + \ldots \tag{7.146}$$

Then, the eigenvalue problem for H can be solved in closed form and leads to energy (mass) formulas in which the eigenvalues are given in terms of the quantum numbers labelling the irreducible representations (irrep) of $G \supset G' \supset G''$ .. e.g.

$$E = \alpha\langle C(G)\rangle + \alpha'\langle C(G')\rangle + \alpha''\langle C(G'')\rangle + \dots . \tag{7.147}$$

where $\langle..\rangle$ are the expectation values. In this case, one says that the system has a dynamical symmetry (e.g. Gell-Mann-Ne'eman's $SU(3)$ mass equation, see also Appendix I).

So, the general problem to be solved becomes:

i) to identify all possible subgroup chains of $U(6)$
ii) to find the irrep for these chains of groups,
iii) to evaluate the expectation values of the Casimir invariants.

Problems (ii) and (iii) are well-defined group theoretical problems, (i) depends on the physics of the problem.

In Fig. 7.43 we illustrate, for a very general situation, how the largely degenerate number of states, characterized by a set of quantum numbers $[N]n_1, n_2$, becomes split through interactions which do not, however, mix the various individual states.

A large number of examples of such dynamical symmetries that hold exactly, or are only broken in a very slight way, are known in physics, and in nuclear and particle physics in particular. Before applying the above method to the *sd*-interacting boson model, we should like to point out (i) the example of nucleons moving in a central harmonic oscillator potential and (ii) the Zeeman splitting of a particle moving in a magnetic, external field. Both are illustrated in Fig. 7.44. On the extreme left, the large degeneracy of fermions moving in a harmonic oscillator potential and characterized for the full quantum state by $|N(l, 1/2)jm\rangle$, is denoted by the thick line. This very large degeneracy can now be split by the successive addition of the $\alpha \boldsymbol{l}^2$ term, lifting the degeneracy of the l value, while still keeping the $l \pm 1/2$ and magnetic degeneracy. Subsequently, a spin-orbit term $\beta, \boldsymbol{l}.\boldsymbol{s}$ lifts the remaining parallel and anti-parallel orbital spin orientation degeneracy, only keeping the m- (magnetic quantum number) degeneracy. This magnetic degeneracy can then be lifted by placing the system in an external field (applying an interaction term proportional to j_z) and is illustrated on the right-hand side of Fig. 7.44.

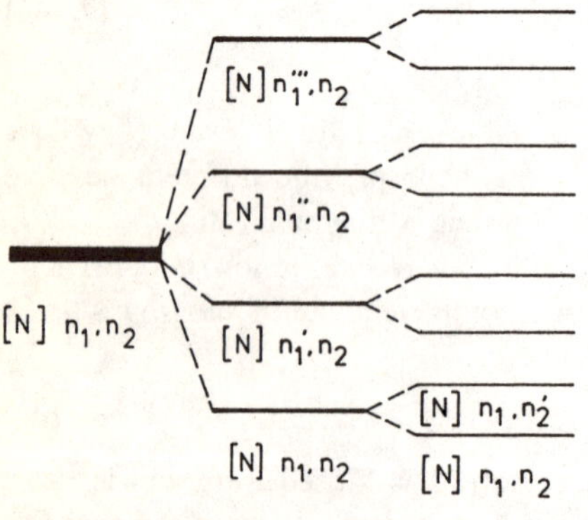

Fig. 7.43. Illustration of how a large number of degenerate levels, characterized by a set of quantum numbers (here denoted by the set $[N]n_1, n_2$), can be split up in sequence through residual interactions which split the degeneracy but keep the symmetry intact (dynamical symmetry)

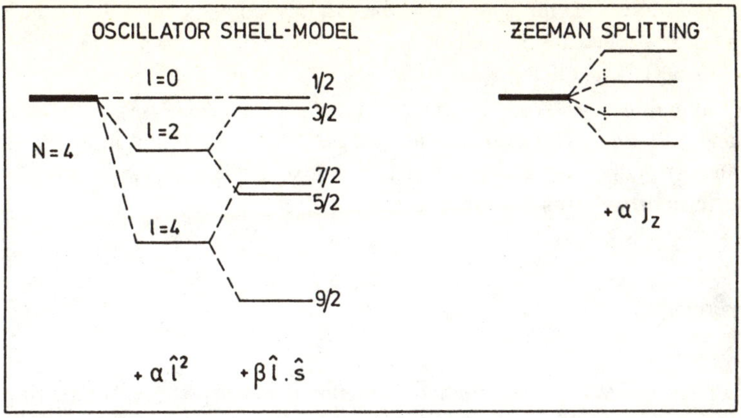

Fig. 7.44. Two examples of dynamical symmetries (left-hand side). The shell-model orbitals where a fermion with intrisic spin 1/2 moves in a harmonic oscillator potential. The large degeneracy of the $N = 4$ shell, denoted by the quantum numbers $|N(l\,1/2)jm\rangle$, is split up throught the $\alpha\boldsymbol{l}^2$ and $\beta\boldsymbol{l}.\boldsymbol{s}$ terms, (right-hand side). Magnetic substate splitting through the interaction of an αj_z term (Zeeman splitting)

7.8.2 Subgroup Chains of $U(6)$

In trying to identify all possible subgroup chains of $U(6)$ such that the angular momentum and its projection quantum number related to the groups $O(3)$ and $O(2)$, respectively, occur, only a few possibilities offer themselves.

Of all 36 generators formed by angular momentum coupling into $(b_l^+\tilde{b}_l)_M^{(L)}$ (with $b_{l=2}^+ = d^+$, $b_{l=0}^+ = s^+$ and $\tilde{b}_{l,m} = (-1)^{l-m}b_{l,-m}$), the 25 generators $(d^+\tilde{d})_M^{(L)}$ close under commutation. They form the subgroup $U(5)$ and one finds $U(6) \supset U(5)$. Furthermore, the three components of $(d^+\tilde{d})_M^{(1)}$ and the seven components of $(d^+\tilde{d})_M^{(3)}$ (10 components in all) also close, forming the generators of the $O(5)$ group. Clearly, the three components $(d^+\tilde{d})_M^{(1)}$ that have a structure similar to the three components of the angular momentum operator L_M close and, finally, a chain ob subgroups results

$$U(6) \supset U(5) \supset O(5) \supset O(3) \supset O(2)\,. \tag{7.148}$$

Two more chains can be constructed, albeit in a less trivial way, with the resulting structure

$$\begin{aligned} &U(6) \supset SU(3) \supset O(3) \supset O(2) \\ &U(6) \supset \;\; O(6) \supset O(5) \supset O(3) \supset O(2)\,. \end{aligned} \tag{7.149}$$

The detailed algebra involved in finding the irrep for these chains and constructing the various Casimir invariant operators, as well as deriving the eigenvalues, is given by Iachello and Arima (1988).

The physics related to the three possible chains is that of anharmonic quadrupole vibrational motion ($U(5)$ chain) rotational motion ($SU(3)$) and γ-unstable rotor motion ($O(6)$).

Hamiltonians more general than those that simply correspond to dynamical symmetries can be studied numerically using diagonalization of the Hamiltonian within the $U(5)$ basis (Scholten, 1980).

Extensive literature exists on the Interacting Boson model (IBM-1, IBM-2, IBM-3,...) and its various extensions. Many references can be found in the book by Iachello and Arima (1988), and in a recent new text on the algebraic properties of boson and fermion systems (editor R. F. Casten, 1993).

7.9 Summary

In the study of nuclei having many nucleons outside the closed shells, the pairing degree of freedom between identical nucleons and the proton-neutron interaction when both valence protons and neutrons are present are the major contributions in establishing the various nuclear structure effects.

In treating the properties of pairing among a large number of identical nucleons before studying the more complex situations of actual single-particle ordering together with realistic nuclear interactions, a number of approximations to the pairing problem have to be studied. We first concentrate on pairing between n nucleons in a single-j-shell orbital. Here, the algebraic properties of seniority (or broken-pair) formalism are pointed out and used to obtain closed expressions for the energy eigenvalues and the corresponding eigenvectors. Subsequently the extension to pairing in non-degenerate levels for the two-particle case is discussed. The eigenvalue pairing equation results in an interesting dispersion relation. The third step of increasing complexity, the handling of n valence particles in non-degenerate orbitals, then leads to a full BCS (Bardeen-Cooper-Schrieffer) treatment of the nuclear many-body problem. The concept of BCS reference state, quasi-particle operators (as linear combinations of particle and hole operators), the gap equations which solve for the optimal nucleon pair distribution (occupation probabilities for the various single-particle orbitals) by minimizing the total energy in the BCS reference state are worked out in some detail. This then leads to a description of the study of excitation modes in single-closed shell nuclei such as the Sn, Pb,... and $N = 50, 82, \ldots$ nuclei. The conditions for obtaining solutions to the BCS gap equations for a given number of particles and given interaction strength are derived and discussed.

As an illustrative example, the BCS method is applied to the $N = 82$ nuclei. A number of feature are discussed for these nuclei: (1) the odd-even mass difference and its relation to the pairing energy (2) the study of energy spectra, including the 0 and 2 qp (quasi-particle) configurations (3) the study of electromagnetic properties and the influence of the specific pairing factors modifying the more simple single-particle transition matrix elements (4) the description of spectroscoping factors for one-nucleon transfer reactions as a test for the specific single-particle character of certain eigenstates and (5) an application of superfluidity to the study of nuclear matter and neutron stars.

This BCS theory is a particle non-conserving theory and particle number is conserved only on average. Therefore, the broken-pair alternative was formulated and is discussed both in relation to BCS theory and, more importantly, as a low-seniority approximation to the more complex shell-model calculation. Some applications, taken from the work of Allaart et al., are used to illustrate this method for single-closed shell nuclei. The extension to nuclei with both an open proton and neutron shell, however, is rather straightforward in that respect and is only briefly presented. This method of broken-pair excitations as an approximation to a much larger shell-model calculation is then used to make contact with one of the most elegant approximations to the full shell-model calculation i.e. that of the Interacting Boson model approximation (IBA).

The IBA is introduced, after a short discussion on symmetries of the nucleus in general, through a simple two-level model containing a one-broken proton and one-broken neutron pair. This model already leads to the concept of low-lying nuclear collective quadrupole motion. The more general case, where a large number of broken pairs (both for protons and/or neutrons) are present, but truncating to just S (spin 0) and D (spin 2) pairs, is described. This SD-shell model space is mapped to a boson space of *sd* bosons (leading to the Interacting Boson model: IBM). The specific procedure for mapping the pairing force and a quadrupole-quadrupole force is outlined in detail. As a result, a typical *sd*-boson IBM Hamiltonian is derived and the major parameters discussed. Some illustrations for these IBM parameters, as well as a numerical study of the even-even Ba and Xe nuclei, are presented. Finally, a section has been added which discussed the basis group theoretical content of the *sd*-boson model. The symmetries of the $U(6)sd$-boson model are pointed out and the three dynamical symmetries (the $U(5)$, the $SU(3)$ and $O(6)$ symmetries) are presented, further details on the $U(5)$ group chain being given. The fundamental aspects of dynamical symmetries, which are used in order to describe the quadrupole collective properties in the nucleus in an elegant way, are stressed in this chapter.

Problems

7.1 Derive the commutation relations for the S-pair fermion creation operator

$$[S_j, S_j^+] = 1 - n_j/\Omega_j \ ,$$

(with $n_j = \sum a_{jm}^+ a_{jm}$ with $\Omega_j \equiv j + 1/2$).
Discuss the approximate boson commutation relation for the fermion pair operator. Show also that

$$\begin{aligned}[H, S_j^+] &= -GS_j^+(\Omega_j - n_j) \\ &= -G(\Omega_j - n_j + 2)S_j^+ \ .\end{aligned}$$

*7.2 Derive the number projected BCS wave function (7.38), starting from the more standard BCS wave function

$$|\tilde{0}\rangle = \prod_{\nu>0}(u_\nu + v_\nu a^+_\nu a^+_{\bar{\nu}})|0\rangle \ .$$

7.3 Show that the condition

$$\frac{\partial}{\partial v_\nu}\langle\tilde{0}|\mathcal{H}|\tilde{0}\rangle$$

indeed leads to the BCS equations for u_ν, v_ν for given values of G and ε_ν.

*7.4 Derive the angular momentum coupled (for constant pairing force strength G) form of the BCS equations (7.56).

7.5 Show that, for a given E_ν spectrum of one-quasi particle energies and given particle number n, the BCS equations can be inverted to give the ε_ν and force strength G (use the Frobenius method).

7.6 Derive a general form of the electromagnetic operator, expressed in quasi-particle operators, for the Biedenharn-Rose (BR) phase convention. How do the pairing reduction factors of Table 7.1 change formally?

7.7 Derive the quadrupole-quadrupole matrix element of (7.116) and the energy eigenvalues when diagonalizing the 2-level energy matrix of (7.117).

7.8 Show that the 0^+ coupled configuration in (7.12) is indeed an eigenstate of the pairing Hamiltonian (7.9) at an energy $E_0 = -G.\Omega$ (relative to the two-particle unperturbed energy).

7.9 Using the fermion to boson mapping as presented in (7.126), derive an expression for (i) ε_s, (ii) u_0, as well as the leading coefficients α_0 and β_0 for the quadrupole operator.

7.10 Starting from the values of α_0 and β_0, derived for a single-j shell, study the variation of ε, κ and χ_ϱ as a function of e.g. the neutron number N_ν.
Hint: N_π, Ω_π give a constant contribution in all cases.
Carry out this calculation for N_ν in the 50–82 neutron shell and for the Ba nuclei.

7.11 Show that $\langle\tilde{0}|\boldsymbol{N}|\tilde{0}\rangle = \sum_\alpha v^2_\alpha$.

*7.12 Use the data for the Sn isotopes, as given in Fig. 7.45. The orbitals around the Fermi level are: $3s_{1/2}$, $2d_{5/2}$, $2d_{3/2}$, $1g_{7/2}$ and $1h_{11/2}$.
a) Assume that U_0 is a smooth function of $\langle\tilde{0}|\boldsymbol{N}|\tilde{0}\rangle$, and read off the one-quasi particle energies E_α for the neutron orbitals in ^{115}Sn and in ^{117}Sn,
b) Assuming that E_α are roughly equal in ^{115,116,117}Sn, where do you expect the excited states to occur in ^{116}Sn?
c) Take the interaction in a schematic form such as

$$V_{\alpha\beta\gamma\delta} = -G.\delta_{\beta\tilde{\alpha}}\delta_{\delta\tilde{\gamma}} \quad \text{with strength } G \ .$$

Prove the following:
* All Δ_α are equal,
* $G = \left\{1/4.\sum_\beta 1/E_\beta\right\}^{-1}$.

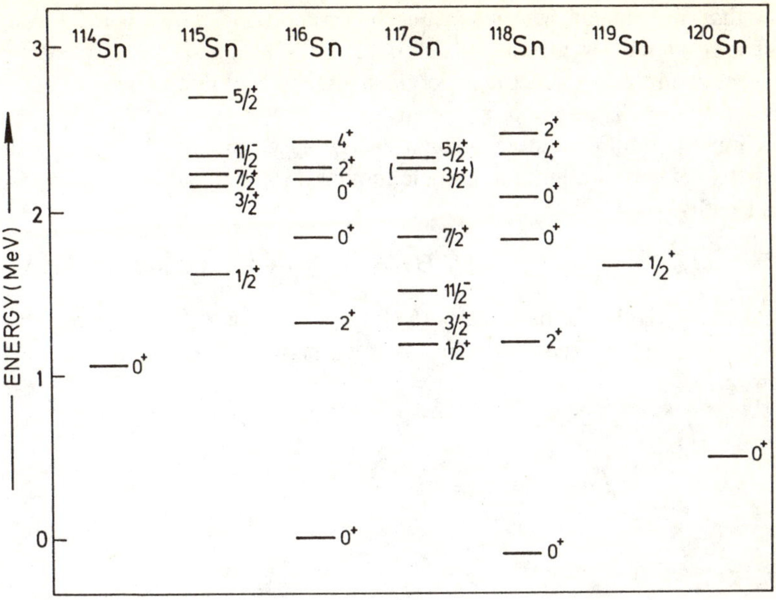

Fig. 7.45

*7.13 Given are the coupled commutators for the $U(6)$ generators

$$\left[\left[b^+_{\lambda_1}\tilde{b}_{\lambda_2}\right]^{(l_1)}_{m_1}, \left[b^+_{\lambda_3}\tilde{b}_{\lambda_4}\right]^{(l_2)}_{m_2}\right] = \sum_{l,m} \sqrt{(2l_1+1)(2l_2+1)}\,\langle l_1 m_1, l_2 m_2 | l m\rangle$$

$$\times \left[\delta_{\lambda_2\lambda_3}(-1)^{\lambda_1-\lambda_4+l}\begin{Bmatrix} l_1 & l_2 & l \\ \lambda_4 & \lambda_1 & \lambda_2 \end{Bmatrix}\left[b^+_{\lambda_1}\tilde{b}_{\lambda_4}\right]^{(l)}_m\right.$$

$$\left. -\delta_{\lambda_1\lambda_4}(-1)^{\lambda_2-\lambda_3+l_1+l_2}\begin{Bmatrix} l_1 & l_2 & l \\ \lambda_3 & \lambda_2 & \lambda_1 \end{Bmatrix}\left[b^+_{\lambda_3}\tilde{b}_{\lambda_2}\right]^{(l)}_m\right], \qquad (*)$$

Show that the vibrational group chain, starting from the 36 generators $[b^+_\lambda \tilde{b}_{\lambda'}]^{(l)}_m$ (with $b^+_{\lambda=2} = d^+$ and $b^+_{\lambda=0} = s^+$), forms the chain

$$U(6) \supset U(5) \supset O(5) \supset O(3) \supset O(2)$$

by evaluating the various closure relations $(*)$.

*7.14 Show that, within the two-level model as discussed in Sect. 7.7, the $E2$ matrix element connecting the 2^+_1 level to the 0^+_1 ground state is a coherent sum of the separate proton and neutron contributions. Show that this coherence remains, even if the two-level model is constructed for $(j_\pi)^{n_\pi}$ and $(j_\nu)^{n_\nu}$ configurations, where n_π (and/or n_ν) denotes a hole configuration.

*7.15 Work out the detailed state labeling and the energy spectrum for the interacting boson $U(5)$ limit with $N = 3$ bosons.

7.16 Show that, in extending the interacting boson model to proton and neutron bosons, besides the symmetric $0^+; 2^+; 0^+, 2^+, 4^+$ states, antisymmetric states $2^+; 1^+, 3^+$ rcsult (antisymmetric under the interchange of the charge label) for a one-proton one-neutron boson system.

7.17 Work out the multiplet structure (and energy spectrum) for the dynamical symmetry related to a fermion particle moving in a harmonic oscillator potential with

$$\boldsymbol{H} = \boldsymbol{H}_{h.0.} + \beta \boldsymbol{l}^2 + \alpha \boldsymbol{l}.\boldsymbol{s} \quad \text{with } \boldsymbol{H}_{h0} = \boldsymbol{p}^2/2m + 1/2.m\omega^2 r^2) ,$$

and characterized by the basis states $|N(l, 1/2); jm\rangle$, for the $N = 4$ multiplet. Consider the βl^2 and the $\alpha \boldsymbol{l}.\boldsymbol{s}$ terms as small perturbations to the oscillator spectrum.

8. Self-Consistent Shell-Model Calculations

8.1 Introduction

In the preceding chapters we have discussed the methods necessary to study the nuclear structure of nuclei throughout the nuclear mass table, excluding strongly deformed nuclei (Bohr, Mottelson 1975) which are outside the scope of the present book. We have discussed the short-range (pairing) properties of atomic nuclei as best illustrated near closed-shell nuclei. We have applied the concept of pairing to nuclei with many valence nucleons outside a single-closed shell nucleus, or even to nuclei with a number of valence protons and neutrons outside doubly-closed shell nuclei (Chap. 7). Also, at closed shells, particle-hole excitations show up as elementary modes of excitation and have been studied in a TDA and RPA approximation (Chap. 6). In the latter chapter, extensive use was made of the methods of second quantization, developed in Chap. 5. In most applications, we started from an average field that was determined in a phenomenological way [harmonic oscillator one-body potential (Chap. 3)] and a residual nucleon-nucleon interaction was used (effective matrix elements, schematic interaction or realistic interaction) which was not self-consistently determined with respect to the one-body potential.

The method of determining the average one-body potential from the nucleon-nucleon interaction (Hartree-Fock method) was briefly outlined in Chap. 3. This method, in principle, allows for a self-consistent study of nuclear structure starting from a given nucleon-nucleon force V_{ij}. This method is a rather ambitious one since it faces the problem of determining at the same time the average nucleonic properties (binding energy, nuclear radii, density distributions, ...) and the excited states in each nucleus (with the constraint of studying spherical nuclei near closed shells only). This ambitious task is schematically drawn in Fig. 8.1 pointing out that three orders of magnitude should be bridged by a single nucleon-nucleon interaction. Because of the need to describe nuclear average properties, the force should establish correct saturation properties, a condition that was not needed for describing the excitation spectra of nuclei with a few valence nucleons outside closed shells. So, the nucleon-nucleon force will be parameterized with the same, relatively small number of parameters, to be determined throughout the nuclear mass region. In the next few sections, we shall outline the construction of a nucleon-nucleon interaction of Skyrme type (Skyrme 1956, Vautherin, Brink 1972, Negele, Vautherin 1972) with a subsequent application to ground-state properties.

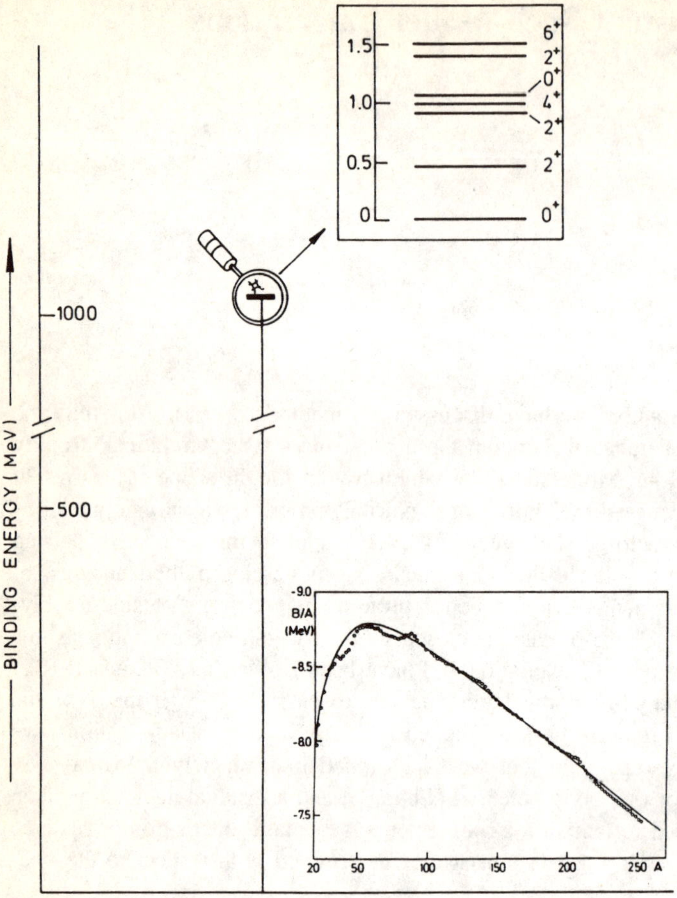

Fig. 8.1. Schematic illustration of a self-consistent calculation where, with a *single* nucleon-nucleon interaction, the global properties (nuclear binding energies, radii, densities, ...) and the local properties (excited states in each nucleus, see insert) are determined. The difference in energy scale of a factor 10^3 is presented on the energy scale with the "magnifier"

Next, we shall illustrate the value of the Skyrme forces as an effective interaction for describing detailed nuclear properties.

The spirit of this last chapter will be slightly different from the early chapters where a fully deductive method with specific and detailed derivations was carried out. Here, we shall discuss and present more the state-of-the-art that can be reached in self-consistent shell-model calculations. This Chap. 8 will rely heavily on developments that emerged in the theory group of the Nuclear Physics Institute in Gent.

8.2 Construction of a Nucleon-Nucleon Force: Skyrme Forces

8.2.1 Hartree-Fock Bogoliubov (HFB) Formalism for Nucleon-Nucleon Interactions Including Three-Body Forces

The most general form of a Hamiltonian where besides two-body forces, also three-body forces occur is

$$H = T + \tfrac{1}{2}\sum_{i,j} V_{i,j} + \tfrac{1}{6}\sum_{i,j,k} W_{i,j,k} , \tag{8.1}$$

with T the kinetic energy, V_{ij} the two-body and $W_{i,j,k}$ the three-body force.

Using second-quantization, this expression (8.1) can be rewritten as (Chap. 5)

$$\begin{aligned} H = &\sum_{\alpha,\gamma} \langle\alpha|T|\gamma\rangle a_\alpha^+ a_\gamma + \tfrac{1}{4}\sum_{\alpha,\beta,\gamma,\delta} \langle\alpha\beta|V|\gamma\delta\rangle_{\text{nas}} a_\alpha^+ a_\beta^+ a_\delta a_\gamma \\ &+ \tfrac{1}{36}\sum_{\substack{\alpha,\beta,\gamma\\ \delta,\varepsilon,\mu}} \langle\alpha\beta\varepsilon|W|\gamma\delta\mu\rangle_{\text{nas}} a_\alpha^+ a_\beta^+ a_\varepsilon^+ a_\mu a_\delta a_\gamma . \end{aligned} \tag{8.2}$$

Here, $\alpha, \beta, \ldots$ denote the single-particle quantum numbers characterizing the orbitals and nas means normalized antisymmetrized matrix elements.

The vacuum of the a_α^+ operators is the state with no particles present $|\,\rangle$. It is sometimes more interesting to define a new reference state (a doubly-closed shell nucleus), a state which acts as a new vacuum state for more general operators than the a_α^+ (c_α^+, being linear, unitary combinations of the a_α^+) with the condition

$$c_\alpha|\tilde{0}\rangle = 0 , \tag{8.3}$$

$$\begin{pmatrix} c_\alpha^+ \\ c_\alpha \end{pmatrix} = \begin{pmatrix} X & Y \\ Y^* & X^* \end{pmatrix} \begin{pmatrix} a_\alpha^+ \\ a_\alpha \end{pmatrix} . \tag{8.4}$$

Using now Wick's theorem (Chap. 5), the Hamiltonian (8.2) can be rewritten as follows (condensed form)

$$H = H^{(0)} + H^{(2)} + H^{(4)} + H^{(6)} , \tag{8.5}$$

with

$$H^{(0)} = \sum_{\alpha,\gamma} \langle\alpha|T + \tfrac{1}{2}U + \tfrac{1}{6}U'|\gamma\rangle \varrho_{\alpha\gamma} + \tfrac{1}{2}\sum_{\alpha,\beta} (\Delta + \Delta')_{\alpha\beta}\kappa^*_{\beta\alpha} , \tag{8.6}$$

$$\begin{aligned} H^{(2)} = &\sum_{\alpha,\gamma} \langle\alpha|T + U + \tfrac{1}{2}U' + P|\gamma\rangle N\left(a_\alpha^+ a_\gamma\right) \\ &+ \tfrac{1}{2}\sum_{\alpha,\beta}\left[(\Delta + \Delta')_{\alpha\beta} N\left(a_\alpha^+ a_\beta^+\right) + \text{hc}\right] , \end{aligned} \tag{8.7}$$

$$H^{(4)} = \tfrac{1}{4} \sum_{\alpha,\beta,\gamma,\delta} \langle\alpha\beta|V + V'|\gamma\delta\rangle N\left(a_\alpha^+ a_\beta^+ a_\delta a_\gamma\right)$$
$$+ \tfrac{1}{12} \sum_{\alpha,\beta,\gamma,\delta} \left[\langle\alpha\beta\varepsilon|W|\gamma\delta\mu\rangle_{\text{nas}} \kappa_{\delta\gamma} N\left(a_\alpha^+ a_\beta^+ a_\varepsilon^+ a_\mu\right) + \text{hc}\right], \tag{8.8}$$

and

$$H^{(6)} = \tfrac{1}{36} \sum_{\substack{\alpha,\beta,\gamma \\ \delta,\varepsilon,\mu}} \langle\alpha\beta\varepsilon|W|\gamma\delta\mu\rangle_{\text{nas}} N\left(a_\alpha^+ a_\beta^+ a_\varepsilon^+ a_\mu a_\delta a_\gamma\right). \tag{8.9}$$

Since $H^{(0)}$ only contains contractions, the expectation value of the Hamiltonian in the new vacuum becomes

$$E^{(0)} = H^{(0)} = \langle\tilde{0}|H|\tilde{0}\rangle . \tag{8.10}$$

In the above expression for the Hamiltonian, we have used the following notation:

$$\varrho_{\alpha\gamma} \equiv a_\alpha^+ a_\gamma , \tag{8.11}$$

$$\kappa_{\beta\alpha}^* \equiv a_\alpha^+ a_\beta^+ \qquad \left(\text{with } \kappa_{\alpha\beta}^* = -\kappa_{\beta\alpha}^*\right), \tag{8.12}$$

$$\langle\alpha|U|\gamma\rangle = \sum_{\beta,\delta} \langle\alpha\beta|V|\gamma\delta\rangle_{\text{nas}} \varrho_{\beta\delta} , \tag{8.13}$$

$$\langle\alpha|U'|\gamma\rangle = \sum_{\beta,\delta} \langle\alpha\beta|V'|\gamma\delta\rangle_{\text{nas}} \varrho_{\beta\delta} , \tag{8.14}$$

which represents a part of the average field coming from the three-body forces, since V' is defined through the relation

$$\langle\alpha\beta|V'|\gamma\delta\rangle_{\text{nas}} = \sum_{\varepsilon,\mu} \langle\alpha\beta\varepsilon|W|\gamma\delta\mu\rangle_{\text{nas}} \varrho_{\varepsilon\mu} , \tag{8.15}$$

and

$$\Delta_{\alpha\beta} = \tfrac{1}{2} \sum_{\gamma,\delta} \langle\alpha\beta|V|\gamma\delta\rangle_{\text{nas}} \kappa_{\delta\gamma} , \tag{8.16}$$

denotes the pairing potential and similarly

$$\Delta'_{\alpha\beta} = \tfrac{1}{2} \sum_{\gamma,\delta} \langle\alpha\beta|V'|\gamma\delta\rangle_{\text{nas}} \kappa_{\delta\gamma} , \tag{8.17}$$

is a contribution to the pairing potential due to the three-body forces. Then

$$\langle\alpha|P|\gamma\rangle = \tfrac{1}{4} \sum_{\beta,\varepsilon,\delta,\mu} \langle\alpha\beta\varepsilon|W|\gamma\delta\mu\rangle_{\text{nas}} \kappa_{\varepsilon\beta}^* \kappa_{\mu\delta} , \tag{8.18}$$

is a pairing contribution, typical for the three-body forces, and makes up part of the single-particle energy.

The HFB method now tries to determine the linear transformation (8.4) which minimizes the HFB ground-state energy $E^{(0)}$. As was shown, in a more simple case in Chap. 7, the term $H^{(2)}$ will become diagonal in the new basis c_α^+.

i) One can now, starting from the given basis a_α^+ and the corresponding known wave functions $\varphi_\alpha(\boldsymbol{r})$ try to determine the transformation quantities X and Y which accomplish the above task (Waroquier 1982). This procedure results in the HFB secular equations

$$\begin{aligned}&\sum_\gamma \langle\alpha|T+U+\tfrac{1}{2}U'+P|\gamma\rangle X_{i\gamma}+\sum_\beta(\Delta+\Delta')_{\alpha\beta}Y_{i\beta}=X_{i\alpha}E_i\,,\\&\sum_\gamma \langle\gamma|T+U+\tfrac{1}{2}U'+P|\alpha\rangle Y_{i\gamma}+\sum_\beta(\Delta+\Delta')^*_{\alpha\beta}X_{i\beta}=-Y_{i\alpha}E_i\,.\end{aligned}\tag{8.19}$$

Solving these equations (including the three-body force through the U', P and Δ' fields), $H^{(2)}$ becomes diagonal, $E^{(0)}$ minimal and $H^{(4)}$ and $H^{(6)}$ are the remaining residual interactions (prove that $E^{(0)}$ is minimal, supposing that the HFB equations (8.19) are fulfilled).

ii) Using the theorem of Bloch and Messiah (Bloch, Messiah 1962), it is possible to show that the general, linear transformation of (8.4) results in pairing effects so that each state is connected via pairing to just *one* other state. In this situation, (8.4) can be reduced to the form

$$c_\alpha^+ = u_\alpha a_\alpha^+ - v_\alpha a_{\overline{\alpha}}\,.\tag{8.20}$$

Now, the single-particle basis α *and* the amplitudes u_α, v_α are unknown and have to be determined through the condition of getting $H^{(2)}$ in diagonal form within the new representation. Using (8.20), the $H^{(2)}$ in (8.7) becomes

$$H^{(2)} = H_{11}^{(2)} + H_{20}^{(2)}\,,\tag{8.21}$$

with

$$H_{11}^{(2)} = \sum_{\alpha,\gamma}\left[\eta_{\alpha\gamma}(u_\alpha u_\gamma - v_\alpha v_\gamma) - (\Delta+\Delta')_{\alpha\overline{\gamma}}(u_\alpha v_\gamma + v_\alpha u_\gamma)\right]c_\alpha^+ c_\gamma\,,\tag{8.22}$$

and

$$H_{20}^{(2)} = \sum_{\alpha,\gamma}\left[\eta_{\alpha\gamma}u_\alpha v_\gamma + \tfrac{1}{2}(\Delta+\Delta')_{\alpha\overline{\gamma}}\cdot(u_\alpha u_\gamma - v_\alpha v_\gamma)\right]\left(c_{\overline{\gamma}}c_\alpha + c_\alpha^+ c_{\overline{\gamma}}^+\right).\tag{8.23}$$

Here, we have introduced the matrix element

$$\eta_{\alpha\gamma} \equiv \langle\alpha|T+U+\tfrac{1}{2}U'+P|\gamma\rangle\,.\tag{8.24}$$

The necessary and sufficient condition for minimizing H with respect to the reference state $|\tilde{0}\rangle$ is that $H_{20}^{(2)}$ disappears. At the same time, $H_{11}^{(2)}$ becomes diagonal, leading to the BCS equations (Chap. 7)

$$\begin{aligned}&\eta_{\alpha\gamma}(u_\alpha u_\gamma - v_\alpha v_\gamma) - (\Delta+\Delta')_{\alpha\overline{\gamma}}(u_\alpha v_\gamma + v_\alpha u_\gamma) = E_\alpha\delta_{\alpha\gamma}\,,\\&\eta_{\alpha\gamma}u_\alpha u_\gamma + \tfrac{1}{2}(\Delta+\Delta')_{\alpha\overline{\gamma}}(u_\alpha u_\gamma - v_\alpha v_\gamma) = 0\,.\end{aligned}\tag{8.25}$$

The solution to these equations (performed self-consistently) diagonalizes $H_{11}^{(2)}$, and takes the matrices $\varrho_{\alpha\gamma}$ and $\kappa^*_{\beta\alpha}$ in their canonical form i.e.

$$\varrho_{\alpha\gamma} = v_\alpha^2 \delta_{\alpha\gamma} \,, \qquad \kappa^*_{\beta\alpha} = u_\alpha v_\alpha \delta_{\beta\overline{\alpha}} \,. \tag{8.26}$$

The ground-state energy $E^{(0)}$ then becomes

$$E^{(0)} = \sum_\alpha v_\alpha^2 \left(T_\alpha + \tfrac{1}{2}\varepsilon_\alpha + \tfrac{1}{3}\varepsilon'_\alpha\right) + \tfrac{1}{2}\sum_\alpha u_\alpha v_\alpha (\Delta + \Delta')_{\alpha\overline{\alpha}} \,, \tag{8.27}$$

with

$$\varepsilon_\alpha = \langle\alpha|U|\alpha\rangle \,, \tag{8.28}$$

$$\varepsilon'_\alpha = \tfrac{1}{2}\langle\alpha|U'|\alpha\rangle \,, \tag{8.29}$$

self-energy corrections.

8.2.2 Application of HFB to Spherical Nuclei

For spherical nuclei near doubly-closed shells, the pair scattering from occupied to the much higher-lying unoccupied orbitals is almost non-existent and the HFB formalism reduces to the spherical HF formalism, for which one has to determine the HF single-particle states $\varphi_\alpha(\boldsymbol{r})$.

Whenever pairing effects are non-negligible, one most often uses a two-step procedure: (i) solve the HF equations with given BCS quantities in order to determine the HF states $\varphi_\alpha(\boldsymbol{r})$ and, (ii) a second minimization is carried out [with the above $\varphi_\alpha(\boldsymbol{r})$] to solve for the new BCS quantities. Using (i) and (ii) now in an iterative way leads to the HF + BCS method.

a) Hartree-Fock (HF) Approximation

Here now, the tensor $\kappa^*_{\beta\alpha}$ disappears and $\varrho_{\alpha\gamma} = \delta_{\alpha\gamma}$ (if α is occupied) and $\varrho_{\alpha\gamma} = 0$ (if α is unoccupied.)

This now leads to

$$H^{(0)} = \sum_{\substack{\alpha \\ \text{(occupied)}}} \langle\alpha|T + \tfrac{1}{2}U + \tfrac{1}{6}U'|\alpha\rangle \,, \tag{8.30}$$

and

$$H^{(2)} = \sum_{\alpha,\gamma} \langle\alpha|T + U + \tfrac{1}{2}U'|\gamma\rangle N\left(a_\alpha^+ a_\gamma\right) . \tag{8.31}$$

We have studied the above equations in Chap. 5 where the new basis reduces to

$$a_\alpha^+ = c_{\overline{\alpha}} \quad (\alpha: \text{ occupied}) \,, \qquad a_\alpha^+ = c_\alpha^+ \quad (\alpha: \text{ unoccupied}) \,. \tag{8.32}$$

The state $\overline{\alpha}$ means the "paired" state to α or the time-reversed state to α (with some extra phase factors, depending on the phase convention used: Condon-Shortley or Biedenharn-Rose) [see (Rowe 1970) and Appendix I].

The term $H^{(2)}$ then becomes

$$H^{(2)} = \sum_{\alpha} \varepsilon_{\alpha}^{\mathrm{HF}} N\left(a_{\alpha}^{+} a_{\alpha}\right) , \tag{8.33}$$

with $\varepsilon_{\alpha}^{\mathrm{HF}}$ the Hartree-Fock energies. The ground-state energy now becomes (using the presence of three-body forces)

$$\begin{aligned} E^{(0)} = & \sum_{\substack{\alpha \\ \text{(occupied)}}} \varepsilon_{\alpha}^{\mathrm{HF}} - \tfrac{1}{2} \sum_{\substack{\alpha,\beta \\ \text{(occupied)}}} \langle \alpha\beta | V | \alpha\beta \rangle_{\mathrm{nas}} \\ & - \tfrac{1}{3} \sum_{\substack{\alpha,\beta,\gamma \\ \text{(occupied)}}} \langle \alpha\beta\gamma | W | \alpha\beta\gamma \rangle_{\mathrm{nas}} . \end{aligned} \tag{8.34}$$

The remaining terms $H^{(4)}$ and $H^{(6)}$ then describe the residual interactions that couple the different particle-hole excitations relative to the vacuum state (a doubly-closed shell nucleus) $|\tilde{0}\rangle$.

b) Solving the HF+BCS Equations in a Spherical Single-Particle Basis

In the most general case, the solution of the HF+BCS equations, in an iterative way, as discussed before is still very complicated (Waroquier et al. 1979, Waroquier 1982). Using now a spherical single-particle basis (and the Condon-Shortley (CS) convention) one has

$$|\overline{\alpha}\rangle = s_{\alpha} | -\alpha \rangle , \tag{8.35}$$

with

$$s_{\alpha} \equiv (-1)^{j_a - m_a} , \tag{8.36}$$

and

$$| -\alpha \rangle = | n_a, l_a, j_a, -m_a \rangle . \tag{8.37}$$

In the BCS equations, using the CS convention, one can choose $v_a > 0$ and $u_a = |u_a|(-1)^{l_a}$ with a resulting qp transformation

$$c_{\alpha}^{+} = u_a a_{\alpha}^{+} - s_{\alpha} v_a a_{-\alpha} . \tag{8.38}$$

Using the above restrictions, the BCS equations can be simplified very much (Waroquier 1982) and solved for the quantities $u_a(\varrho)$, $v_a(\varrho)$, ... with $\varrho = \pi, \nu$ (proton, neutron) in the HF + BCS iterative way.

8.2.3 The Extended Skyrme Force

Trying to construct a nucleon-nucleon force that is apt to describe both the nuclear global and local properties, using a parameterization that remains constant over the whole nuclear mass region is a most challenging task. Thereby we start from a Skyrme type force (Skyrme 1956) as discussed in a basic article of Vautherin and Brink (Vautherin, Brink 1972). Here, however, only nuclear ground-state properties are concentrated on. Going beyond these global aspects of the nucleus, one needs to cope with the specific pairing correlations in order to describe excited states in nuclei. Moreover, one has to treat this short-range aspect of the force in a way consistent with the other, saturation aspects of the nucleon-nucleon force. A number of calculations were performed with this aspect in mind (Vautherin, Brink 1972, Beiner et al. 1975, Liu, Brown 1976, Krewald et al. 1977).

One can now try to add extra terms, with one or two extra parameters to be fixed in order to reproduce pairing aspects properly and lead to realistic two-body matrix elements.

The suggested extension has two-body and three-body interactions: the two-body part contains an extra zero-range density-dependent term and in the three-body part, velocity-dependent terms are added. This becomes, in a schematic way

$$\text{SkE}\begin{cases}\nearrow \text{2-body part: } V(\boldsymbol{r}_1,\boldsymbol{r}_2) = V^{(0)} + V^{(1)} + V^{(2)} \\ \qquad\qquad + V^{(ls)} + V_{\text{Coul.}} + (1-x_3)V_0 \,, \\ \searrow \text{3-body part: } W(\boldsymbol{r}_1,\boldsymbol{r}_2,\boldsymbol{r}_3) = x_3 W_0(\boldsymbol{r}_1,\boldsymbol{r}_2,\boldsymbol{r}_3) \\ \qquad\qquad + W_1(\boldsymbol{r}_1,\boldsymbol{r}_2,\boldsymbol{r}_3,\ \boldsymbol{k}_1,\boldsymbol{k}_2,\boldsymbol{k}_3) \,,\end{cases} \tag{8.39}$$

where $V^{(0)}$, $V^{(1)}$, $V^{(2)}$ and $V^{(ls)}$ have the same structure as the original Skyrme force (Vautherin, Brink 1972):

$$\begin{aligned} &\text{i)} \quad V^{(0)} = t_0(1 + x_0 P_\sigma)\delta(\boldsymbol{r}_1 - \boldsymbol{r}_2) \,, \\ &\text{ii)} \quad V^{(1)} = \tfrac{1}{2}t_1\left[\delta(\boldsymbol{r}_1 - \boldsymbol{r}_2)\boldsymbol{k}^2 + \boldsymbol{k}'^2\delta(\boldsymbol{r}_1 - \boldsymbol{r}_2)\right] , \\ &\text{iii)} \quad V^{(2)} = t_2 \boldsymbol{k}' \cdot \delta(\boldsymbol{r}_1 - \boldsymbol{r}_2)\boldsymbol{k} \,, \\ &\text{iv)} \quad V^{(ls)} = iW_0'(\boldsymbol{\sigma}_1 + \boldsymbol{\sigma}_2) \cdot \left(\boldsymbol{k}' \times \delta(\boldsymbol{r}_1 - \boldsymbol{r}_2)\boldsymbol{k}\right) . \end{aligned} \tag{8.40}$$

Here, $\boldsymbol{k}$ denotes the momentum operator, acting to the right

$$\boldsymbol{k} = \frac{1}{2i}(\nabla_1 - \nabla_2) \,, \tag{8.41}$$

and

$$\boldsymbol{k}' = -\frac{1}{2i}(\nabla_1 - \nabla_2) \,, \tag{8.42}$$

acting to the left. The spin-exchange operator P_σ has been defined in Chap. 3 and the Coulomb force has its standard form. The density-dependent zero-range force V_0 reads

$$V_0 = \tfrac{1}{6}t_3(1 + P_\sigma)\varrho((\boldsymbol{r}_1 + \boldsymbol{r}_2)/2)\delta(\boldsymbol{r}_1 - \boldsymbol{r}_2) \,. \tag{8.43}$$

In the three-body part, one has the original term W_0

$$W_0 = t_3\delta(\boldsymbol{r}_1 - \boldsymbol{r}_2)\delta(\boldsymbol{r}_1 - \boldsymbol{r}_3) , \tag{8.44}$$

to which a velocity dependent zero-range term W_1 is added, given as

$$\begin{aligned} W_1 = \tfrac{1}{6}t_4\Big[\left(\boldsymbol{k}_{12}'^2 + \boldsymbol{k}_{23}'^2 + \boldsymbol{k}_{31}'^2\right)\delta(\boldsymbol{r}_1 - \boldsymbol{r}_2)\delta(\boldsymbol{r}_1 - \boldsymbol{r}_3) \\ + \delta(\boldsymbol{r}_1 - \boldsymbol{r}_2)\delta(\boldsymbol{r}_1 - \boldsymbol{r}_3)\left(\boldsymbol{k}_{12}^2 + \boldsymbol{k}_{23}^2 + \boldsymbol{k}_{31}^2\right)\Big] . \end{aligned} \tag{8.45}$$

We point out that only a fraction x_3 of the original three-body term W_0 is retained, but a fraction $(1 - x_3)V_0$ of a density-dependent two-body force is added.

One can show that both interactions V_0 and W_0 contribute in the same way to the binding energy in "time-reversal invariant" systems (even-even nuclei). Thereby, the parameter x_3 remains as a "free" parameter relative to the nucleon ground-state properties and has to be determined such that properties of excited states can be well described.

One can express the ground-state energy corresponding to the Skyrme force given above in (8.39) in terms of a number of elementary density functions as

$$E_0^{\mathrm{HF}} = \int H(\boldsymbol{r})\, d\boldsymbol{r} , \tag{8.46}$$

where $H(\boldsymbol{r})$ describes the Hamilton density [expressed in terms of $\varrho_q(\boldsymbol{r})$, $\tau_q(\boldsymbol{r})$, $\boldsymbol{M}_q(\boldsymbol{r})$, $\boldsymbol{S}_q(\boldsymbol{r})$, $\boldsymbol{V}_q(\boldsymbol{r})$ and $\boldsymbol{J}_q(\boldsymbol{r})$: the nucleon density, the kinetic energy density, the current density, the spin density, the spin-kinetic density and the spin-current density function, respectively] (Waroquier 1982).

Minimization of the HF ground-state energy against independent variations of the single-particle wave functions $\varphi_\alpha(\boldsymbol{r})$ leads to

$$\delta H(\boldsymbol{r}) = 0 \quad \text{for} \quad \delta\varphi_\alpha(\boldsymbol{r}) \quad (\text{all} \quad \alpha) , \tag{8.47}$$

and results in a differential equation for the $\varphi_\alpha(\boldsymbol{r})$. For spherical nuclei, the radial part can be separated from the standard angular part and results in a radial differential equation for the part $\varphi_\alpha(r)$.

For spherical even-even nuclei (time-reversal invariant systems), the densities $\boldsymbol{S}_q(\boldsymbol{r})$, $\boldsymbol{M}_q(\boldsymbol{r})$ and $\boldsymbol{V}_q(\boldsymbol{r})$ disappear and the contribution of V_0 and W_0 to the Hartree-Fock ground-state energy becomes equal i.e.

$$E_0^{\mathrm{HF}}(V_0) = E_0^{\mathrm{HF}}(W_0) = \tfrac{1}{4}t_3 \int \varrho_p(\boldsymbol{r})\varrho_{\mathrm{tot}}(\boldsymbol{r})\, d\boldsymbol{r} , \tag{8.48}$$

with $\varrho_p(\boldsymbol{r})$, $\varrho_{\mathrm{tot}}(\boldsymbol{r})$ the proton and total nucleon density, respectively. This does not at all imply that both interactions V_0 and W_0 are identical with respect to evaluating the corresponding two-body matrix elements.

We give, in detail, the Hamiltonian density as a function of the remaining densitiy $\varrho_q(\boldsymbol{r})$, $\tau_q(\boldsymbol{r})$ and $\boldsymbol{J}_q(\boldsymbol{r})$:

$$
\begin{aligned}
&\text{kinetic energy} \to \frac{\hbar^2}{2m}\tau(\boldsymbol{r}) \,, \\
&V^{(0)} \to \tfrac{1}{2}t_0\left[(1+x_0/2)\varrho_{\text{tot}}^2 - \left(\tfrac{1}{2}+x_0\right)\left(\varrho_p^2+\varrho_n^2\right)\right] \\
&V^{(1)} \to \tfrac{1}{8}t_1\left[2\varrho_{\text{tot}}\tau_{\text{tot}} - \varrho_p\tau_p - \varrho_n\tau_n\right] - \tfrac{3}{32}t_1 \\
&\qquad\quad \times\left[2\varrho_{\text{tot}}\boldsymbol{\nabla}^2\varrho_{\text{tot}} - \varrho_p\boldsymbol{\nabla}^2\varrho_p - \varrho_n\boldsymbol{\nabla}^2\varrho_n\right] + \tfrac{1}{16}t_1\left[\boldsymbol{J}_p^2+\boldsymbol{J}_n^2\right] \,, \\
&V^{(2)} \to \tfrac{1}{8}t_2\left[2\varrho_{\text{tot}}\tau_{\text{tot}} + \varrho_p\tau_p + \varrho_n\tau_n\right] + \tfrac{1}{32}t_2 \\
&\qquad\quad \times\left[2\varrho_{\text{tot}}\boldsymbol{\nabla}^2\varrho_{\text{tot}} + \varrho_p\boldsymbol{\nabla}^2\varrho_p + \varrho_n\boldsymbol{\nabla}^2\varrho_n\right] - \tfrac{1}{16}t_2\left[\boldsymbol{J}_p^2+\boldsymbol{J}_n^2\right] \,, \\
&V^{(ls)} \to -\tfrac{1}{2}W_0'\left[\varrho_{\text{tot}}\boldsymbol{\nabla}\cdot\boldsymbol{J}_{\text{tot}} + \varrho_p\boldsymbol{\nabla}\cdot\boldsymbol{J}_p + \varrho_n\boldsymbol{\nabla}\cdot\boldsymbol{J}_n\right] \,, \\
&V_{\text{Coul}} \to \tfrac{1}{2}\varrho_p V^c(\boldsymbol{r}) \,, \\
&(1-x_3)V_0 \to \tfrac{1}{4}(1-x_3)t_3\varrho_p\varrho_n\varrho_{\text{tot}} \,, \\
&x_3W_0 \to \tfrac{1}{4}x_3t_3\varrho_p\varrho_n\varrho_{\text{tot}} \,, \\
&W_1 \to \frac{t_4}{24}\Bigg[-\varrho_p\varrho_n\boldsymbol{\nabla}^2\varrho_{\text{tot}} - \frac{1}{2}\varrho_p^2\boldsymbol{\nabla}^2\varrho_n - \frac{1}{2}\varrho_n^2\boldsymbol{\nabla}^2\varrho_p \\
&\qquad + 2\tau_{\text{tot}}\varrho_p\varrho_n + \tau_n\varrho_p^2 + \tau_p\varrho_n^2 + \tfrac{1}{2}\varrho_{\text{tot}}\boldsymbol{\nabla}\varrho_p\cdot\boldsymbol{\nabla}\varrho_n \\
&\qquad + \tfrac{1}{4}\varrho_p\left(\boldsymbol{\nabla}\varrho_n\right)^2 + \tfrac{1}{4}\varrho_n\left(\boldsymbol{\nabla}\varrho_p\right)^2 + \tfrac{1}{2}\boldsymbol{J}_p^2\varrho_n + \tfrac{1}{2}\boldsymbol{J}_n^2\varrho_p\Bigg] \,,
\end{aligned}
$$

and

$$
V^c(\boldsymbol{r}) = \int \varrho_p(\boldsymbol{r}')\frac{e^2}{|\boldsymbol{r}-\boldsymbol{r}'|}d\boldsymbol{r}' \,. \tag{8.49}
$$

The stationary condition for the ground-state energy now leads to

$$
\delta\left(E_0^{\text{HF}}\right) = \delta(\langle\text{HF}|H|\text{HF}\rangle) = \delta\int H(\boldsymbol{r})\,d\boldsymbol{r} = 0 \,. \tag{8.50}
$$

Since single-particle wave functions need to be normalized, (8.50) leads to a constrained extremum problem

$$
\delta\int\left[H(\boldsymbol{r}) - \sum_{\alpha_q,q}\varepsilon_{\alpha_q}^{\text{HF}}\varrho_{\alpha\alpha}^{(q)}\varrho_{\alpha\alpha}^{(q)}(\boldsymbol{r})\right]d\boldsymbol{r} = 0 \,. \tag{8.51}
$$

The specific form of the Skyrme force (zero-range terms) makes this variational problem more tractable since we know $H(\boldsymbol{r})$ as a function of the elementary density function. This results in

$$
\begin{aligned}
\delta H(\boldsymbol{r}) &= f\left(\delta\varrho_q, \delta\tau_q, \delta\boldsymbol{J}_q\right) \\
&= \sum_q\left[\frac{\hbar^2}{2m_q^*(\boldsymbol{r})}\delta\tau_q(\boldsymbol{r}) + U_q(\boldsymbol{r})\delta\varrho_q(\boldsymbol{r}) + \boldsymbol{W}_q(\boldsymbol{r})\cdot\delta\boldsymbol{J}_q(\boldsymbol{r})\right] \,,
\end{aligned} \tag{8.52}
$$

where the coefficients $m_q^*(\boldsymbol{r})$, $U_q(\boldsymbol{r})$ and $\boldsymbol{W}_q(\boldsymbol{r})$ define the nucleon effective mass, the potential $U_q(\boldsymbol{r})$ and spin-orbit potential $\boldsymbol{W}_q(\boldsymbol{r})$, respectively. Here, we only give the effective mass in detail

$$\frac{\hbar^2}{2m_q^*(\boldsymbol{r})} = \frac{\hbar^2}{2m_q} + \frac{1}{4}\left(t_1 + t_2\right)\varrho_{\text{tot}}(\boldsymbol{r}) + \frac{1}{8}\left(t_2 - t_1\right)\varrho_q(\boldsymbol{r}) + \frac{1}{24}t_4\left(\varrho_{\text{tot}}^2(\boldsymbol{r}) - \varrho_q^2(\boldsymbol{r})\right) . \tag{8.53}$$

The other terms $U_q(\boldsymbol{r})$ and $\boldsymbol{W}_q(\boldsymbol{r})$ are given in (Waroquier 1982).

Finally, one obtains the differential equation

$$-\boldsymbol{\nabla} \cdot \left[\frac{\hbar^2}{2m_q^*(\boldsymbol{r})}\boldsymbol{\nabla}\varphi_{\alpha_q}(\boldsymbol{r})\right] + \left[U_q(\boldsymbol{r}) - i\boldsymbol{W}_q(\boldsymbol{r}) \cdot (\boldsymbol{\nabla} \times \boldsymbol{\sigma})\right] \varphi_{\alpha_q}(\boldsymbol{r}) = \varepsilon_{\alpha_q}^{\text{HF}}\varphi_{\alpha_q}(\boldsymbol{r}) . \tag{8.54}$$

This equation resembles very much a one-body Schrödinger equation with effective mass $m_q^*(\boldsymbol{r})$. Using the spherical shell-model basis for expressing $\varphi_{\alpha_q}(\boldsymbol{r})$, a simplified radial equation results where now, the functions $\hbar^2/2m_q^*$, U_q and $\boldsymbol{W}_q$ become functions of the radial variable r, only.

Since we have now

$$\varrho_q(\boldsymbol{r}) = \frac{1}{4\pi}\sum_{a_q}(2j_a + 1)v_{a_q}^2\varphi_{a_q}^2(r) = \varrho_q(r) , \tag{8.55}$$

$$\tau_q(\boldsymbol{r}) = \frac{1}{4\pi}\sum_{a_q}(2j_a + 1)v_{a_q}^2 \left[\left(\frac{d\varphi_{a_q}}{dr}\right)^2 + l_a\left(l_a + 1\right)/r^2 \cdot \varphi_{a_q}^2(r)\right] = \tau_q(r) , \tag{8.56}$$

$$\boldsymbol{J}_q(\boldsymbol{r}) = -\frac{1}{4\pi}\sum_{a_q}(2j_a + 1)v_{a_q}^2 \left[l_a\left(l_a + 1\right) + \frac{3}{4} - j_a\left(j_a + 1\right)\right] \times \frac{1}{r}\varphi_{a_q}^2(r)\mathbb{1}_r = \boldsymbol{J}_q(r) , \tag{8.57}$$

the purely radial equation for φ_{a_q} becomes finally

$$\frac{d^2\varphi_{a_q}(r)}{dr^2} + \left[\frac{2m_q^*(r)}{\hbar^2}\frac{d}{dr}\left(\frac{\hbar^2}{2m_q^*(r)}\right) + \frac{2}{r}\right]\frac{d}{dr}\varphi_{a_q}(r) + \left[\frac{2m_q^*(r)}{\hbar^2}\left(\varepsilon_{a_q}^{\text{HF}} - U_q(r) + \frac{1}{r}W_q(r)\left[l_a\left(l_a + 1\right) + \frac{3}{4} - j_a\left(j_a + 1\right)\right]\right) - \frac{1}{r^2}l_a\left(l_a + 1\right)\right]\varphi_{a_q}(r) = 0 , \tag{8.58}$$

with

$$\boldsymbol{W}_q(r) = W_q(r)\mathbb{1}_r .$$

Before coming to a discussion on how to determine the parameters in the SkE-forces, we should like to make the following points.

We denote, in a shorthand notation, the extended Skyrme force as

$$\text{(I)}\quad \begin{array}{|c|}\hline V + (1 - x_3)V_0 \\ \hline x_3 W_0 + W_1 \\ \hline \end{array}\,, \tag{8.59}$$

with

$$\begin{aligned} V_0 &= \tfrac{1}{6} t_3 (1 + P_\sigma)\varrho\big((\boldsymbol{r}_1 + \boldsymbol{r}_2)/2\big)\delta(\boldsymbol{r}_1 - \boldsymbol{r}_2)\,, \\ W_0 &= t_3 \delta(\boldsymbol{r}_1 - \boldsymbol{r}_2)\delta(\boldsymbol{r}_1 - \boldsymbol{r}_3)\,. \end{aligned} \tag{8.60}$$

The original Skyrme force looks like

$$\text{(II)}\quad \begin{array}{|c|}\hline V \\ \hline W_0 \\ \hline \end{array}\,. \tag{8.61}$$

The big difference between (I) and (II) is:

- the addition of a velocity-dependent three-body force,
- the introduction of the fraction parameter x_3.

In time-reversal invariant systems, V_0 and W_0 contribute the same energy to E_0^{HF} and so, (I) and (III) give rise to the same HF equations, with (III) defined as the force

$$\text{(III)}\quad \begin{array}{|c|}\hline V \\ \hline W_0 + W_1 \\ \hline \end{array}\,. \tag{8.62}$$

This observation does not at all imply that the force parameterizations (I) and (III) would be identical. On the contrary, the x_3 parameter plays a major role in determining, in case (I), correct pairing two-body matrix elements.

8.2.4 Parameterization of Extended Skyrme Forces: Nuclear Ground-State Properties

We shall determine the parameters that appear in the SkE parameterization of (8.39) by insisting on a good reproduction of both nuclear matter and ground-state properties in even-even nuclei.

Nuclear matter is described as a medium with infinite dimensions, a uniform density and an equal number of non-interacting protons and neutrons, without distinction (the Coulomb force turned off).

The nuclear matter binding energy per nucleon (E/A) amounts to $-16\,\text{MeV}$. Variations in the density ϱ around the equilibrium density, as a function of the quantity $\alpha \equiv (N - Z)/A$, can be written as

$$E/A(\varrho, \alpha^2) = (E/A)_{\text{eq.}} + \frac{1}{2}K' \left(\frac{\varrho - \varrho_F}{\varrho_F}\right)^2 + a_\tau \cdot \alpha^2 . \tag{8.63}$$

The coefficient K' is the nuclear incompressibility, defined by

$$\begin{aligned} K' &= \varrho_F^2 \left(\frac{\partial^2 E/A}{\partial \varrho^2}\right)_{\varrho=\varrho_F} \\ &= \frac{1}{9}k_F^2 \left(\frac{\partial^2 E/A}{\partial k^2}\right)_{k=k_F} = \frac{1}{9}K , \end{aligned} \tag{8.64}$$

where k_F denotes the Fermi momentum.

The numerical value of K is uncertain and lies between 100 and 300 MeV. The coeffcient a_τ is called the isospin symmetry energy term. An experimental value around 25 to 30 MeV is estimated. In the Fermi theory of nuclear matter a_τ can be expressed as $\left(\varepsilon_F = (\hbar^2/2m)k_F^2\right)$

$$a_\tau = \tfrac{1}{3}\varepsilon_F , \tag{8.65}$$

with a resulting value of $a_\tau = 12.3\,(\pm 0.51)$ MeV.

In extending the non-interacting Fermi nuclear matter theory into a theory of an interacting system, the Landau description of nuclear matter is obtained (Landau 1956). Using the SkE forces of (8.39), a value of the binding energy per nucleon is obtained as

$$\begin{aligned} \left(\frac{E}{A}\right)_{\text{nm}} &= \frac{\hbar^2}{2m}\frac{3}{5}k_F^2 + \frac{3}{8}t_0\varrho_F + \frac{3}{80}(3t_1 + 5t_2)\varrho_F k_F^2 \\ &\quad + \frac{1}{16}t_3\varrho_F^2 + \frac{3}{160}t_4\varrho_F^2 k_F^2 . \end{aligned} \tag{8.66}$$

The saturation condition for the nucleon-nucleon force leads to an extra condition on the SkE parameters

$$\begin{aligned} &\frac{\hbar^2}{2m}\frac{6}{5}k_F + \frac{3}{4\pi^2}t_0 k_F^2 + \frac{1}{8\pi^2}(3t_1 + 5t_2)k_F^4 \\ &\quad + \frac{1}{6\pi^4}t_3 k_F^5 + \frac{1}{15\pi^4}t_4 k_F^7 = 0 . \end{aligned} \tag{8.67}$$

The nuclear compressibility is now given by

$$\begin{aligned} K &= \frac{\hbar^2}{2m}\frac{6}{5}k_F^2 + \frac{3}{2\pi^2}t_0 k_F^3 + \frac{1}{2\pi^2}(3t_1 + 5t_2)k_F^5 \\ &\quad + \frac{5}{6\pi^4}t_3 k_F^6 + \frac{7}{15\pi^4}t_4 k_F^8 . \end{aligned} \tag{8.68}$$

We have now three equations with the unknown quantities t_0, $(3t_1 + 5t_2)$, t_3 and t_4 for a given set of nuclear matter properties k_F, $(E/A)_{\text{nm}}$ and K.

The nucleon effective mass becomes

$$\frac{\hbar^2}{2m^*} = \frac{\hbar^2}{2m} + \frac{1}{16}(3t_1 + 5t_2)\varrho_F + \frac{1}{32}t_4\varrho_F^2 . \tag{8.69}$$

If $t_4 \neq 0$, the parameters K and (m^*/m) can vary independently of each other (this property can be used to reproduce the correct level density near the Fermi surface in finite nuclei). So, given $(E/A)_{\mathrm{nm}}$, K and (m^*/m) in nuclear matter, the quantities t_0, t_3, t_4 and $(3t_1+5t_2)$ can be determined uniquely. Separate values t_1, t_2 can only be determined making use of properties in finite nuclei. In a Thomas-Fermi model for the nuclear density, one can derive the expression for the nuclear mean-square radius $\langle r^2 \rangle$ as

$$\langle r^2 \rangle_{T-F} = \frac{3}{5}\left(\frac{9\pi A}{8}\right)^{2/3} \frac{1}{k_F^2}\left[1 - \frac{7}{81}\pi^2\left(\frac{8}{9\pi A}\right)^{2/3} k_F^2 (9t_1 - 5t_2)/t_0\right] . \tag{8.70}$$

Finally, the spin-exchange parameter x_0 can be determined from the isospin symmetry energy coefficient a_τ since

$$a_\tau = \frac{\hbar^2}{2m}\frac{1}{3}k_F^2 - \frac{t_0 k_F^3}{12\pi^2}(1+2x_0) + \frac{t_2 k_F^5}{9\pi^2} - \frac{t_3}{36\pi^4}k_F^6 - \frac{t_4}{135\pi^4}k_F^8 . \tag{8.71}$$

Combining the above results, it is clear that the SkE forces have eight parameters t_0, t_1, t_2, t_3, t_4, x_0, x_3 and the strength of the spin-orbit force W_0'. These parameters should determine a large number of experimental quantities (finite nuclei, nuclear matter).

In a *first step*, the quantities t_0, t_3, t_4 and $(3t_1+5t_2)$ are extracted from nuclear matter properties ($(E/A)_{\mathrm{nm}}$, k_F, K and m^*).

In a *second step*, the quantity $9t_1 - 5t_2$ is fitted using properties of finite nuclei, such as the nuclear radii in a Thomas-Fermi model.

In a *third step*, the parameter x_0 is determined from the isospin symmetry energy a_τ.

In a *fourth step*, self-consistent HF calculations for ^{16}O, ^{40}Ca, ^{48}Ca, ^{90}Zr, ^{132}Sn and ^{208}Pb are carried out. From these calculations, a value for W_0' is extracted as $W_0' = 120\,\mathrm{MeV\,fm}^5$. In this step, a very good reproduction of (E/A) is imposed and, to a minor extent, the reproduction of the HF single-particle energies, more in particular in the vicinity of the Fermi level.

One performs steps 1 to 4 a number of times in order to obtain an idea of the influence of the nuclear matter properties on the Hartree-Fock self-consistent aspects of finite nuclei. In this way, parameter regions can be delimited. Finally, four parameterizations SkE1, SkE2, SkE3 and SkE4 (Table 8.1) are retained, giving good Hartree-Fock results (Table 8.2). In Fig. 8.2, we show a nomogram, illustrating the parameter interval for constant $k_F = 1.33\,\mathrm{fm}^{-1}$ and $(E/A) = -16\,\mathrm{MeV}$. For given value of m^* and K, independent of each other, t_0, t_3, t_4 and $(3t_1+5t_2)$ can be uniquely determined. The dashed line illustrates the SkE2 parameter set. We also illustrate, in Figs. 8.3 and 8.4, the HF single-particle energies in ^{16}O and ^{208}Pb. In the above tables and figures, we also give the results of the SkIII force, the Skyrme force used originally (with $x_3 = 1$ and $t_4 = 0$) (Beiner et al. 1975). From the above results, a number of rather general remarks can be made:

Table 8.1. Illustration of a number of parameterizations (SkE2 and SkE4) yielding suitable values of nuclear matter and ground-state quantities for doubly-closed shell nuclei. We also compare with the original Skyrme interaction SkIII (Beiner et al. 1975, Waroquier et al. 1983a)

	t_0 (MeV fm^3)	t_1 (MeV fm^5)	t_2 (MeV fm^5)	t_3 (MeV fm^6)	x_0	W_0' (MeV fm^5)
SkE2	−1299.30	802.41	−67.89	19558.96	0.270	120
SkE4	−1263.11	692.55	−83.76	19058.78	0.358	120
SkIII	−1128.75	395.0	−95.0	14000.0	0.45	120
	t_4 (MeV fm^8)	K (MeV)	$(E/A)_{\mathrm{nm}}$ (MeV)	k_F (fm^{-1})	m^*/m	a_τ (MeV)
SkE2	−15808.79	200	−16.0	1.33	0.72	29.7
SkE4	−12258.97	250	−16.0	1.31	0.75	30.0
SkIII	0.0	356	−15.87	1.29	0.76	28.2

Table 8.2. Binding energies per nucleon E/A (MeV), proton and neutron points rms radii r_p and r_n (fm) and charge rms radii r_c (fm), corresponding to various Skyrme parameterizations after self-consistent Hartree-Fock calculations (Waroquier et al. 1983a)

	E/A	r_p	r_n	r_c		E/A	r_p	r_n	r_c
		^{16}O						^{90}Zr	
SkE2	−7.92	2.63	2.60	2.68	SkE2	−8.67	4.17	4.24	4.21
SkE4	−7.96	2.65	2.62	2.70	SkE4	−8.71	4.22	4.29	4.26
SkIII	−8.03	2.64	2.61	2.70	SkIII	−8.69	4.26	4.31	4.30
exp	−7.98			2.71[a]	exp	−8.71			4.27[c]
		^{40}Ca						^{132}Sn	
SkE2	−8.56	3.37	3.31	3.42	SkE2	−8.36	4.62	4.84	4.66
SkE4	−8.59	3.40	3.35	3.46	SkE4	−8.36	4.68	4.89	4.71
SkIII	−8.57	3.41	3.36	3.46	SkIII	−8.36	4.73	4.90	4.78
exp	−8.55		3.36[e]	3.48[b]	exp	−8.36			
		^{48}Ca						^{208}Pb	
SkE2	−8.63	3.39	3.56	3.44	SkE2	−7.87	5.41	5.57	5.45
SkE4	−8.65	3.43	3.59	3.47	SkE4	−7.87	5.47	5.62	5.50
SkIII	−8.69	3.46	3.60	3.50	SkIII	−7.87	5.52	5.64	5.56
exp	−8.67		3.54[e]	3.48[b]	exp	−7.87			5.50[d]

[a] (Sick, Mc Carthy 70) [b] (Wohlfart et al. 78)
[c] (Alkhazov et al. 76) [d] (Sick 73)
[e] (Shlomo, Schaeffer 79)

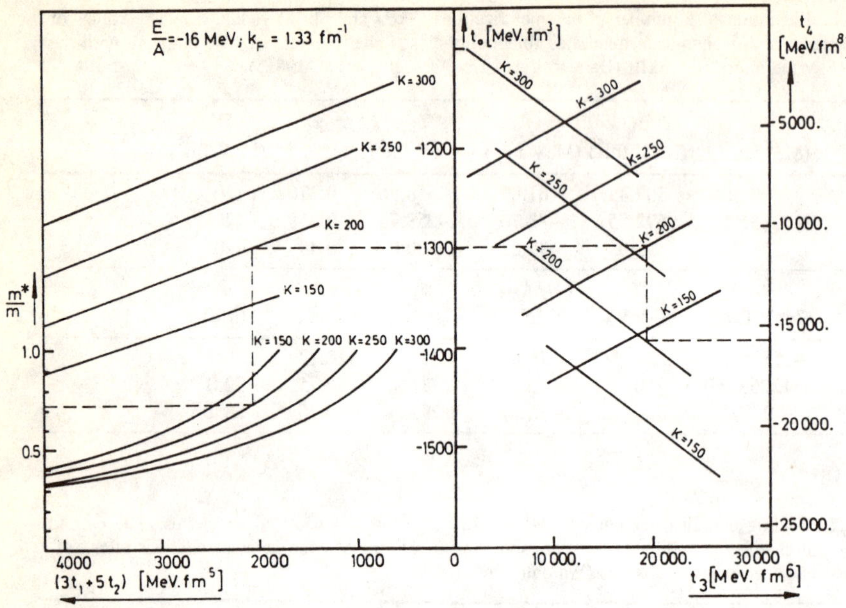

Fig. 8.2. Nomogram illustrating the range of parameters in m^*/m, $(3t_1 + 5t_2)$, t_3 and t_4 for a constant value $k_F = 1.33\,\text{fm}^{-1}$ and $(E/A)_{nm} = -16\,\text{MeV}$. The dashed line indicates the parameterization SkE2 (K is expressed in units of MeV) (Waroquier et al. 1979)

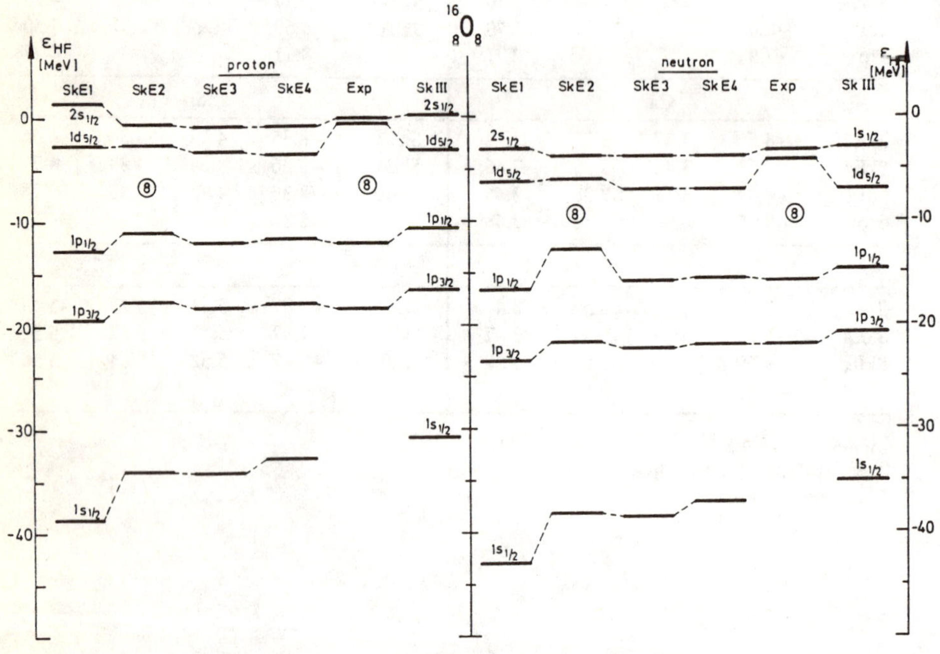

Fig. 8.3. Single-particle states in ^{16}O from a self-consistent Hartree-Fock calculation using various Skyrme force parameterizations (Waroquier 1982)

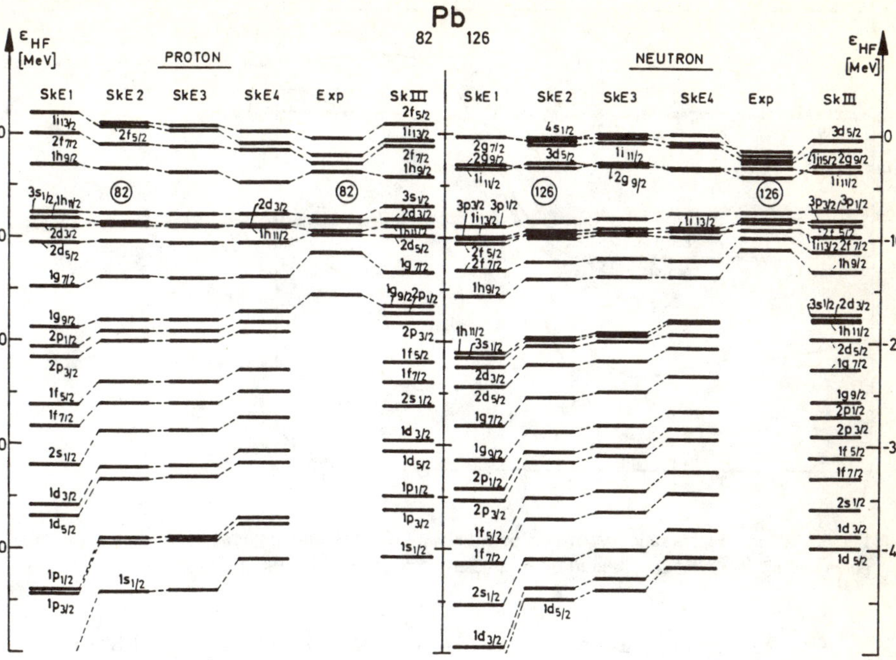

Fig. 8.4. Same caption as Fig. 8.3, but for ^{208}Pb (Waroquier 1982)

i) Deep-lying orbitals are more bound, using SkE1, compared to the other forces.

ii) The level density around the Fermi energy is almost equal in the SkE2, SkE3 and SkE4 parameterization, pointing out that k_F and K can compensate each other's influence on the level density.

iii) All four parameterizations give a good reproduction of self-consistent binding energies. The force SkE4 gives the better results, however.

iv) Saturation properties can best be studied in comparing the nucleon density distributions (see Fig. 3.17). The density, in particular for heavy nuclei, approaches the equilibrium density ϱ_F in nuclear matter ($\varrho_F = 0.159\,\text{fm}^{-3}$ for SkE2).

v) The Skyrme force is a non-local force. The most important consequence comes about in a large variation of the effective mass ratio m^*/m (Fig. 8.5): a low-value in the nucleus, even up to the nuclear surface. This non-locality is needed to reproduce the deep-lying single-particle orbitals. The level-density near the Fermi energy is too low in general. To reproduce correctly the single-particle spacings, one needs a value $m^*/m \cong 1$ near the Fermi level and $m^*/m \cong 0.6$ for deep-lying orbitals. Figure 8.5 illustrates this behaviour. Since experimental single-particle energies have been extrapolated from one-nucleon transfer reactions, not taking into account coupling of the single-particle motion to the nuclear surface collective modes of motion, an exact reproduction should not be imposed.

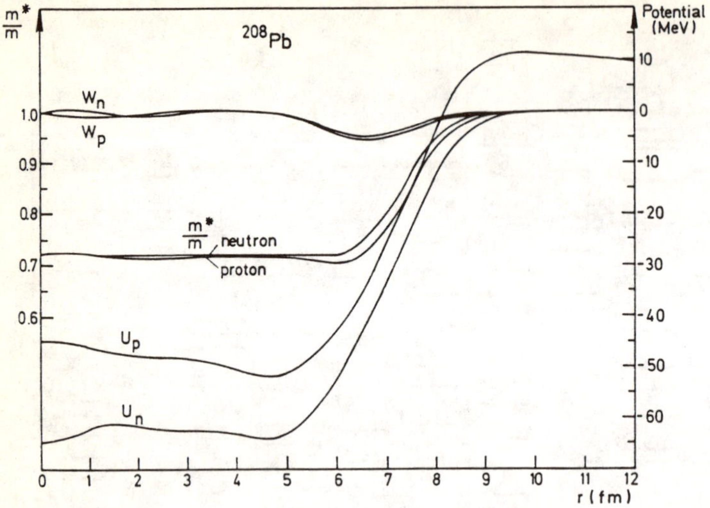

Fig. 8.5. Effective mass ratio m^*/m, the potential U_q (MeV) (q =proton or neutron) and the spin-orbit potential W_q in ^{208}Pb, using the SkE2 parameterization (Waroquier 1982)

As a general conclusion, we find that SkE2, SkE4 and the original SkIII force reproduce nuclear matter and *global* aspects of finite nuclei very well. A later discrimination shall have to be made when comparing *local* aspects in finite nuclei (Sect. 8.3).

8.3 Excited-State Properties of SkE Forces

Having constructed an extended Skyrme force in order to describe nuclear matter and ground-state properties in doubly-closed shell nuclei (binding energy, radii, single-particle spectra, saturation properties), we have determined only seven of the eight free parameters and retained two rather good parameterizations (SkE2 and SkE4).

In the present Sect. 8.3, we shall use the extra parameter x_3 for determining the forces such that correct pairing properties result throughout the whole nuclear mass region starting from the known two-particle spectra in nuclei like ^{18}O, ^{42}Ca, ^{50}Ca, ^{134}Te, ... Since these nuclei are mainly studied in a small model space consisting of an open shell with just two valence nucleons, effective forces and effective operators have to be determined (Chaps. 3, 4). Using specific forces that are fitted to such a small model space (using schematic interactions or using empirical two-body matrix elements as discussed in Sect. 3.2.2) a good reproduction can quite often be obtained. These forces, however, can vary very much when used in different parts of the nuclear mass region. Realistic forces on the other hand, determined from the free nucleon-nucleon scattering properties, generate "bare" two-body matrix elements that are almost always not very well adapted to be used in a small model space. The realistic force has to be modified into a corresponding effective force by considering (through perturbation theory mainly) effects of configurations outside of the model space ($3p-1h$, $4p-2h$ excitations, ...). Also the SkE2 and SkE4 forces have to be corrected for this "core-polarization" in order to be apt for describing nuclei with just two valence nucleons in a small model space.

So, we shall study subsequently how to determine the fraction parameter x_3 from particle-particle correlations (Sect. 8.3.1), the particle-hole excitations in doubly-closed shell nuclei (Sect. 8.3.2) with applications to giant multipole resonances and finally (Sect. 8.3.3) we shall study some effects due to rearrangement in the density-dependence of the residual interaction.

8.3.1 Particle-Particle Excitations: Determination of x_3

The case of two particles outside a closed shell has been studied in detail in Sect. 3.2 (in coordinate space) and in Sect. 7.3 (second quantization). Here, we discuss briefly how the effect of configurations outside the chosen model space of two-particle configurations can be taken into account in an effective way in the two-particle model space, thereby "renormalizing" the original force and single-particle properties.

Relative to the situation with a Fermi level, in the single-particle energy spectrum (Fig. 8.6) and denoting the single-closed shell as the A-nucleon reference state $|\mathrm{HF}\rangle$, the realistic eigenvector for the $A+2$ nucleus can be written as

$$
\begin{aligned}
|A+2;\alpha\rangle = &\sum_{m<n} c_{mn}(\alpha) a_m^+ a_n^+ |\mathrm{HF}\rangle \\
&+ \sum_{\substack{m<n<r \\ i}} c_{mnr,i}(\alpha) a_m^+ a_n^+ a_r^+ a_i |\mathrm{HF}\rangle \\
&+ \sum_{\substack{m<n<r<s \\ i<j}} c_{mnrs,ij}(\alpha) a_m^+ a_n^+ a_r^+ a_s^+ a_i a_j |\mathrm{HF}\rangle + \dots \,.
\end{aligned}
\tag{8.72}
$$

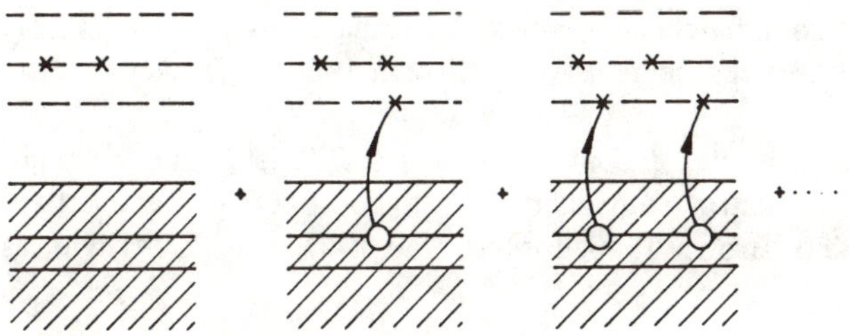

Fig. 8.6. The different components in the wave function of (8.72) for a nucleus with two valence nucleons outside a closed shell configuration: besides the two-particle component, the $3p-1h$, $4p-2h, \ldots$ components are also indicated

If we now force the model space to contain only the specific two-particle configurations, relative to $|\mathrm{HF}\rangle$, the model eigenvector has to be represented by

$$|A+2;\alpha\rangle_{\mathrm{Model}} = \sum_{m<n}{}' c_{mn}^{M}(\alpha) a_m^+ a_n^+ |\mathrm{HF}\rangle \,, \tag{8.73}$$

where $'$ indicates that we only sum over two-particle states with expansion coefficients $c_{mn}^{M}(\alpha)$. We can define a projection operator that projects onto the two-particle model space such that

$$|A+2;\alpha\rangle_{\mathrm{Model}} = P|A+2;\alpha\rangle \,. \tag{8.74}$$

Using this operator, the eigenvalue equation in the small model space can be expressed formally as

$$\begin{aligned}\left[E_\alpha - H^{(0)} - \left(\varepsilon_m + \varepsilon_n\right)\right] c_{mn}^{M}(\alpha) \\ = \sum_{m'<n'}{}' \langle mn|V_{\mathrm{eff}}\left(E_\alpha\right)|m'n'\rangle_{\mathrm{nas}} c_{m'n'}^{M}(\alpha) \,,\end{aligned} \tag{8.75}$$

with

$$V_{\mathrm{eff}}\left(E_\alpha\right) = V + VQ\frac{1}{E_\alpha - H^{(0)} - QVQ}QV \,, \tag{8.76}$$

and $(Q + P = 1)$.

Now, the "effective" interaction becomes dependent on the energy E_α itself. Using an expansion of (8.76), the effective interaction can be rewritten as

$$V_{\mathrm{eff}}(E) = V + V\frac{Q}{E - H^{(0)}}V_{\mathrm{eff}}(E) \,, \tag{8.77}$$

or

$$V_{\mathrm{eff}}\left(E_\alpha\right)|A+2;\alpha\rangle_{\mathrm{Model}} = V|A+2;\alpha\rangle \,, \tag{8.78}$$

which is an equivalent expression.

Using the SkE forces, we have to introduce the construction of such an effective model interaction in order to reproduce the two-particle spectra and determine the parameter x_3.

At present we do not go into details of how to determine from (8.75) and (8.77) the appropriate two-body matrix elements (Kuo, Brown 1966, 1968, Brown, Kuo 1967, Kuo 1974, 1981).

It occurs that intermediate $3p-1h$ configurations in (8.77) determine the major corrections to the bare SkE matrix elements in order to construct an effective two-particle model space (Kuo, Brown 1966, Brown, Kuo 1967, Kuo 1974) with as major characteristics:

i) a large increase of the pairing matrix elements and, to a lesser extent, of the $J^\pi = 2^+$ matrix elements,
ii) a slight reduction of the high J^π matrix elements.

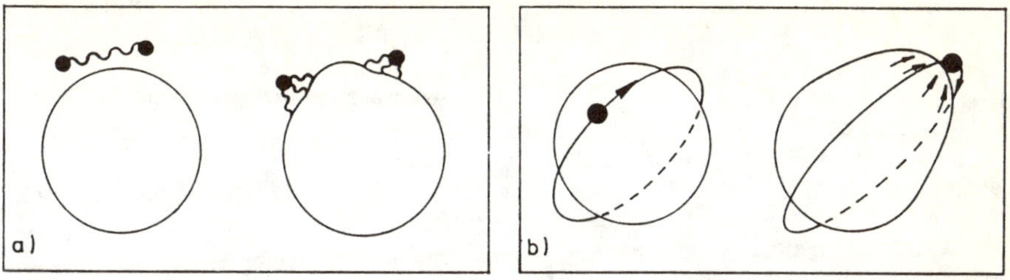

Fig. 8.7. (**a**) Schematic illustration of the "bare" two-nucleon interaction and the "polarized" two-nucleon interaction where long-range correlations in the core are induced, modifying the original, "bare" interaction matrix elements (shown in the figure as a macroscopic deviation of the core from its original, spherical shape). (**b**) The "bare" single particle motion in an average field and a "dressed" single particle, dragging in its motion the polarization effect in the core (shown in the figure as a macroscopic deviation of the core, from its spherical shape following the extra nucleon in its motion through the nucleus)

One can depict this $3p-1h$ intermediate state as a polarization of the nuclear core. This effect induces a long-range attractive part [quite similar to the effect of the $P_2\,(\cos\theta_{12})$ force] acting over distances of the order of the nuclear dimensions (Fig. 8.7a). Detailed calculations of the $3p-1h$ correction indeed show that the intermediate $J^\pi = 2^+$ states contribute most of the correction.

The effective force can then be specified by the two-body matrix elements

$$\begin{aligned}\langle ab; JM|V_{\text{eff}}|cd; JM\rangle_{\substack{\text{nas}\\ qq}} =& \langle ab; JM|V+V'|cd; JM\rangle_{\substack{\text{nas}\\ qq}}\\ &+\langle ab; JM|V_{3p-1h}|cd; JM\rangle_{\substack{\text{nas}\\ qq}}\\ &+\langle ab; JM|V_{4p-2h}|cd; JM\rangle_{\substack{\text{nas}\\ qq}}\ . \end{aligned} \tag{8.79}$$

Evaluation of the matrix elements occurring in (8.79) is discussed in (Kuo, Brown 1968, Waroquier 1982).

We have to remember that, due to the use of an effective model space, the zero-order Hamiltonian also becomes modified into

$$H_{\text{eff}}^{(0)} = H^{(0)} + \sum_{m>F}{}' \varepsilon_m^{\text{eff}} a_m^+ a_m\ , \tag{8.80}$$

where $\varepsilon_m^{\text{eff}} \neq \varepsilon_m^{\text{HF}}$. The energies are mainly determined from fits to nuclei with one nucleon outside the closed-shell nucleus under consideration. We call these nucleons, "dessed" nucleons that interact in the model space via the effective interaction V_{eff} (see Fig. 8.7b).

Using the above method and the self-consistently determined single-particle orbitals and matrix elements one adjusts x_3 such that, after taking into account the above "renormalization" effects, energy spectra in closed-shell nuclei ± 2 particles are well reproduced.

In a first step, we study some general properties of two-body matrix elements, using the SkE4 force for the $(1d_{5/2})^2$ configuration (which applies to ^{18}O). The pure three-body version ($x_3 = 1$) yields repulsive matrix elements (Fig. 8.8). The situation $x_3 = 0$ gives too wide a spectrum compared to the ^{18}O data. For a suitable x_3

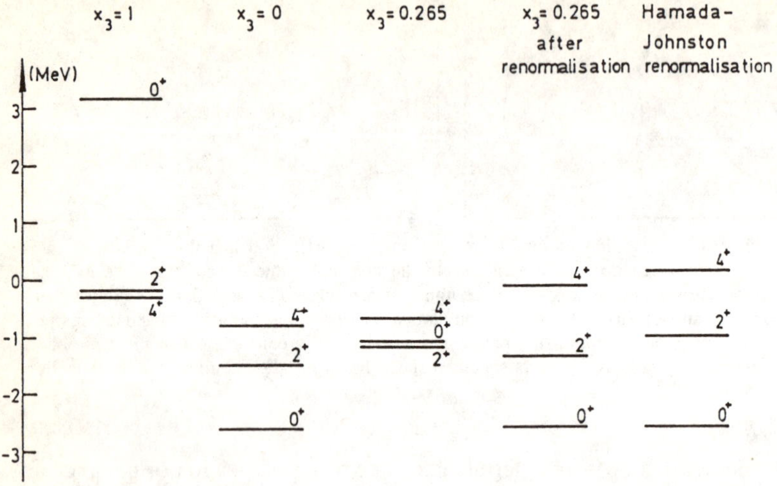

Fig. 8.8. Antisymmetric and normalized neutron two-body matrix element in ^{18}O, using the SkE4 parameterization $\langle (1d_{5/2})^2; JM|V + V'|(1d_{5/2})^2; JM\rangle_{\text{nas}}$ (Waroquier et al. 1983b)

value, a spectrum which is very much comparable to a realistic Hamada-Johnston (Hamada, Johnston 1962) potential results. A similar conclusion is obtained after diagonalizing in a full two-particle space and taking into account renormalization effects ($3p - 1h$, $4p - 2h$). It is now of the utmost importance to use a constant x_3 parameter value throughout the whole mass region. This can indeed be accomplished using effective (dressed) single-particle energies instead of the self-consistent HF energies.

Finally, a set of values

$x_3 = 0.43$	SkE2
$x_3 = 0.265$	SkE4

results.

Before discussing some typical examples, using these forces and comparing to other forces, we point out that all calculations are performed in a self-consistent way (except that in a number of cases adjusted single-particle energies were used). We point out once again the successive steps in the calculations:

i) For given $A(N, Z)$, the spherical HF calculations are carried out determining $\varepsilon_\alpha^{\text{HF}}$, $\varphi_\alpha(\boldsymbol{r})$,
ii) We calculate the SkE particle-particle "two-body" matrix elements using the above self-consistent wave functions,
iii) We renormalize the two-body matrix elements,
iv) We construct the energy matrix, diagonalize it in the model space and determine energy spectra, transition probabilities, etc.... .

Using the above outline, shell-model calculations have been carried out for most nuclei with two proton (-neutron) particles (-holes) outside a doubly-closed shell configuration i.e. ^{18}O, ^{42}Ca, ^{50}Ca, ^{58}Ni, ^{134}Te, ^{130}Sn (Waroquier 1982, Waroquier et al. 1983b). Besides the typical influence of the x_3 parameter in SkE force, one quite often uses the SkE force with an empirical set of single-particle energies, denoted by an asterisk (SkE*). An extensive discussion of the ^{18}O, ^{42}Ca, and ^{130}Sn spectra is carried out in (Waroquier 1982, Waroquier et al. 1983b). The combined energy spectra are shown in Fig. 8.9, using the SkE*2 force (x_3 = 0.43). Only positive parity states are retained. The levels in ^{42}Ca, drawn with dashed lines have mainly a $4p - 2h$ character and are outside the two-particle model spaces, discussed here. This figure gives a good indication of the overall agreement for a large number of nuclei throughout the nuclear mass region.

As a typical illustration, we give the detailed comparison for ^{134}Te where the two protons move in the 50–82 proton valence model space. The bare SkE two-body matrix elements are corrected for the $3p - 1h$ core polarization effects.

Here, we compare the SkE results with results obtained using an effective Gaussian two-body force. Such an effective force has been used in the study of the N = 82 nuclei with good results (Waroquier 1970, Waroquier 1971, Heyde 1971). In Fig. 8.10, we compare the pairing two-body matrix elements, using the bare and renormalized SkE2 force and a Gaussian force ($V_0 = -39$ MeV, $t = +0.2$ and $t = -0.7$) where t denotes the triplet to singlet spin ratio. We remark that:

- the behaviour is very similar for all orbitals,
- the Gaussian matrix elements, using a value $t = -0.7$, are remarkably similar to the SkE2 matrix elements and thus the SkE2 force approximates very well effective interactions constructed to study these N = 82 nuclei more particularly.

In Fig. 8.11, we compare the ^{134}Te energy spectra and, apart from a somewhat too large $0^+ - 2^+$ energy separation using the SkE forces, a very detailed similarity between all interactions (SkE2, SkE4, Gaussian with $t = +0.2$) results. Moreover, when comparing, in Table 8.3, the $|1g_{7/2}\, 2d_{5/2}; JM\rangle$ matrix elements ($1^+ \leq J^\pi \leq 6^+$), one observes that the relative energy differences are all very similar (although different in absolute value). Only the Tabakin interaction gives a spectrum that is too compressed and the Gaussian force with $t = -0.7$ approximates best the SkE matrix elements.

As a general conclusion on the study of the SkE force as a particle-particle interaction it has been pointed out that the fraction parameter x_3 could be determined. This value was fixed at 0.43 and 0.265 for SkE2 and SkE4, respectively. It was found that x_3 tends to a constant value throughout the whole nuclear mass region. At this stage, *all* parameters of the SkE force are determined. One might hope that the forces SkE2 and SkE4 also give correct results in odd-mass nuclei.

We have, moreover, indicated that the SkE forces behave very much as "realistic" interactions: specific core-polarization effects renormalize the "bare" matrix

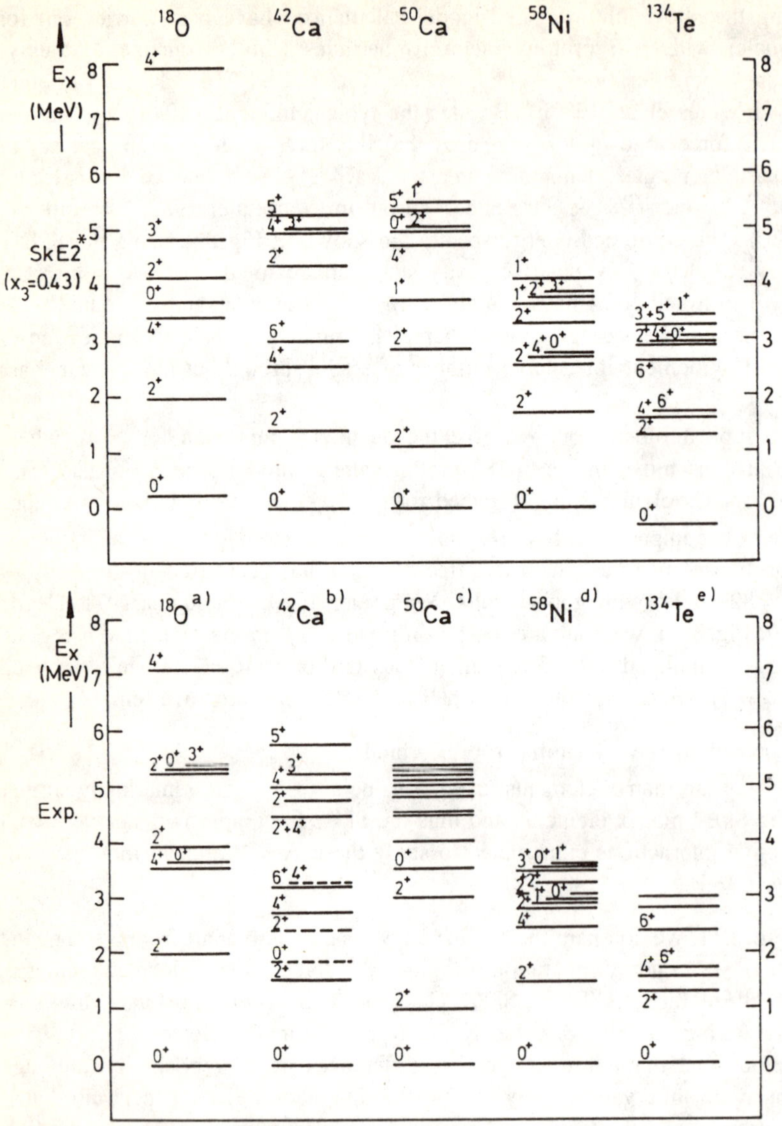

Fig. 8.9. Two-particle spectra of some (doubly-closed shell + two nucleons) nuclei, using the SkE2*(x_3 = 0.43) force. Only positive parity states are retained. The dashed lines in the ^{42}Ca spectrum, in the lower part of the figure, denote mainly $4p - 2h$ excitations not observed in the theoretical $2p$ shell-model calculations. SkE2* points towards the use of adjusted single-particle energies (Waroquier et al. 1983b): [a] (Fortune 1978), [b] (Vold et al. 1978a, 1978b), [c] (Bjerregaard et al. 1967, Broglia 1968, Tape 1975), [d] (Bertin et al. 1969, Start 1970), [e] (Kerek et al. 1972)

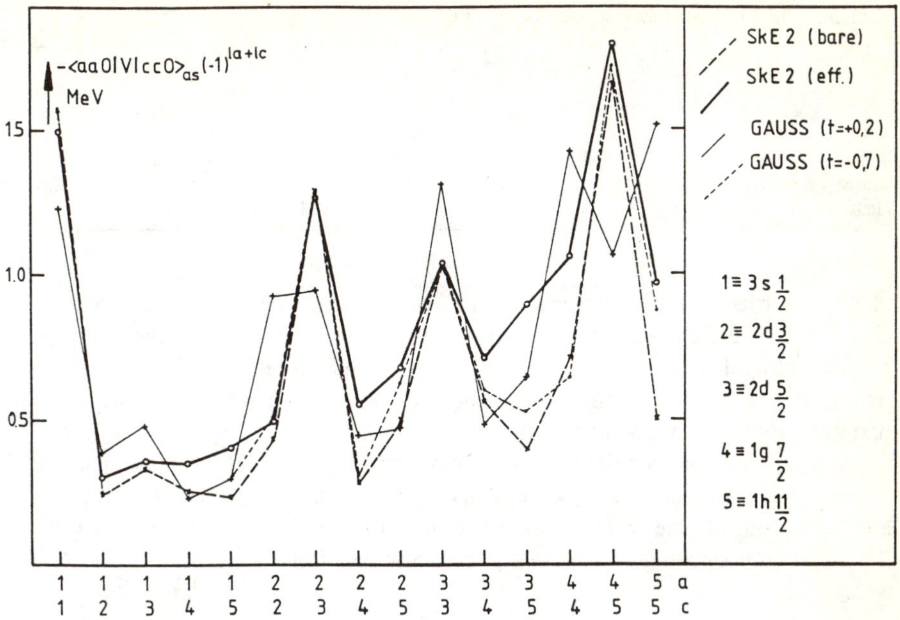

Fig. 8.10. The proton pairing matrix elements in the orbitals for the 50 – 82 $N = 4$ harmonic oscillator shell and for ^{134}Te. We compare the "bare" and renormalized SkE4 matrix elements with two-body matrix elements for a Gaussian, effective interaction (using $V_0 = -39$ MeV and spin triplet to singlet ratio $t = +0.2$ and -0.7) (Waroquier 1982)

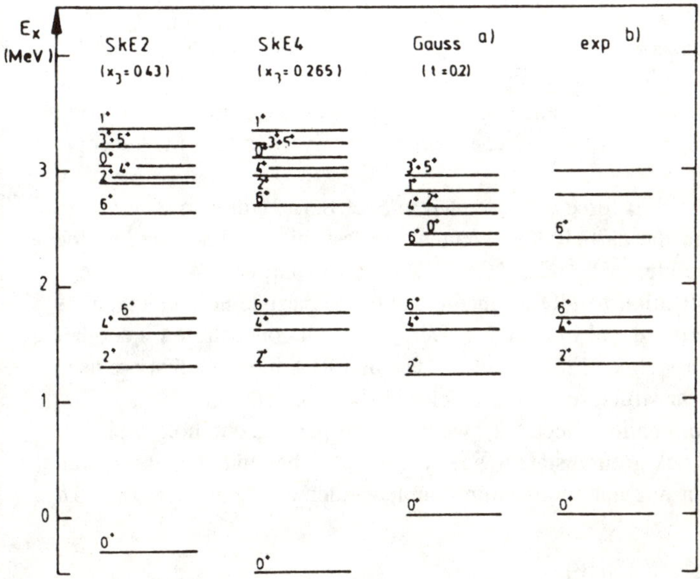

Fig. 8.11. Positive parity states in ^{134}Te. We compare the data (*b*), taken from (Kerek et al. 1972) with calculations using an effective Gaussian interaction (*a*) (Sau et al. 1980) and with self-consistent calculations using the SkE2 and SkE4 parameterizations (Waroquier 1982)

Table 8.3. Diagonal two-body matrix elements for the multiplet $(1g_{7/2}2d_{5/2})$ in ^{134}Te

$\langle 1g_{7/2}2d_{5/2}; JM\|V\|1g_{7/2}2d_{5/2}JM\rangle_{\text{nas}}$	1^+	2^+	3^+	4^+	5^+	6^+
SkE2 ($x_3 = 0.43$)	+0.52	+0.12	+0.37	+0.15	+0.38	−0.20
SkE4 ($x_3 = 0.265$)	+0.46	+0.11	+0.37	+0.16	+0.40	−0.20
Tabakin	−0.07	+0.05	−0.12	−0.03	−0.11	−0.28
Gauss ($t = 0.2$)	−0.20	−0.19	−0.11	−0.21	−0.10	−0.61
Gauss ($t = -0.7$)	+0.70	+0.37	+0.24	+0.04	+0.11	−0.35

elements for use in a highly truncated shell-model space. The $3p-1h$ corrections, especially in light nuclei, are very important in order to provide the necessary long-range quadrupole component for the renormalized SkE forces. It is also remarkable that such a good agreement with phenomenological interactions, adjusted to a particular mass region, results.

It is obvious that for light nuclei, not *all* low-lying levels in the two-particle shell-model calculations can be reproduced. There are a number of states that are mainly outside the small two-particle model space. Such levels contain large deformed components and are mainly from a $4p-2h$ origin.

8.3.2 The Skyrme Interaction as a Particle-Hole Interaction

a) General Introduction

In the present section, we shall study the above (Sect. 8.3.1) Skyrme force parameterizations SkE2 and SkE4 for use in particle-hole calculations in doubly-closed shell nuclei.

Besides the microscopic picture of a particle-hole interaction, one can also take a more macroscopic point of view. A linear combination of particle-hole states in a doubly-closed shell nucleus can be seen as an excitation of the core: density fluctuations appear in the average field which give rise to a macroscopic particle-hole interaction.

We expect that the SkE forces, determined in Sect. 8.3.1, will also give rise to a correct behaviour in the particle-hole channel. To test this, we have studied the doubly-closed shell nuclei ^{16}O, ^{40}Ca, ^{48}Ca, ^{90}Zr, ^{132}Sn, ^{146}Gd and ^{208}Pb in a self-consistent way. Application to ^{16}O as an illustration of particle-hole calculations was already discussed in detail in Sect. 6.4. Here, we shall concentrate on another nucleus ^{40}Ca, but first we recall the standard TDA and RPA methods that are used to calculate the nuclear structure in doubly-closed shell nuclei.

In the TDA approximation (Sect. 6.2), we use a one-particle one-hole basis relative to the Hartree-Fock ground state to span the basis of the related configuration space. Thereby, in an angular momentum coupled scheme, we use the $1p-1h$ operators

$$|ph^{-1}; JM\rangle \equiv A^{+}_{ph}(JM|\text{HF}\rangle$$
$$= \sum_{m_p,m_h} \langle j_p m_p, j_h - m_h|JM\rangle$$

$$\times (-1)^{j_h - m_h} a^+_{j_p m_p} a_{j_h m_h} |\mathrm{HF}\rangle \,. \tag{8.81}$$

In the RPA approach, a more general ground state (containing particle-hole excitations) is considered for which one defines collective excited states by (Sect. 6.3)

$$Q^+_{J,\alpha}|\tilde{0}\rangle = |\Psi(J\alpha)\rangle \,, \tag{8.82}$$

and the RPA vacuum state through the condition that

$$Q_{J,\alpha}|\tilde{0}\rangle = 0 \,, \tag{8.83}$$

for all (J,α). In this way, each excited state can be well approximated by the explicit operator

$$\begin{aligned} Q^+_{JM\alpha} = \sum_i \{ & X_{p_i h_i}(J,\alpha) A^+_{p_i h_i}(JM) \\ & - Y_{p_i h_i}(J,\alpha)(-1)^{J-M} A_{p_i h_i}(J-M)\} \,. \end{aligned} \tag{8.84}$$

The corresponding RPA secular equation becomes

$$HQ^+_{J,\alpha}|\tilde{0}\rangle = \hbar\omega_{J,\alpha} Q^+_{J,\alpha}|\tilde{0}\rangle \,, \tag{8.85}$$

or

$$\langle\tilde{0}|Q_{J,\beta} H Q^+_{J,\alpha}|\tilde{0}\rangle = \hbar\omega_{J,\alpha}\delta_{\alpha\beta} \,. \tag{8.86}$$

This problem cannot be solved in general, since one does not know the exact structure of the RPA ground-state $|\tilde{0}\rangle$. For a given interaction, however, one could improve on this by calculating the ground-state admixtures. Thereby, improved or extended RPA calculations could be carried out (Waroquier 1983c). We only know that $|\tilde{0}\rangle$ is a vacuum for the $Q^+_{J,\alpha}$ operators.

One can rewrite (8.86) using the commutator expression

$$\langle\tilde{0}|\left[Q_{J,\beta}, \left[H, Q^+_{J,\alpha}\right]\right]|\tilde{0}\rangle = \hbar\omega_{J,\alpha}\langle\tilde{0}|\left[Q_{J,\beta}, Q^+_{J,\alpha}\right]|\tilde{0}\rangle \,, \tag{8.87}$$

with the normalization condition of

$$\langle\tilde{0}|\left[Q_{J,\beta}, Q^+_{J,\alpha}\right]|\tilde{0}\rangle = \delta_{\alpha\beta} \,. \tag{8.88}$$

With the following approximations:

i) Use the form of (8.84) for the $Q^+_{J,\alpha}$ operator,

ii) Replace in (8.87) the RPA ground state $|\tilde{0}\rangle$ by the Hartree-Fock ground state $|\mathrm{HF}\rangle$. One can then express the secular equation in (8.87) in a more explicit way

$$\begin{aligned} &\begin{pmatrix} Q+E & R \\ R^* & (Q+E)^* \end{pmatrix} \begin{pmatrix} X(J,\alpha) \\ Y(J,\alpha) \end{pmatrix} \\ &= \hbar\omega_{J,\alpha} \begin{pmatrix} \mathbb{1} & 0 \\ 0 & -\mathbb{1} \end{pmatrix} \begin{pmatrix} X(J,\alpha) \\ Y(J,\alpha) \end{pmatrix} \,, \end{aligned} \tag{8.89}$$

where Q and R are the angular momentum coupled representation of the Q and R matrix elements of Sect. 6.3 and are given by

$$\begin{aligned} Q_{i,j} &= \langle\tilde{0}| \left[\left[A_{p_i h_i}(JM), H^{(4)}\right], A^{+}_{p_j h_j}(JM)\right] |\tilde{0}\rangle \,, \\ R_{i,j} &= -\langle\tilde{0}| \left[\left[A_{p_i h_i}(JM),\, H^{(4)}\right], (-1)^{J-M} A_{p_j h_j}(J-M)\right] |\tilde{0}\rangle \,, \end{aligned} \tag{8.90}$$

where $H^{(4)}$ is given in (8.8).

Using the RPA approximation conditions

$$\begin{aligned} &\langle\tilde{0}|a^{+}_{p_i} a_{p_j}|\tilde{0}\rangle \to 0 \,, \\ &\langle\tilde{0}|a^{+}_{h_i} a_{h_j}|\tilde{0}\rangle \to \delta_{ij} \,, \end{aligned} \tag{8.91}$$

the angular momentum Q and R matrix elements become

$$\begin{aligned} Q_{ij} = &\sum_{J_1} (2J_1+1) \begin{Bmatrix} j_{p_i} & j_{h_j} & J_1 \\ j_{p_j} & j_{h_i} & J \end{Bmatrix} (-1)^{j_{h_i}+j_{p_j}-J_1} \\ &\times \langle p_i h_j; J_1 | V + V' | h_i p_j; J_1 \rangle_{\text{nas}} \,, \end{aligned} \tag{8.92}$$

and

$$\begin{aligned} R_{ij} = &\sum_{J_1} (2J_1+1) \begin{Bmatrix} j_{p_i} & j_{p_j} & J_1 \\ j_{h_j} & j_{h_i} & J \end{Bmatrix} (-1)^{j_{h_i}+j_{p_j}+J-J_1} \\ &\times \langle p_i p_j; J_1 | V + V' | h_i h_j; J_1 \rangle_{\text{nas}} \,. \end{aligned} \tag{8.93}$$

The RPA secular equations can be rewritten as

$$\begin{pmatrix} Q+E & R \\ -R^* & -(Q+E)^* \end{pmatrix} \begin{pmatrix} X(J,\alpha) \\ Y(J,\alpha) \end{pmatrix} = \hbar\omega_{J,\alpha} \begin{pmatrix} X(J,\alpha) \\ Y(J,\alpha) \end{pmatrix} , \tag{8.94}$$

This is the standard RPA secular equation for a non-hermitian matrix. If $\hbar\omega_{J,\alpha}$ is an eigenvalue, then $-\hbar\omega^*_{J,\alpha}$ is also an eigenvalue with eigenvectors

$$\begin{pmatrix} X(J,\alpha) \\ Y(J,\alpha) \end{pmatrix} \quad \text{and} \quad \begin{pmatrix} Y^*(J,\alpha) \\ X^*(J,\alpha) \end{pmatrix} ,$$

respectively.

Before concentrating on some applications, we discuss briefly (i) the elimination of the spurious centre-of-mass motion, and (ii) the electromagnetic decay properties in single-closed shell nuclei.

i) The nuclear A-body Hamiltonian can be separated into a part that describes the motion of the centre of mass of the A-nucleon system *and* the $A-1$ internal coordinates. The Hartree-Fock ground state of a doubly-closed shell nucleus has the centre-of-mass motion in its ground state. An excited state of the centre-of-mass motion consists of a linear combination of $1p-1h$ excitations and is admixed with the internal $1p-1h$ excitations (with the centre-of-mass motion in its ground state). One has now to project out this spurious centre-of-mass motion which, for the $1p-1h$ space, results in a single 1^- state that can be constructed explicitly.

ii) For the electromagnetic decay properties, we use the expressions as derived in Chap. 4. Within the RPA, this results in the transition probability (reduced) from an excited state (J,α) to the 0^+ ground state

$$B\left(\text{el.}; J,\alpha \to 0^+_{\text{gs}}\right) = \frac{1}{2J+1}\delta_{JL}\Big| \sum_i \Big[X_{p_i h_i}(J,\alpha) + (-1)^J Y_{p_i h_i}(J,\alpha)\Big] \langle p_i \|\mathbf{0}(\text{el.}; L)\| h_i\rangle\Big|^2 , \tag{8.95}$$

with the reduced single-particle matrix element for the $\mathbf{0}(\text{el.}; L)$ operator defined in Sect. 4.3.

Similarly, the more complicated expression between an excited state (J,α) and (J,β) can be derived.

We also mention briefly the non-energy weighted (NEW) and energy-weighted (EW) sum rule since they give information on the collectivity of certain excited states.

a) The NEW sum rule is defined as

$$\begin{aligned} S_{\text{NEW}}(\mathbf{0}L) &= \sum_\alpha B\left(\mathbf{0}L; 0^+_{\text{gs}} \to J,\alpha\right) \\ &= \delta_{JL}\sum_M \langle \Psi_0 | 0^*_{JM} 0_{JM} | \Psi_0\rangle , \end{aligned} \tag{8.96}$$

with

$$0^*_{JM} = (-1)^M 0_{J,-M} , \tag{8.97}$$

where $|\Psi_0\rangle$ denotes the ground-state. According to the appropriate approximation this stands for $|\text{HF}\rangle$ or $|\tilde{0}\rangle$.

Within TDA, this leads to

$$S^{\text{TDA}}_{\text{NEW}}(\text{el.}; L) = \sum_i |\langle p_i \|\mathbf{0}(\text{el.}; L)\| h_i\rangle|^2 , \tag{8.98}$$

with the TDA fulfilling the "non-energy-weighted" sum rule.

Within the RPA, this gives, on the other hand

$$\begin{aligned} S^{\text{RPA}}_{\text{NEW}}(\text{el.}; L) = \sum_{i,\alpha} \Big| \Big[X^*_{p_i h_i}(J,\alpha) + (-1)^J Y^*_{p_i h_i}(J,\alpha)\Big] \\ \times \langle p_i \|\mathbf{0}(\text{el.}; L)\| h_i\rangle\Big|^2 \delta_{JL} . \end{aligned} \tag{8.99}$$

Here, the RPA does not fulfil the NEW sum rule. Within an identical model space, one has

$$S^{\text{RPA}}_{\text{NEW}}(\text{el.}; L) \leq S^{\text{TDA}}_{\text{NEW}}(\text{el.}; L) . \tag{8.100}$$

b) The energy-weighted (EW) sum rule is defined as

$$S_{\mathrm{EW}}(\mathrm{el.}; L) = \sum_{\alpha} \hbar\omega_{J,\alpha} B\left(\mathrm{el.}\ L; 0^+_{\mathrm{gs}} \to J, \alpha\right) \tag{8.101}$$

and can be rewritten as

$$S_{\mathrm{EW}}(\mathrm{el.}; L) = \frac{1}{2}\delta_{JL} \sum_{M} \langle \Psi_0 | \left[\left[0^*_{JM}, H \right], 0_{JM} \right] | \Psi_0 \rangle \,, \tag{8.102}$$

with $|\Psi_0\rangle$ defined as in (8.96).

Within TDA, only a part of the Hamiltonian H [the part Q + E of (8.89)] is taken into account. The TDA does not fulfil the EW sum rule. In the RPA, however, one can prove (Waroquier 1982) that the EW sum rule is fulfilled.

b) Description of Low-lying Collective Vibrations

Having recalled the necessary elements and expressions for testing the quality of the particle-hole character of the Skyrme forces, we shall now discuss some applications to the doubly-closed shell nuclei ^{16}O, ^{40}Ca. For ^{16}O. We refer to Chap. 6.

We now concentrate on ^{40}Ca: a nucleus with an analogous structure to ^{16}O. In Figs. 8.12, 13 we present detailed TDA and RPA calculations, using the SkE4* interaction for the 3^-, 5^- and 0^-, 1^-, 2^-, 4^- levels, respectively.

One observes that the low-lying collective 3^- and 5^- levels are mainly $T = 0$ states. For the other states, the percentage of isospin character T is presented in all cases (Figs. 8.12, 13).

For $1\hbar\omega$ $1p-1h$ excitations, the hole moves in the $N = 2$ shells ($2s1d$ orbitals) and the particle in the $N = 3$ shell ($1f2p$ orbitals). This restricted space is denoted by the indices (7–10) in Fig. 8.12. Extending to $2\hbar\omega$ $1p-1h$ excitations, we also consider hole states in the $N = 1$ shell ($1p$ orbitals) and particle states in the $N = 4$ shell ($1g2d3s$ orbitals). This space is denoted by the indices (7–16) in Fig. 8.12. One clearly observes the influence of this model space increase on excitation energies for 3^- and 5^- levels. The influence is negligible, except for the collective 3^- and 5^- levels. Some states of more complex character ($3p-3h$) do, however, occur at low excitation energy, levels that cannot be described within the model space chosen here.

Comparing TDA with RPA, the most clear-cut difference is the imaginary eigenvalue of the lowest 3^-, $T = 0$ state in RPA. This results for SkE2 and SkE4 both in the small and larger model space. For all other states, even the collective 5^- level, the TDA-RPA differences are minor. Also, use of the realistic Tabakin interaction (Dieperink et al. 1968) results in an imaginary root for the 3^-, $T = 0$ level. This deficiency in solving the RPA secular equations is a constant problem in the description of low-lying ($E_x \gtrsim 4\,\mathrm{MeV}$) strongly collective states. A possible explanation can be found in the neglect of ground-state correlations

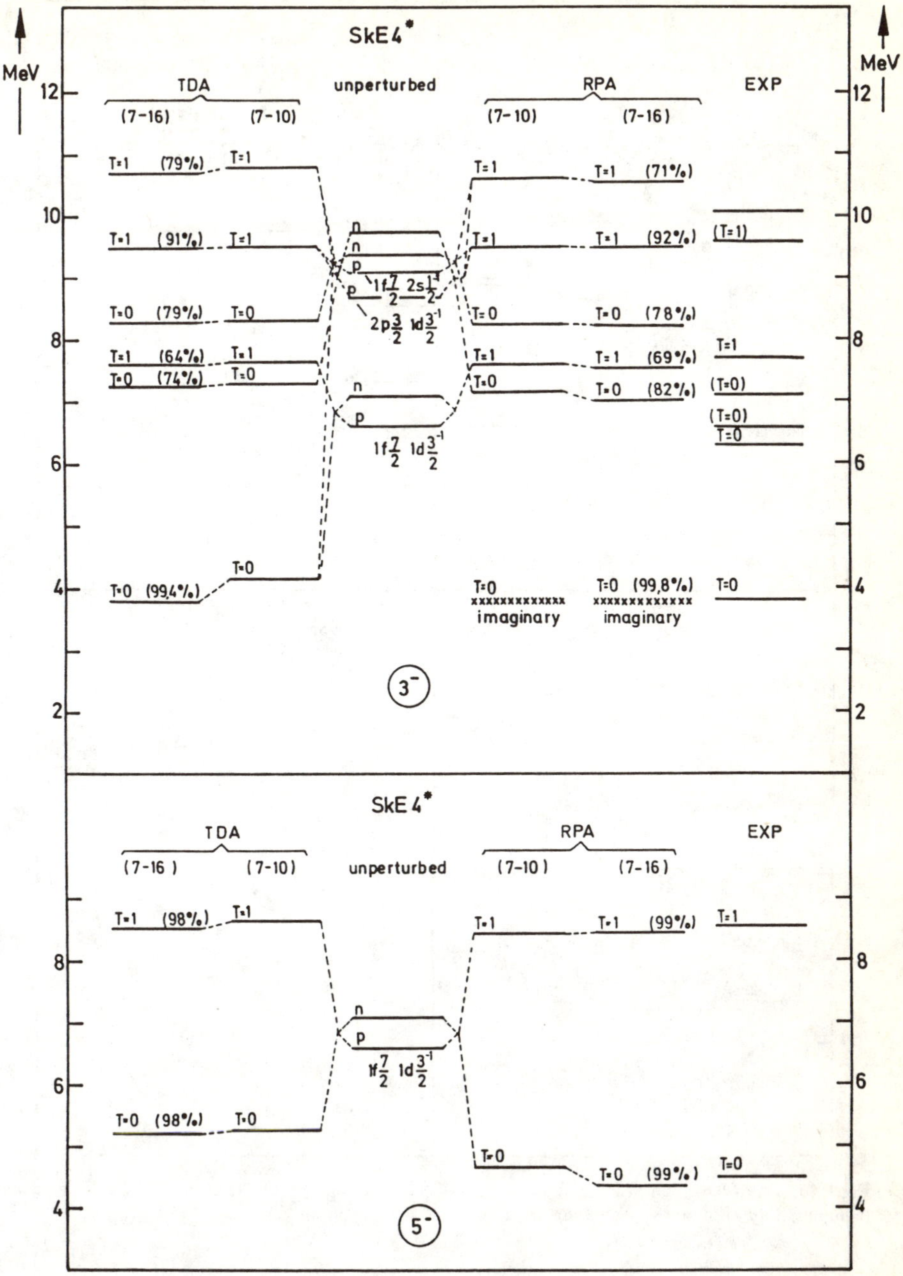

Fig. 8.12. The $J^\pi = 3^-$ and 5^- states in ^{40}Ca. We compare the TDA and RPA results, both in a small ($1\hbar\omega$ excitations with the particle in the $1f2p$ shell) and a large ($3\hbar\omega$ excitations with the particle in the $1g2d3s$ shell) configuration space. The RPA eigenvalue for the strongly collective $3_i^- \, T = 0$ state becomes imaginary (Waroquier 1982)

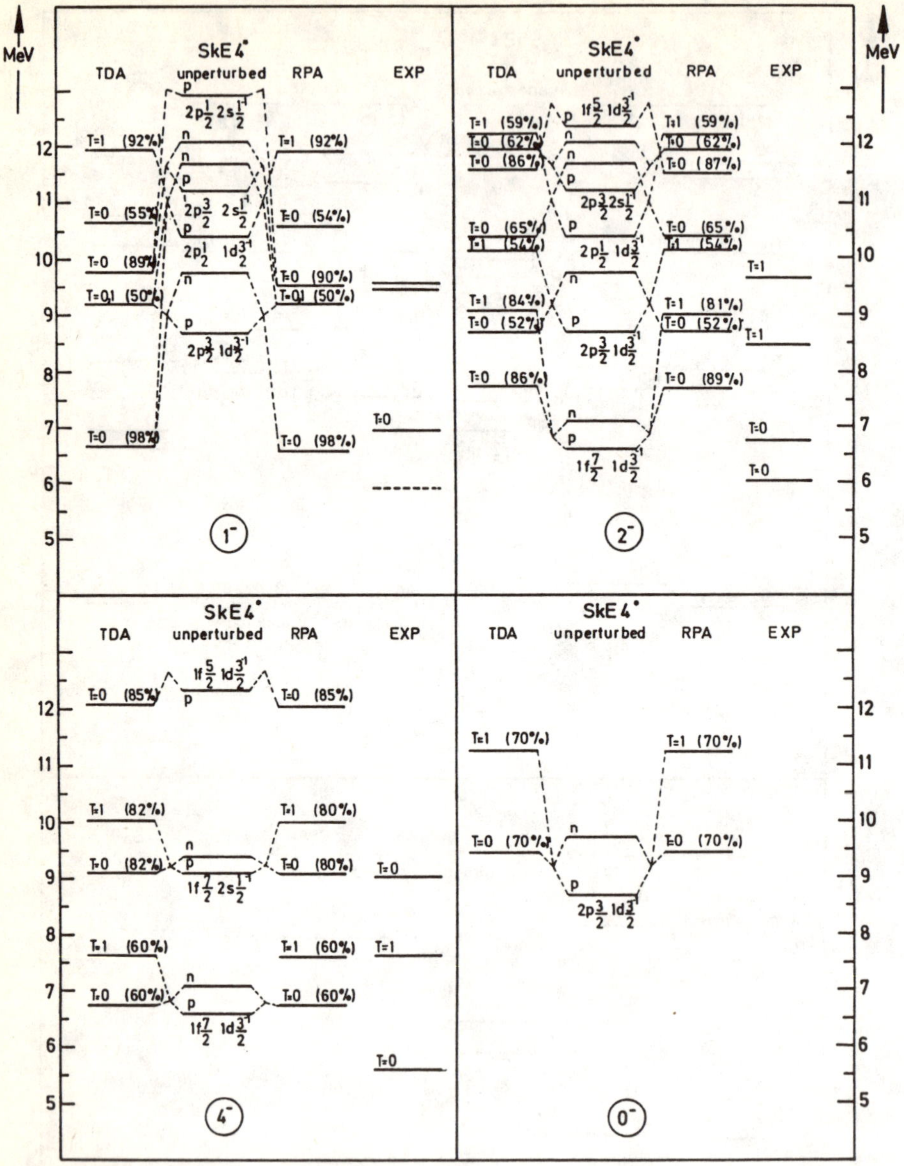

Fig. 8.13. The $J^\pi = 0^-, 1^-, 2^-$ and 4^- states in ^{40}Ca. We compare the TDA and RPA calculations, including $3\hbar\omega$ excitations (Waroquier 1982)

Table 8.4. The single-particle energies used for the $1p - 1h$ calculations in ^{40}Ca. We compare with other sets used in calculations by (Dieperink et al. 1968) (I), (Hsieh et al. 1975) (II) and (Gloeckner, Lawson 1975) (III)

	SkE4 self-consistent		SkE4* adjusted		I	II	III	
	p	n	p	n			p	n
$1p_{3/2}$	−24.72	−25.38	−26.72	−27.38				
$1p_{1/2}$	−20.98	−21.54	−22.98	−23.54				
$1d_{5/2}$	−12.48	−12.84	−13.48	−13.84	−13.30	−15.67	−13.31	−13.63
$1d_{3/2}$	−5.95	−6.12	−6.60	−7.10	−7.30	−6.37	−7.27	−7.24
$2s_{1/2}$	−7.03	−7.34	−9.10	−9.40	−9.80	−9.17	−9.82	−9.91
$1f_{7/2}$	0.0	0.0	0.0	0.0	0.0	0.0	0.0	0.0
$1f_{5/2}$	+7.74	+8.33	+5.74	+6.33	+6.2	+6.5	+6.5	+6.5
$2p_{3/2}$	+4.09	+4.67	+2.09	+2.67	+2.0	+2.1	+1.89	+2.18
$2p_{1/2}$	+5.81	+6.65	+3.81	+4.65	+4.1	+3.9	+3.59	+4.06
$1g_{9/2}$	+11.76	+12.52	+9.76	+10.52				

that are present in the RPA ground-state $|\tilde{0}\rangle$ when deriving the RPA secular equations. This is corroborated independently by the observation of a very low-lying 0^+ state in ^{40}Ca (E_x = 3.35 MeV) which is of a $2p - 2h$ origin (pairing vibration).

There are now a number of possibilities to remedy for the above deficiency in the RPA method:

i) One could enlarge the unperturbed $(1f_{7/2}\,1d_{3/2}^{-1})$ $1p - 1h$ energy by 1.4 MeV (Table 8.4). The 3^- lowest eigenvalue now becomes real (at E_x = 1.64 MeV) pointing towards a highly critical situation in determining the correct excitation energy for these low-lying collective states. This is clear in Sect. 6.3 (Figs. 6.8, 9) where the RPA eigenvalues are presented as a function of $1/\chi$. A slight change in χ can render the lowest root real or imaginary. Also, a slight change in the lowest unperturbed $1p - 1h$ energy can make this change in character (real, imaginary) of the lowest root possible.

ii) As a direct consequence of (i), the x_3 fraction parameter which determines the magnitude of the two-body matrix elements can make the above change in character of the lowest collective 3^- root possible. The precise influence [in the small model space (7–10)] of x_3 is presented in Fig. 8.14. An increase in x_3 (an increase in the importance of the three-body force W_0) lowers the collective character of the interaction. At a value $x_3 = 0.45$ (SkE4) the 3^-, $T = 0$ eigenvalue becomes real again. One can, however, not generally conclude that with increasing x_3 value, the agreement between theory and experiment improves substantially.

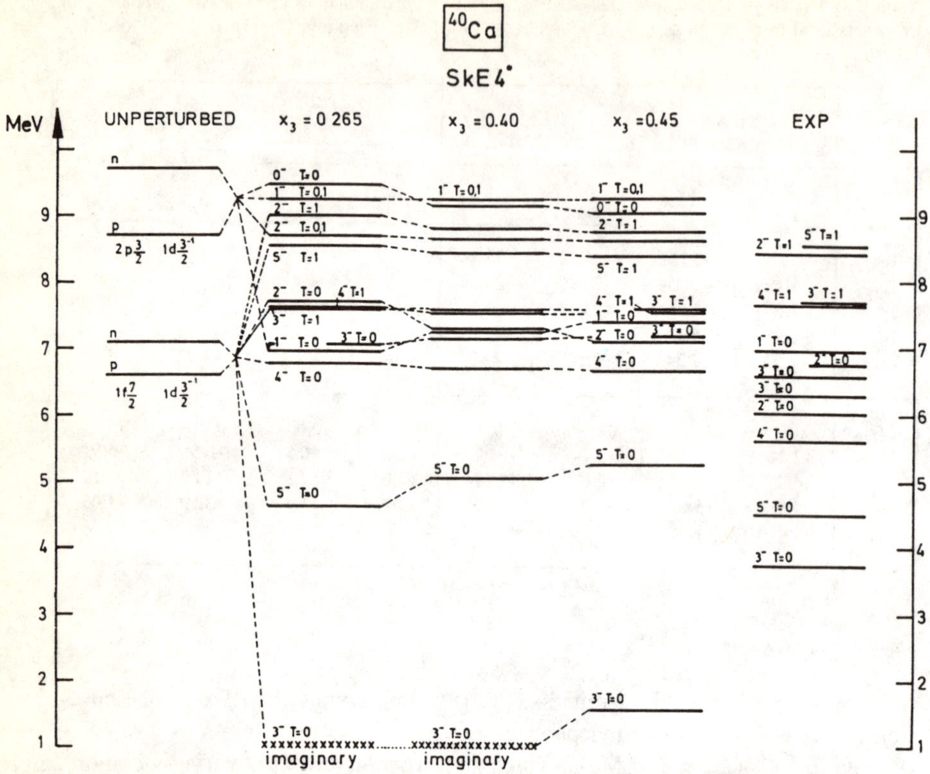

Fig. 8.14. Influence of the fraction parameter x_3 on the low-lying $1p - 1h$ excitations in ^{40}Ca as obtained from the RPA calculations using the small ($1\hbar\omega$) configuration space (Waroquier 1982)

iii) The above two suggestions can be used when calculating the ^{40}Ca energy spectrum with a phenomenological or schematic interaction for which the strength is fitted so as to obtain good agreement for a number of nuclei in a specific model space. Using realistic forces, however, (such as SkE2 and SkE4) this freedom does not exist so that the method of approximation (TDA, RPA, ...) itself has to be corrected (Sect. 8.3.3).

c) Giant Multipole Resonances

Having studied the low-lying collective excitations (vibrations) and non-collective excitations in doubly-closed shell nuclei, an extension towards a description of giant multipole resonances, using the SkE effective interactions, is attempted. Since we always consider bound-state wave functions, widths cannot be calculated. However, the problem of location of the dipole resonance energies and strengths can well be studied. In the light of the results in Sect. 8.3.2b, such a calculation is most interesting.

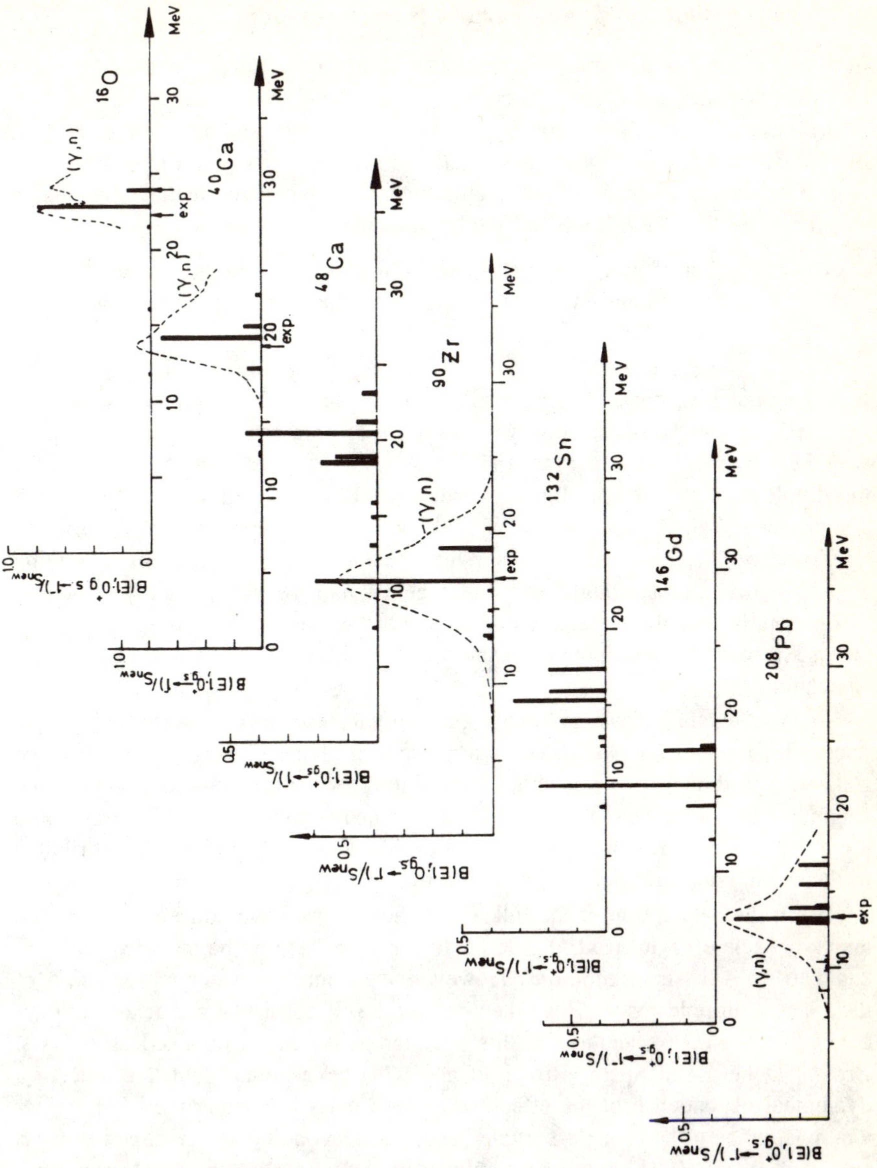

Fig. 8.15. Giant dipole resonances (GDR) in doubly-closed shell-nuclei as obtained from a self-consistent RPA calculation using the SkE4 force parameterization (Waroquier et al. 1983b)

As a first application we show the dipole strength distribution, using SkE4, in Fig. 8.15 for ^{16}O, ^{40}Ca, ^{48}Ca, ^{90}Zr, ^{132}Sn, ^{146}Gd and ^{208}Pb. The excitation energy is very well reproduced in *all* cases using the RPA method with the SkE4 interaction, in a self-consistent way. For a detailed discussion of both the giant dipole resonance and other, higher-multipole resonances in ^{16}O-^{208}Pb, we refer to (Waroquier 1982, Waroquier et al. 1983b).

d) General Outlook: SkE as a Particle-Hole Interaction

In the preceding subsections, a detailed study of the adequacy of the SkE force as a particle-hole interaction was carried out. It was shown that after having fully determined the SkE interaction by fitting the fraction parameter x_3 to spectra of two-particle (two-hole) valence nucleon configurations, good agreement had been obtained in describing doubly-closed shell nuclei. Many features have been satisfactorily described. We recall the main results:

i) Description of isospin purity in the collective low-lying vibration $J^\pi = 3^-$ and 5^-, $T = 0$ states and in the giant multipole resonances in spite of the use of self-consistent orbitals.

ii) The truncation to a $1p - 1h$ model space is generally sufficient to describe low-lying negative-parity collective vibrations as well as positive-parity states (the latter in heavy nuclei only). However, one needs an enlargement of the model space with $3p - 3h$ configurations to satisfy the energy-weighted sum rules (except for the dipole strength, which almost exhausts the EWSR), to describe higher-lying resonances ($\cong 40\,\mathrm{MeV}$ in ^{16}O, $\cong 20\,\mathrm{MeV}$ in ^{208}Pb), to reproduce *all* experimentally observed negative-parity states in light nuclei (^{16}O, ^{40}Ca ...). Some low-lying $3p-3h$ states exhibit strongly rotational features, and are predominantly composed of the coupling of the collective octupole with the quadrupole vibration. The energy position of the multipole resonances is mainly unaffected by this model-space truncation.

iii) The excitation energy of the giant dipole resonance is fairly well reproduced. In light nuclei one observes a separation of the dipole strength into *two* states due to the effect of the spin-orbit splitting, while a much more pronounced fragmentation of the strength is encountered in heavy nuclei, still concentrated into a restricted energy region ($\Delta E_x \cong 5\,\mathrm{MeV}$). The latter can probably be attributed to the momentum dependence of the Skyrme interaction considered.

iv) In the description of the full level schemes, we have noticed that adjusted single-particle energies (SkE*) give the better agreement in the model space truncated to $1p - 1h$ configurations. However, we remark a better reproduction of the giant multipole resonances when complete self-consistency is taken into account (SkE). An explanation of this peculiar behaviour is obtained as follows. Brown, Dehesa and Speth (Brown et al. 1979) have pointed out that, due to a dynamical dependence of the effective nucleon mass in nuclear particle-hole excitations, unperturbed single-particle levels corresponding to an effective mass $m^*/m \cong 0.60 - 0.64$ should be employed in order to push the dipole state to a sufficiently high energy, whereas for low-lying excited states a value $m^*/m \cong 1$ seems more appropriate. Since the SkE interaction, inside the nuclear interior, corresponds to low values $m^*/m \cong 0.7$ (Fig. 8.16), it is quite consistent that the self-consistent HF single-particle energies give a much better description of the GDR excitation energies, whereas adjustments (SkE*) need to be introduced for reproducing low-lying $1p-1h$ excitations in doubly-closed shell nuclei, corresponding more to $m^*/m \cong 1$. In conclusion, a different set of unperturbed energies in RPA calculations, dependent on each J^π state being calculated, is desirable.

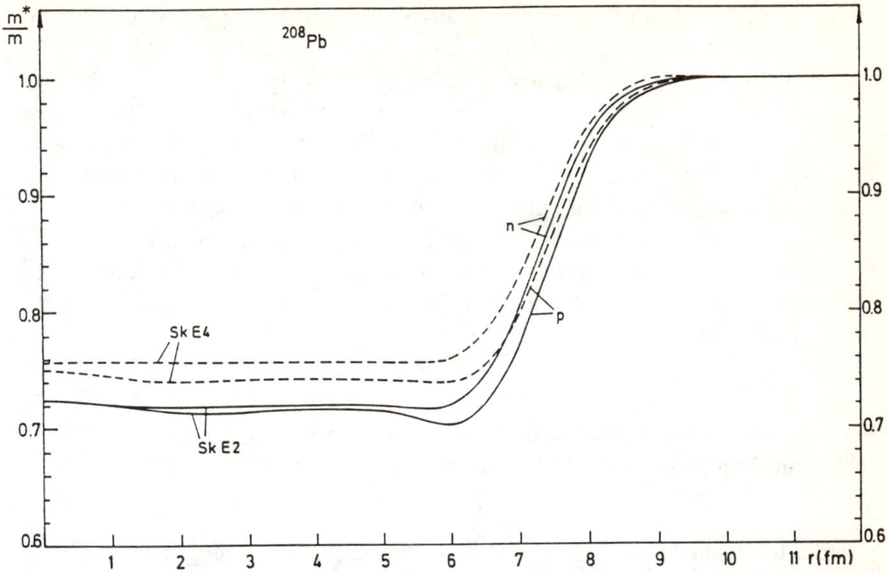

Fig. 8.16. The effective m^*/m ratio in ^{208}Pb using the SkE force, for protons (p) and neutrons (n) separately (Waroquier et al. 1983b)

v) The unnatural-parity states ($J^\pi = 2^-, 4^-$) are less satisfactorily described than the natural-parity levels using SkE. This feature is even more manifest when using realistic interactions.

vi) A serious defect of the SkE interaction remains the existence of imaginary solutions in the standard RPA. It should be emphasized that use of the realistic Tabakin potential also gives rise to an imaginary solution for the low-lying collective octupole state. When using phenomenological residual interactions, the force strength acts as a "renormalization" parameter, which is adjusted in order to reproduce excitation energies according to the approximation used (TDA or RPA). Using the TDA adjusted values for the force parameters in the RPA calculation, one should also obtain an imaginary solution for the collective $J_j^\pi = 3_1^-$, $T = 0$ state. When calculating the particle-hole interaction starting from a realistic potential or from an effective force such as SkE, one has no such additional parameters to adjust. Thus, there is a danger of attributing the defect of imaginary solutions to the neglect of explicitly taking into account ground-state correlations in the derivation of the standard RPA secular equation. This argument is supported by studies of Rowe (Rowe 1968, 1970, Rowe et al. 1971). In order to account for possible $2p-2h$ correlations in the physical ground state $|\Psi_0\rangle$, Rowe deduced renormalized single-particle energies and densities, which reduce the matrix elements of the standard RPA secular equation. This renormalized RPA method clearly indicates that the imaginary solution becomes real and that the other solutions remain almost unaltered. Moreover, the $J_i^\pi = 3_1^-$ excitation energy thus obtained approaches the TDA value. Rowe's results support the possibility that the SkE interaction could give rise to a real and satisfactory eigenvalue for the lowest octupole vibrational excitation, if the explicit ground state $|\Psi_0\rangle$ is considered, without altering significantly the other higher-lying states and modes. Part of such a calculation has already been performed (Waroquier et al. 1983c) with the SkE interaction by evaluating the $2p-2h$ correlations in the ground state of ^{16}O and ^{40}Ca. The conclusions of this study strongly invalidate the Hartree-Fock ground state for the physical correlated ground state in solving the equations of motion, when describing low-lying collective excitations with realistic potentials or effective interactions, adjusted in self-consistent Hartree-Fock calculations. The appearance of imaginary solutions can therefore be attributed to a defect of the standard RPA, rather than to a defect of the SkE interaction.

8.3.3 Rearrangement Effects for Density-Dependent Interactions and Applications for SkE Forces

In order to obtain correct saturation of the nuclear binding energy and density in a self-consistent Hartree-Fock calculation, it has been shown (Moszkowski 1970, Negele 1970, Vautherin, Brink 1972, Campi, Spring 1972, Beiner et al. 1975, Köhler 1976, Dechargeé, Gogny 1980, Waroquier et al. 1983a) that the effective interaction should contain a density-dependent two-body force *or* a three-body term, which, for spin-saturated even-even nuclei behaves like a density-dependent two-body force. The effect of including a density dependence is most significant at the nuclear surface, but becomes significantly suppressed in the nuclear interior. All density-dependent effective nucleon-nucleon interactions used in the literature exhibit density-dependent parts which bear a striking resemblance to each other and which are of zero or very short-range nature. When exciting the nuclear A-body system a subsequent variation in the nucleon density results. Through the density dependence of the nucleon-nucleon interaction, this implies also a modification of the average field in which the interacting nucleons move. This extra term, modifying the average Hartree-Fock field is called the *rearrangement potential energy correction.* This extra term will also modify the HFB equations (8.19, 25).

We shall discuss the importance of the rearrangement terms, using the SkE forces in a number of numerical applications: they drastically effect the collective states in e.g. doubly-closed shell nuclei.

Since the two-body interaction V shows an explicit density-dependence, the interaction matrix elements $\langle\alpha\beta|V[\varrho]|\gamma\delta\rangle_{\text{nas}}$ will show an extra dependence on the modifications of the average field through the rearrangement terms implied by exciting the nuclear system. A detailed discussion on how to evaluate any interaction matrix element occurring in shell-model calculations of an even-even nucleus, including these rearrangement terms, is discussed by Waroquier et al. (Waroquier et al. 1987).

When applied to the SkE force, we use an expression with the three-body component

$$W_0 = x_3 t_3 \delta(\boldsymbol{r}_1 - \boldsymbol{r}_2)\delta(\boldsymbol{r}_1 - \boldsymbol{r}_3) , \tag{8.103}$$

and a density-dependent two-body force

$$V_0 = (1 - x_3)\frac{1}{6}t_3(1 + P_\sigma)\delta(\boldsymbol{r}_1 - \boldsymbol{r}_2)\varrho\left(\frac{\boldsymbol{r}_1 + \boldsymbol{r}_2}{2}\right) . \tag{8.104}$$

The advantage of the partitioning in a fraction x_3 and $(1-x_3)$ for the three-body and density-dependent two-body force has been discussed in Sect. 8.2. Here we present how important the rearrangement effects are on the particle-hole matrix elements of doubly-closed shell nuclei, and how they affect the collective properties in an RPA calculation. Due to the particular structure of the above density-dependent SkE force $(1 - x_3)V_0$, rather simple expressions for the $T = 0$ and $T = 1$ Q and R matrix elements in the RPA secular equations result [(8.92–8.94); (Waroquier et al. 1987)]. Using the assumption for the densities $\varrho_p = \varrho_n = \frac{1}{2}\varrho_{\text{tot}}$, which is not entirely correct, one simplifies the discussion of rearrangement effects in the RPA considerably. The $T = 1$ matrix elements remain unaffected while in the $T = 0$ channel only natural parity states acquire an additional repulsive contribution ($t_3 > 0$). This correction is not negligible, as shown in Fig. 8.17. Here, we depict the Q- and R-matrix elements of the major particle-hole configurations in ^{40}Ca, as a function of the fraction parameter x_3. The strongly attractive $Q_{T=0}$ matrix elements and the repulsive $R_{T=0}$ matrix elements are reduced considerably, reducing the collective features of the $T = 0$ states. In Sect. 8.3, we observed a number of problems with the lowest $T = 0$, 3^- eigenvalue when solving the RPA equations. In Fig. 8.18, we illustrate the effect the inclusion of rearrangement has on the low-lying $T = 0$, 3^- and 5^- levels. The collective 3^- level now obtains a real eigenvalue and good agreement for ^{40}Ca is obtained in a self-consistent way (no adjustment of single-particle energies). One can conclude that the defect attributed to the SkE force parameterization (Waroquier et al. 1983b) is resolved by taking into account rearrangement effects properly. Further we stress that excitation energies and strength distributions for giant multipole resonances are hardly affected so that all conclusions concerning SkE as given in Sect. 8.3 remain.

It is also so that rearrangement effects, proven to be important in the particle-hole channel for doubly-closed shell nuclei, have to be taken into account when discussing the particle-particle aspects of the SkE forces. We therefore apply the above methods to a discussion of ^{18}O and ^{42}Ca (Fig. 8.19). No attempt has been made to adjust single-particle spectra in an optimal way. Only a slightly smaller x_3 value was taken in the SkE2 parameterization. The higher x_3 values lead to more compressed energy spectra (Sect. 8.3.1). The rearrangement terms reduce core-polarization corrections and induce a serious compression in the two-particle energy spectra, shown in Fig. 8.19. This effect is counterbalanced by choosing a somewhat smaller x_3 value in the force. This explains why good results could be obtained in Sect. 8.3.1 in the absence of rearrangement effects. In the particle-hole RPA no such compensating effects occur and here, rearrangements are really needed to improve the comparison between experiment and calculations.

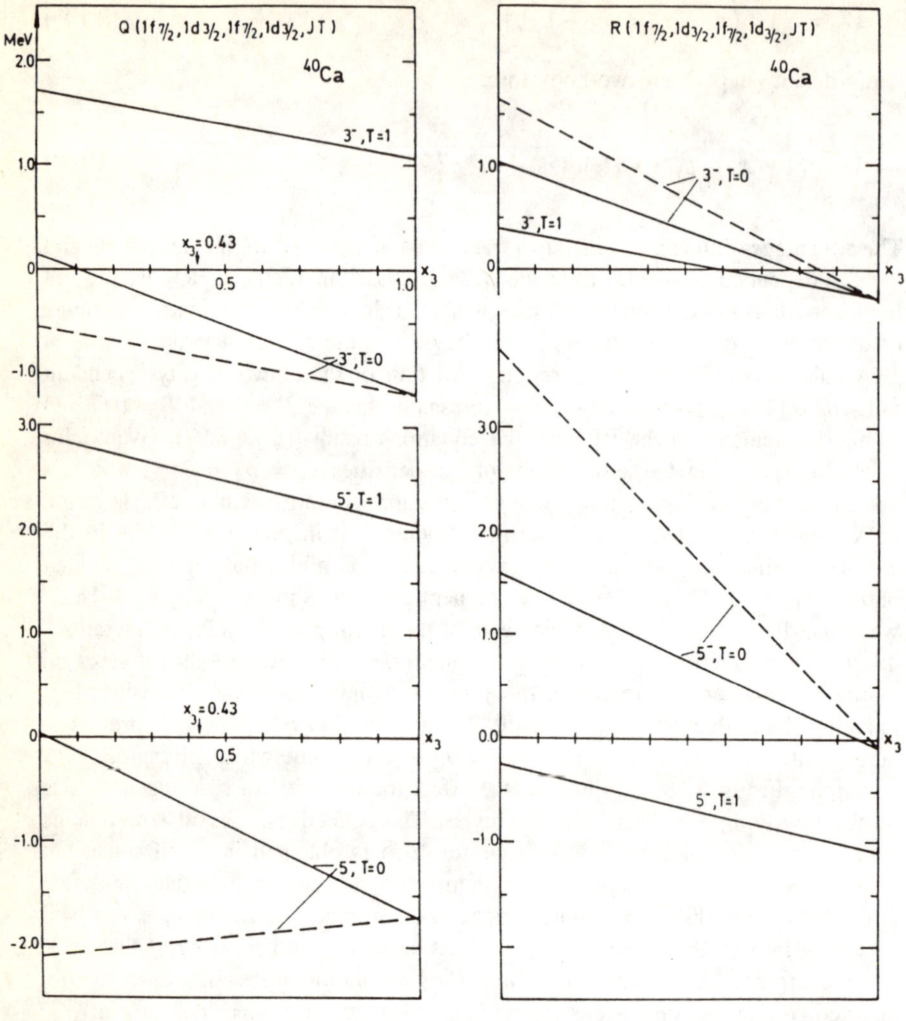

Fig. 8.17. Influence of rearrangement on a particular Q- and R matrix element in ^{40}Ca in the J^π = 3^- and 5^- states using the SkE2 parameterization. The dashed line represents the situation without rearrangement corrections (Waroquier et al. 1987)

We finally would like to report on a very extensive and state-of-the art self-consistent calculation, using all of the above ingredients, in the single-closed shell even-even nucleus ^{116}Sn. The model space here is taken to contain rather complex configurations of $4qp$ character. We take this nucleus ^{116}Sn since a detailed set of experiments was performed including ^{116}Sn(p,p'), ^{116}Sn(e,e'), ^{115}In(^{3}He,d) and ^{115}In(α,t) (Van der Werf 1986).

In the present calculation [see also (Waroquier et al. 1987) for more details on the formalism] we consider neutron $2qp$, proton $1p-1h$ across $Z = 50$ and coupled $(1p-h)_\pi \otimes (2qp)_\nu$ configurations. These calculations are performed

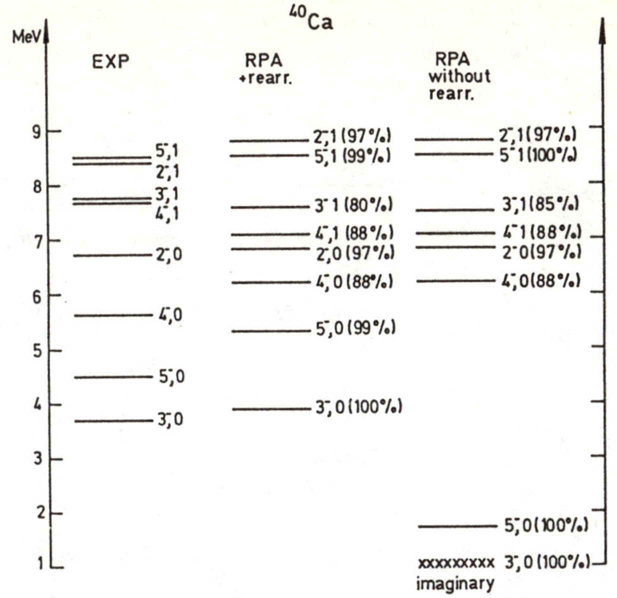

Fig. 8.18. Low-lying negative parity (J^{π}, T) states in ^{40}Ca. The theoretical levels correspond to a fully self-consistent RPA calculation using the SkE2 (x_3 = 0.43) force. The numbers between brackets denote the isospin purity (Waroquier et al. 1987)

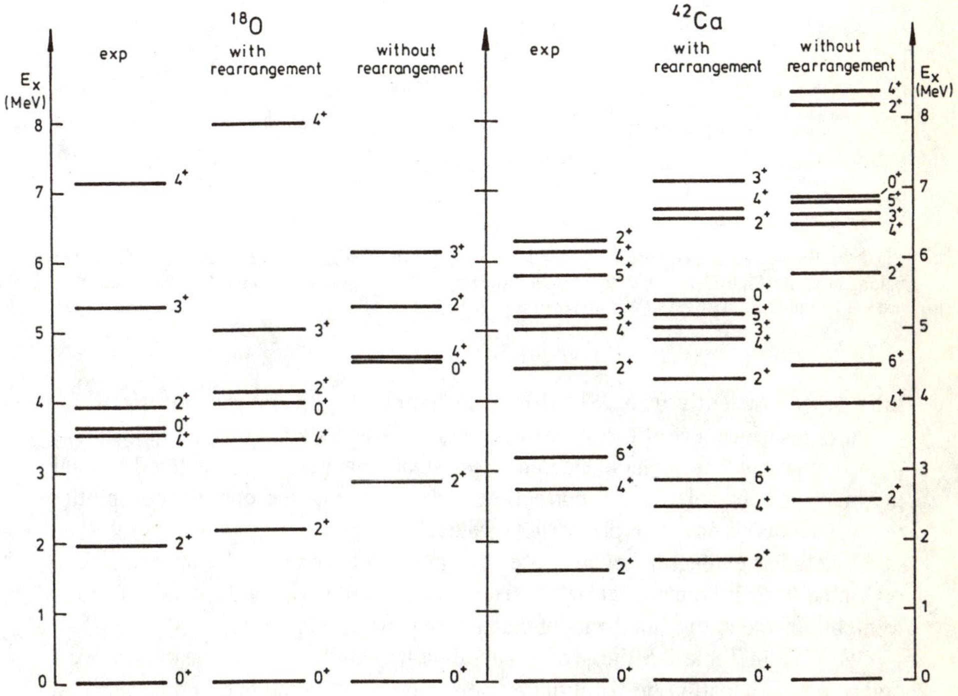

Fig. 8.19. Two-particle spectra in ^{18}O and ^{42}Ca using the SkE2 (x_3 = 0.37) force. Only positive parity states are retained. The dominant low-lying $4p - 2h$ excitations, observed in the experimental spectra, are not inserted in the figure as they could not be reproduced in a two-particle model space (Waroquier et al. 1987)

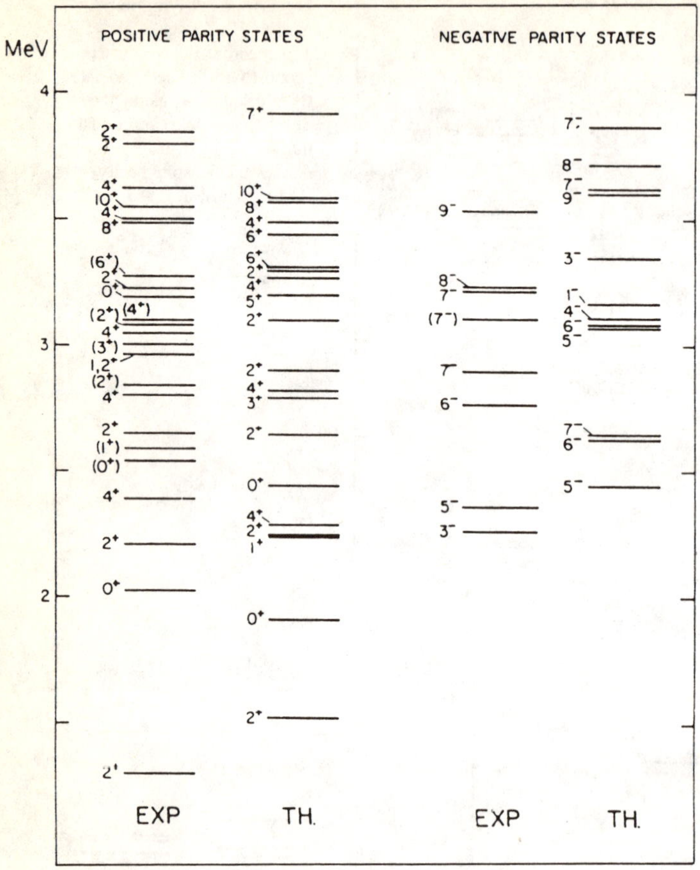

Fig. 8.20. Experimental and calculated low-lying positive and negative parity states for ^{116}Sn. Levels belonging to the "intruder" band, based on a proton $2p-2h$ excitation across the $Z=50$ closed shell, are not included in the figure (Waroquier et al. 1987)

fully self-consistently using SkE forces including also rearrangement effects and provides a serious test of the above interaction. Only a slight change in the single-particle energies has been considered: these small changes can be justified through the missing core-polarization corrections on the level of the one-nucleon motion in the nucleus (dressed single-particle states).

A detailed evaluation of all necessary matrix elements in the above basis is presented in (Waroquier et al. 1987). Here, we present some of the salient features to highlight the above shell-model calculations, using Figs. 8.20, 21.

We give in Table 8.5 the proton single-particle and neutron quasi-particle energies (and occupation probabilities v^2) that were used in the determination of the ^{116}Sn spectrum of Fig. 8.20. Here, low-lying positive and negative parity states are displayed separately to make comparison easier. The data on the deformed proton $2p-2h$, which are fully outside the present model space, have been left

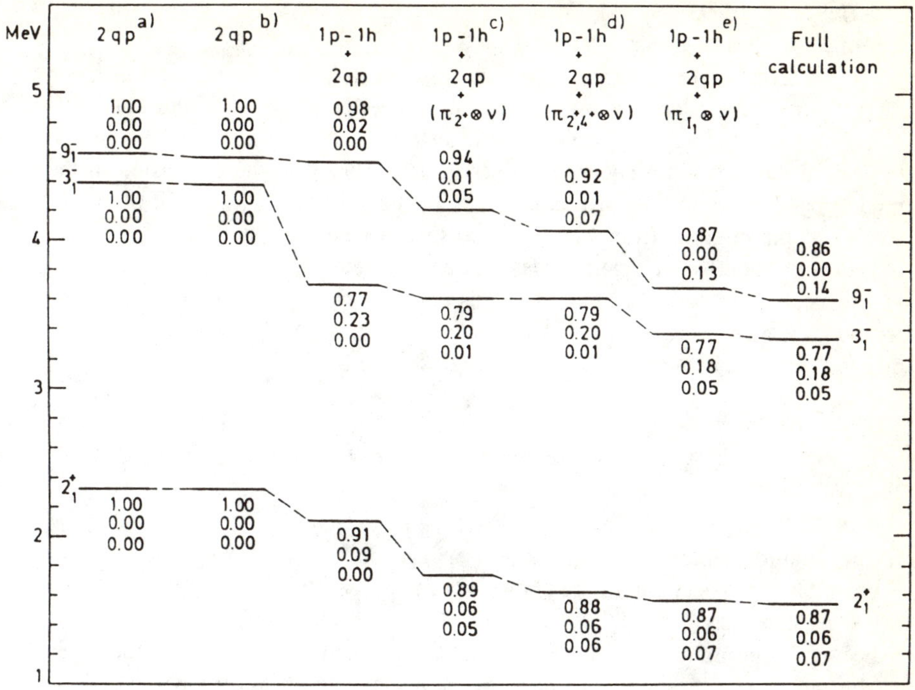

Fig. 8.21. Excitation energies of the $J_i^\pi = 2_1^+, 3_1^-$ and 9_1^- levels as a function of the dimension of the model space: (*a*) only neutron 2qp excitations in the 50 – 82 shell are included, (*b*) space of (*a*) but including also the $1g_{9/2}$ orbital, (*c*) the coupled configurations are included with the proton $1p - 1h$ configurations coupled to $I^\pi = 2^+$, (*d*) space of (*c*) but now with J^π = all natural parity spins. The numbers near each level refer to the wave function square amplitudes of (from top to bottom) 2qp, $1p - 1h$ and $(1p - 1h)_\pi \otimes (2\text{qp})_\nu$ configurations (Waroquier et al. 1987)

Table 8.5. Proton single-particle, neutron one quasi-particle energies (in MeV) and occupation probabilities

	$\varepsilon(p)$	$E(n)$	v^2
$1f_{7/2}$	−5.00		
$1f_{5/2}$	−3.00		
$2p_{3/2}$	−2.00		
$2p_{1/2}$	−1.00		
$1g_{9/2}$	0.00	5.00	0.99
$1g_{7/2}$	4.30	2.15	0.84
$2d_{5/2}$	4.40	2.20	0.92
$1h_{11/2}$	5.90	2.10	0.12
$2d_{3/2}$	5.20	1.40	0.25
$3s_{1/2}$	5.40	1.00	0.55

out (Heyde et al. 1983, 1985). There is a general good agreement though a detailed identification has not been attempted.

It is interesting to discuss in some more detail the influence of the dimension of the model space chosen on the excitation energy of some of the collective states (2_1^+, 3_1^-) and on the "stretched" 9_1^- level (Fig. 8.21). Here, it is shown that in general the excitation energy is lowered, proportional to the dimension of the model space improving the agreement with the data. For the 9_1^-, a shift of almost 1 MeV occurs in going from a neutron $2qp$ space to the full space, an effect which is very important to get a correct description of states in ^{116}Sn.

In calculating spectroscopic strengths for one-proton transfer reactions one should have a description of the ^{115}In nucleus within a model space, similar to the one in ^{116}Sn. Such calculations are in progress (Waroquier 1989).

8.4 Summary

In this chapter, which is somewhat less pedagogical in spirit, we point out how one can take into account all the above aspects of nuclear structure, but, in a fully self-consistent way.

The starting point here is the construction of a single nucleon-nucleon interaction which has the ability to describe both the ground-state properties of the atomic nucleus (binding energy, saturation properties, charge and mass radii, ...) and the specific properties of excited states equally well. To this end, the powerful properties of zero-range density-dependent forces have been shown to indicate that these interactions are valid candidates for a self-consistent approach to nuclear structure.

With the use of Skyrme forces, problems arose mainly in connection with the pairing properties: we have been able to devise a force which represents both correct pairing (short-range) characteristics and saturating properties. The presence of three-body terms is instrumental in that respect and the effect of three-body forces on Hartree-Fock-Bogoliubov (HFB) theory is first discussed in a rather general way. The method simplifies considerably for spherical doubly-closed and single-closed shell nuclei. An extended Skyrme force (SkE force), which fulfills all the above requirements remarkably well, is derived and discussed. The parameters are derived from a simultaneous description of the ground-state properties and the pairing characteristics in finite nuclei.

Detailed applications of the SkE force to doubly-closed shell nuclei throughout the nuclear mass table relating to binding energy and radii are presented. Moreover, the single-particle Hartree-Fock spectra for doubly-closed shell nuclei are given.

In the next sections, excited state properties are discussed using the extended Skyrme force. The construction of a renormalized effective force, taking into account excitations outside the model space and using perturbation theory, is performed for various open shell nuclei with two-particle (or two-hold) configurations. We even compare the salient features of the two-body matrix elements evaluated using simple, schematic forces, with those from the extended Skyrme force. Similarly, we study the $1p - 1h$ excitations as observed in doubly-closed shell nuclei and thereby test the behavior of the SkE force as a good particle-hole interaction. Besides some general discussions on the evaluation of transition matrix elements and sum rules, we concentrate on the reproduction of collective, low-lying states in doubly-closed shell nuclei. Both the isoscalar (low-lying 3^-, 2^+, ..., states) as well as the higher-lying isovector giant multipole resonances are discussed, in greater detail for excitations in ^{40}Ca. The variation in position and strength of the giant electric dipole mode throughout the whole nuclear mass table is also presented.

We finally discuss some interesting aspects related to the self-consistent method of performing shell-model calculations such as the question of the general validity of using density-dependent interactions, and the consequences of rearrangement effects. For the latter point in particular, we study in detail the effect of rearrangement on the position of the lowest collective root in ^{40}Ca. To conclude, an application to an open shell nucleus from the Sn series is given, showing the potential of the methods discussed.

9. Some Computer Programs

In this chapter we discuss briefly a set of simple computer programs that allow, at the level of assimilating the present material, calculations that go beyond some analytical evaluations or academic questions. Even though this collection is not very high-brow, the necessary Wigner $3nj$ symbols are included. We give a code to evaluate the Slater integrals and the matrix elements of a δ-residual interaction. These ingredients allow for the construction of some simple secular equations. To this end we include a diagonalization code, based on the Jacobi method for diagonalizing matrices of small dimensions. We furthermore include, for the calculation of transition rates the computation of radial integrals of any power of $r(r^L)$ weighted with the harmonic oscillator wave functions. A BCS code, using a constant pairing force is included for finding the occupation probabilities in a given number of single-particle orbitals. We also include a code to convert particle-hole into particle-particle matrix elements. Finally, we include a more extended code where, at the same time, an apllication to collective excitation is incorporated. We discuss particle-core coupling, and present the major codes for evaluating energies and transition rates. The library package includes some interesting subroutines for e.g. the calculation of the reduced EL and ML transition matrix elements.

In many cases, as well as the source programs, a typical input command file is given. For the latter, particle-core coupling code, a typical output file is also given in order to allow for precise checking when implementing the code on a given computing system.

9.1 Clebsch-Gordan Coefficients

Here, with the function program VN02BA (j_1, j_2, j, m_1, m_2, m), we evaluate the Clebsch-Gordan coefficient

$$\langle j_1 m_1,\, j_2 m_2 | j m\rangle \,, \tag{9.1}$$

with parameters j_1, j_2, j, m_1, m_2, m in the function to be used having double the physical value of the angular momentum quantum number. This particular convention is also used in Sects. 9.2 and 9.3, when evaluating $6j$ and $9j$-symbols.

A small main program is added which includes the input set. In this main program, a common block /VN06FC/FCT(40) is generated which contains the quantities FCT(I) = (I$-$1)!/10^{I}, values to be used in the calculation of other Wigner

nj-invariants too. It is the dimension of this array FCT(DIMENSION) which constrains the angular momenta to be used in evaluating the Clebsch-Gordan coefficient.

PROGRAM

```
      FUNCTION VN02BA(J1,J2,J,M1,M2,M)
      INTEGER Z,ZMIN,ZMAX,FASE
      COMMON/VN06FC/FCT(50)
      CC=0.0
      IF(M1+M2-M)20,1,20
1     IF(IABS(M1)-IABS(J1))2,2,20
2     IF(IABS(M2)-IABS(J2))3,3,20
3     IF(IABS(M)-IABS(J))4,4,20
4     IF(J-J1-J2)5,5,20
5     IF(J-IABS(J1-J2))20,6,6
6     ZMIN=0
      IF(J-J2+M1)7,8,8
7     ZMIN=-J+J2-M1
8     IF(J-J1-M2+ZMIN)9,10,10
9     ZMIN=-J+J1+M2
10    ZMAX=J1+J2-J
      IF(J2+M2-ZMAX)11,12,12
11    ZMAX=J2+M2
12    IF(J1-M1-ZMAX)13,14,14
13    ZMAX=J1-M1
14    DO 15 Z=ZMIN,ZMAX,2
      JA=Z/2+1
      JB=(J1+J2-J-Z)/2+1
      JC=(J1-M1-Z)/2+1
      JD=(J2+M2-Z)/2+1
      JE=(J-J2+M1+Z)/2+1
      JF=(J-J1-M2+Z)/2+1
      FASE=((-1)**(Z/2))
      F2=FASE
15    CC=CC+F2/(FCT(JA)*FCT(JB)*FCT(JC)*FCT(JD)*FCT(JE)*FCT(JF))
      JA=(J1+J2-J)/2+1
      JB=(J1-J2+J)/2+1
      JC=(-J1+J2+J)/2+1
      JD=(J1+M1)/2+1
      JE=(J1-M1)/2+1
      JF=(J2+M2)/2+1
```

```
      JG-(J2-M2)/2+1
      JH-(J+M)/2+1
      JI-(J-M)/2+1
      JJ-(J1+J2+J+2)/2+1
      F1-J+1
      CC-SQRT(F1*FCT(JA)*FCT(JB)*FCT(JC)*FCT(JD)
     1*FCT(JE)*FCT(JF)*FCT(JG)*FCT(JH)*FCT(JI)/FCT(JJ))*CC
20    VN02BA-CC/SQRT(10.0)
      RETURN
      END
```

PROGRAM

```
      COMMON/VN06FC/FCT(40)
      FCT(1)-1.
      DO 400 I-2,40
400   FCT(I)-FCT(I-1)*(I-1.)/10.
4     READ(1,2)J1,J2,J3,M1,M2,M3,ICON
2     FORMAT(7I2)
      X-VN02BA(J1,J2,J3,M1,M2,M3)
      WRITE(6,3)J1,J2,J3,M1,M2,M3,X
3     FORMAT(1H ,6I4,E22.8)
      IF(ICON.EQ.0)GOTO5
      GOTO 4
5     STOP
      END
```

9.2 Wigner 6j-Symbol

Here, with the function program VN02B9 $(j_1, j_2, j_3, l_1, l_2, l_3)$, we evaluate the Wigner 6j-symbol

$$\begin{Bmatrix} j_1 & j_2 & j_3 \\ l_1 & l_2 & l_3 \end{Bmatrix}, \tag{9.2}$$

where again, double the physical value of the angular momentum quantum numbers are used.

This function VN02B9 uses the function VN02BB (j_1, j_2, j_3, FCT) in order to calculate the Wigner 6j-symbol.

As with the calculation of the Clebsch-Gordan coefficient, the array FCT(40) within the common block /VN06FC/FCT(40) has to be used.

PROGRAM

```
      REAL*8 FUNCTION VN02B9(J1,J2,J3,L1,L2,L3)
      IMPLICIT REAL*8 (A-H,O-Z)
      COMMON/VN06FC/FCT(40)
      CC=0.0
      IF(J1+J2-J3) 20,1,1
 1    IF(IABS( J1-J2)-J3) 2,2,20
 2    IF (J1+J2+J3-2*((J1+J2+J3)/2)) 20,3,20
 3    IF(J1+L2-L3) 20,4,4
 4    IF (IABS(J1-L2)-L3) 5,5,20
 5    IF (J1+L2+L3-2*((J1+L2+L3)/2)) 20,6,20
 6    IF(L1+J2-L3) 20,7,7
 7    IF(IABS(L1-J2)-L3) 8,8,20
 8    IF(L1+J2+L3-2*((L1+J2+L3)/2)) 20,9,20
 9    IF(L1+L2-J3) 20,10,10
 10   IF(IABS(L1-L2)-J3) 11,11,20
 11   IF(L1+L2+J3-2*((L1+L2+J3)/2)) 20,12,20
 12   OMEGA=0.0
      IF(J3) 37,38,37
 37   IF(L3) 40,39,40
 38   VN02B9=(-1.)**((J1+L2+L3)/2)/SQRT((FLOAT(J1)+1.)*(FLOAT(L2)+1.))
      GO TO 41
 39   VN02B9=(-1.)**((J1+J2+J3)/2)/SQRT((FLOAT(J1)+1.)*(FLOAT(J2)+1.))
      GO TO 41
 40   IWMIN=J1+J2+J3
      IF(IWMIN-J1-L2-L3) 13,14,14
 13   IWMIN=J1+L2+L3
 14   IF(IWMIN-L1-J2-L3) 15,16,16
 15   IWMIN=L1+J2+L3
 16   IF(IWMIN-L1-L2-J3) 17,18,18
 17   IWMIN=L1+L2+J3
 18   IWMAX=J1+J2+L1+L2
      IF(IWMAX-J2-J3-L2-L3) 22,22,23
 23   IWMAX=J2+J3+L2+L3
 22   IF (IWMAX-J1-J3-L1-L3) 24,24,25
 25   IWMAX=J1+J3+L1+L3
 24   IF(IWMIN-IWMAX) 26,26,20
 26   DO 701 IW = IWMIN,IWMAX,2
      IW1=IW/2+2
      IW2=(IW-J1-J2-J3)/2+1
```

```
      IW3=(IW-J1-L2-L3)/2+1
      IW4=(IW-L1-J2-L3)/2+1
      IW5=(IW-L1-L2-J3)/2+1
      IW6=(J1+J2+L1+L2-IW)/2+1
      IW7=(J1+J3+L1+L3-IW)/2+1
      IW8=(J2+J3+L2+L3-IW)/2+1
      IF(IW-4*(IW/4)) 30,31,30
31    PH=1.0
      GO TO 35
30    PH=-1.0
35    OMEGA=OMEGA+PH*FCT(IW1)/FCT(IW2)/FCT(IW3)/FCT(IW4)/FCT(IW5)
     1/FCT(IW6)/FCT(IW7)/FCT(IW8)
701   CONTINUE
      CC=OMEGA*VN02BB(J1,J2,J3,FCT)*VN02BB(J1,L2,L3,FCT)*VN02BB(L1,J2,L3
     1,FCT)*VN02BB(L1,L2,J3,FCT)
20    VN02B9=CC*10.0
41    RETURN
      END
      REAL*8 FUNCTION VN02BB(J1,J2,J3,FCT)
      IMPLICIT REAL*8 (A-H,O-Z)
      DIMENSION FCT(40)
      IW1=(J1+J2-J3)/2+1
      IW2=(J1-J2+J3)/2+1
      IW3=(-J1+J2+J3)/2+1
      IW4=(J1+J2+J3+2)/2+1
      FDELTA=SQRT(FCT(IW1)*FCT(IW2)*FCT(IW3)/FCT(IW4))
      VN02BB=FDELTA/3.16227765
      RETURN
      END
```

PROGRAM

```
      IMPLICIT REAL*8 (A-H,O-Z)
      COMMON/VN06FC/FCT(40)
      FCT(1)=1.
      DO 400 I=2,40
400   FCT(I)=FCT(I-1)*(I-1.)/10.
4     READ(1,2)J1,J2,J3,L1,L2,L3,ICON
2     FORMAT(7I2)
      X=VN02B9(J1,J2,J3,L1,L2,L3)
      WRITE(6,3)J1,J2,J3,L1,L2,L3,X
```

```
3     FORMAT(1H ,6I4,D22.12)
      IF(ICON.EQ.0)GOTO5
      GOTO 4
5     STOP
      END
```

9.3 Wigner 9j-Symbol

Below, we give the function program that evaluates the Wigner 9j-symbol, WINEJ ($j_{11}, j_{12}, j_{13}, j_{21}, j_{22}, j_{23}, j_{31}, j_{32}, j_{33}$) and is used to evaluate

$$\left\{\begin{matrix} j_{11} & j_{12} & j_{13} \\ j_{21} & j_{22} & j_{23} \\ j_{31} & j_{32} & j_{33} \end{matrix}\right\} . \tag{9.3}$$

The input parameters are double the physical value of the corresponding angular momentum quantum numbers. The evaluation uses a summation over products of three Wigner 6j coefficients. Again, the array FCT(40) has to be defined in a main program via the common block, /VN06FC/FCT(40).

PROGRAM

```
      FUNCTION WINEJ(J11,J12,J13,J21,J22,J23,J31,J32,J33)
      COMMON/VN06FC/FCT(40)
      WINEJ-0.
      IF(J11+J21-J31)2,1,1
1     IF(IABS(J11-J21)-J31)3,3,2
3     IF(J11+J21+J31-2*((J11+J21+J31)/2))2,4,2
4     IF(J21+J22-J23)2,5,5
5     IF(IABS(J21-J22)-J23)6,6,2
6     IF(J21+J22+J23-2*((J21+J22+J23)/2))2,7,2
7     IF(J31+J32-J33)2,8,8
8     IF(IABS(J31-J32)-J33)9,9,2
9     IF(J31+J32+J33-2*((J31+J32+J33)/2))2,10,2
10    IF(J11+J12-J13)2,11,11
11    IF(IABS(J11-J12)-J13)12,12,2
12    IF(J11+J12+J13-2*((J11+J12+J13)/2))2,13,2
13    IF(J12+J22-J32)2,14,14
14    IF(IABS(J12-J22)-J32)15,15,2
15    IF(J12+J22+J32-2*((J12+J22+J32)/2))2,16,2
16    IF(J13+J23-J33)2,17,17
17    IF(IABS(J13-J23)-J33)18,18,2
```

```
18   IF(J13+J23+J33-2*((J13+J23+J33)/2))2,19,2
19   KMIN=MAX0(IABS(J11-J33),IABS(J32-J21),IABS(J23-J12))
     KMAX=MIN0(J11+J33,J32+J21,J23+J12)
     DO 28 K=KMIN,KMAX,2
28   WINEJ=WINEJ+(-1)**K*(FLOAT(K)+1.)*VN02B9(J11,J21,J31,J32,J33,K)*
    1VN02B9(J12,J22,J32,J21,K,J23)*VN02B9(J13,J23,J33,K,J11,J12)
2    RETURN
     END
```

9.4 Calculation of Table of Slater Integrals

In the present program, a list of Slater integrals F^0 (see (3.79))

$$F^0 = \frac{1}{4\pi}\int u_{n_1 l_1}(r)u_{n_2 l_2}(r)u_{n_3 l_3}(r)u_{n_4 l_4}(r)r^{-2}dr\ , \tag{9.4}$$

is evaluated, Slater integrals that occur for a $\delta(\boldsymbol{r}_2 - \boldsymbol{r}_1)$ residual interaction.

The quantum numbers are introduced as $n(i), l(i)$ with n the radial quantum number ($n = 1, 2, \ldots$) and l the orbital angular momentum. Even though in the actual integral (9.4) the radial quantum number n varies as $n = 0, 1, 2, \ldots$; in the input to the program the more standard notation $n' \equiv n + 1 = 1, 2, 3, \ldots$ is used. According to a given maximal number of radial wave functions, given by the variable IMAX, all Slater integrals are evaluated automatically. The strength of the interaction is determined by the variable VEFF and the harmonic oscillator range by the variable $NU(\equiv \nu)$ with as definition

$$\nu = m\omega/2\hbar\ , \tag{9.5}$$

or in more easy-to-handle units

$$\nu = \frac{41A^{-1/3}mc^2}{2(\hbar c)^2}\ , \tag{9.6}$$

where mc^2 is the nucleon rest mass and $\hbar c$ the combination 197 MeV fm such that ν has indeed the dimension of fm^{-2}.

The radial integral is evaluated via an analytical expression, in closed form, in the function program SLATER. This function uses a number of coefficients, calculated in the function VN07AL, needed to evaluate the Slater integrals.

The expressions used are

$$\begin{aligned} F^0 =& \frac{1}{8\pi} N_{n_1 l_1} N_{n_2 l_2} N_{n_3 l_3} N_{n_4 l_4} (2\nu)^{-(l_1+l_2+l_3+l_4)/2-3/2} \\ &\times \int_0^\infty x^{(l_1+l_2+l_3+l_4)/2+1/2} e^{-2x} v_{n_1 l_1}(x) v_{n_2 l_2}(x) \\ &\times v_{n_3 l_3}(x) v_{n_4 l_4}(x) dx\ . \end{aligned} \tag{9.7}$$

with

$$N_{nl} = \left[\frac{2^{l-n+2}(2\nu)^{l+3/2} \cdot (2l+2n+1)!!}{\sqrt{\pi}[(2l+1)!!]^2 n!} \right]^{1/2} , \tag{9.8}$$

$$v_{nl}(x) = \sum_{k=0}^{n} (-1)^k \cdot 2^k \frac{n!}{(n-k)!k!} \frac{(2l+1)!!}{(2l+2k+1)!!} x^k , \tag{9.9}$$

and

$$u_{nl}(x) = N_{nl} \cdot r^{l+1} e^{-\nu r^2} v_{nl}(2\nu r^2) . \tag{9.10}$$

Using the integrals

$$\begin{aligned} &\int_0^\infty x^p e^{-2x} dx = 2^{-(p+1)} p! \\ &\int_0^\infty x^{p+1/2} e^{-2x} dx = 2^{-(2p+5/2)} \sqrt{\pi} (2p+1)!! \end{aligned} \tag{9.11}$$

with p a positive number, one finally gets

$$\begin{aligned} F^0 =& \pi^{-2} \nu^{3/2} \prod_i 2^{-(n_i+l_i)/2} \begin{Bmatrix} \sqrt{2} \\ \sqrt{\pi} \end{Bmatrix} \left[\prod_i (2l_i + 2n_i + 1)!! n_i \right]^{1/2} \\ &\times \sum_{k_1=0}^{n_1} \sum_{k_2=0}^{n_2} \sum_{k_3=0}^{n_3} \sum_{k_4=0}^{n_4} \frac{(-1)^K \cdot 2^{-K} (2K+2L)!!}{\prod_i \{(n_i - k_i)!! k_i!! (2l_i + 2k_i + 1)!!\}} . \end{aligned} \tag{9.12}$$

Here we have defined

$$\begin{aligned} K &= k_1 + k_2 + k_3 + k_4 , \\ L &= \frac{(l_1 + l_2 + l_3 + l_4)}{2} + \frac{1}{2} . \end{aligned} \tag{9.13}$$

The numbers between accolades

$$\begin{Bmatrix} \sqrt{2} \\ \sqrt{\pi} \end{Bmatrix}$$

mean that one uses $\sqrt{2}$ or $\sqrt{\pi}$ when L is integer or half-integer, respectively.

PROGRAM

```
      PROGRAM SLATMAT
      REAL NU
      REAL INT1,INT2
      DIMENSION N(10),L(10)
      COMMON/VN07CB/NU,AKNL(4,4,8),ANL(4,8),INT1(20),INT2(20)
C
C SLATMAT MAKES A LIST OF SLATER INTEGRALS
```

```
C
      CALL VN07AL
      WRITE(3,360)NU
  360 FORMAT(1H ,'NU=',F10.8/)
      READ(1,1)IMAX,VEFF
  1   FORMAT(I2,F7.2)
      WRITE(3,7)IMAX,VEFF
  7   FORMAT(1H ,'MAX',I4,'VEFF',F10.5/)
      DO 2 I=1,IMAX
      READ(1,3)N(I),L(I)
      WRITE(3,4)N(I),L(I)
  3   FORMAT(2I2)
  4   FORMAT(1H ,'N',I4,'L',I4)
  2   CONTINUE
C
C
C
      DO 5 I=1,IMAX
      DO 5 J=I,IMAX
      DO 5 K=1,IMAX
      DO 5 M=K,IMAX
      IF(10*I+J.GT.10*K+M)GOTO 5
      IF((-1)**(L(I)+L(J)).NE.(-1)**(L(K)+L(M)))GOTO 5
      NA=N(I)-1
      NB=N(J)-1
      NC=N(K)-1
      ND=N(M)-1
      LA=L(I)
      LB=L(J)
      LC=L(K)
      LD=L(M)
      X=SLATER(NA,LA,NB,LB,NC,LC,ND,LD)*VEFF
      WRITE(3,6)N(I),LA,N(J),LB,N(K),LC,N(M),LD,X
  6   FORMAT(1H ,'<',4I3,'|V|',4I3,'>',F10.5)
  5   CONTINUE
      STOP
      END
      FUNCTION SLATER(NI,LI,NII,LII,NK,LK,NKK,LKK)
C
C FUNCTION SUBPROGRAM CALCULATES SLATER-INTEGRALS FOR A DELTA FORCE
C
```

```
C  USING FOUR RADIAL HARMONIC OSCILLATOR WAVEFUNCTIONS WITH QUANTUM-
C  NUMBERS (NI,LI),(NII,LII),(NK,LK),(NKK,LKK) WHERE N=0,1,2,3.
C
C  (NI,LI),(NII,LII),(NK,LK),(NKK,LKK): THE FOUR SETS OF QUANTUM
C  NUMBERS OF THE 4 RADIAL WAVE FUNCTIONS WITH N=0,1,2,3 AND
C  L=0,1,2,3,4,5,6,7
C  INT: THE SLATER-INTEGRAL (INCLUDING DIVISION BY 4*PI )
C
      COMMON/VN07CB/NU,AKNL(4,4,8),ANL(4,8),INT1(20),INT2(20)
      REAL NU
      REAL INT1,INT2
      ROM=0.
      L=LI+LII+LK+LKK
      NI1=NI+1
      NII1=NII+1
      NK1=NK+1
      NKK1=NKK+1
      LI1=LI+1
      LII1=LII+1
      LK1=LK+1
      LKK1=LKK+1
      L1=(L+1)/2
      IF(L/2.EQ.L/2.) GO TO 1
      DO 2 I=1,NI1
      DO 2 II=1,NII1
      DO 2 K=1,NK1
      DO 2 KK=1,NKK1
      L2=L1+I+II+K+KK-4
    2 ROM=ROM+AKNL(I,NI1,LI1)*AKNL(II,NII1,LII1)*AKNL(K,NK1,LK1)*AKNL(KK
     1,NKK1,LKK1)*INT1(L2)
      GO TO 3
    1 DO 4 I=1,NI1
      DO 4 II=1,NII1
      DO 4 K=1,NK1
      DO 4 KK=1,NKK1
      L2=L/2+I+II+K+KK-3
    4 ROM=ROM+AKNL(I,NI1,LI1)*AKNL(II,NII1,LII1)*AKNL(K,NK1,LK1)*AKNL(KK
     1,NKK1,LKK1)*INT2(L2)
    3 SOM=ROM
      SLATER=ANL(NI1,LI1)*ANL(NII1,LII1)*ANL(NK1,LK1)*ANL(NKK1,LKK1)*SOM
     1*(2*NU)**1.5/78.95683523
```

```
      RETURN
      END
      SUBROUTINE VN07AL
C
C  SUBROUTINE THAT CALCULATES A NUMBER OF COEFFICIENTS NECESSARY WHEN
C  CALCULATING THE SLATER INTEGRALS
C
C  FAK1(N),N=1,4: FAK1(N)=(N-1)!
C  FAK2(N),N=1,11: FAK2(N)=(2N-1)!!
C  ANL(N,L),N=1,4, L=1,8:
C     ANL(N,L)=SQRT(2**(L-N+2)*(2*NU)**(L+1/2)*(2L+2N-3)!!)/SQRT(SQRT(PI)
C               *((2L-1)!!)**2*(N-1)!)
C  AKNL(K,N,L),K=1,4, N=1,4, L=1,8:
C     AKNL(K,N,L)=(-2)**(K-1)*(N-1)!*(2L-1)!!/(K-1)!*(N-K)!*(2L+2K-3)!!)
C  INT1(I),I=1,20: INT1(I)=I!/2**(I+1)
C  INT2(I),I=1,20: INT2(I)=SQRT(PI)*(2I-1)!!/2**(2I+1/2)
C  NU=M*OMEGA/2*H-BAAR
C
      REAL NU
      REAL INT1,INT2
      COMMON/VN07CB/NU,AKNL(4,4,8),ANL(4,8),INT1(20),INT2(20)
      DIMENSION FAK1(21),FAK2(20)
      READ(1,1) NU
    1 FORMAT(F10.8)
      FAK1(1)=1.
      DO 2 IR=1,20
    2 FAK1(IR+1)=FAK1(IR)*IR
      FAK2(1)=1.
      DO 3 IR=1,19
    3 FAK2(IR+1)=FAK2(IR)*(2*IR+1)
      DO 4 N=1,4
      DO 4 L=1,8
      LNS=L+N
      F1=FAK2(LNS-1)/(FAK2(L)**2*FAK1(N))
      ANL(N,L)=SQRT(2.**(L-N+2.)*F1)
      DO 4 K=1,N
      KNS=N-K
      LKS=L+K
    4 AKNL(K,N,L)=(-2)**(K-1)*FAK1(N)*FAK2(L)/(FAK1(K)*FAK1(KNS+1)*FAK2(
     1LKS-1))
      DO 5 I=1,20
```

```
      INT1(I)=FAK1(I+1)/2**(I+1)
    5 INT2(I)=1.25331413731550*FAK2(I)/4.**I
      RETURN
      END
```

```
$ KRIS
$ ASSIGN SYS$INPUT FOR001
$ ASSIGN SYS$OUTPUT FOR003
$ R SLATMAT.EXE
0.1
10 100.00
 1 0
 1 1
 2 0
 1 2
 2 1
 1 3
 3 0
 2 2
 1 4
 1 5
```

9.5 Calculation of δ-matrix Element

Making use of the above program SLATER, the two-body matrix elements using a residual interaction of the type

$$V = -V(\text{eff})(1 - \alpha + \alpha \boldsymbol{\sigma}_1 \cdot \boldsymbol{\sigma}_2)\delta(\boldsymbol{r}_1 - \boldsymbol{r}_2) , \tag{9.14}$$

are calculated. We evaluate proton-neutron matrix elements as

$$\begin{aligned} &\langle (n_p l_p j_p)(n_n l_n j_n);\, JM|V|(n_p' l_p' j_p')(n_n' l_n' j_n');\, JM\rangle \\ &= \hat{j}_n \hat{j}_n' \hat{j}_p \hat{j}_p' (-1)^{j_p + j_p' + l_p + l_p'} (2J+1)^{-1} \langle j_p \tfrac{1}{2}, j_n - \tfrac{1}{2}|J0\rangle \langle j_p' \tfrac{1}{2} j_n' - \tfrac{1}{2}|J0\rangle \\ &\quad \times \left[\tfrac{1}{2}\left\{1 - (-1)^{j_n + j_n' + l_p + l_p'} S(j_n j_p, J) S(j_n' j_p', J)(4J(J+1))^{-1}\right\}\right. \\ &\quad \left. -\alpha\left[1 + (-1)^{l_n + l_p + J}\right]\right] V(\text{eff}) F^0 , \end{aligned} \tag{9.15}$$

with

$$S(j_p j_n, J) = (2j_p + 1) + (-1)^{j_p + j_n + J}(2j_n + 1) . \tag{9.16}$$

Here, we used the coupling order $\boldsymbol{l} + \frac{1}{2} = \boldsymbol{j}$. Note that the order of coupling the angular momenta is important.

The choice of other quantities is discussed in Sect. 9.4 where the evaluation of Slater integrals has been discussed.

A typical example of an input file is given where we have the $1h_{9/2}$ proton and $1i_{13/2}$ neutron orbital as configurations that are used in order to determine the matrix elements

$$\langle 1h_{9/2}1i_{13/2};\, JM|V|1h_{9/2}1i_{13/2};\, JM\rangle (2 \leq J \leq 11)\,. \tag{9.17}$$

Remark: The code can even be used to calculate two-body matrix elements for a δ-interaction in the specific case of $\langle j^2; JM|V|j'^2; JM\rangle$ matrix elements. In that case, the variable IDENT has to be chosen equal to 1.

PROGRAM

```
      PROGRAM DELT
      REAL NU,NORM1,NORM2
      REAL INT1,INT2
      INTEGER PAR
      COMMON/SHELL/E(20),N(20),L(20),J(20),IT(20),NUM(20)
      COMMON/PARAM/ALFA,VEFF,IDENT
      COMMON/VN06FC/FCT(40)
      COMMON/VN07CB/NU,AKNL(4,4,8),ANL(4,8),INT1(20),INT2(20)
C
C      MATRIX ELEMENTS IN ORDER <J(P)J(N),JJ|V|J(P')J(N'),JJ>
C (L+S)J IN THAT ORDER
C R(N,L) POSITIVE AT THE ORIGIN
C N-1,2,3,....AS INPUT ( N-0,1,2  in code)
C JA-2*J   DOUBLE PHYSICAL VALUE
C LA  SINGLE PHYSICAL VALUE
C V--VEFF*(1-ALPHA+ALPHA*SIGMA(P)*SIGMA(N))*DELTA FUNCTION
C NU-(M*OMEGA)/(2.*HBAR)
C
C  GOOD ESTIMATES ARE  VEFF-400;ALPHA-0.2
C IDENT-1  IDENTICAL PARTICLES
C
      CALL VN07AL
      WRITE(3,360)NU
 360  FORMAT(1H ,'NU-',F10.8/)
      READ(1,400)ALFA,VEFF,IDENT
 400  FORMAT(2F8.4,I2)
      WRITE(3,365)ALFA,VEFF
 365  FORMAT(1H ,'ALFA-',F8.4,'VEFF-',F8.4)
      FCT(1)-1.
```

```
      DO 50 I=2,40
      FCT(I)=FCT(I-1)*(I-1.)/10.
50    CONTINUE
      READ(1,31)IMAX
31    FORMAT(I2)
      DO 38 I=1,IMAX
      READ(1,32)E(I),N(I),L(I),J(I),IT(I),NUM(I)
32    FORMAT(F5.3,5I4)
      WRITE(3,33)E(I),N(I),L(I),J(I),IT(I),NUM(I)
33    FORMAT(1H ,'E=',F5.3,' N L J=',3I4,'/2  IT=',I2,'NUM=',I4)
38    CONTINUE
60    READ(1,34)I1,I2,I3,I4,JP,ICON
34    FORMAT(6I2)
      X=DELTA(I1,I2,I3,I4,JP)
      WRITE(3,35)X
35    FORMAT(1H ,'MATRIX EL.=',E20.8)
       IF(ICON.EQ.0)GOTO 60
      STOP
      END
      FUNCTION DELTA(I1,I2,I3,I4,JP)
      COMMON/PARAM/ALFA,VEFF,IDENT
      COMMON/SHELL/E(20),N(20),L(20),J(20),IT(20),NUM(20)
       REAL NORM1,NORM2
      NA=N(I1)-1
      NB=N(I2)-1
       NC=N(I3)-1
      ND=N(I4)-1
      A1=0.5
      IF(IDENT.EQ.1)A1=1.
      IF(JP.EQ.0.AND.IDENT.NE.1)GOTO 106
      IF(JP.GE.1.AND.IDENT.NE.1)GOTO 101
106    ZL=0.
      GOTO 102
101    ZL=S(J(I2),J(I1),JP)*S(J(I4),J(I3),JP)/(4.*JP*(JP+1.))
102   XL=SQRT((J(I1)+1.)*(J(I2)+1.)*(J(I3)+1.)*(J(I4)+1.))*(-1)**((J(I1)
     1+J(I3))/2+L(I1)+L(I3))/(2.*JP+1.)*
     2VN02BA(J(I2),J(I1),2*JP,1,-1,0)*
     3VN02BA(J(I4),J(I3),2*JP,1,-1,0)*
     4(A1*(1.-(-1)**((J(I2)+J(I4))/2+L(I1)+L(I3))*ZL)-
     5ALFA*(1.+(-1)**(L(I1)+L(I2)+JP)))*VEFF*
     6SLATER(NA,L(I1),NB,L(I2),NC,L(I3),ND,L(I4))
```

```
      NORM1=1.
      NORM2=1.
      IF(IDENT.NE.1)GOTO 104
       IF(I1.EQ.I2)NORM1=1./SQRT(2.)
      IF(I3.EQ.I4)NORM2=1./SQRT(2.)
 104  XL=XL*NORM1*NORM2
      DELTA=XL
      RETURN
      END
      FUNCTION S(IA,IB,IC)
C HULPFUNKTIE
      S=IA+1+(IB+1)*((-1)**((IA+IB)/2+IC))
      RETURN
      END
      FUNCTION VN02BA(J1,J2,J,M1,M2,M)
      COMMON/VN06FC/FCT(40)
       INTEGER Z,ZMIN,ZMAX,FASE
C VN02BA DENOTES THE CLEBSCH-GORDAN COEFFICIENT
C  <J1,M1,J2,M2 I J1,J2,J,M>
C
      CC=0.
      IF(M1+M2-M)20,1,20
1      IF(IABS(M1)-IABS(J1))2,2,20
2     IF(IABS(M2)-IABS(J2))3,3,20
3     IF(IABS(M)-IABS(J))4,4,20
4     IF(J-J1-J2)5,5,20
5      IF(J-IABS(J1-J2))20,6,6
6     ZMIN=0
       IF(J-J2+M1)7,8,8
7      ZMIN=-J+J2-M1
8     IF(J-J1-M2+ZMIN)9,10,10
9     ZMIN=-J+J1+M2
10    ZMAX=J1+J2-J
       IF(J2+M2-ZMAX)11,12,12
11     ZMAX=J2+M2
12     IF(J1-M1-ZMAX)13,14,14
13     ZMAX=J1-M1
14     DO 15 Z=ZMIN,ZMAX,2
      JA=Z/2+1
       JB=(J1+J2-J-Z)/2+1
       JC=(J1-M1-Z)/2+1
```

```
      JD=(J2+M2-Z)/2+1
      JE=(J-J2+M1+Z)/2+1
      JF=(J-J1-M2+Z)/2+1
      FASE=((-1)**(Z/2))
      F2=FASE
15    CC=CC+F2/(FCT(JA)*FCT(JB)*FCT(JC)*FCT(JD)*FCT(JE)*
     1FCT(JF))
      JA=(J1+J2-J)/2+1
      JB=(J1-J2+J)/2+1
      JC=(-J1+J2+J)/2+1
      JD=(J1+M1)/2+1
      JE=(J1-M1)/2+1
      JF=(J2+M2)/2+1
      JG=(J2-M2)/2+1
      JH=(J+M)/2+1
      JI=(J-M)/2+1
      JJ=(J1+J2+J+2)/2+1
      F1=J+1
      CC=SQRT(F1*FCT(JA)*FCT(JB)*FCT(JC)*FCT(JD)*FCT(JE)*
     1FCT(JF)*FCT(JG)*FCT(JH)*FCT(JI)/FCT(JJ))*CC
 20   VN02BA=CC /SQRT(10.)
      RETURN
      END
      FUNCTION SLATER(NI,LI,NII,LII,NK,LK,NKK,LKK)
C
C  FUNCTION SUBPROGRAM CALCULATES SLATER INTEGRALS FOR A DELTA FORCE
C
C  USING FOUR RADIAL HARMONIC OSCILLATRO WAVEFUNCTIONS WITH QUANTUM
C  NUMBERS (NI,LI),(NII,LII),(NK,LK),(NKK,LKK) WHERE N=0,1,2,3.
C
C  (NI,LI),(NII,LII),(NK,LK),(NKK,LKK): THE FOUR SETS OF QUANTUM
C  NUMBERS OF THE 4 RADIAL WAVEFUNCTIONS WITH N=0,1,2,3 AND
C  L=0,1,2,3,4,5,6,7
C  INT: THE SLATER-INTEGRAL (DIVISION BY 4*PI INCLUDED)
C
      COMMON/VN07CB/NU,AKNL(4,4,8),ANL(4,8),INT1(20),INT2(20)
      REAL NU
      REAL INT1,INT2
      ROM=0.
      L=LI+LII+LK+LKK
      NI1=NI+1
```

```
      NII1=NII+1
      NK1=NK+1
      NKK1=NKK+1
      LI1=LI+1
      LII1=LII+1
      LK1=LK+1
      LKK1=LKK+1
      L1=(L+1)/2
      IF(L/2.EQ.L/2.) GO TO 1
      DO 2 I=1,NI1
      DO 2 II=1,NII1
      DO 2 K=1,NK1
      DO 2 KK=1,NKK1
      L2=L1+I+II+K+KK-4
    2 ROM=ROM+AKNL(I,NI1,LI1)*AKNL(II,NII1,LII1)*AKNL(K,NK1,LK1)*AKNL(KK
     1,NKK1,LKK1)*INT1(L2)
      GO TO 3
    1 DO 4 I=1,NI1
      DO 4 II=1,NII1
      DO 4 K=1,NK1
      DO 4 KK=1,NKK1
      L2=L/2+I+II+K+KK-3
    4 ROM=ROM+AKNL(I,NI1,LI1)*AKNL(II,NII1,LII1)*AKNL(K,NK1,LK1)*AKNL(KK
     1,NKK1,LKK1)*INT2(L2)
    3 SOM=ROM
      SLATER=ANL(NI1,LI1)*ANL(NII1,LII1)*ANL(NK1,LK1)*ANL(NKK1,LKK1)*SOM
     1*(2*NU)**1.5/78.95683523
      RETURN
      END
      SUBROUTINE VN07AL
C
C  SUBROUTINE CALCULATES A NUMBER OF COEFFICIENTS NECESSARY FOR THE CAL
C  CULATIONS OF THE SLATER INTEGRALS
C
C  FAK1(N),N=1,4: FAK1(N)=(N-1)!
C  FAK2(N),N=1,11: FAK2(N)=(2N-1)!!
C  ANL(N,L),N=1,4,_L=1,8:
C     ANL(N,L)=SQRT(2**(L-N+2)*(2*NU)**(L+1/2)*(2L+2N-3)!!)/SQRT(SQRT(PI)
C                *((2L-1)!!)**2*(N-1)!)
C  AKNL(K,N,L),K=1,4, N=1,4, L=1,8:
C     AKNL(K,N,L)=(-2)**(K-1)*(N-1)!*(2L-1)!!/(K-1)!*(N-K)!*(2L+2K-3)!!)
```

```
C   INT1(I),I=1,20: INT1(I)=I!/2**(I+1)
C   INT2(I),I=1,20: INT2(I)=SQRT(PI)*(2I-1)!!/2**(2I+1/2)
C   NU=M*OMEGA/2*H-BAAR
C
      REAL NU
      REAL INT1,INT2
      COMMON/VN07CB/NU,AKNL(4,4,8),ANL(4,8),INT1(20),INT2(20)
      DIMENSION FAK1(21),FAK2(20)
      READ(1,1) NU
    1 FORMAT(F10.8)
      FAK1(1)=1.
      DO 2 IR=1,20
    2 FAK1(IR+1)=FAK1(IR)*IR
      FAK2(1)=1.
      DO 3 IR=1,19
    3 FAK2(IR+1)=FAK2(IR)*(2*IR+1)
      DO 4 N=1,4
      DO 4 L=1,8
      LNS=L+N
      F1=FAK2(LNS-1)/(FAK2(L)**2*FAK1(N))
      ANL(N,L)=SQRT(2.**(L-N+2.)*F1)
      DO 4 K=1,N
      KNS=N-K
      LKS=L+K
    4 AKNL(K,N,L)=(-2)**(K-1)*FAK1(N)*FAK2(L)/(FAK1(K)*FAK1(KNS+1)*FAK2(
     1LKS-1))
      DO 5 I=1,20
      INT1(I)=FAK1(I+1)/2**(I+1)
    5 INT2(I)=1.25331413731550*FAK2(I)/4.**I
      RETURN
      END
```

- -

```
    $ ASSIGN SYS$INPUT FOR001
    $ ASSIGN SYS$OUTPUT FOR003
    R DELTA.EXE
    0.083
      0.5000300.0000
     2
    0.000    1    5    9    1    1
    0.000    1    6   13    0    2
     1 2 1 2 2 0
```

```
1 2 1 2 3 0
1 2 1 2 4 0
1 2 1 2 5 0
1 2 1 2 6 0
1 2 1 2 7 0
1 2 1 2 8 0
1 2 1 2 9 0
1 2 1 210 0
1 2 1 211 1
```

9.6 Matrix Diagonalization

Here, we include a short program to diagonalize real symmetric matrices using the Jacobi method. In the present option, the dimensions have been put at 20×20. These numbers can be easily enlarged. However, since the Jacobi method is mainly used for small matrices, it is advised not to enlarge dimensions beyond 100×100.

Besides the subroutine VN0108 (N), we include a main program that introduces the matrix A to be diagonalized in lower diagonal form (see input example).

In the output *all* eigenvalues and corresponding eigenvectors are provided.

PROGRAM

```
      PROGRAM DIAGOT
C     TEST OF DIAGONALISATION OF EIGENVALUES AND
C     EIGENVECTORS OF SMALL SYMMETRIC REAL MATRICES
C     UP TO DIMENSION 20
      COMMON/VN01B2/A(20,20),IDIM,S(20,20)
      READ(1,1)IDIM
 1    FORMAT(I2)
      WRITE(3,2)IDIM
 2    FORMAT(1H ,' DIMENSION OF MATRIX=',I4/)
      DO 5 I=1,IDIM
      READ(1,3)(A(I,J),J=1,I)
 3    FORMAT(10F8.5/10F8.5)
      WRITE(3,4)(A(I,J),J=1,I)
 4    FORMAT(1H ,'MATRIX',10F8.5/10F8.5)
 5    CONTINUE
      DO 10 I=1,IDIM
      DO 10 J=1,I
      A(J,I)=A(I,J)
```

```
10    CONTINUE
      CALL VN0108(IDIM)
      DO 20 I=1,IDIM
      WRITE(3,6)A(I,I)
      WRITE(3,7)(S(J,I),J=1,IDIM)
7     FORMAT(1H ,'EIGENVECTOR'/10F9.5/10F9.5)
6     FORMAT(1H ,'EIGENVALUE',F9.5)
20    CONTINUE
      STOP
      END
```

```
$ASSIGN SYS$INPUT FOR001
$ASSIGN SYS$INPUT FOR003
R DIAGOT.EXE
10
 3.00000
 0.05000 2.80119
 0.00000 0.00000 2.80830
 0.00000 0.00000 0.00000 2.98050
 0.00000 0.00000 0.00000 0.00000 2.98561
 0.00000 0.00000 0.00000 0.00000 0.00000 2.99318
 0.00000 0.00000 0.00000 0.00000 0.00000 0.00000 2.07925
 0.00000 0.00000 0.00000 0.00000 0.00000 0.00000 0.00000 2.09859
 0.00000 0.00000 0.00000 0.00000 0.00000 0.00000 0.00000 0.00000 2.12036
 0.00000 0.00000 0.00000 0.00000 0.00000 0.00000 0.00000 0.00000 0.00000 2.59965
```

PROGRAM

```
   SUBROUTINE VN0108(N)
C  VN0108 DIAGONALISES SQUARE MATRICES UP TO THE ORDER OF 20*20
C
      COMMON/VN01B2/A(20,20),IDIM,S(20,20)
       IF(N-1)21,21,22
22    DO 1 I=1,N
      S(I,I)=1.
      I1=I+1
       IF(I1.GT.N)GOTO 60
       DO 1 J=I1,N
      S(I,J)=0.
1     S(J,I)=S(I,J)
60    NUL=0
      NBMAX=1000
      EPS=1.0E-3
```

```
      IN=3
      EF=10.
       DO 2 I=1,IN
      EPS=EPS/EF
       DO 2 NB=NUL,NBMAX
      IF(NB)4,5,4
4     IF(IDR+IDI)2,2,5
5      IDR=0
      IDI=0
      DO 33 II=2,N
       JI=II-1
      DO 33 JJ=1,JI
      C=A(II,JJ)+A(JJ,II)
      D=A(II,II)-A(JJ,JJ)
      IF(ABS(C)-EPS)6,7,7
6     CC=1.
      SS=0.
      GOTO 8
7      CC=D/C
       IF(CC)10,9,9
9      SIG=1.
       GOTO 12
10    SIG=-1.
12    COT=CC+SIG*SQRT(1.+CC*CC)
      SS=SIG/SQRT(1.+COT*COT)
      CC=SS*COT
      IDR=IDR+1
8     E=A(II,JJ)-A(JJ,II)
      CH=1.
      SH=0.
      IF(ABS(E)-EPS)14,13,13
13     CO=CC*CC-SS*SS
       SI=2.*SS*CC
      H=0.
       G=0.
       HJ=0.
      DO 15 K=1,N
      IF(K-II)16,15,16
16     IF(K-JJ)18,15,18
18     H=H+A(II,K)*A(JJ,K)-A(K,II)*A(K,JJ)
      S1=A(II,K)*A(II,K)+A(K,JJ)*A(K,JJ)
       S2=A(JJ,K)*A(JJ,K)+A(K,II)*A(K,II)
```

```
      G=G+S1+S2
      HJ=HJ+S1-S2
15    CONTINUE
      D=D*CO+C*SI
      H=2.*H*CO-HJ*SI
      F=(2.*E*D-H)/(4.*(E*E+D*D)+2.*G)
      IF(ABS(F)-EPS)14,17,17
17    CH=1./SQRT(1.-F*F)
      SH=F*CH
      IDI=IDI+1
14    C1=CH*CC-SH*SS
      C2=CH*CC+SH*SS
      S1=CH*SS+SH*CC
      S2=SH*CC-CH*SS
      IF(ABS(S1)+ABS(S2))20,33,20
20    DO 30 L=1,N
      A1=A(L,II)
      A2=A(L,JJ)
      A(L,II)=C2*A1-S2*A2
      A(L,JJ)=C1*A2-S1*A1
      A1=S(L,II)
      A2=S(L,JJ)
      S(L,II)=C2*A1-S2*A2
30    S(L,JJ)=C1*A2-S1*A1
      DO 31 L=1,N
      A1=A(II,L)
      A2=A(JJ,L)
      A(II,L)=C1*A1+S1*A2
31    A(JJ,L)=C2*A2+S2*A1
33    CONTINUE
2     CONTINUE
      TAUSQ=0.
      N1=N-1
      DO 150 I=1,N1
      II=I+1
      DO 150 J=II,N
150   TAUSQ=TAUSQ+(A(I,J)+A(J,I))**2
      RETURN
21    S(1,1)=1.
      RETURN
      END
```

9.7 Radial Integrals Using Harmonic Oscillator Wave Functions

Here, we evaluate the radial integrals, using the harmonic oscillator wave functions, as discussed in (9.10) of this chapter. The integral, however, is evaluated using dimensionless variables.

The function that evaluates these integrals is
VN0105(N1, L1, LAMB, N2, L2)
where N1, L1 (N2, L2) denote the radial (starting at N1 = 0, 1, ...) and the orbital quantum number, respectively.

The integral becomes

$$\int u_{n'l'}(x)x^{\lambda}u_{nl}(x)dx = \left[\frac{\Gamma(n+1)\Gamma(n'+1)}{\Gamma(n+1+t-\tau)\Gamma(n'+1+t-\tau')}\right]^{1/2} \tau!\tau'! \times \sum_{\sigma} \frac{\Gamma(t+\sigma+1)}{\sigma!(n-\sigma)!(n'-\sigma)!(\sigma+\tau-n)!(\sigma+\tau'-n')!}, \tag{9.18}$$

with

$$\begin{aligned} \tau &= \tfrac{1}{2}(l'-l+\lambda) \\ \tau' &= \tfrac{1}{2}(l-l'+\lambda) \\ t &= \tfrac{1}{2}(l+l'+\lambda+1) . \end{aligned} \tag{9.19}$$

The sum on σ is restricted via the condition

$$\left.\begin{matrix} n \\ n' \end{matrix}\right\} \geq \sigma \geq \left\{\begin{matrix} n-\tau \\ n'-\tau' \end{matrix}\right. . \tag{9.20}$$

Some integrals are

$$\begin{aligned} \langle n,l-1|r|n,l\rangle &= (n+l+\tfrac{1}{2})^{1/2}, \\ \langle n-1,l+1|r|n,l\rangle &= n^{1/2}, \\ \langle n+1,l-1|r|n,l\rangle &= (n+1)^{1/2}, \\ \langle n,l+1|r|n,l\rangle &= (n+l+\tfrac{3}{2})^{1/2}, \end{aligned} \tag{9.21}$$

$$\begin{aligned} \langle n,l|r^2|n,l\rangle &= 2n+l+\tfrac{3}{2} = N+\tfrac{3}{2}, \\ \langle n-1,l|r^2|n,l\rangle &= (n(n+l+\tfrac{1}{2}))^{1/2}, \\ \langle n,l-2|r^2|n,l\rangle &= ((n+l+\tfrac{1}{2})(n+l-\tfrac{1}{2}))^{1/2}, \\ \langle n+1,l-2|r^2|n,l\rangle &= (n(n-1))^{1/2}, \\ \langle n+2,l-2|r^2|n,l\rangle &= ((n+1)(n+2))^{1/2}. \end{aligned} \tag{9.22}$$

PROGRAM

```
      FUNCTION VN0105(N1,L1,LAMB,N2,L2)
C     VN0105(N1,L1,LAMB,N2,L2) =(N1,L1/RHO**LAMB/N2,L2)
C     WHERE /N,L) IS THE RADIAL PART OF THE WAVE FUNCTION
C     ASSOCIATED WITH THE SYMMETRICAL HARMONICAL
      OSCILLATOR POTENTIAL AND
      WHERE RHO IS THE REDUCED R-COORDINATE
 1    NUL=0
      IF(N2-N1)2,3,3
2      NA=N1
      N1=N2
      N2=NA
      NA=L1
      L1=L2
      L2=NA
 3    S=0.
       NN1=2*N1
       NN2=2*N2
       DO 9 K=1,N1,1
      KK=2*K
       T1=VN0107(NN1)/VN0107(KK)
      M1=2*(N1-K+1)
      M2=2*(N2-N1+K)
       T2=VN0107(NN2)/(VN0107(M1)*VN0107(M2))
       M3=2*(N1-K+1)+L1+L2+LAMB+1
       T3=VN0107(M3)
      T4=1.
      T5=1.
      IF(K-1)31,5,31
31     AL=L1-L2-LAMB
      KA=K-2
       DO 4 I=NUL,KA,1
      T4=T4*(AL/2.+FLOAT(I))
4      CONTINUE
5      IF((K.EQ.1).AND.(N1.EQ.N2))GOTO 7
      BL=L2-L1-LAMB
      KB=K-2+N2-N1
       DO 6 J=NUL,KB,1
      T5=T5*(BL/2.+FLOAT(J))
6      CONTINUE
7      S=S+(T1*T2*T3*T4*T5)*10.**(FLOAT(LAMB)/2.
```

```
     1+FLOAT(-2*K+N1-N2+2))
 9    CONTINUE
11     N5=2*(N1+L1)+1
       N6=2*(N2+L2)+1
10     SQ=VN0107(NN1)*VN0107(NN2)*VN0107(N5)*VN0107(N6)
       SQ1=SQRT(SQ)
12     VN0105=(S*FLOAT((-1)**(N1+N2)))/SQ1
      RETURN
       END
      FUNCTION VN0107(N)
C     VN0107(N)=GAMMA(N/2)/(10**(N/2-1))
C        WHERE GAMMA IS THE WELL KNOWN GAMMA-FUNCTION AND
C        WHERE N IS AN INTEGER
C     FCT(M)=(M-1)!/(10**(M-1))
C       WHERE M IS A POSITIVE INTEGER
      COMMON/VN06FC/FCT(40)
10     I=0
  1   K=N/2
      IF(K*2-N)3,2,12
  2   IF(N)15,15,21
 21   GA=FCT(K)
      IF(I)90,90,13
 3    K=N/2
      IF(K)31,6,31
31     K1=2*K
       GA=(FCT(K1)*SQRT(31.4159))/(FCT(K)*2.**(2*K-1))
      IF(I)90,90,13
  6   GA=SQRT(31.4159)
       IF(I)90,90,13
 12   I=1
      N=N*(-1)+2
       GOTO 3
13     G=GA
       N=N*(-1)+2
       FA=(-1)**((1-N)/2)
       GA=(FA*31.4159)/G
      GOTO 90
90    VN0107=GA
15     RETURN
       END
```

9.8 BCS Equations with Constant Pairing Strength

In the present, simple BCS code, we solve the two coupled algebraic equations,

$$\begin{aligned} &\sum_j \left[(\varepsilon_j - \lambda)^2 + \Delta^2\right]^{-1/2} = \frac{2}{G} , \\ &\sum_j \left(j + \frac{1}{2}\right) \left[1 - \frac{\varepsilon_j - \lambda}{[(\varepsilon_j - \lambda)^2 + \Delta^2]^{1/2}}\right] = n , \end{aligned} \tag{9.23}$$

where G and n are input values, G the pairing strength and n the number of identical valence nucleons outside the closed-shell configurations. For G, we can use estimates as discussed in Sect. 7.5.1.

Here, in (9.23), the sum on j goes over *all* single-particle orbitals.

From solving the BCS equations which is done in the program BCS one gets λ and Δ as output values and subsequently, one easily derives v_j^2, and u_j^2 and E_j.

In solving for the BCS equations, a typical input set is given with

i) STREN, NUM1, IMAX
STREN: the pairing strength G
NUM1: the number of valence nucleons n
IMAX: the number of single-particle orbitals considered.

ii) $J(I), E(I)$
We give as input $2 \times j$, ε_j for the different single-particle orbitals with j the angular momentum and εj the single-particle energy corresponding to given j.

iii) FER, DEL
We have to give as input an estimate of the Fermi level energy λ and of the gap Δ. If, incidentally, the increments $\delta\lambda$ and $\delta\Delta$ are such that the solution diverges (i.e. the new value of Δ becomes negative), a new set of starting values for λ, Δ has to be used as an input.

The present example starts from the set

$$2f_{7/2},\ 1h_{9/2},\ 1i_{13/2},\ 3p_{3/2},\ 2f_{5/2} \text{ and } 3p_{1/2} ,$$

respectively, with energies −0.075, 0.599, 1.392, 3.156, 3.530 and 4.700 (MeV), respectively.

PROGRAM

```
      DIMENSION E(20),J(20),VKW(20),EE(20)
300   READ(1,1)STREN,NUM1,IMAX,IWRITE,ICON1,ICON2
 1    FORMAT(F6.3,5I2)
      WRITE(6,51)STREN,NUM1,IMAX
 51   FORMAT(1H ,'STRENGTH=',F6.3,'NUMBER',I4,'LEVELS',I3/)
      IF(ICON1.EQ.1)GOTO 200
      DO 3 I=1,IMAX
      READ(1,4)J(I),E(I)
      WRITE(6,5)J(I),E(I)
 4    FORMAT(I2,F6.3)
 5    FORMAT(1H ,I2,'/2   E =',F6.3/)
 3    CONTINUE
 200  READ(1,6)FER,DEL
 6    FORMAT(2F6.3)
      KT=0
      FX=0.
      FY=0.
50    F1=-2.0/STREN
      F2=(-1)*FLOAT(NUM1)
      F1L=0.
      F2L=0.
      F1D=0.
      F2D=0.
      DO 7 I=1,IMAX
      Z1=((E(I)-FER)**2+DEL**2)**0.5
      Z2=Z1*((E(I)-FER)**2+DEL**2)
      F1D=F1D-DEL*(J(I)+1.)/2./Z2
      F1L=F1L+(E(I)-FER)*(J(I)+1.)/2./Z2
      F2D=F2D+(E(I)-FER)*DEL*(J(I)+1.)/2./Z2
      F2L=F2L+DEL**2*(J(I)+1.)/2./Z2
      F1=F1+(J(I)+1.)/2/Z1
      F2=F2+(1.-(E(I)-FER)/Z1)*(J(I)+1.)/2
      IF(IWRITE.EQ.0)GOTO 7
      WRITE(6,52)Z1,Z2,F1D,F1L,F2D,F2L,F1,F2
52    FORMAT(1H ,8E16.6)
7     CONTINUE
      A1=F1L*F2D-F1D*F2L
      A2=F2*F1D-F1*F2D
      A3=F1*F2L-F2*F1L
      DLAM=A2/A1
```

```
      DDEL=A3/A1
      FER=FER+DLAM
      DEL=DEL+DDEL
      IF(IWRITE.EQ.0)GOTO 11
      WRITE(6,53)A1,A2,A3,DLAM,DDEL,FER,DEL,KT
 53   FORMAT(1H ,7E16.6,I4/)
 11   PREC=0.001
      IF(ABS(F1).LT.PREC.AND.ABS(F2).LT.PREC)GOTO 40
      KT=KT+1
      FVER1=ABS(FX-F1)
      FVER2=ABS(FY-F2)
      FX=F1
      FY=F2
      IF(IWRITE.EQ.0)GOTO 13
      WRITE(6,54)FVER1,FX,F1,FVER2,FY,F2
 54   FORMAT(1H ,6E16.6/)
 13   IF(DEL.LT.0)GOTO 200
      GOTO 50
C  CALCULATE  VKW  AND QP ENERGY FROM FER AND DEL
C
 40   WRITE(6,56)FER,DEL
 56   FORMAT(1H ,'FERMI LEVEL',F8.5,'GAP   ',F8.5///)
      DO 15 I=1,IMAX
      ZA=((E(I)-FER)**2+DEL**2)**0.5
      ZB=E(I)-FER
      VKW(I)=0.5*(1.-ZB/ZA)
      EE(I)=ZA
      WRITE(6,55)VKW(I),EE(I)
 55   FORMAT(1H ,'VKW=',F8.5,'1QP ENERGY=',F8.5,I2,'/2 '//)
 15   CONTINUE
      IF(ICON2.EQ.1.AND.ICON1.EQ.1)GOTO 80
      IF(ICON2.EQ.0)GOTO 300
 80   STOP
      END
```

```
$ ASSIGN SYS$INPUT FOR001
$ ASSIGN SYS$OUTPUT FOR006
$ R BCS
 0.195 1 6 1
 7-0.075
 9 0.599
```

```
13 1.392
 3 3.156
 5 3.530
 1 4.700
 1.000 1.000
 1.000 0.200
```

9.9 Evaluation of Particle-Hole Matrix Elements

Using techniques from Chap. 5, where both particle and hole creation operators were introduced, one is able to relate the particle-hole angular momentum coupled matrix element to the particle-particle angular momentum coupled matrix elements. The theoretical expression relation for both becomes

$$\begin{aligned}\langle j_a j_b^{-1}; JM|V|j_c j_d^{-1}; JM\rangle &= -\sum_{J'}(2J'+1)\begin{Bmatrix} j_a & j_d & J' \\ j_c & j_b & J \end{Bmatrix} \\ &\times \langle j_a j_d; JM'|V|j_c j_b; J'M'\rangle \ , \end{aligned} \tag{9.24}$$

and this evaluated in the simple code PANDYA.

On input, the particle-particle matrix elements are needed and are placed into the array RMAT(I), starting from a minimal to a maximal value of J'. Then, the quantum numbers j_a, j_b, j_c, j_d and J and a control variable are read as input i.e.

a) JMIN,JMAX
b) RMAT(I): the two-body particle-particle matrix elements
c) JA,JB,JC,JD,JJ,ICON: the angular momentum quantum numbers (twice the physical values), total angular momentum and a control variable.

9.10 Particle-Core Coupling

Before discussing the code in some detail, we give the basic physical ideas behind the coupling of particles and collective core vibrational excitations (9.10.1). We discuss the part evaluating the energy eigenvalues and eigenvectors (9.10.2), the part for evaluating transition rates and static moments (9.10.3) as well as some features of the subroutine library (9.10.4).

This application is a much larger-scale computer application, making contact between typical shell-model and collective phenomena, and is an application with many possibilities. We shall discuss the main line of particle(or – hole or even one-quasi – particle)-core coupling, although the code contains even more extended options such as: treating the core in an anharmonic way, using a two-core system for treating $1p$ and $2p - 1h$ (or $1h$ and $2h - 1p$) excitations in a single nucleus. Extensive discussions on the particle-core coupling model are presented in Bohr and Mottelson (1975), where many references to the original literature are also given.

9.10.1 Formalism

In the concept of particle-vibration coupling (PVC), an odd-particle (-hole or quasi-particle) moving in the field of a vibrating, collective nucleus becomes coupled to these collective vibrations in a very natural way. In describing the nucelear surface through a multipole expansion as

$$R(\theta,\varphi) = R_0\left(1 + \sum_{\lambda,\mu} \alpha^*_{\lambda,\mu} Y^\mu_\lambda(\theta,\varphi)\right) + O(\alpha^2)\,, \tag{9.25}$$

the time-dependent, collective variables $\alpha^*_{\lambda,\mu}$ will form the basic quantities for describing small amplitude oscillations around the spherical equilibrium shape.

The potential felt by a nucleon moving in this vibrating field at the point (r,θ,φ) then becomes

$$\overline{V}(r,\theta,\varphi) = V(r) - r\,dV/dr.\sum_{\lambda,\mu} \alpha^*_{\lambda,\mu} Y^\mu_\lambda\,, \tag{9.26}$$

where higher-order terms are neglected in expanding around the point r.

Therefore, the total Hamiltonian describing the collective and the single-particle motion, in interaction, can be depicted as (for weak and intermediate coupling strength)

$$\boldsymbol{H} = \boldsymbol{H}_{\text{s.p.}} + \boldsymbol{H}_{\text{coll.}} + \boldsymbol{H}_{\text{int.}}\,, \tag{9.27}$$

with a collective Hamiltonian

$$\boldsymbol{H}_{\text{coll.}} = 1/2.\sum_{\lambda,\mu} \left(B_\lambda |\dot{\alpha}_{\lambda\mu}|^2 + C_\lambda |\alpha_{\lambda,\mu}|^2\right)\,. \tag{9.28}$$

Using the standard quantization procedure, one can introduce creation and annihilation operators as

$$b^+_{\lambda,\mu} = \sqrt{\frac{C_\lambda}{2\hbar\omega_\alpha}}\left\{\alpha_{\lambda,\mu} - \frac{i\omega_\lambda}{C_\lambda}(-1)^\mu \pi_{\lambda,-\mu}\right\}\,,$$

$$b_{\lambda,\mu} = \sqrt{\frac{C_\lambda}{2\hbar\omega_\lambda}}\left\{(-1)^\mu \alpha_{\lambda,-\mu} + \frac{i\omega_\lambda}{C_\lambda}\pi_{\lambda,\mu}\right\}\,. \tag{9.29}$$

Thereby, the collective Hamiltonian can be rewritten as

$$\boldsymbol{H}_{\text{coll.}} = \sum_{\lambda,\mu} \hbar\omega_\lambda \left(b^+_{\lambda\mu} b_{\lambda,\mu} + 1/2\right)\,, \tag{9.30}$$

and, subsequently, the coupling Hamiltonian becomes

$$\boldsymbol{H}_{\text{int.}} = -\sum_{\lambda,\mu}\left(\frac{\pi}{2\lambda+1}\right)^{1/2} \xi_\lambda \hbar\omega_\lambda \left(b_{\lambda\mu} + (-1)^\mu b^+_{\lambda,-\mu}\right) Y^\mu_\lambda(\hat{r})\,, \tag{9.31}$$

where we use the shorthand notation for the coupling strength parameters

$$\xi_\lambda = \langle r\,dV(r)/dr\rangle_{\alpha_{\lambda,\mu=0}} \left(\frac{2\lambda+1}{2\pi\hbar\omega_\lambda C_k}\right)^{1/2}\,. \tag{9.32}$$

and

$$\hbar\omega_\lambda = \hbar\left(C_\lambda/B_\lambda\right)^{1/2} . \tag{9.33}$$

In constructing now a basis of $\boldsymbol{H}_{\text{coll.}} + \boldsymbol{H}_{\text{s.p.}}$ as the basis for diagonalizing the interaction Hamiltionian, and denoting the core collective eigenstates by the number of quanta (N) and its total angular momentum (R), characterizing the single-particle motion by j ($\equiv n, l, j$), we obtain the following energy eigenvalues, calling $\boldsymbol{H}_0 = \boldsymbol{H}_{\text{coll.}} + \boldsymbol{H}_{\text{s.p.}}$

$$\boldsymbol{H}_0|j(NR); JM\rangle = \left\{\varepsilon_j + \sum_\lambda N_\lambda \hbar\omega_\lambda\right\}|j(NR); JM\rangle , \tag{9.34}$$

where we have used a shorthand notation for the full collective eigenstate. In the event of our only considering quadrupole and octupole vibrations ($\lambda = 2$ and $\lambda = 3$), which is the case in the computer code, the collective eigenstates read

$$\left|(N_{\text{q}}, R_{\text{q}})\,(N_{\text{o}}, R_{\text{o}}); RM\right\rangle , \tag{9.35}$$

where $N_{\text{q}}, R_{\text{q}}(N_{\text{o}}, R_{\text{o}})$ denote the number of quadrupole (octupole) phonons and collective quadrupole (octupole) angular momentum with, moreover, $\boldsymbol{R} = \boldsymbol{R}_q + \boldsymbol{R}_0$.

The non-diagonal matrix elements of $\boldsymbol{H}_{\text{int.}}$ (9.31) can now couple various basis states, although the final energy matrix becomes rather sparse. This is caused by the selection rules $|\Delta N_{\text{q}}| = 1$, $|\Delta N_{\text{o}}| = 1$ for non-vanishng matrix elements. For quadrupole vibrations only, the various elements of the energy matrix read

$$\begin{aligned}
&\langle j', N'R'; JM|\boldsymbol{H}_0|j, NR; JM\rangle \\
&\quad = \{\varepsilon_j + (N + 5/2)\hbar\omega_2\}\,\delta_{jj'}\delta_{NN'}\delta_{RR'} ,
\end{aligned} \tag{9.36}$$

$$\begin{aligned}
&\langle j', N'R'; JM|\boldsymbol{H}_{\text{int.}}|j, NR; JM\rangle \\
&\quad = -1/2.\xi_2\hbar\omega_2(-1)^{J+1/2}\hat{j}\hat{j}'\begin{Bmatrix} j & R & J \\ R' & j' & 2 \end{Bmatrix}\begin{pmatrix} j' & 2 & j \\ -1/2 & 0 & 1/2 \end{pmatrix} \\
&\qquad \times \left\{(-1)^{R'}\langle N'R'\|b_2^+\|NR\rangle + (-1)^R\langle NR\|b_2^+\|N'R'\rangle\right\} .
\end{aligned} \tag{9.37}$$

The expression become somewhat more complicated if octupole and quadrupole vibrations are also considered, but there is no change in the basic structure. The extension to the coupling of a single-hole (or a quasi-particle) configuration only slightly changes the above expressions. The energy eigenvalues ε_j then becomes $\tilde{\varepsilon}_j$ or E_j $\left(\equiv \sqrt{(\varepsilon_j - \lambda)^2 + \Delta^2}\right)$. In the case of quasi-particle coupling, various pairing factors are introduced, but this is discussed in Bohr and Mottelson (1975). There, the evaluation of the boson reduced matrix elements, needed as input numbers, are also discussed.

In Sects. 9.10.2, 3 and 4, the practical implementation is outlined briefly (major input and output structures).

Of course, electromagnetic properties need to be evaluated and, besides the well-known single-particle EL and ML operators discussed at length in Chap. 4, collective contributions are also present. For a uniform charge density of the oscillating nucleus, one obtains for the EL operators

$$O(\text{el.}, LM)_{\text{coll.}} = \int \rho_{\text{charge}}(\boldsymbol{r}) r^L Y_L^M(\hat{r})\, d\boldsymbol{r} \;, \tag{9.40}$$

$$= 3/4\pi . Ze\, R_0^L \sqrt{\frac{\hbar\omega_L}{2C_L}} \left(b_{L,M}^{+} + (-1)^M b_{L,-M} \right) \;, \tag{9.41}$$

where higher-order terms are again left out. The evaluation of the collective, magnetic part is more involved (Bohr, Mottelson (1975)) and we only give, without derivation, the result for the collective dipole part as

$$O(\text{mag.}, 1M)_{\text{coll.}} = g_R R_M \mu_N \;. \tag{9.42}$$

9.10.2 Description of the Core-coupling Code: Energy Eigenvalues and Eigenvectors

In this section, the major input and output options are discussed. A number of input variables have 'default' values that are not relevant in carrying out particle (-hole, one quasi-particle) core coupling. Those variables or output lines then refer to either (i) the use of an anharmonic core, (ii) the use of a two-core coupling calculation and are indicated by an asterisk. At the end, some of the major quantities are discussed.

INPUT

(1)IKIES,MULT (2I2)

(2)IS,IT,IN (3I2)

Here, the boson reduced matrix elements are read in as input for calculating up to three quadrupole and up to two octupole phonons. Boson coefficients-of-fractional parentage (cfp) are read in as

$$\langle \| b^{+} \| \rangle = \text{IS}\sqrt{\text{IT}/\text{IN}}$$

*(3)IDIAGONAL (Default value of '0'). The value 1 can allow for anharmonicities to be taken into account.

(4)IAAN,IKI,IKIA,IREAD,ISPIN,IANHA (6I2)

IKI,IKIA = 1 standard core-coupling calculation,

IKI,IKIA = 2 two-core coupling calculation,

IAAN: Total number of single-particle (-hole) orbitals,

IREAD = 1 Wave functions for different J values are read (via unit 14) on disk space using the direct access mode,

ISPIN ($2J$) Upper spin value (as double- valued) until the point where wave functions are read to disk space

(5)IPRIN, IBASIS,NOG,ITEST,MEER (5I2)

These are test variables that can generate control output for testing. Normally, all values are set to '0' as the default value.

(6)IORDEN,KIES,ISINP,NORM,MEER (5I2)

IORDEN = 1 Energy eigenvalues are put in increasing order. Important for easy use,

KIES = 0 Default value, otherwise extra control output is produced,

ISINP = 0 Default value. If put to '1', then only the first component for each J wave function is given on output,

NORM = 0 Default value. If put to '1', then eigenvalues, for each J value, are separately normalized to 0.0,

NUMBER: The number of eigenvectors given on output (Default value of '5').

*(7)IWAF

IJI(I),ILI(I),I=1,IWAF

AMPL(I), I=1,IWAF

Only for the use of a two-core coupling calculation. This option is not discussed in detail in the present program comment.

(8)J(IP),L(IP),IP=1,IAAN (24I2)

For all orbitals used in the particle-core coupling calculation, $2j$ (J(IP) and l (L(IP)) are read. It is easy to organize the levels with increasing energy (e.g. $2d_{5/2}$, $1g_{7.2}$, ... for the light Sb nuclei).

(9)E(IP),VKW(IP),IP=1,IAAN (16F5.3/8F5.3)

The single-particle (or single-hole or one-quasi particle) energies in the sequence corresponding to the input of line (8). Relative values are most easily used. The character (particle, hole, quasi-particle) is characterized by VKW(IP) being 0, 1, or $0 \leq v^2 \leq 1$, respectively.

*(10)EUNP(1),EUNP(2) (2F5.3)

Only for use in a two-core coupling calculation.

(11)ELIM,KSI2(1),KSI2(2),HW2(1),HW2(2) (F6.3,4F5.3)

ELIM: The limiting energy, putting a truncation in the construction of the basis for a given J value. Configurations are considered if

$$\varepsilon_j + N_q \hbar\omega_2 + N_o \hbar\omega_3 \leq \text{ELIM} ,$$

KSI2,HW2: Coupling strength ξ_2 and phonon energy $\hbar\omega_2$ used in the core-coupling calculation. Only the values for (1) are relevant. The other set (2) goes with a second core, if used

(12)ELIM,KSI3(1),KSI3(2),HW3(1),HW3(2) (F6.3, 4F5.3)

See the explanation for input line (11), but now for the octupole coupling with strength ξ_3 and $\hbar\omega_3$ values.

(13)OPHON(1),OPHON(2),QPHON(1),QPHON(2) (4I2)

OPHON: The number of octupole phonons considered in the calculations, increased by 1 (Default value of '3')

QPHON: The number of quadrupole phonons considered in the calculations, increased by 1 (Default value of '4')

The index (1) refers to the regular core-coupling; (2) is for use only with a two-core coupling calculation. This also holds for input lines (14) and (15).

(14)IPI(1),IPI(2),IPF(1),IPF(2) (4I2)

IPI(1),IPF(1): The number (rank number) corresponding to the initial and final orbital considered in the core-coupling calculation.

(15)IOCT(1),IOCT(2),IQUAD(1),IQUAD(2) (4I2)

IOCT(1),IQUAD(1) = 1 (Default value). This means that both octupole and quadrupole phonon excitations are taken into account. Putting these values to '0' excludes a certain multipolarity from the basis construction.

*(16)ANHAR(1),ANHAR(2) (2F6.3)

Quantities relevant to the use of anharmonic quadrupole vibrations.

(17)I,IPAR

I(2J): Angular momentum for particle-core coupling calculation,

IPAR: Parity (+1, −1) for the angular momentum J^π,

The set $(0,0)$ stops the calculation in a normal and correct way.

OUTPUT

On output, the code gives, besides an unambigous repetition of input variables and control variables the basis configuration space for each J^π value. The basis is given in the order

$$\left|(N_o R_o, N_q R_q)\, R, j; J\right\rangle ,$$

where o(q) means octupole (quadrupole), R denotes the total collective angular momentum, j the single-particle angular momentum and J the total angular momentum. If the maximal values of N_q and N_o are (as default) 3 and 2, the various combinations N_q, R_q are $(00; 12; 20, 22, 24; 30, 32, 33, 34, 36)$ and for N_o, R_o $(00; 13; 20, 22, 24, 26)$.

Then, all eigenvalues are printed (in units MeV) on an absolute scale. This means that to construct the excitation spectrum, values relative to the lowest eigenvalue, over all J^π values, have to be constructed.

The corresponding eigenvectors (with a default value of '5') are given and the various amplitudes are ordered in the same way as the basis configurations. If a single-particle contribution for the given J^π value exists, this is always the first amplitude, corresponding to the $|(00,00)0, j; J = j\rangle$ basis state.

We now briefly discuss how, for a given region, the basic parameters such as $\xi_\lambda, \hbar\omega_\lambda$ can be determined. For $\hbar\omega_\lambda$, one can start from the 0_1^+ to 2_1^+ or 0_1^+ to

3_1^- separation (for quadruple and octuple phonons, respectively) in the adjacent even-even nucleus. The coupling strength ξ_λ can be estimated from known $B(E\lambda)$ values in the adjacent even-even nucleus. Using a harmonic approximation, one finds that

$$B(E\lambda; 0 \rightarrow \lambda) = \left(3/4\pi . Ze . R_o{}^\lambda\right)^2 \beta_\lambda^2 ,$$

and, since one has

$$\xi_\lambda = \langle r\, dV/dr \rangle_{\alpha_{\lambda\mu=0}} \beta_\lambda \frac{1}{\sqrt{\pi\hbar\omega_\lambda}} ,$$

one obtain a good starting estimate as

$$\xi_\lambda = \langle r\, dV/dr \rangle \frac{B(E; 0 \rightarrow \lambda)^{1/2}}{3/4\pi . ZeR_o{}^\lambda \cdot \hbar\omega_\lambda \sqrt{\pi}} .$$

One can determine $\langle \ldots \rangle$ numerically. A good, average value for medium-heavy and heavy nuclei amounts to $\simeq 50\,\text{MeV}$. Some variation in the ξ_λ values is allowed for an optimal search, to obtain a good fit to the experimental spectrum.

9.10.3 Transition Rate and Moment Calculation

In evaluating magnetic and electric moments and transition rates, starting from the wave functions as determined (see Sect. 9.10.2), a large part of the input variables is identical to the description (and default values) as given in that section. Therefore, if nothing important changes, the input list will be given without explanation.

INPUT

(1)IKIES,MULT (2I2)

(2)IS,IT,IN (3I2)

(3)IAAN,IKI,IKIA,IREAD,ISPIN,IBCS (6I2)

If IREAD=1, wave functions are read (for a given J^π value) from disk space at the appropriate moment,

IBCS: Default value ‘1’. This includes the specific pairing factors for the EL and ML matrix element calculations.

(4)IPRIN,GRENS (I1,F6.4)

IPRIN = 0 (Default values). The choice ‘1’ produces much extra output,

GRENS: This is a limiting number such that, if the product of the initial and final state amplitudes $|c^i(j, NR; J_i) c^f(j', N'R'; J_f)| <$ GRENS, this contribution is neglected in the calculation of the particular matrix element. One has to take convergence into consideration when putting in too large a value for the variable GRENS.

*(5)IWAF (12)

IJI(II),ILI(II),II=1,IWAF (20I2)

AMPL(II),II=1,IWAF (10F6.3)

Only for use in a two-core coupling calculation.

(6)N(IP),J(IP),L(IP),IP=1,IAAN (36I2)

The quantum numbers for the various orbitals (conform to the order of those in the energy calculation) $n, 2j, l$. It should be noted that the radial quantum number starts from $n = 1, 2, \ldots$

(7)E(IP),VKW(IP),IP=1,IAAN (16F5.3/8F5.3)

*(8)EUNP(1),EUNP(2) (2F5.3)

(9)ELIM,HW2(1),HW2(2),HW3(1),HW3(2) (F6.3,4F5.3)

(10)OPHON(1),OPHON(2),QPHON(1),QPHON(2) (4I2)

(11)IPI(1),IPI(2),IPF(1),IPF(2) (4I2)

(12)IOCT(1),IOCT(2),IQUAD(1),IQUAD(2) (4I2)

Input for the electromagnetic operators now occurs.

(13)GS,GL,GR (3F6.3)

The intrinsic, orbital and collective g-factors. For the collective part, we use the standard estimate of $g_R = Z/A$. Also, for the intrinsic part, a typical quenching relative to the free values i.e. $g_s^{\text{eff.}} = 0.7\, g_s^{\text{free}}$, is considered.

(14)Z,AMA,EP,C2(1),C2(2),C3(1),C3(2),ICON

Z: The atomic charge,

AMA: The atomic mass number,

EP: The effective charge of the single-particle (-hole or quasi-particle). Quite standard values are $e_p^{\text{eff.}} = 1.5\, e$, $e_n^{\text{eff.}} = 0.5\, e$. The collective charges, given by the C2 and C3 values, refer to the rigidity of the core and good estimates can be obtained starting from the known $B(E)$ values, since one has the relations

$$B\,(E\lambda; 0 \rightarrow \lambda) = \left(3/4\pi . ZeR_0^{\lambda}\right)^2 \beta_\lambda^2\,,$$

and

$$\beta_\lambda = \left((2\lambda + 1)\hbar\omega_\lambda / 2C_\lambda\right)^{1/2}\,.$$

Otherwise, knowing the ξ_λ value, a good estimate is achieved through the relation

$$\beta = \xi_\lambda \sqrt{\pi}\, \hbar\omega_\lambda / \langle r\, dV/dr \rangle$$

ICON: Default value '0'

(15)LAM,LAPPA,MOM,ICON1,ICON2 (5I2)

LAM = L, the multipolarity of the moment or transition,

LAPPA = +1(−1), for electric (magnetic) moments or transitions,

MOM = 1, for a static moment calculation, '0' for transition rates,

ICON1,ICON2 control variables. The standard options are '1,0' but, to end a sequence of transitions or moment calculations, the option '0,0' has to be used. An example will be given later on.

(16)JII,IPARI,NRI (3I2)

Denotes the angular momentum ($2J$), the parity and the rank number for the initial state e.g. 9 1 1 denotes the first $9/2_1^+$ state.

If MOM=1, then the two input lines are not considered in the calculation.

(17)JFF,IPARF,NRF (3I2)

The same as for input line (16), but now for the final state.

(18)GAMMA (F5.3)

The gamma energy (in MeV) for the transition between initial and final state. Experimental energies are preferred if one wishes to compare lifetimes in the calculation.

An example for a sequence of an $E2$ transition and an $M1$ (with $E_\gamma = 1\,\text{MeV}$) between the $9/2_1^+$ and 7.2_1^+ levels are on input (for the lines 15–18):

```
2 1 0 1 0
9 1 1
7 1 1
1.000
1-1 0 0 0
9 1 1
7 1 1
1.000
```

OUTPUT

On output, the following points are important, before even discussing details of the output format.

Only $M1$ transitions (and moments) for magnetic properties are active at present and $E1, \ldots E3$ for the electric properties. Extension to $E4, \ldots$ is quite trivial. This is not the case for the other magnetic $M\lambda$ operators, since the general, magnetic collective part is, at present, not built in.

First, after the standard recapitulation of input values, the wave function for initial ($J_i^{\pi_i}$) and final ($J_f^{\pi_f}$) states are printed. This gives an easy control if the correct wave functions were read from disk space.

Then comes

S.P.CONTR. :

COLL.EL." -1 :

COLL.EL." -2 :

COLL.MAG. :

TOTAL MATRIX EL. :

MOMENT (in units μ_N of e.fm^L),

TRANSITIONS

MAG.DIPOLE TR. PROB. (s^{-1})	ELECTRIC L TR.PROB. (s^{-1})
LIFETIME (s)	LIFETIME (s)
REDUCED TR. PROB. $(\mu_N)^2$	REDUCED L TR.PROB. (e^2.fm^{2L}).

The single-particle contribution represents that part coming from the initial and final parts of the wave functions having $(N, R) \equiv (0, 0)$. The collective contribution is the total matrix element, excluding the contribution coming from the $(N, R) \equiv (0, 0)$ part. Here, the main effect normally comes from collective one-photon $2^+ \to 0^+$ (or $3^- \to 0^+$) quadrupole (octupole) transitions. The collective part-2 only becomes relevant for a two-core coupling calculation. Similar comments apply for the collective, magnetic dipole part. In all of these, the effective charges and g-factors are already included. The total matrix element is simply the sum of the above values.

For the electric matrix elements $\langle J_f^{\pi_f} \| O(\text{el.}, L) \| J_i^{\pi_i} \rangle$ and $B(EL)$, the units are e.fm^L and e^2.fm^{2L} for the reduced transition probabilities. For the magnetic dipole matrix element, the untis are μ_N and μ_N^2 for the reduced $M1$ transition probability. The total electric (magnetic) transition probability (s^{-1}) $T(EL)$ and $T(ML)$ includes the energy and remaining factors (see Chap. 4) and the lifetime (s) ($\ln 2/T(M1)$) or $\ln 2/T(E2)$).

9.10.4 Library for Particle-Core Coupling

We finally discuss the various functions needed to perform the particle-core coupling calculations. Only if a quasi-particle core-coupling calculation is to be carried out does one first have to perform a quasi-particle calculation to determine the one-quasi particle energies and the various orbital occupation probabilities. This is explained in Sect. 9.8 for a constant pairing force G. Moreover, a large number of ingredients (calculations of Clebsch-Gordan coefficients, Wigner $6j$-symbol, radial integrals, diagonalization code,...) have already been given in the present chapter. The other parts are not explained in great detail.

1) The subroutine VN06DD(N,NBMAX,IN,EPS,EF,TAUSQ) diagonalizes the matrix [A] which is passed through the common block /VN06CE/ A(80,80), S(80,80). The eigenvectors are inserted into the matrix [S] per column. The quantities NBMAX,IN,EPS and EF relate to the number of rotations needed to obtain, to within a given accuracy, a diagonal matrix. The value TAUSQ contains a measure of $\sum_{i,j} |a(i,j)|^2$ at the final moment of diagonalization. Finally, N denotes the dimension of the matrix.

2) The subroutine VN06DE(NORM) orders the energy eigenvalues of A in increasing order and, at the same time, orders the corresponding eigenvectors.

3) The subroutine VN06P2(IK) is a subroutine controlling a large number of the output. For IPRIN=1, the boson reduced matrix elements $\langle \| b_2^+ \| \rangle$ and $\langle \| b_3^+ \| \rangle$ are printed. If IK=1, the basis is additionally given on output, if IK=2 and IPRIN=1, the full energy matrix [H] is printed and for IK=3, only the dimensions of the various energy matrices H are printed.

4) Function VN06TR(IK,IKA,IP,IPA,LAM,LAPPA).

Here, one evaluates the single-particle EL and $M1$ matrix elements, eventually including the respective pairing factors if IBCS=1, if a quasi-particle is coupled to the core vibrations. LAM denotes the multipolarity and LAPPA=1(-1) denotes that electric (magnetic) properties are studied.

5) Function VN02BA: Clebsch-Gordan coefficient,

6) Function VN02B9: Wigner $6j$-symbol,

7) Function VN02BB: Help function for evaluation of Wigner $6j$-symbol,

8) Function VN06PH: Here, the basis (configuration space) for given J^π is constructed for the transition code. The quantum numbers are put into vectors I1, I2 and I3.

9) Function VN0105: Calculation of the radial integrals over harmonic oscillator wave functions,

10) Function VN0107: Calculation of the Gamma function,

11) Subroutine VN06P1.

Here, the basis (configuration space) for a given J^π is constructed in the particle-core coupling calculation. The quantum numbers are put into the vectors I1, I2 and I3 with the vector EEE containing the unperturbed energies.

12) Subroutine VN06P3(IAAN)

Here, we construct the energy matrix according to eqs. (9.36) and (9.37). The energy matrices to be diagonalized are placed in the matrix [HIN].

9.11 Running the Codes

The diskette contains the source for the various programs discussed in Sects. 9.1 to 9.10. In almost all cases, a simple main program is also present for evaluating the function programs. These parts are self-explanatory and the source codes are easily converted into an executable program. The various codes presented run on a MICROVAX 3500 system, but no special features are used or implemented in these various codes. The FORTRAN standard used is the VAX FORTRAN running under the VMS 5.1-1 version. Therefore, the codes presented should run on any computer compatible with this FORTRAN compiler. The codes have not been tested for other systems but should, in general, cause no specific problems as far as conversion is concerned.

In a number of cases, a typical command file has been added to run the appropriate code (Table of Slater integrals, evaluating the δ-interaction matrix elements, diagonalizing hermitian matrices, a BCS pairing model code,...).

Only the particle-core coupling package is more extensive in scope. Here, it is more interesting to compile the library into an object file and link with the energy eigenvalue, eigenvector calculation or the transition (and moment) calculation as two separate executable programs. For each, a typical input and corresponding output file are given which should allow for easy control and installation of the various parts in particle-core coupling on one's own computing system.

CLEBSCH.FOR: Contains the function program VN02BA as well as a test program to evaluate a given Clebsch-Gordan coefficient.

SIXJ.FOR: Contains the function program VN02B9, the help function VN02BB and a test program to evaluate a given Wigner $6j$-symbol.

WINEJ.FOR: Contains the function program WINEJ in order to evaluate a given Wigner $9j$-symbol.

SLATMAT.FOR: Contains the main program SLATMAT for evaluating a table of Slater integrals. In this file all necessary function programs are also added i.e. SLATER and VN07AL. A small command file for running the executable file SLATMAT.EXE obtained by linking the object programs SLATMAT, SLATER and VN07AL.

DELTA.FOR: Contains the program DELT for evaluating a given two-body matrix element using a zero-range delta interaction. The necessary function programs are added: DELTA,S,VN02BA,SLATER and VN07AL. A small command file for running the executable file DELTA.EXE, obtained by linking the object programs DELT, DELTA, S,VN02BA,SLATER and VN07AL is present.

DIAGOT.FOR: Contains the program DIAGOT in order to diagonalize a given $n \times n$ matrix (symmetric and real). The function program VN0108 is added, as well as a small command file for running the executable file DIAGOT.EXE obtained by linking the object programs corresponding to DIAGOT and VN0108.

RADIAL.FOR: Contains the program VN0105 in order to evaluate radial integrals of r^λ using harmonic oscillator wave functions. The necessary function program VN0107 for calculating the gamma function is also added. An executable file can be made by linking the object programs corresponding to VN0105 and VN0107.

BCS.FOR: Contains the program BCS for performing a BCS calculation which contains a constant pairing strength G. A command file is added in order to run the executable file BCS.EXE obtained by linking the object program BCS.

PANDYA.FOR: Contains the programs PANDYA and VN02B9 and VN02BB

PCOREEN.FOR: Contains the main program for carrying out a particle-core coupling calculation for the energy eigenvalues and corresponding eigenvectors.

PCORETRA.FOR: Contains the main program for carrying out a calculation of transition rates and static moments, starting from wave functions obtained using the former program PCOREEN.FOR.

PCOREBIB.FOR: Library which contains the various function and subroutine programs which go together with the main codes PCOREEN.FOR and PCORETRA.FOR. This library contains, in sequence VN06DD, VN06DE, VN06DE,

VN06P2, VN06TR, VN06PH, VN02BA, VN02B9, VN02BB, VN0105, VN0107, VN06P1 and VN06P3.

PCOREEN.COM: Input command file for ${}^{133}_{57}\mathrm{Sb}$

PCORTRA.COM: Input command file for ${}^{133}_{51}\mathrm{Sb}$

The executable file PCOREODD.EXE and TRANS.EXE are obtained by linking the object files for PCOREEN.FOR with the library and PCORETRA.FOR with the library, respectively.

PCOREEN.LOG: Output file with energy eigenvalues and eigenvectors for ${}^{133}_{51}\mathrm{Sb}$

PCORETRA.LOG: Output file for moments and transition rates for ${}^{133}_{51}\mathrm{Sb}$.

9.12 Summary

In this chapter, which is intended to allow the reader to test the theoretical knowledge presented in the earlier chapters in 'real' situations, a number of programs for evaluating various nuclear structure properties are presented.

Here, we present in each section a short explanation of the particular application: short notes on formalism, typical information relating to the organization of the input and output options for running the codes,... On the MS-DOS-diskette, the various programs are given in the sequence in which they are discussed in the text. Therefore, the order followed is: (1) a code for evaluating Clebsch-Gordan coefficients, (2) a code for evaluating Wigner $9j$-symbols, (4) a code which allows for the evaluation of a table of Slater radial integrals, (5) a code for calculating any particular two-body matrix element using a zero-range δ-interaction with the inclusion of spin exchange, (6) a code for diagonalizing symmetric and real matrices using the Jacobi method. Here, a typical command file is added. (7) a code for calculating radial integrals of powers of the radial coordinate r^L using harmonic oscillator wave functions, (8) a code which solves the BCS gap equations for the occupation probabilities and one-quasi particle energies, starting from a constant pairing force for a given number of interacting particles n distributed over i single-particle orbitals, (9) a code for evaluating the particle-hole angular momentum coupled matrix element in terms of the particle-particle angular momentum coupled matrix elements, and (10) a more extended package on particle (-hole, -one-quasi particle)-core coupling.

The latter package contains, first, a concise write-up of the basic formalism of particle core coupling. Then, the two major codes for (a) evaluating the energy eigenvalues and eigenvectors and, (b) calculating electromagnetic transition matrix elements and static moments are presented with respect to input and output format. The full library for this particle-core coupling package is discussed and, on the diskette, not only the source programs for these various codes but also a typical input, as well as the corresponding output files, arc presented in order to facilitate the implementation of these codes on one's own computing system.

The various codes have been written in FORTRAN and run on a MICROVAX 3500 system. No specific system options are in use and, therefore, no complications should arise when installing the various programs.

10. Shell Structure: A Major Theme in the Binding of Various Physical Systems

10.1 Introduction

The nuclear shell-model description, in order to account for the stability of the various nuclei as well as of the peculiar shell-structures observed in the energy spectra of many nuclei, has basically been copied from the much older Bohr model for the atom. Whereas, in this Bohr model, the atomic nucleus plays the natural role for the singular character ($\frac{1}{r}$) of the Coulomb potential that the electrons feel in their motion, the nuclear case is not so straightforward. As was discussed at length in Chap. 3, it is, however, possible to determine to a good approximation an average one-body potential, using self-consistent Hartree-Fock calculations starting from the original nucleon-nucleon two-body interactions.

Many other systems exhibit similar shell structures. Some were anticipated rather early, but experimental measuring conditions often delayed a clear-cut manifestation, while others came as rather more of a surprise. The typical extrapolations of the Bohr atom, whereby the bound electron is replaced by e.g. the negative muon (μ^-) or pion (π^-) leading towards muonic, pionic, mesonic atoms, have been observed. In these systems, since for the radius the mass of the bound particle appears in the denominator, this mass (m_{e^-}, m_{μ^-}, m_{π^-}, $\cdots$) leads to a natural "scale" for the bound system with these particles ($\mu^-, \pi^-, \cdots$) moving very close to, and almost inside, the extended nuclear charge distributions. The corresponding emission of radiation also has its scale modified into the keV and MeV region. Moreover, the more strongly bound levels acquire a width through the strong interaction coupling the mesons directly to the nucleon and thus, in many cases, the electromagnetic transitions at the lower end of the decay chain are no longer observed (see Fig. 10.1). We shall not discuss these mesonic atoms in detail at present (Backenstoss, 1993). We concentrate, though, on a number of systems going from two-body systems (e^-e^+ positronium, $q\bar{q}$ quark-anti quark combination or quarkonium) towards the rather complex mesoscopic cluster systems where shell- structures were shown to reveal some of the basic properties in those metallic compounds.

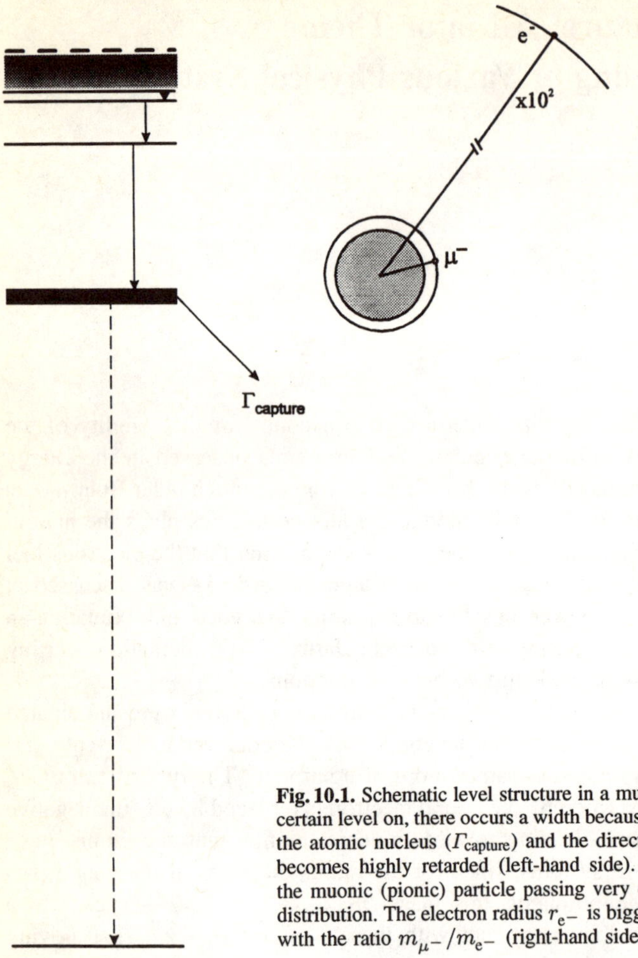

Fig. 10.1. Schematic level structure in a muonic (pionic) atom. From a certain level on, there occurs a width because of capture possibilities by the atomic nucleus (Γ_{capture}) and the direct electromagnetic transition becomes highly retarded (left-hand side). Pictorial representation of the muonic (pionic) particle passing very close to the nuclear charge distribution. The electron radius r_{e^-} is bigger than e.g. the muonic one with the ratio m_{μ^-}/m_{e^-} (right-hand side)

10.2 Bound Two-Body Systems: Positronium and Quarkonium

10.2.1 Positronium

Positronium exhibits the characteristics of a bound two-fermion system through the combined effect of the Coulomb and magnetic (hyperfine) forces. This is basically a relativistic problem needing the full use of quantum electrodynamics (QED) but, in order to gain insight, we can use the non-relativistic Schrödinger equation as a basic guide (Bloom, Feldman, 1982), where the mass for a single particle is replaced by the reduced mass i.e. $m_{e/2}$. The angular momentum structure will lead to the combination of $\ell = 0,1,2$, (S,P,D, $\cdots$) states with the S=0 (singlet) and S=1 (triplet) intrinsic spin combinations which we shall denote by the notation $^{2S+1}\ell_J$. The exact Coulomb potential will lead to a spectrum given on the extreme left of

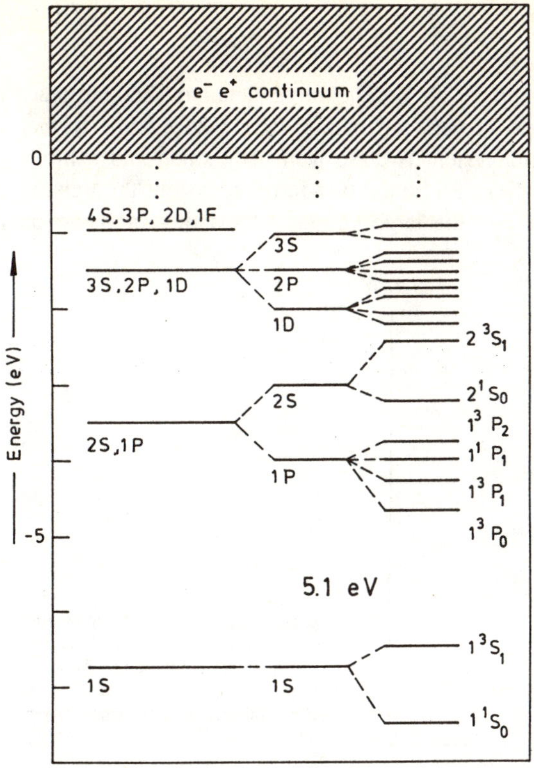

Fig. 10.2. Energy level scheme of positronium. Both the continuum and bound state system (schematically) are drawn. The effect of Coriolis forces and magnetic (spin-spin) forces is drawn in an enlarged way (adapted from Isgur and Karl (1983))

Fig. 10.2. Because of the importance of magnetic forces, a spin-dependent form could be used to express the major effect of relativistic motion, i.e. one can use a form (Frauenfelder, Henley, 1991)

$$V_{Mag.} = a\,\boldsymbol{\sigma}_{e^-} \cdot \boldsymbol{\sigma}_{e^+} \delta\left(\boldsymbol{r}_{e^-} - \boldsymbol{r}_{e^+}\right) . \tag{10.1}$$

This spin dependence will cause a splitting between the S=0 and S=1 states. Moreover, spin-orbit effects are present and these will lift the degeneracy that otherwise remains for various orbital angular momentum states such as e.g. 2S, 1P; 3S, 2P, 1D with the first number, the radial quantum number n. The major term (excluding spin-orbit and spin-spin effects) gives the binding energy in the e^- e^+ systems as (Perkins, 1987)

$$E_n = -\frac{1}{4n^2}\alpha^2 m_e c^2 , \tag{10.2}$$

(with $\alpha = \frac{e^2}{4\pi\epsilon_0}\frac{1}{\hbar c}$ the fine-structure comitant).

This amounts to the energy difference $E_2 - E_1 = 5.1\,\mathrm{eV}$ for Positronium as observed (Fig. 10.2). For the ground-state magnetic (spin-spin interaction) splitting, Rigorous QED calculations give the result

$$\Delta E(1^3S_1 - 1^1S_0) = m_e c^2 . \frac{7}{12} \alpha^4 , \tag{10.3}$$

which results in a value of 8.45×10^{-4} eV. The more detailed shell-structure relating to the e^-e^+ bound system is depicted in Fig. 10.3, where some further results from QED are shown. This system is quite well understood and exhibits in a very nice way the characteristics of a bound two-fermion system through the Coulomb force. We shall not go into the large amount of existing literature on positronum in further detail.

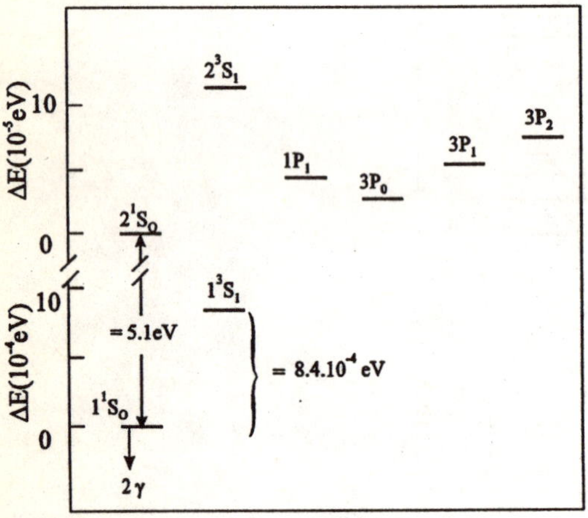

Fig. 10.3. The energy levels in positronium. A different energy scale is used in the upper as compared to the lower part in order to best illustrate the hyperfine splitting (adapted from Perkins (1987))

10.2.2 Quarkonium: Bound Quark-Antiquark Systems

Observational Properties. A whole new era of physics was opened up through the experiments carried out by Richter el al. at SLAC (Stanford) (Richter, 1977) and Ting et al. (Brookhaven National Laboratory) (Ting, 1977) with the discovery of what turned out to be a bound state of the charmed quark (c) and its antiquark ($\bar{c}$), the now famous J/ψ particle. The width of the state for decay was only about 70 keV, in very marked contrast to other particles at such high energies of $\simeq 3$ GeV. Later on, a number of "excited" $c\bar{c}$ states were, shown to exist, leading to a large family exhibiting a structure that is largely reminiscent of positronium (e^-e^+). The data on $c\bar{c}$ are presented in Fig. 10.4.

The precise form of the potential acting between the quark and its antiquark is not well known. Since, in the field theory describing quark-quark interactions, the mediating fields are described through the exchange of massless vector gluons (just as in QED the exchange proceeds through the massless photon), a potential very much like the Coulomb potential can be considered (Quig, Rosner, 1979). However, at large separation, quarks are subject to a confining force that also requires the presence of a potential which grows linearly with r. Therefore, a typical QCD potential will be of the form (Perkins, 1987)

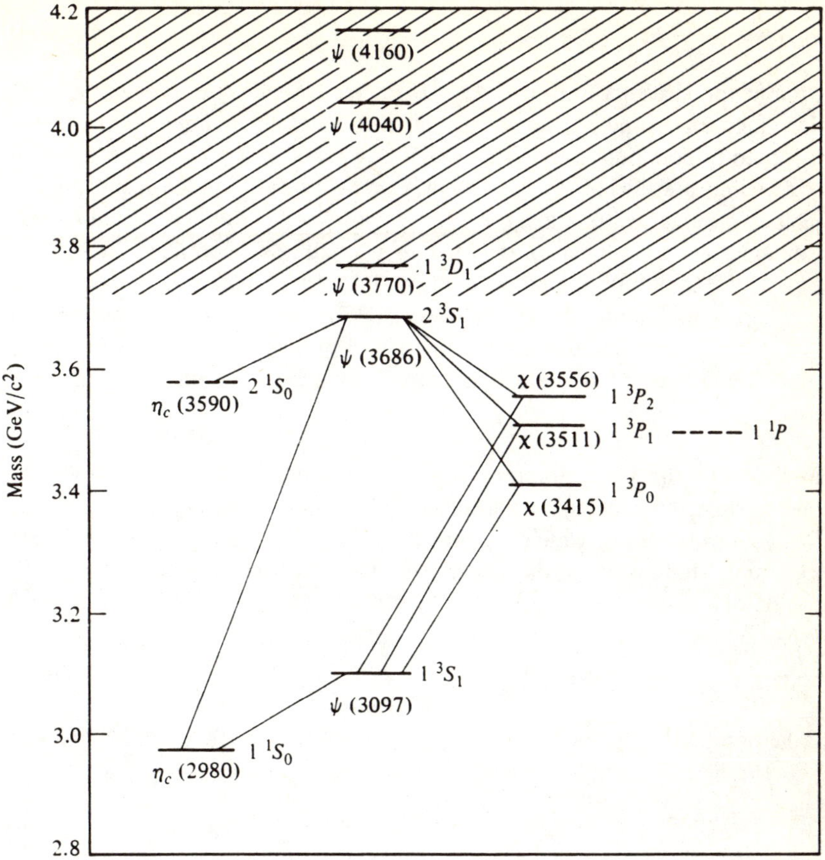

Fig. 10.4. Energy spectrum of charmonium. The solid lines indicate the experimentally well-known states ; dashed lines present less known or expected states. The connecting lines indicate transition photon lines. The shaded region is the continuum for decay into the $D\bar{D}$ mesons. (taken from Frauenfelder and Henley (1991))

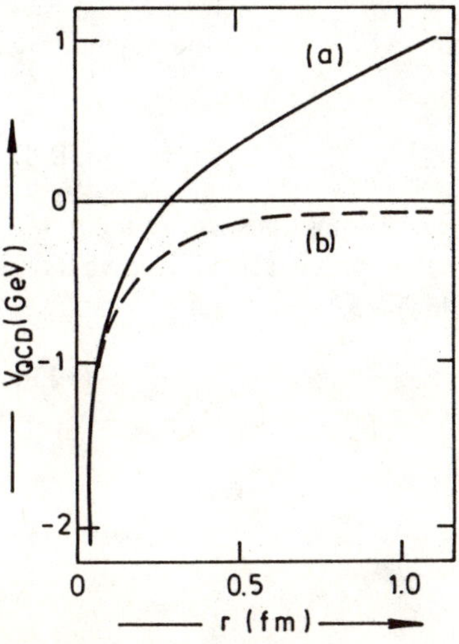

Fig. 10.5. Illustration of the possible $q\bar{q}$ binding potential, known as V_{QCD}. Both the full form (eq. 10.4) and the pure Coulomb part (dashed line) are drawn. The parameters used are $\alpha_s = 0.3$ and $k = 1$ GeV.fm^{-1}

$$V_{\text{QCD}} = -\frac{4}{3}\frac{\alpha_s}{r} + kr \; . \tag{10.4}$$

Here, the factor $\frac{4}{3}$ results from the fact that three quarks couple through the octet of color gluon states and α_s then shows up as a coupling strength, similar to the α fine-structure coupling strength in QED. A plot of this potential is given in Fig. 10.5, using a value of $\alpha_s = 0.3$ and $k = 1\,\text{GeV fm}^{-1}$. Such a potential is amenable to a numerical treatment in a non-relativistic Schrödinger equation. It is clear, though, that a large part of the binding at small r is determined through the structure of the Coulomb-like potential $(1/r)$. In order to understand the structure of the full potential somewhat better, we shall (for the large separation regime) solve the Schrődinger equation for the linear part of the potential. This extension will give further insight into shell-structures that appear for the class of linear potentials.

Potential Model for Charmonium. Separating out the relative angular momentum and intrinsic spin parts for the quark and antiquark, we are led to a radial equation (non-relativistic) which contains the mass of the quark m_c (in the COM frame). Using straight away the linear part, the equation becomes (for $\ell = 0$) (Flűgge, 1974)

$$-\frac{\hbar^2}{m_c}\frac{d^2u(r)}{dr^2} + (kr - E)u(r) = 0 \; . \tag{10.5}$$

Using the notation

$$\ell^3 = \frac{1}{k}\frac{\hbar^2}{m_c E} \; ; \quad \lambda = \ell^2 \frac{m_c E}{\hbar^2} \; , \tag{10.6}$$

we can transform equation (10.5), using as the new radial coordinate $\xi = \frac{r}{\ell} - \lambda$, into the form

$$\frac{d^2u}{d\xi^2} - \xi u = 0 \; . \tag{10.7}$$

The regular boundary conditions at the origin $u(r) = 0$ and the asymptotic behaviour lead to the conditions

$$u(-\lambda) = 0 \; ; \quad u(\infty) \to 0 \; . \tag{10.8}$$

The differential equation (10.7) is solved by Bessel functions of the order 1/3 and, incorporating the correct boundary conditions as expressed by (10.8), we have to use the Airy function solution (Abramowiz, 1964)

$$u(\xi) = C.Ai(\xi) \; , \tag{10.9}$$

with

$$Ai(\xi) = \frac{1}{\pi}\sqrt{\frac{\xi}{3}}K_{1/3}\left(\frac{2}{3}\xi^{2/3}\right) (\xi > 0) \; . \tag{10.10}$$

The asymptotic form is obtained from the general formula

$$K_\nu(z) \to \sqrt{\frac{\pi}{2z}}\, e^{-z} \quad \text{for} \quad z \to \infty\,. \tag{10.11}$$

In this way, we obtain

$$u(\xi) \to \frac{C}{2}\sqrt{\frac{3}{\pi}}\,\xi^{-1/4}\, e^{-\frac{2}{3}\xi^{3/2}} \quad \text{for} \quad \xi \to \infty\,, \tag{10.12}$$

with a correct asymptotic behaviour.

In the classically allowed interval i.e. for negative ξ values, the Airy function is presented by Bessel functions as

$$Ai(-\xi) = \frac{1}{3}\sqrt{\xi}\left\{J_{1/3}\left(\frac{2}{3}\xi^{3/2}\right) + J_{-1/3}\left(\frac{2}{3}\xi^{3/2}\right)\right\}\,, \tag{10.13}$$

or

$$u(\xi) = \frac{C}{3}\sqrt{|\xi|}\left\{J_{1/3}\left(\frac{2}{3}|\xi|^{3/2}\right) + J_{-1/3}\left(\frac{2}{3}|\xi|^{3/2}\right)\right\} \quad \text{for} \quad \xi > 0\,. \tag{10.14}$$

Now, because of the boundary condition $u(-\lambda) = 0$, this results in the energy eigenvalue equation

$$Ai(-\lambda) = 0\,.$$

The roots λ_n will lead to the corresponding eigenvalues E_n through the relation

$$E_n = \left(\frac{(\hbar k)^2}{m_c}\right)^{1/3} \cdot \lambda_n\,. \tag{10.15}$$

The above energies have to be added to the energy of the unperturbed $c\bar{c}$ pair such that the mass of the nth excited state of charmonium becomes

$$M_n = 2m_c + \frac{1}{c^2}\left(\frac{(\hbar k)^2}{m_c}\right)^{1/3} \lambda_n\,. \tag{10.16}$$

Since two input parameters occur (k and m_c), one can use the two masses $M_1 = 3.097$ GeV and $M_2 = 3.686$ GeV in order to determine these as $k = (0.458\ \text{GeV})^2$ and $m_c c^2 = 1.155$ GeV (Greiner, 1989). With this knowledge and the various roots of the Airy functions, given in Table 10.1, one can easily derive the masses of some of the lowest radially excited $c\bar{c}$ states (see Table 10.2).

There exists an interesting relation for higher levels (corresponding to $\lambda \gg 1$) since we can then use the asymptotic relations (Flűgge, 1974)

$$J_{1/3}(z) \to \sqrt{\frac{2}{\pi z}}\cos\left(z - \frac{5\pi}{12}\right)$$

$$J_{-1/3}(z) \to \sqrt{\frac{2}{\pi z}}\cos\left(z - \frac{\pi}{12}\right)\,, \tag{10.17}$$

Table 10.1. Lowest roots of Airy function $Ai(-\lambda_n) = 0$

n	λ_n
1	2.338
2	4.088
3	5.521
4	6.787
5	7.944
6	9.023
7	10.040
8	11.009
9	11.936
10	12.829

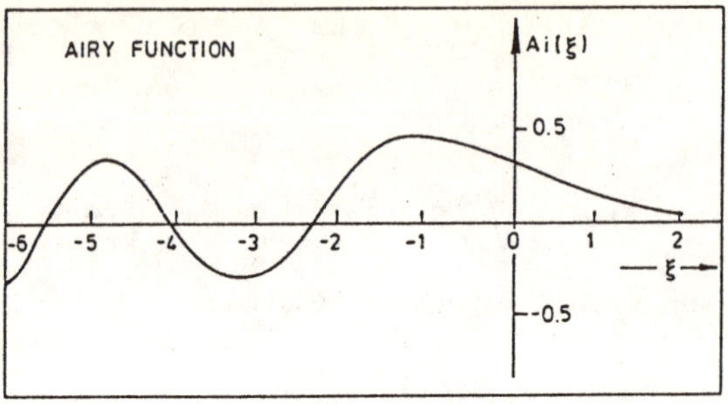

Table 10.2. The lowest four roots of the Airy function (n, λ_n), the corresponding mass M_n, derived from eq. (10.16), a relativistic correction $\Delta M_{\text{rel.}}$, important for $n > 2$, and the experimental mass M_{exp}.

n	λ_n	M_n/GeV	ΔM_{rel}/GeV	M_n/GeV	M_{exp}/GeV
1	2.338	3.097*	–	3.097	3.097
2	4.088	3.686*	–	3.686	3.686
3	5.521	4.17	−0.15	4.02	4.040
4	6.787	4.59	−0.23	4.36	4.415

* The lowest two mass states Ψ (3.097 GeV) and Ψ' (3.686 GeV) were used to determine the two unknown parameters k and m_c in eq. (10.16)

or

$$u(\xi) \to \frac{C}{\sqrt{3\pi}} |\xi|^{-1/4} \cos\left(\frac{2}{3}|\xi|^{3/2} - \frac{\pi}{4}\right) , \tag{10.18}$$

for large negative values of ξ and, thus, the eigenvalue condition becomes

$$\frac{2}{3}\lambda_n^{3/2} = \left(2n - \frac{1}{2}\right)\frac{\pi}{2} , \tag{10.19}$$

supplementing the values given in Table 10.1. The latter relation leads to the energy eigenvalues (for $n \gg 1$)

$$E_n = \left(\frac{(\hbar k)^2}{m_c}\right)^{1/3} \left(\frac{3}{4\pi}\left(2n - \frac{1}{2}\right)\right)^{2/3} . \tag{10.20}$$

In comparing the actual data for the $c\bar{c}$ spectrum (Fig. 10.4) with those for positronium (Fig. 10.3) where only the Coulomb part $(1/r)$ acts, one observes a relative lowering of the 1P-states relative to the 2S states. This effect can also be explained, but one then has to solve the more complicated equation (10.5) by including the angular momentum centrifugal effect. Since this is caused by the linear part, one can obtain a rough feeling of the relative changes in nL states by comparing the exact results of the Coulomb $(1/r)$ and harmonic oscillator potential (r^2). This is illustrated in Fig. 10.6, where an interpolation versus a possible $c\bar{c}$ situation is also indicated.

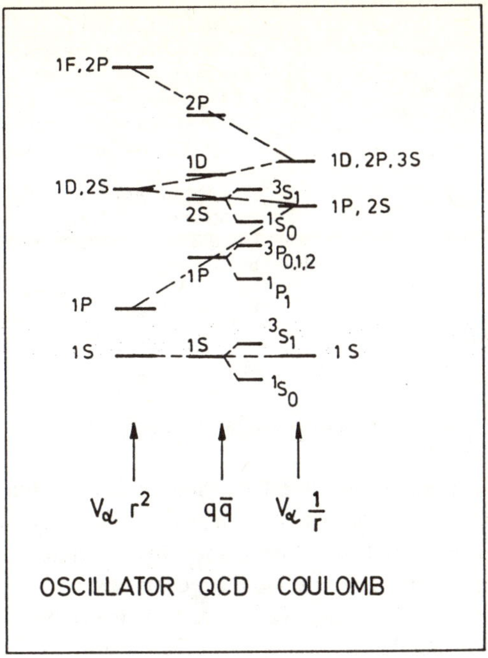

Fig. 10.6. The lowest energy levels for a harmonic oscillator potential (left-hand part) and for a pure Coulomb potential (right-hand part). The notation used is $n\ell$ = 1S, 1P, ... An interpolation between both extremes might correspond to the actual situation for binding the $q\bar{q}$ system and is positioned half-way between the exact solvable potentials. The spin-spin splitting is also indicated

10.3 Shell Structure in Metal Clusters

10.3.1 Experimental Situation

Mass spectra for a particular class of metallic clusters were produced some time ago (Knight, 1984) and shown to exhibit large abundance peaks at $N = 8, 20, 40, 58, 92, 138, \ldots$. This very peculiar observation for Na clusters, which was later also shown to exist in various other cluster combinations (Ag, Au, Cs) points towards the extra stability associated with the motion of independent, delocalized atomic 3s electrons that are bound in a spherical symmetric potential well. At present, we shall not discuss the precise details for producing the various metal clusters (de Heer et al., 1987). In the expansion of an inert gas (Argon or Xenon) through a 0.1–0.2 mm diameter nozzle, random thermal motion is converted into a uniform translational motion, thereby causing a cooling of the inert gas. Introducing atomic Na vapor into this system will cause the Na atoms to combine into large clusters with an overall broad uniform size distribution (Fig. 10.7).

It became clear that a most unexpected feature, i.e. the observation of quantal effects in systems of atoms, established a basic connection between fields as distinct as the quantum mechanics of few body systems (electronic structure in an atom, nucleonic motion in the nucleus) and the study of atomic cluster formation.

Various techniques have subsequently been developed in order to study not only rather small N aggregates ($N = 2, 8, 20, 40$), but also to find evidence for

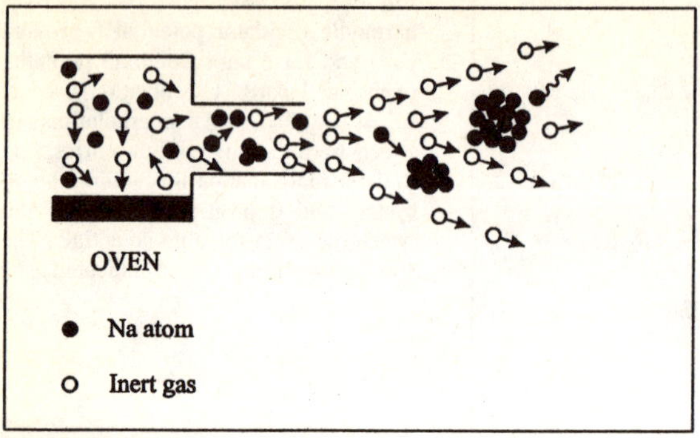

Fig. 10.7. Production of Na clusters. In the expansion of an inert gas (Ar, Xe, ...) through a fine nozzle, random thermal motion is converted into translational motion, resulting in a strong cooling of the gas (denoted by open circles ∘). Introducing atomic Na vapor into this system (circles: •), the Na atoms will aggregate into clusters (adapted from Bjørnholm (1992))

very large N combinations (up to $N \simeq 3000$ atoms) (Bjørnholm, 1990, 1992), (Nishioka, 1990).

In Fig. 10.8 (top part) one clearly observes the overall distribution in the intensity distribution I_N versus the number of atoms N in the cluster. The quantity N is also equal to the number of conduction electrons in the atomic cluster. On top of the smooth background distributions a number of saw-tooth irregularities at magic numbers are clearly observed. A more convenient way to display these points of increased stability (and increased abundance) is to plot the relative intensity changes as

$$\begin{aligned} \Delta(\ell n\, I_N) &= \ell n \left(\frac{I_N + 1}{I_N} \right) \\ &\approx 2 \frac{(I_{N+1} - I_N)}{(I_{N+1} + I_N)} \,. \end{aligned} \tag{10.21}$$

This quantity, plotted in Fig. 10.8 (lower part) expresses the magic numbers in an even more dramatic way, whereby the closed shells at $N = 8, 20, 40, 58, 92, 138$, 196 clearly stand out. These numbers correspond fairly well to the shell structure of closed shell configurations using a square-well potential without spin-orbit interactions (see Chap. 3) and more elaborate calculations, using either a Woods-Saxon potential approximation (Nishioka, 1990)

$$U(r) = \frac{V_0}{1 + e^{\frac{r - r_0}{a_0}}} \,, \tag{10.22}$$

with

$$V_0 = -6.0\,\text{eV} \,, \quad r_0 = R_0 . N^{1/3} \,, \quad R_0 = 2.25\,\text{Å} \,, \quad a_0 = 0.74\,\text{Å} \,,$$

or the even more realistic Local Density Approximation (LDA) potential (Brack et al., 1991a, 1991b) very nicely reproduce the stable structures (see Table 10.3).

In the plot of the experimental data, the higher N combinations become progressively less pronounced (Fig. 10.8 lower parts) due to noise in the finite counting statistics. Using correctly weighted logarithmic derivatives for properly spaced points N, Bjørnholm pointed out that the expression

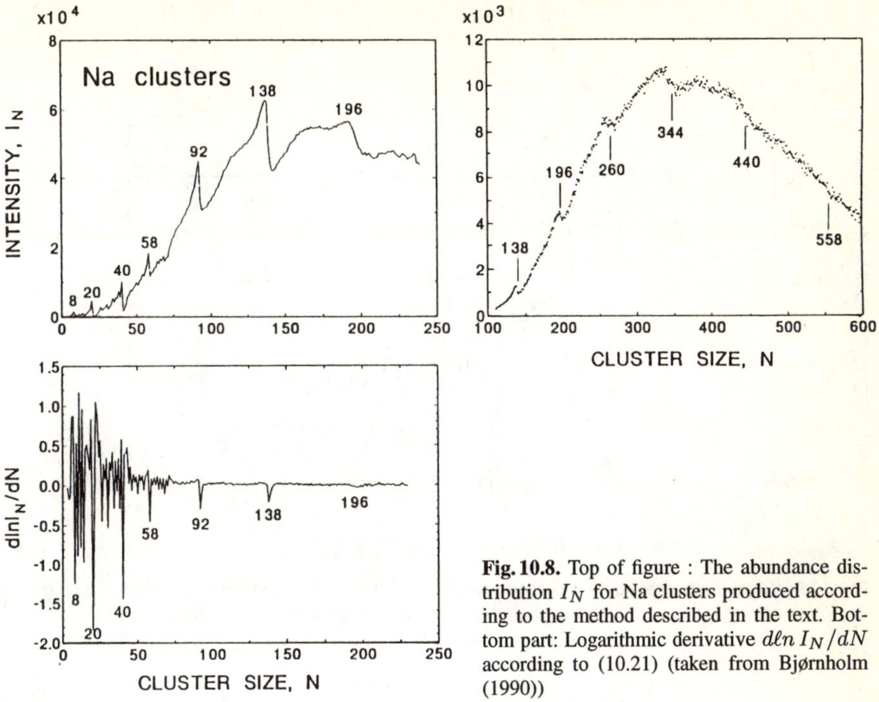

Fig. 10.8. Top of figure : The abundance distribution I_N for Na clusters produced according to the method described in the text. Bottom part: Logarithmic derivative $d\ell n\, I_N/dN$ according to (10.21) (taken from Bjørnholm (1990))

Table 10.3. Magic numbers describing spherical shells in metal clusters

Shell	Na cluster[1)]	W.S.[2)]	LDA[3)]
0	2	2	2
1	8	8	8
2	20	20	18/20
3	40	40	34/40
4	58	58	58
5	92	92	92
6	138	138	138
7	196	198	186/196
8	260 ± 4	254/268	254
9	344 ± 4	338	338
10	440 ± 2	440	440
11	558 ± 8	562	556
12		694	676
13		832	832
14		1012	

[1)] Bjørnholm, 1990; [2)] Nishioka, 1990; [3)] Brack, 1991a, 1991b

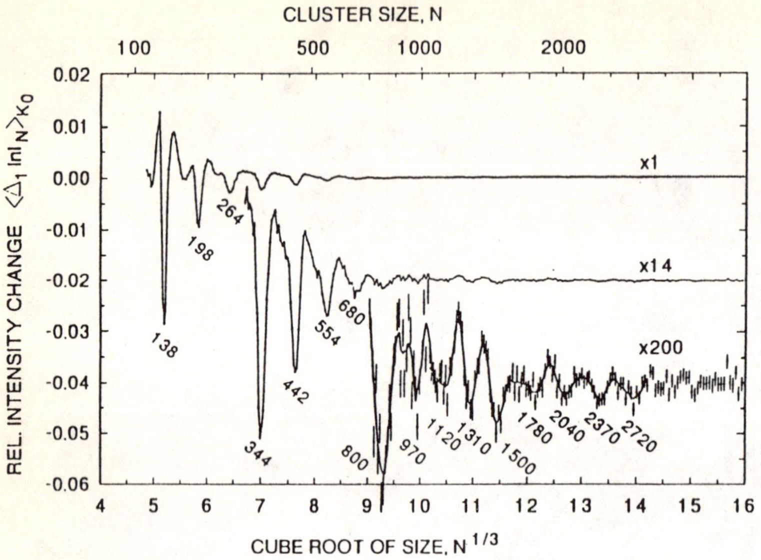

Fig. 10.9. Relative changes in cluster abundances, using the weighted logarithmic derivative as given in eq. (10.23). The above quantity is plotted against $N^{1/3}$. (Taken from Bjørnholm (1992))

$$\langle \Delta(\ell n I_N) \rangle_{\kappa_0} = \frac{\sum_{\kappa=2\kappa_0/3}^{\kappa_0} \frac{2(I_{N+1+\kappa} - I_{N-\kappa})(2\kappa+1)}{I_{N+1+\kappa} + I_{N-\kappa}}}{\sum_{\kappa=2\kappa_0/3}^{\kappa_0} (2\kappa+1)^2} , \tag{10.23}$$

on choosing a value of $\kappa_0 = \frac{3}{100}N$, and so sampling over intervals of $2\kappa_0 + 1$, i.e. $\pm 3\%$ of the actual size, stands out much more clearly for the higher N magic numbers do (Fig. 10.9).

More detailed information about metal clusters has been obtained and an extensive reference list can be found in (de Heer, 1987), (Kresin, 1992) for detailed studies of all salient features related to this unexpected illustration of shell structures. Therefore, periodicities that are unambigously due to quantum shell structure effects have been found to exist in a large number of cluster abundance measurements.

10.3.2 Theoretical Description

Here, we shall briefly present some of the many facets that show up in the quantum mechanical many-body problem posed by the metal clusters. We indicate some of the approximations that make the problem tractable with at one extreme, the complex features of "ab initio" quantum mechanical studies and, at the other, the single-particle static square well potential without spin orbit term. A detailed discussion, including a large reference list is given by (Brack, 1992).

The Full Quantum-Mechanical Many-Body Problem. The Hamiltonian describing a neutral cluster, consisting of N nuclei with Z electrons each, reads

$$
\begin{aligned}
\boldsymbol{H} = \sum_{i=1}^{N} \Bigg\{ \frac{\boldsymbol{P}_i^2}{2M} + \sum_{k=1}^{Z} \left(\frac{\boldsymbol{p}_{i,k}^2}{2m} - \frac{Ze^2}{|\boldsymbol{r}_{i,k} - \boldsymbol{R}_i|} \right) \\
+ \frac{1}{2} \sum_{j=1}^{N} \left(\frac{(Ze)^2}{|\boldsymbol{R}_i - \boldsymbol{R}_j|} + \sum_{k,\ell=1}^{Z} \frac{e^2}{|\boldsymbol{r}_{i,k} - \boldsymbol{r}_{j,\ell}|} \right) \Bigg\} ,
\end{aligned}
\tag{10.24}
$$

where $M, \boldsymbol{P}_i, \boldsymbol{R}_i$ $(\boldsymbol{P}_j, \boldsymbol{R}_j)$ denote the mass, momenta and coordinates of the nuclei and $m, \boldsymbol{p}_{i,k}, \boldsymbol{r}_{i,k}(\boldsymbol{p}_{j,\ell}, \boldsymbol{r}_{j,\ell})$ the corresponding quantities in the nucleus $i(j)$. The above Hamiltonian is fully determined through the Coulomb forces, but is too general to be solved exactly. Because of the different mass scales for nuclei and electrons, the motion of the nuclei can be taken out and treated classically. Electrons have to be treated using quantum mechanics.

In simple metals such as Ag, Al, Na,... the separation into valence electrons and core electrons (well bound and localized) leads to a further simplification and can be depicted by a Hamiltonian containing N interacting ions and $x.N$ interacting electrons (with x the number of valence electrons per atom) moving in the field caused by the ions. This results in

$$
\boldsymbol{H} = \boldsymbol{H}_N + \boldsymbol{H}_{e\ell} , \tag{10.25}
$$

with

$$
\boldsymbol{H}_N = \sum_{i=1}^{N} \left(\frac{\boldsymbol{P}_i^2}{2M} + \frac{1}{2} \sum_{i<j=1}^{N} \frac{(xe)^2}{|\boldsymbol{R}_i - \boldsymbol{R}_j|} \right) , \tag{10.26}
$$

$$
\boldsymbol{H}_{e\ell} = \sum_{k=1}^{xN} \left(\frac{\boldsymbol{p}_k^2}{2m} + V_I(\boldsymbol{r}_k) + \frac{1}{2} \sum_{k<\ell=1}^{xN} \frac{e^2}{|\boldsymbol{r}_k - \boldsymbol{r}_\ell|} \right) , \tag{10.27}
$$

and the ionic potential $V_I(\boldsymbol{r}_i)$ defined as

$$
V_I(\boldsymbol{r}_k) = -\sum_{i=1}^{N} \frac{xe^2}{|\boldsymbol{r}_k - \boldsymbol{R}_i|} . \tag{10.28}
$$

The latter potential can be replaced by some pseudo-potential, as is quite conveniently done in solid state physics.

The remaining electron-electron interactions, described by (10.24) to (10.28), still pose a huge problem and an approximation which is quite often used considers the introduction of a mean field. Then the electron wave functions describe non-interacting fermions that can be derived from a variational principle leading to the familiar Hartree-Fock equations (Chaps. 3, 8). An alternative to the mean field is obtained from a density functional formulation in which correlations can be included approximately in a local average field (see Chap. 8 and methods discussed therein).

This mean field concept explains most of the electronic structure effects, and many other properties of metal clusters.

Approximating the Many-Body Problem. "Ab initio" calculations, starting from the full Hamiltonian of (10.24), and treating all the electronic degrees of freedom, can be carried out within a Born-Oppenheimer approximation. The separation between nuclear and electron degrees of freedom, and the use of a H.F. approximation in order to determine the electron wave functions can be performed for small cluster systems ($N \gtrless 20$).

Some of the dynamics related to the nuclear motion have been incorporated into the so-called "molecular dynamics" calculations but the results are not very different from the "ab initio" calculations.

A next step is to completely ignore the nuclear motion, using only the electronic Hamiltonian of (10.27). The ionic structure is then represented by some external potential V_I (cf. (10.28)), or a corresponding pseudo-potential. The solution of this mean field problem is still very complicated and an optimization of the ion geometry, entering through the potential V_I, actually implies some form of molecular dynamics. Therefore, a rather dramatic, but efficient, approximation consists of averaging out the ionic structure and replacing the corresponding charge distribution by a "constant" background charge in a finite (spherical, deformed, vibrating and/or rotating, ...) volume. This is basically the origin of the favoured *jellium* model, as used in the description of metallic bulk and surface properties (Brack et al., 1991a, 1991b, 1992).

Many studies have been carried out (see e.g. Brack, 1992) for finite clusters, and the common outcome is that a self-consistent jellium model can account qualitatively, and sometimes even in great detail, for the experimentally observed metallic cluster properties (Nishioka (1990), Bjørnholm (1990), Brack (1992)).

Thc justification for the jellium model (see Fig. 10.10 as an illustration of the density and self-consistent potential for the neutral Na_{198} cluster (Brack, 1992)) will remain an object of much debate and research for some time to come. Its big virtue is, of course, its applicability to large cluster systems containing up to a few thousand atoms. The most pertinent and beautiful example is the correct explanation of shells and supershells in large alkali cluster systems.

The Extreme Simplification: The Square-well Potential. As can be observed from the self-consistent total external potential the remaining valence electrons feel (see Fig. 10.10), a wide square-well with finite depth V_0 which seems a very good approximation to the jellium model. The study of the stability of cluster systems then reduces to the shell-structure calculation within such a square-well potential with finite depth and so, to the radial equation

$$-\frac{\hbar^2}{2m}\frac{d^2u}{dr^2} + \left[\frac{\ell(\ell+1)\hbar^2}{2mr^2} - V_0\right] u = Eu \,. \qquad (10.29)$$

This problem is very well known and leads to the spherical Bessel function solutions with the eigenvalue equation leading to a quantization of the energy. The latter becomes particularly simple for an infinite square-well extension which should hold for the deeply-bound electron states. By taking up the lowest roots of the various $j_\ell(kr)$ solutions (with k related to the energy E by the expression

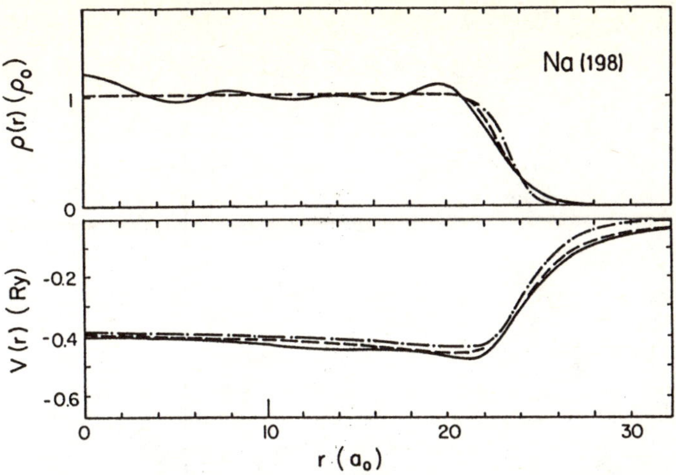

Fig. 10.10. Variational density $\rho(r)$ (in units of ρ_0) and the self-consistent total potential $V(r)$ for the neutral cluster Na_{198} (taken from Brack (1992))

$E_{n\ell} = \frac{\hbar^2}{2mR_0^2}(X_{n,\ell}^2))$ (see Table 10.4) one can already obtain a good description of the lowest stability configurations as observed in the I_N distributions (see also Fig. 10.11).

Table 10.4. Lowest roots of $\ell = 0, \ldots 10$ spherical Bessel functions (less than 19.00)

$\ell = 0$		$\ell = 1$		$\ell = 2$	
$n = 1$	3.1416	$n = 1$	4.4934	$n = 1$	5.7634
2	6.2832	2	7.7252	2	9.0950
3	9.4248	3	10.9041	3	12.3229
4	12.5664	4	14.0661	4	15.5146
5	15.7079	5	17.2207	5	18.6890
6	18.8495				
$\ell = 3$		$\ell = 4$		$\ell = 5$	
$n = 1$	6.9879	$n = 1$	8.1825	$n = 1$	9.3558
2	10.4171	2	11.7049	2	12.9665
3	13.6980	3	15.0396	3	16.3547
4	16.9236	4	18.3012		
$\ell = 6$		$\ell = 7$		$\ell = 8$	
$n = 1$	10.5128	$n = 1$	11.6570	$n = 1$	12.7908
2	14.2074	2	15.4312	2	16.6410
3	17.6480	3	18.9229		
$\ell = 9$		$\ell = 10$			
$b = 1$	13.9158	$n = 1$	15.0334		
2	17.8386				

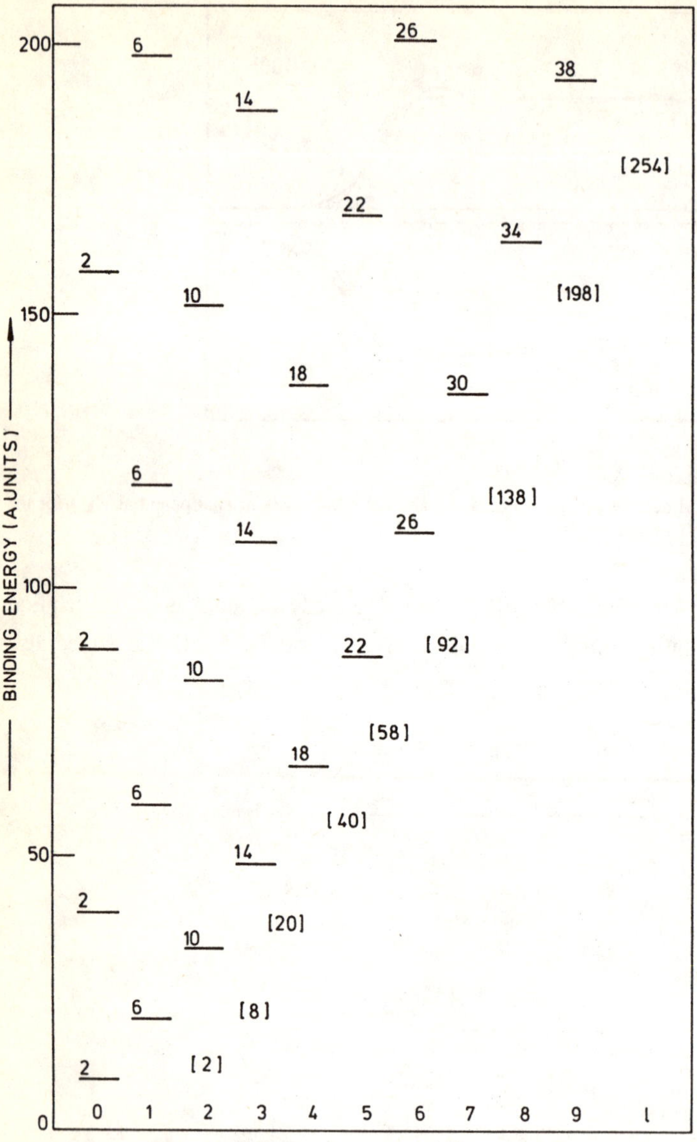

Fig. 10.11. Square-well level structure (n, ℓ) for the energy eigenvalues $E_{n,\ell}$. The bound states up to $\ell = 9$ $(n = 1)$ and with a cut-off near the $\ell = 6$ $(n = 2)$ eigenvalue are given. Both partial and cumulative occupation (between square brackets) numbers are given. The energy scale is in arbitrary units

10.4 Summary

In the present Chapter, it was shown that in systems ranging from a rather simple bound two-body character (e^+e^- positronium, $q\bar{q}$ quarkonium) up to highly complex many-body metal clusters, organized shell-structures appear.

In the case of positronium, a shell-structure highly reminescent of the Hydrogen atom governed by the Coulomb force ($1/r$ dependence) occurs. As well as the highly-degenerate orbital momentum symmetries (2S, 1P; 3S, 2P, 1D;...), various QED effects (spin-orbit splitting, hyperfine interactions) are needed in order to account for the finer details in the bound electron-positron system.

The analogous quark-antiquark charmonium ($c\bar{c}$) "bound" system gives rise to a number of clearly discernible resonances, known as the J/ψ particle(s) and the family of its excited states. Here, the "binding" potential is not very well known. The confinement idea of quarks permanently bound suggests that a linear (r) potential could give a good guideline. The non-relativistic one-body Schrödinger equation was studied in this linear potential model and energy eigenvalues and wave functions studied connected with the Airy-function properties. Here, too, one has to include spin-orbit and spin-spin interactions in order to account for the family of $c\bar{c}$ states as observed.

At the other extreme end, it was observed that in the mass abundance spectra for certain metallic cluster systems (Ag, Au, Cs, Na,...), extra stability arose for certain stable configurations. It very soon became clear that the quantum mechanics could be simplified to the study of the possible stable configurations for the "delocalized" electrons within the background of a positively charged "jellium" potential. This resembles the motion of an electron in an almost square-well potential giving rise in a very natural way to the extra stability associated with the $N = 2, 8, 20, 40, 58, 92, \ldots$ configurations. In Chap. 10, we presented some of the peculiarities of the mass abundance distributions, as well as theoretical attempts to describe such highly-complex many-body interacting systems subject to the Coulomb force as the major agent in causing the extra stability of the metal clusters.

These two extremes (two-body systems, many-body systems) reflect the rich structure presented in the organizational possibilities of the one-body Schrödinger equation for simple potentials (Coulomb: $1/r$ in positronium, atoms, metal clusters, oscillator r^2 in the atomic nucleus; linear potential r in the quark-antiquark bound system).

Problem

10.1 Starting from an infinitely-high square-well potential, construct the energy eigenvalue spectrum. Show that this simple potential model can account for the lowest stable configurations in metal clusters. Study the modifications that occur when the square-well potential is changed into a finite well.

Appendix

A. The Angular Momentum Operator in Spherical Coordinates

The transformation from cartesian to spherical coordinates for the angular momentum operator components (L_x, L_y and L_z) can be evaluated in detail. Starting from Chap. 1, Fig. 1.1, one has the relations

$$\begin{aligned} r &= (x^2 + y^2 + z^2)^{1/2} , \\ \varphi &= \mathrm{tg}^{-1} \frac{y}{x} , \\ \theta &= \mathrm{tg}^{-1} \left(\frac{x^2 + y^2}{z^2} \right)^{1/2} . \end{aligned} \tag{A.1}$$

The derivatives of r, φ and θ with respect to the cartesian coordinates x, y, z become, respectively

$$\begin{aligned} \frac{\partial r}{\partial x} &= \sin\theta \cdot \cos\varphi , \\ \frac{\partial r}{\partial y} &= \sin\theta \cdot \sin\varphi , \\ \frac{\partial r}{\partial z} &= \cos\theta , \end{aligned} \tag{A.2}$$

$$\begin{aligned} \frac{\partial \varphi}{\partial x} &= -\frac{1}{r} \frac{\sin\varphi}{\sin\theta} , \\ \frac{\partial \varphi}{\partial y} &= \frac{1}{r} \frac{\cos\varphi}{\sin\theta} , \\ \frac{\partial \varphi}{\partial z} &= 0 , \end{aligned} \tag{A.3}$$

and

$$\begin{aligned} \frac{\partial \theta}{\partial x} &= \frac{1}{r} \cos\theta \cdot \cos\varphi , \\ \frac{\partial \theta}{\partial y} &= \frac{1}{r} \cos\theta \cdot \sin\varphi , \\ \frac{\partial \theta}{\partial z} &= -\frac{1}{r} \sin\theta . \end{aligned} \tag{A.4}$$

Now, we calculate

$$L_z = \alpha \left(x\frac{\partial}{\partial y} - y\frac{\partial}{\partial x} \right) , \tag{A.5}$$

or, written out in detail

$$\begin{aligned} L_z = \alpha \Big\{ & r \sin\theta \cos\varphi \left[\sin\theta \sin\varphi \frac{\partial}{\partial r} + \frac{1}{r} \cos\theta \sin\varphi \frac{\partial}{\partial \theta} + \frac{1}{r} \cdot \frac{\cos\varphi}{\sin\theta} \frac{\partial}{\partial \varphi} \right] \\ & - r \sin\theta \sin\varphi \left[\sin\theta \cos\varphi \frac{\partial}{\partial r} + \frac{1}{r} \cos\theta \cos\varphi \frac{\partial}{\partial \theta} - \frac{1}{r} \frac{\sin\varphi}{\sin\theta} \frac{\partial}{\partial \varphi} \right] \Big\} , \end{aligned} \tag{A.6}$$

or

$$\begin{aligned} L_z = \alpha \Big(& r \sin^2\theta \cos\varphi \sin\varphi \frac{\partial}{\partial r} + \sin\theta \cos\theta \sin\varphi \cos\varphi \frac{\partial}{\partial \theta} \\ & + \cos^2\varphi \frac{\partial}{\partial \varphi} - r \sin^2\theta \cos\varphi \sin\varphi \frac{\partial}{\partial r} \\ & - \sin\theta \cos\theta \sin\varphi \cos\varphi \frac{\partial}{\partial \theta} + \sin^2\varphi \frac{\partial}{\partial \varphi} \Big) , \end{aligned} \tag{A.7}$$

or

$$L_z = \alpha \frac{\partial}{\partial \varphi} . \tag{A.8}$$

Completely similar calculations can be carried out to determine the other components L_y and L_z. Subsequently, one easily determines the value of $\boldsymbol{L}^2 \equiv L_x^2 + L_y^2 + L_z^2$.

B. Explicit Calculation of the Transformation Coefficients for Three-Angular Momentum Systems

An explicit calculation of the transformation (1.119) for a system of three independent angular momenta goes as follows.

First, we uncouple the coupled basis states $|j_1(j_2j_3)J_{23}; JM\rangle$ explicitly via the expression

$$\begin{aligned} |j_1(j_2j_3)J_{23}; JM\rangle = \sum_{\substack{m_1, m_2, \\ m_3, M_{23}}} & \langle j_2m_2, j_3m_3|J_{23}M_{23}\rangle \\ & \times \langle j_1m_1, J_{23}M_{23}|JM\rangle |j_1m_1\rangle |j_2m_2\rangle |j_3m_3\rangle . \end{aligned} \tag{B.1}$$

Coupling now the angular momenta $(j_1j_2)J_{12}$ and then $(J_{12}j_3)J$ in order to recover the transformation (1.119) one gets

$$\begin{aligned} |j_1(j_2j_3)J_{23}; JM\rangle = \sum_{\substack{m_1, m_2, m_3, \\ M_{12}, M_{23}, J_{12}}} & \langle j_2m_2, j_3m_3|J_{23}M_{23}\rangle \\ & \times \langle j_1m_1, J_{23}M_{23}|JM\rangle \langle j_1m_1, j_2m_2|J_{12}M_{12}\rangle \\ & \times \langle J_{12}M_{12}, j_3m_3|JM\rangle |(j_1j_2)J_{12}j_3; JM\rangle . \end{aligned} \tag{B.2}$$

In the latter expression (B.2), we can also sum over the magnetic quantum number M, and since the transformation coefficients are independent of M, one divides by the factor $(2J+1)$.

Rewriting (B.2) in terms of the Wigner $3j$-symbols we get the following expression for the transformation (1.119) (we make use of the notation $\hat{\jmath} \equiv (2j+1)^{1/2}$ throughout the appendices, as well as in the text)

$$
\begin{aligned}
|j_1(j_2j_3)J_{23}; JM\rangle = \sum_{J_{12}} \Bigg\{ & \sum_{\substack{\text{all magn.}\\ \text{qn}}} (2J+1)^{-1} \hat{J}_{12}\hat{J}_{23}\hat{J}\hat{J} \\
& \times (-1)^{j_1-j_2-M_{12}}(-1)^{J_{12}-j_3-M} \\
& \times (-1)^{j_2-j_3-M_{23}}(-1)^{j_1-J_{23}-M} \\
& \times \begin{pmatrix} j_1 & j_2 & J_{12} \\ m_1 & m_2 & -M_{12} \end{pmatrix} \begin{pmatrix} J_{12} & j_3 & J \\ M_{12} & m_3 & -M \end{pmatrix} \\
& \times \begin{pmatrix} j_2 & j_3 & J_{23} \\ m_2 & m_3 & -M_{23} \end{pmatrix} \begin{pmatrix} j_1 & J_{23} & J \\ m_1 & M_{23} & -M \end{pmatrix} \Bigg\} \\
& \times |(j_1j_2)J_{12}j_3; JM\rangle \ .
\end{aligned} \tag{B.3}
$$

Now, making use of the quantity between curly brackets, where the sum over *all* magnetic quantum numbers over a product of four Wigner $3j$-symbols is observed, and (1.121), one finally gets the result (1.119) back as

$$
\begin{aligned}
|j_1(j_2j_3)J_{23}; JM\rangle = \sum_{J_{12}} & (-1)^{j_1+j_2+j_3+J} \hat{J}_{12}\hat{J}_{23} \\
& \times \begin{Bmatrix} j_1 & j_2 & J_{12} \\ j_3 & J & J_{23} \end{Bmatrix} |(j_1j_2)J_{12}j_3; JM\rangle \ .
\end{aligned} \tag{B.4}
$$

Thereby, we have proved the transformation relation and obtain the result in terms of a Wigner $6j$-symbol.

C. Tensor Reduction Formulae for Tensor Products

Here, we give an explicit derivation of the reduction formula (2.106). The method used is rather general and can also be used for deriving other reduction formulae.

We start from the standard matrix element to which we apply the Wigner-Eckart theorem

$$
\begin{aligned}
M \equiv & \langle \alpha_1 j_1, \alpha_2 j_2; JM | T^{(k)}_{\kappa} | \alpha'_1 j'_1, \alpha'_2 j'_2; J'M' \rangle \\
= & (-1)^{J-M} \begin{pmatrix} J & k & J' \\ -M & \kappa & M' \end{pmatrix} \\
& \times \langle \alpha_1 j_1, \alpha_2 j_2; J || \boldsymbol{T}^{(k)} || \alpha'_1 j'_1, \alpha'_2 j'_2; J' \rangle \ .
\end{aligned} \tag{C.1}
$$

We can now work out the left-hand part of (C.1) by uncoupling the wave functions for the combined system and also uncoupling the tensor operator $T_\kappa^{(k)}$ into its uncoupled form. One gets

$$\begin{aligned} M = &\sum \langle j_1 m_1, j_2 m_2 | JM \rangle \langle j_1' m_1', j_2' m_2' | J'M' \rangle \\ &\times \langle k_1 \kappa_1, k_2 \kappa_2 | k\kappa \rangle \langle \alpha_1 j_1 m_1 | T_{\kappa_1}^{(k_1)} | \alpha_1' j_1' m_1' \rangle \\ &\times \langle \alpha_2 j_2 m_2 | T_{\kappa_2}^{(k_2)} | \alpha_2' j_2' m_2' \rangle \,. \end{aligned} \tag{C.2}$$

If we apply the Wigner-Eckart theorem again in (C.2), one obtains for M, the expression, in short-hand

$$\begin{aligned} M = &\sum \langle \ldots | \ldots \rangle \langle \ldots | \ldots \rangle \langle \ldots | \ldots \rangle (-1)^{j_1 - m_1} \begin{pmatrix} j_1 & k_1 & j_1' \\ -m_1 & \kappa_1 & m_1' \end{pmatrix} \\ &\times (-1)^{j_2 - m_2} \begin{pmatrix} j_2 & k_2 & j_2' \\ -m_2 & \kappa_2 & m_2' \end{pmatrix} \langle \alpha_1 j_1 || \boldsymbol{T}^{(k_1)} || \alpha_1' j_1' \rangle \\ &\times \langle \alpha_2 j_2 || \boldsymbol{T}^{(k_2)} || \alpha_2' j_2' \rangle \,. \end{aligned} \tag{C.3}$$

In (C.2) and (C.3), the sum goes over the magnetic quantum numbers $\{m_1, m_2, m_1', m_2', \kappa_1, \kappa_2\}$. We rewrite the above expressions using the Wigner-$3j$-symbols and make use of the orthogonality relations for $3j$-symbols to bring the $3j$-symbol onto the right-hand side of (C.1) so as to calculate the reduced matrix element itself. We so obtain

$$\begin{aligned} &\langle \alpha_1 j_1, \alpha_2 j_2; J || \boldsymbol{T}^{(k)} || \alpha_1' j_1', \alpha_2' j_2'; J' \rangle \\ &= \sum_{\text{all } m_i} \begin{pmatrix} j_1 & j_2 & J \\ m_1 & m_2 & -M \end{pmatrix} \begin{pmatrix} j_1' & j_2' & J' \\ m_1' & m_2' & -M' \end{pmatrix} \begin{pmatrix} k_1 & k_2 & k \\ \kappa_1 & \kappa_2 & -\kappa \end{pmatrix} \\ &\quad \times \begin{pmatrix} j_1 & k_1 & j_1' \\ -m_1 & \kappa_1 & m_1' \end{pmatrix} \begin{pmatrix} j_2 & k_2 & j_2' \\ -m_2 & \kappa_2 & m_2' \end{pmatrix} \begin{pmatrix} J & k & J' \\ -M & \kappa & M' \end{pmatrix} \\ &\quad \times \hat{J} \hat{J}' \hat{k} (-1)^{j_1 - m_1} (-1)^{j_2 - m_2} \\ &\quad \times (-1)^{j_1 - j_2 + M} (-1)^{j_1' - j_2' + M'} (-1)^{k_1 - k_2 + \kappa} (-1)^{J - M} \\ &\quad \times \langle \alpha_1 j_1 || \boldsymbol{T}^{(k_1)} || \alpha_1' j_1' \rangle \langle \alpha_2 j_2 || \boldsymbol{T}^{(k_2)} || \alpha_2' j_2' \rangle \,, \end{aligned} \tag{C.4}$$

$$\begin{aligned} &= \sum_{\text{all } m_i} \begin{pmatrix} j_1 & j_2 & J \\ m_1 & m_2 & -M \end{pmatrix} \begin{pmatrix} j_1' & j_2' & J' \\ -m_1' & -m_2' & M' \end{pmatrix} \begin{pmatrix} k_1 & k_2 & k \\ -\kappa_1 & -\kappa_2 & \kappa \end{pmatrix} \hat{J} \hat{J}' \hat{k} \\ &\quad \times \begin{pmatrix} j_1 & j_1' & k_1 \\ m_1 & -m_1' & -\kappa_1 \end{pmatrix} \begin{pmatrix} j_2 & j_2' & k_2 \\ m_2 & -m_2' & -\kappa_2 \end{pmatrix} \begin{pmatrix} J & J' & k \\ -M & M' & \kappa \end{pmatrix} \\ &\quad \times (-1)^{\text{Phase}} \langle \alpha_1 j_1 || \boldsymbol{T}^{(k_1)} || \alpha_1' j_1' \rangle \langle \alpha_2 j_2 || \boldsymbol{T}^{(k_2)} || \alpha_2' j_2' \rangle \,, \end{aligned} \tag{C.5}$$

where

$$\begin{aligned} \text{Phase} = &j_1 - j_2 + j_1' - j_2' + k_1 - k_2 + J + j_1 + j_2 + j_1' + j_2' + J' \\ &+ k_1 + k_2 + k + J + J' + k - M - M' - \kappa - M - m_1 - m_2 \,. \end{aligned} \tag{C.6}$$

Since we know how the Wigner $9j$-symbol is defined in terms of a sum over all magnetic quantum numbers m_i, we use the definition

$$\begin{Bmatrix} j_{11} & j_{12} & j_{13} \\ j_{21} & j_{22} & j_{23} \\ j_{31} & j_{32} & j_{33} \end{Bmatrix} = \sum_{\text{all } m} \begin{pmatrix} j_{11} & j_{12} & j_{13} \\ m_{11} & m_{12} & m_{13} \end{pmatrix} \begin{pmatrix} j_{21} & j_{22} & j_{23} \\ m_{21} & m_{22} & m_{23} \end{pmatrix}$$

$$\times \begin{pmatrix} j_{31} & j_{32} & j_{33} \\ m_{31} & m_{32} & m_{33} \end{pmatrix} \begin{pmatrix} j_{11} & j_{21} & j_{31} \\ m_{11} & m_{21} & m_{31} \end{pmatrix}$$

$$\times \begin{pmatrix} j_{12} & j_{22} & j_{32} \\ m_{12} & m_{22} & m_{32} \end{pmatrix} \begin{pmatrix} j_{13} & j_{23} & j_{33} \\ m_{13} & m_{23} & m_{33} \end{pmatrix} . \tag{C.7}$$

So, from (C.6), one finally obtains

$$\langle \alpha_1 j_1, \alpha_2 j_2; J || \boldsymbol{T}^{(k)} || \alpha_1' j_1', \alpha_2' j_2'; J' \rangle$$

$$= \hat{J}\hat{J}'\hat{k} \begin{Bmatrix} j_1 & j_2 & J \\ j_1' & j_2' & J' \\ k_1 & k_2 & k \end{Bmatrix} \langle \alpha_1 j_1 || \boldsymbol{T}^{(k_1)} || \alpha_1' j_1' \rangle \langle \alpha_2 j_2 || \boldsymbol{T}^{(k_2)} || \alpha_2' j_2' \rangle . \tag{C.8}$$

D. The Surface-Delta Interaction (SDI)

The SDI interaction has some very interesting properties which make it quite a "realistic" interaction (Faessler 1967) although, at first glance, it looks just like any other effective interaction:

i) Interaction Between Nucleons at the Nuclear Surface. The local kinetic energy T for a nucleon within the nuclear potential $U(r)$ is much larger in the central region as compared to the nuclear surface region (see Fig. D.1) since, for a nucleon moving at the nuclear fermi surface, the total energy E is $E = T + U(r)$, a constant value. The nucleon scattering process is mainly a $l = 0$ wave (S-wave) process of nucleons that interact in a "frontal" way. The relative kinetic energy for nucleon-nucleon scattering in the nuclear interior is now much larger as compared to the nuclear surface region. It appears that, for S-wave scattering, the phase shift for ${}^1S_0$ scattering (${}^{2S+1}l_J$) increases with decreasing collision energy. From the Born approximation theory for nucleon-nucleon scattering, one has (Fetter, Walecka 1971)

$$\text{tg}\delta = -\frac{mk}{\hbar^2} \int_0^\infty V(r) j_l^2(kr) r^2 dr \, , \tag{D.1}$$

with

$$E = \frac{\hbar^2 k^2}{m} \, ,$$

the energy available in the relative system at which the phase parameters have to be evaluated. The $\delta({}^1S_0)$ phase is illustrated in Fig. D.2 and illustrates indeed the importance of scattering at the lower energy (nuclear surface region) as compared to the higher energy (nuclear interior) scattering process.

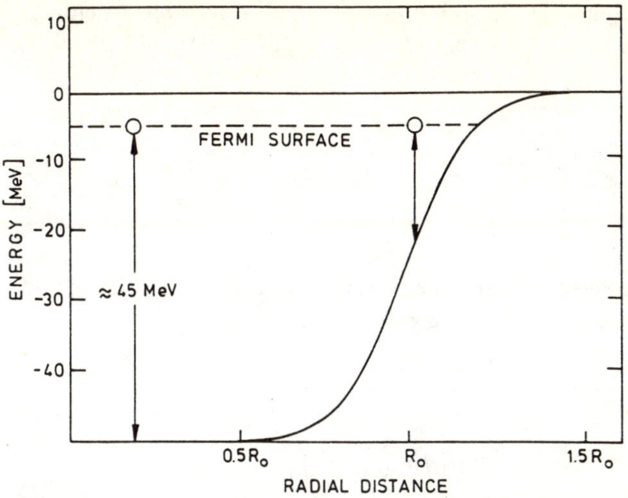

Fig. D.1. A Woods-Saxon potential is used to illustrate the relative kinetic energy. The well depth is $\cong 50$ MeV. The diffuseness parameter is $a_0 = 0.64$ fm and the radius parameter is 1.25 fm. The mass number is $A = 185$. The local kinetic energy of a particle near the Fermi surface is about 45 MeV. At the nuclear surface, the kinetic energy is appreciably lower (taken from (Faessler, Plastino 67))

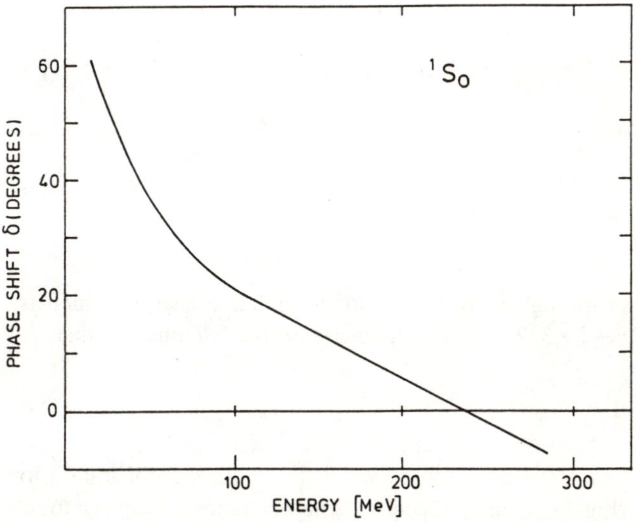

Fig. D.2. The singlet $l = 0$ wave ($^1S_0$) phase shift for proton-proton scattering (in a qualitative way) as a function of the laboratory bombarding energy

Of course the Born-integrals are not yet the relative integrals of the nuclear relative two-body interaction $V(r)$ (with the nucleon-nucleon relative coordinate) since in the integral the $j_l(kr)$ spherical Bessel functions show up and not the relative harmonic oscillator wave functions.

ii) Relation Between Spherical Bessel Functions and Harmonic Oscillator Functions (Abramowitz 1964). One can write the harmonic oscillator functions as

$$R_{nl}(r) = \left[\frac{2^{l-n+2}(2l+2n+1)!!\nu^{3/2}}{\pi^{1/2}\cdot n![(2l+1)!!]^2}\right]^{1/2} e^{-\nu r^2/2}(\nu r^2)^{l/2} \times {}_1F_1\left(-n; l+\frac{3}{2}; \nu r^2\right), \tag{D.2}$$

with ${}_1F_1(\alpha, \beta, x)$ the confluent, hypergeometric series and $\nu = m\omega/2\hbar$. There now exists a relation (for small r-values and large $n(n \to \infty)$) that

$$\begin{aligned} {}_1F_1\left(-n; l+\frac{3}{2}; \nu r^2\right) &= \Gamma\left(l+\frac{3}{2}\right) e^{\nu r^2/2} \\ &\times \left(\frac{1}{2}\left(2n+l+\frac{3}{2}\right)\nu r^2\right)^{-\frac{l}{2}-\frac{1}{4}} \\ &\times \left(\frac{2}{\pi}\right)^{1/2}\left(2\left(2n+l+\frac{3}{2}\right)\nu r^2\right)^{1/4} \\ &\times j_l((\nu\nu')^{1/2}r) \qquad \text{(D.3)} \\ &= \frac{(2l+1)!!}{2^{l/2}} e^{\nu r^2/2}\left(2n+l+\frac{3}{2}\right)^{-l/2} \\ &\times (\nu r^2)^{-l/2} j_l((\nu\nu')^{1/2}r)\,. \qquad \text{(D.4)} \end{aligned}$$

This now gives the relation

$$R_{nl}(r) = N_{nl}\frac{(2l+1)!!}{(2(2n+l+3/2))^{l/2}} j_l(kr)\,, \tag{D.5}$$

with $k^2 \equiv \nu\nu'$.

This relation follows through identification of the relative energy, which is, $\hbar^2k^2/m$ and also $\hbar\omega(2n+l+3/2)$, such that, since $\nu = m\omega/2\hbar$, one has that

$$\frac{k^2}{\nu} = 2\left(2n+l+\frac{3}{2}\right) = \nu'\,. \tag{D.6}$$

The relation (D.5) has been tested in the mass region $A \cong 140$. We illustrate it for $R_{1s}(r)$, R_{2s} and $j_0(kr)$ (Fig. D.3), and, indeed, for small r-values (clearly up to the nuclear radius $R_0 = 1.2\,A^{1/3}$ fm) this relation (D.5) is well fulfilled. Agreement increases with increasing n, and for equal n with increasing l.

The reduced integrals which appear in the calculation of Sect. 3.2.3d, when we couple to the relative spin wave function $|(1/2\,1/2); SM_S\rangle$ lead to

$$\langle(nl, S); \mathcal{J}T|V(r)|(nl, S); \mathcal{J}T\rangle = \int_0^\infty R_{nl}^2(r)V(r)r^2dr\,, \tag{D.7}$$

or

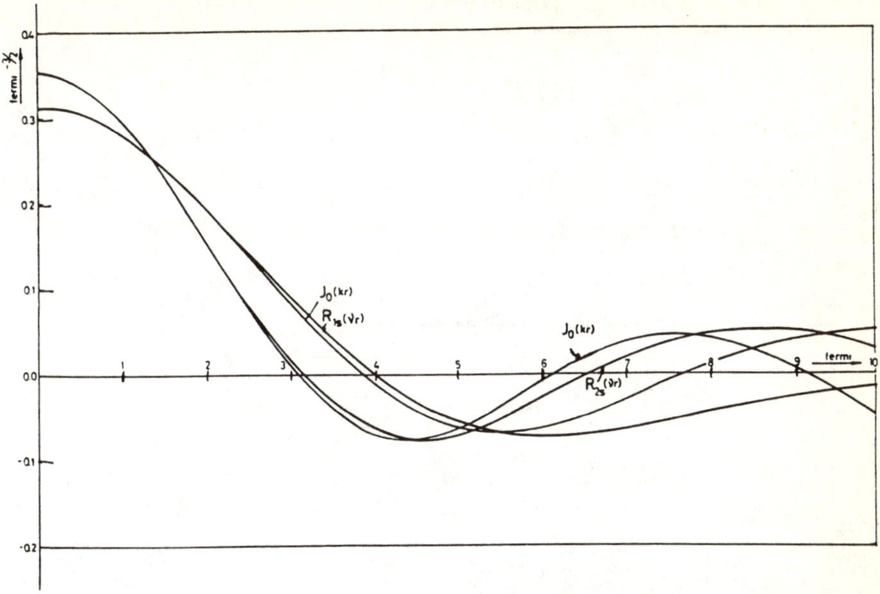

Fig. D.3. Comparison of the radial harmonic oscillator wave functions $R_{1s}(r)$, $R_{2s}(r)$ with the related spherical Bessel function $j_0(kr)$. The precise relation between both functions is expressed by the relation (D.5). The ordinate is in units $\mathrm{fm}^{-3/2}$

$$\langle (nl,S);\mathcal{J}T|V(r)|(nl,S);\mathcal{J}T\rangle = \frac{2^{l-n+1}(2l+2n+1)!!\hbar\omega\cdot km}{\pi^{1/2}\cdot n!(\nu')^{l+1/2}\hbar^2} \times \int_0^\infty V(r)j_l^2(kr)r^2dr\,. \tag{D.8}$$

Using the Born-approximation, relating the integral in (D.8) to the phases describing scattering in the $(l,S);\mathcal{J}T$ channel, one can finally write the radial integrals describing nuclear structure as

$$\langle (nl,S);\mathcal{J}T|V(r)|(nl,S);\mathcal{J}T\rangle = -\frac{2^{l-n+1}(2l+2n+1)!!\hbar\omega}{\pi^{1/2}\cdot n!(\nu')^{l+1/2}}\mathrm{tg}\delta(E)\,. \tag{D.9}$$

From (D.6) (giving ν'), the energy for scattering $E=\hbar^2k^2/m$ (the energy that is present in the relative nucleon-nucleon motion) and $\hbar\omega = 41\,A^{-1/3}$ MeV, one can now determine the relative integrals over the nuclear, relative 2-body interaction $V(r)$. One can use the tabulated values of Arndt, Wright and McGregor (see Table D.1) for $\delta(E)$ (McGregor 68a).

iii) A Delta-Force Expression. The free nucleon-nucleon interaction indeed has a short range. Even though the detailed form of the nucleon-nucleon interaction in the nuclear medium probably deviates from the free nucleon-nucleon interaction, the short-range property remains a major characteristic. This can be expressed most easily using a δ-function force $\delta(\boldsymbol{r}_1-\boldsymbol{r}_2)$.

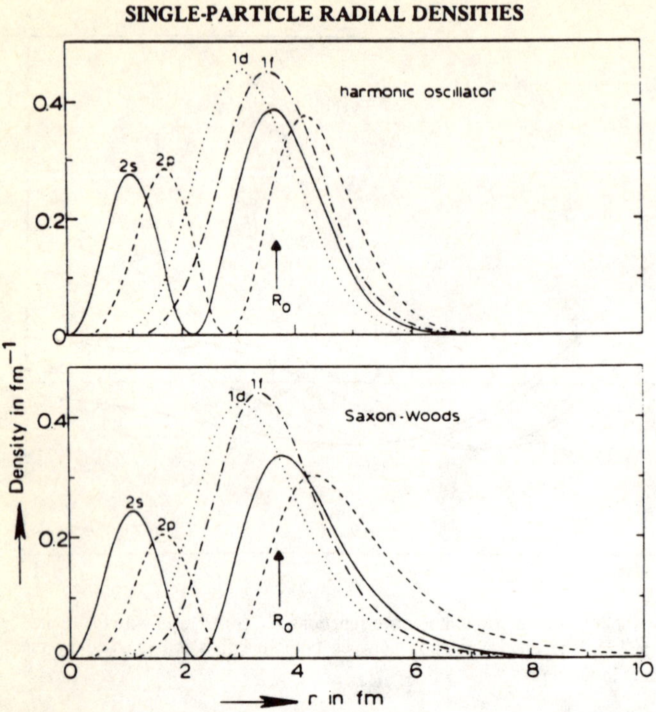

Fig. D.4. The probability density $r^2 R_{nl}^2(r)$ for a neutron in the $1d$, $2s$, $1f$ or $2p$ single-particle state calculated for the nucleus ^{29}Si. The plots illustrate that for both the harmonic oscillator and Woods-Saxon potential, the probabilities near the nuclear surface $r = R_0$ do not depend strongly (state-independence) on n *nor* l (taken from (Brussaard, Glaudemans 77))

iv) State Independence of $R_{nl}(r = R_0)$. If one calculates the probability of finding a nucleon between the distance $r, r+dr$, one gets the expression $4\pi r^2\, dr\, R_{nl}^2(r)$. For $r < R_0$, this function is strongly dependent on n and l. For $r \cong R_0$, however, this probability is almost constant (see Fig. D.4), and thus the radial integral, using the surface delta force $\delta(\boldsymbol{r}_1 - \boldsymbol{r}_2)\delta(r_1 - R_0)$, simplifies to

$$R_{n_1 l_1}(R_0) R_{n_2 l_2}(R_0) R_{n_3 l_3}(R_0) R_{n_4 l_4}(R_0) R_0^2 \, , \tag{D.10}$$

and can even be approximated by a constant (non state-dependent) quantity A (Brussaard, Glaudemans 1977).

E. Multipole Expansion of $\delta(\boldsymbol{r}_1 - \boldsymbol{r}_2)$ Interaction

In obtaining a multipole expansion of the delta effective interaction $\delta(\boldsymbol{r}_1 - \boldsymbol{r}_2)$, we start from the integral representation

Table D.1 Phase-shift analysis (with error bars) for a laboratory energy interval 5 MeV $\leq E_{\text{lab}} \leq$ 400 MeV. The phase shifts are given in degrees (taken from (Mac Gregor et al. 68a))

Lab energy (MeV)	$^1S_0$	$^1D_2$	$^1G_4$	$^3P_0$	$^3P_1$	$^3P_2$	ϵ_2
5	54.65 ± 0.03	0.06 ± 0.00	0.00 ± 0.00	1.77 ± 0.02	−1.09 ± 0.01	0.29 ± 0.01	−0.06 ± 0.00
10	54.97 ± 0.07	0.20 ± 0.00	0.00 ± 0.00	3.83 ± 0.04	−2.32 ± 0.01	0.80 ± 0.02	−0.23 ± 0.00
15	53.01 ± 0.09	0.38 ± 0.00	0.01 ± 0.00	5.61 ± 0.07	−3.41 ± 0.02	1.41 ± 0.03	−0.44 ± 0.00
20	50.75 ± 0.11	0.57 ± 0.01	0.03 ± 0.00	7.09 ± 0.10	−4.36 ± 0.03	2.07 ± 0.04	−0.66 ± 0.01
25	48.51 ± 0.11	0.77 ± 0.01	0.05 ± 0.00	8.28 ± 0.12	−5.20 ± 0.04	2.75 ± 0.04	−0.87 ± 0.01
30	46.36 ± 0.11	0.98 ± 0.01	0.07 ± 0.00	9.23 ± 0.14	−5.95 ± 0.05	3.43 ± 0.05	−1.08 ± 0.01
40	42.37 ± 0.12	1.38 ± 0.02	0.12 ± 0.00	10.54 ± 0.18	−7.28 ± 0.05	4.76 ± 0.05	−1.45 ± 0.02
50	38.78 ± 0.13	1.77 ± 0.02	0.17 ± 0.00	11.25 ± 0.20	−8.45 ± 0.06	6.02 ± 0.05	−1.76 ± 0.02
60	35.55 ± 0.14	2.15 ± 0.03	0.23 ± 0.00	11.51 ± 0.22	−9.51 ± 0.06	7.19 ± 0.05	−2.03 ± 0.03
70	32.62 ± 0.16	2.53 ± 0.04	0.29 ± 0.00	11.45 ± 0.23	−10.51 ± 0.07	8.27 ± 0.05	−2.25 ± 0.03
80	29.94 ± 0.18	2.89 ± 0.04	0.35 ± 0.00	11.13 ± 0.23	−11.47 ± 0.07	9.26 ± 0.05	−2.43 ± 0.04
90	27.48 ± 0.20	3.25 ± 0.05	0.41 ± 0.01	10.62 ± 0.23	−12.39 ± 0.07	10.17 ± 0.05	−2.57 ± 0.04
100	25.21 ± 0.21	3.60 ± 0.05	0.47 ± 0.01	9.97 ± 0.23	−13.29 ± 0.07	10.99 ± 0.05	−2.68 ± 0.04
120	21.08 ± 0.23	4.27 ± 0.06	0.59 ± 0.01	8.36 ± 0.23	−15.02 ± 0.07	12.41 ± 0.06	−2.83 ± 0.04
140	17.38 ± 0.25	4.91 ± 0.07	0.71 ± 0.02	6.49 ± 0.24	−16.70 ± 0.09	13.58 ± 0.06	−2.91 ± 0.04
160	13.96 ± 0.27	5.52 ± 0.08	0.82 ± 0.03	4.50 ± 0.26	−18.33 ± 0.11	14.53 ± 0.07	−2.91 ± 0.05
180	10.71 ± 0.29	6.10 ± 0.10	0.93 ± 0.03	2.44 ± 0.30	−19.91 ± 0.13	15.30 ± 0.08	−2.87 ± 0.06
200	7.58 ± 0.31	6.66 ± 0.11	1.04 ± 0.04	0.38 ± 0.34	−21.46 ± 0.16	15.91 ± 0.09	−2.79 ± 0.08
220	4.51 ± 0.34	7.19 ± 0.12	1.15 ± 0.05	−1.65 ± 0.40	−22.96 ± 0.20	16.39 ± 0.10	−2.68 ± 0.09
240	1.46 ± 0.38	7.69 ± 0.14	1.25 ± 0.06	−3.64 ± 0.46	−24.43 ± 0.24	16.77 ± 0.12	−2.55 ± 0.12
260	−1.57 ± 0.43	8.17 ± 0.16	1.35 ± 0.07	−5.57 ± 0.53	−25.86 ± 0.27	17.04 ± 0.15	−2.39 ± 0.14
280	−4.62 ± 0.50	8.63 ± 0.17	1.45 ± 0.08	−7.43 ± 0.60	−27.26 ± 0.30	17.24 ± 0.18	−2.23 ± 0.17
300	−7.67 ± 0.59	9.07 ± 0.19	1.55 ± 0.09	−9.22 ± 0.69	−28.62 ± 0.34	17.37 ± 0.21	−2.05 ± 0.20
320	−10.75 ± 0.71	9.49 ± 0.21	1.65 ± 0.10	−10.93 ± 0.76	−29.94 ± 0.37	17.44 ± 0.24	−1.86 ± 0.23
340	−13.85 ± 0.84	9.89 ± 0.23	1.74 ± 0.11	−12.57 ± 0.83	−31.23 ± 0.40	17.46 ± 0.27	−1.67 ± 0.26
360	−16.97 ± 0.99	10.28 ± 0.25	1.83 ± 0.12	−14.12 ± 0.90	−32.49 ± 0.43	17.44 ± 0.31	−1.47 ± 0.29
380	−20.11 ± 1.15	10.64 ± 0.27	1.92 ± 0.13	−15.61 ± 0.97	−33.72 ± 0.46	17.38 ± 0.35	−1.27 ± 0.32
400	−23.27 ± 1.33	11.00 ± 0.29	2.01 ± 0.14	−17.02 ± 1.04	−34.91 ± 0.49	17.28 ± 0.39	−1.07 ± 0.35

$$\delta(\boldsymbol{r}_1 - \boldsymbol{r}_2) = \frac{1}{(2\pi)^3} \int \mathrm{e}^{i\boldsymbol{k}\cdot(\boldsymbol{r}_1 - \boldsymbol{r}_2)} d\boldsymbol{k} \,, \tag{E.1}$$

$$= \frac{1}{(2\pi)^3} \int \mathrm{e}^{i\boldsymbol{k}\cdot\boldsymbol{r}_1} \cdot \mathrm{e}^{-i\boldsymbol{k}\cdot\boldsymbol{r}_2} d\boldsymbol{k} \,. \tag{E.2}$$

We use the plane-wave expansions

$$\mathrm{e}^{i\boldsymbol{k}\cdot\boldsymbol{r}_1} = \sum_{l,m} (4\pi) i^l j_l(kr_1) Y_l^{m^*}(\hat{\boldsymbol{k}}) Y_l^m(\hat{\boldsymbol{r}}_1) \,, \tag{E.3}$$

and

$$\mathrm{e}^{-i\boldsymbol{k}\cdot\boldsymbol{r}_2} = \sum_{l',m'} (4\pi) i^{l'} (-1)^{l'} j_{l'}(kr_2) Y_{l'}^{m'}(\hat{\boldsymbol{k}}) Y_{l'}^{m'^*}(\hat{\boldsymbol{r}}_2) \,. \tag{E.4}$$

Combining these results, one obtains

$$\delta(\boldsymbol{r}_1 - \boldsymbol{r}_2) = \frac{1}{(2\pi)^3} \sum_{l,l',m,m'} \int (16\pi^2) i^{l+l'} (-1)^{l'} j_l(kr_1) j_{l'}(kr_2)$$

$$\times Y_l^m(\hat{\boldsymbol{r}}_1) Y_{l'}^{m'^*}(\hat{\boldsymbol{r}}_2) Y_l^{m^*}(\hat{\boldsymbol{k}}) Y_{l'}^{m'}(\hat{\boldsymbol{k}}) k^2 dk \, d\hat{\boldsymbol{k}} \,. \tag{E.5}$$

Table D.1 (continued)

Lab energy (MeV)	$^3F_2$	$^3F_3$	$^3F_4$	ϵ_4	$^3H_4$	$^3H_5$	$^3H_6$
5	0.00 ± 0.00	−0.01 ± 0.00	0.00 ± 0.00	0.00 ± 0.00	0.00 ± 0.00	0.00 ± 0.00	0.00 ± 0.00
10	0.01 ± 0.00	−0.04 ± 0.00	0.00 ± 0.00	0.00 ± 0.00	0.00 ± 0.00	0.00 ± 0.00	0.00 ± 0.00
15	0.04 ± 0.00	−0.10 ± 0.00	0.01 ± 0.00	−0.02 ± 0.00	0.00 ± 0.00	0.00 ± 0.00	0.00 ± 0.00
20	0.07 ± 0.00	−0.17 ± 0.00	0.01 ± 0.00	−0.03 ± 0.00	0.00 ± 0.00	−0.01 ± 0.00	0.00 ± 0.00
25	0.10 ± 0.00	−0.26 ± 0.00	0.02 ± 0.00	−0.06 ± 0.00	0.00 ± 0.00	−0.02 ± 0.00	0.00 ± 0.00
30	0.14 ± 0.00	−0.35 ± 0.00	0.04 ± 0.00	−0.08 ± 0.00	0.01 ± 0.00	−0.03 ± 0.00	0.00 ± 0.00
40	0.22 ± 0.00	−0.55 ± 0.00	0.08 ± 0.00	−0.14 ± 0.00	0.02 ± 0.00	−0.06 ± 0.00	0.00 ± 0.00
50	0.30 ± 0.01	−0.75 ± 0.01	0.13 ± 0.00	−0.21 ± 0.00	0.03 ± 0.00	−0.10 ± 0.00	0.01 ± 0.00
60	0.37 ± 0.01	−0.94 ± 0.01	0.18 ± 0.01	−0.28 ± 0.00	0.04 ± 0.00	−0.14 ± 0.00	0.01 ± 0.00
70	0.43 ± 0.02	−1.12 ± 0.02	0.25 ± 0.01	−0.35 ± 0.00	0.06 ± 0.00	−0.19 ± 0.00	0.02 ± 0.00
80	0.48 ± 0.02	−1.28 ± 0.02	0.32 ± 0.01	−0.42 ± 0.01	0.08 ± 0.00	−0.24 ± 0.01	0.03 ± 0.00
90	0.53 ± 0.03	−1.44 ± 0.03	0.40 ± 0.01	−0.49 ± 0.01	0.10 ± 0.01	−0.29 ± 0.00	0.04 ± 0.00
100	0.56 ± 0.04	−1.58 ± 0.04	0.48 ± 0.02	−0.55 ± 0.01	0.12 ± 0.01	−0.35 ± 0.01	0.05 ± 0.00
120	0.61 ± 0.06	−1.83 ± 0.06	0.65 ± 0.03	−0.66 ± 0.01	0.16 ± 0.01	−0.46 ± 0.02	0.08 ± 0.01
140	0.64 ± 0.08	−2.04 ± 0.08	0.83 ± 0.03	−0.77 ± 0.02	0.20 ± 0.02	−0.57 ± 0.04	0.11 ± 0.01
160	0.64 ± 0.10	−2.22 ± 0.11	1.01 ± 0.04	−0.86 ± 0.03	0.25 ± 0.03	−0.68 ± 0.05	0.15 ± 0.01
180	0.62 ± 0.13	−2.37 ± 0.13	1.20 ± 0.05	−0.93 ± 0.04	0.29 ± 0.04	−0.78 ± 0.07	0.19 ± 0.02
200	0.59 ± 0.15	−2.50 ± 0.15	1.39 ± 0.06	−1.00 ± 0.05	0.34 ± 0.05	−0.88 ± 0.09	0.23 ± 0.03
220	0.55 ± 0.17	−2.61 ± 0.18	1.58 ± 0.07	−1.06 ± 0.06	0.38 ± 0.07	−0.98 ± 0.11	0.27 ± 0.03
240	0.49 ± 0.20	−2.71 ± 0.20	1.77 ± 0.08	−1.11 ± 0.07	0.42 ± 0.08	−1.08 ± 0.14	0.32 ± 0.04
260	0.43 ± 0.22	−2.79 ± 0.23	1.96 ± 0.09	−1.15 ± 0.08	0.46 ± 0.10	−1.17 ± 0.16	0.36 ± 0.05
280	0.37 ± 0.24	−2.85 ± 0.25	2.14 ± 0.10	−1.19 ± 0.09	0.50 ± 0.12	−1.26 ± 0.19	0.41 ± 0.06
300	0.30 ± 0.27	−2.91 ± 0.27	2.32 ± 0.11	−1.22 ± 0.10	0.54 ± 0.14	−1.34 ± 0.22	0.46 ± 0.06
320	0.23 ± 0.29	−2.96 ± 0.30	2.50 ± 0.12	−1.24 ± 0.11	0.58 ± 0.16	−1.43 ± 0.25	0.51 ± 0.07
340	0.15 ± 0.31	−3.01 ± 0.32	2.67 ± 0.13	−1.26 ± 0.12	0.62 ± 0.18	−1.51 ± 0.28	0.57 ± 0.08
360	0.08 ± 0.34	−3.04 ± 0.34	2.85 ± 0.14	−1.28 ± 0.14	0.65 ± 0.20	−1.58 ± 0.31	0.62 ± 0.09
380	0.00 ± 0.36	−3.07 ± 0.37	3.01 ± 0.15	−1.29 ± 0.15	0.69 ± 0.22	−1.66 ± 0.34	0.67 ± 0.10
400	−0.08 ± 0.38	−3.10 ± 0.39	3.18 ± 0.16	−1.30 ± 0.16	0.72 ± 0.24	−1.73 ± 0.38	0.72 ± 0.11

This can be integrated using the angular element $d\hat{\boldsymbol{k}}$ to give

$$\delta(\boldsymbol{r}_1 - \boldsymbol{r}_2) = \frac{16\pi^2}{8\pi^3} \sum_{l,m} \int j_l(kr_1) j_l(kr_2) k^2 dk Y_l^m(\hat{\boldsymbol{r}}_1) Y_l^{m^*}(\hat{\boldsymbol{r}}_2) . \tag{E.6}$$

The integral $\int \ldots dk$ gives the result $(\pi/2)\delta(r_1 - r_2)/(r_1 r_2)$,, and one obtains finally

$$\delta(\boldsymbol{r}_1 - \boldsymbol{r}_2) = \sum_{l,m} \frac{\delta(r_1 - r_2)}{r_1 r_2} Y_l^m(\hat{\boldsymbol{r}}_1) Y_l^{m^*}(\hat{\boldsymbol{r}}_2) , \tag{E.7}$$

or

$$\delta(\boldsymbol{r}_1 - \boldsymbol{r}_2) = \sum_{l} \frac{\delta(r_1 - r_2)}{r_1 r_2} \frac{2l+1}{4\pi} P_l(\cos\theta_{12}) , \tag{E.8}$$

which leads to

$$v_l(r_1, r_2) = \frac{2l+1}{4\pi} \frac{\delta(r_1 - r_2)}{r_1 r_2} . \tag{E.9}$$

F. Calculation of Reduced Matrix Element $\langle(1/2l)j||Y_k||(1/2l')j'\rangle$ and Some Important Angular Momentum Relations

Here, besides a calculation of the reduced matrix elements for the spherical harmonics taken between single-particle states (n,l,j), we also discuss some of the necessary angular momentum coefficient relations, needed in the calculation of the two-body matrix elements for a δ-interaction (see Sect. 3.2.3).

We start from the recoupling expressions, transforming among the equivalent three-angular momentum basis states,

$$|(j_1j_2)J_{12}j_3;JM\rangle = \sum_{J_{23}}(-1)^{j_1+j_2+j_3+J}\hat{J}_{12}\hat{J}_{23} \times \begin{Bmatrix} j_1 & j_2 & J_{12} \\ j_3 & J & J_{23} \end{Bmatrix} |j_1(j_2j_3)J_{23};JM\rangle\ . \tag{F.1}$$

This now can be rewritten as

$$\sum_{M_{12}}\left\{\sum_{\text{all } m_i}\langle j_1m_1,j_2m_2|J_{12}M_{12}\rangle\langle J_{12}M_{12},j_3m_3|JM\rangle\right\} \times |j_1m_1\rangle|j_2m_2\rangle|j_3m_3\rangle$$
$$= \sum_{J_{23},M_{23}}\left\{\sum_{\text{all } m_i}\langle j_2m_2,j_3m_3|J_{23}M_{23}\rangle\langle j_1m_1,J_{23}M_{23}|JM\rangle\hat{J}_{12}\hat{J}_{23} \times(-1)^{j_1+j_2+j_3+J}\begin{Bmatrix} j_1 & j_2 & J_{12} \\ j_3 & J & J_{23} \end{Bmatrix}\right\}|j_1m_1\rangle|j_2m_2\rangle|j_3m_3\rangle\ . \tag{F.2}$$

Herefrom, by re-ordering the summation symbols, one obtains

$$\sum_{M_{12}}\langle j_1m_1,j_2m_2|J_{12}M_{12}\rangle\langle J_{12}M_{12},j_3m_3|JM\rangle = \sum_{J_{23},M_{23}}(-1)^{j_1+j_2+j_3+J}\hat{J}_{12}\hat{J}_{23}\begin{Bmatrix} j_1 & j_2 & J_{12} \\ j_3 & J & J_{23} \end{Bmatrix} \times\langle j_2m_2,j_3m_3|J_{23}M_{23}\rangle\langle j_1m_1,J_{23}M_{23}|JM\rangle\ . \tag{F.3}$$

By putting everything in terms of Wigner $3j$-symbols, one finally gets the particular relation

$$\sum_{J_{23},M_{23}}(-1)^{J_{23}}\begin{pmatrix} j_3 & j_2 & J_{23} \\ m_3 & m_2 & -M_{23} \end{pmatrix}\begin{pmatrix} j_1 & J & J_{23} \\ m_1 & -M & M_{23} \end{pmatrix} \times\begin{Bmatrix} j_1 & j_2 & J_{12} \\ j_3 & J & J_{23} \end{Bmatrix}(2J_{23}+1)$$

$$= \sum_{M_{12}} (-1)^{J_{12}+m_1+m_3} \begin{pmatrix} j_1 & j_2 & J_{12} \\ m_1 & m_2 & -M_{12} \end{pmatrix}$$
$$\times \begin{pmatrix} j_3 & J & J_{12} \\ m_3 & -M & M_{12} \end{pmatrix} . \tag{F.4}$$

Using the orthogonality relations for Wigner $3j$-symbols, the summation on (J_{23}, M_{23}) can be carried out by multiplying both sides of (F.4) with the $3j$-symbol

$$\begin{pmatrix} j_1 & J & J_{23} \\ m_1 & -M & M_{23} \end{pmatrix}$$

and summing over $(m_1, -M)$. Then one gets the relation

$$\begin{pmatrix} j_3 & j_2 & J_{23} \\ m_3 & m_2 & -M_{23} \end{pmatrix} \begin{Bmatrix} j_3 & j_2 & J_{23} \\ j_1 & J & J_{12} \end{Bmatrix}$$
$$= \sum_{M,m_1,M_{12}} (-1)^{j_1+J+J_{12}-M-m_1-M_{12}} \begin{pmatrix} j_1 & j_2 & J_{12} \\ m_1 & m_2 & -M_{12} \end{pmatrix}$$
$$\times \begin{pmatrix} j_3 & J & J_{12} \\ m_3 & -M & M_{12} \end{pmatrix} \begin{pmatrix} j_1 & J & J_{23} \\ -m_1 & M & -M_{23} \end{pmatrix} . \tag{F.5}$$

Putting the above results together, relations of the following type have been obtained, relations which we denote in a symbolical way by

$$\begin{Bmatrix} \cdot & \cdot & \cdot \\ \cdot & \cdot & \cdot \end{Bmatrix} = \sum_{\text{all } m_i} (-1)^{\text{Phase}} \begin{pmatrix} \cdot & \cdot & \cdot \\ \cdot & \cdot & \cdot \end{pmatrix} \begin{pmatrix} \cdot & \cdot & \cdot \\ \cdot & \cdot & \cdot \end{pmatrix} \begin{pmatrix} \cdot & \cdot & \cdot \\ \cdot & \cdot & \cdot \end{pmatrix} \begin{pmatrix} \cdot & \cdot & \cdot \\ \cdot & \cdot & \cdot \end{pmatrix} , \tag{F.6}$$

$$\begin{Bmatrix} \cdot & \cdot & \cdot \\ \cdot & \cdot & \cdot \end{Bmatrix} \begin{pmatrix} \cdot & \cdot & \cdot \\ \cdot & \cdot & \cdot \end{pmatrix} = \sum_{3(m_i)} (-1)^{\text{Phase}'} \begin{pmatrix} \cdot & \cdot & \cdot \\ \cdot & \cdot & \cdot \end{pmatrix} \begin{pmatrix} \cdot & \cdot & \cdot \\ \cdot & \cdot & \cdot \end{pmatrix} \begin{pmatrix} \cdot & \cdot & \cdot \\ \cdot & \cdot & \cdot \end{pmatrix} , \tag{F.7}$$

$$\sum_{J',M'} \begin{Bmatrix} \cdot & \cdot & \cdot \\ \cdot & \cdot & J' \end{Bmatrix} \begin{pmatrix} \cdot & \cdot & J' \\ \cdot & \cdot & M' \end{pmatrix} \begin{pmatrix} \cdot & \cdot & J' \\ \cdot & \cdot & -M' \end{pmatrix} (2J'+1)$$
$$= \sum_{M} \begin{pmatrix} \cdot & \cdot & \cdot \\ \cdot & \cdot & -M \end{pmatrix} \begin{pmatrix} \cdot & \cdot & \cdot \\ \cdot & \cdot & M \end{pmatrix} . \tag{F.8}$$

Calculation of $\langle (\frac{1}{2}l)j || \boldsymbol{Y}_k || (\frac{1}{2}l')jj' \rangle$

i) Using the reduction formula type (I) of (2.106) and since $\boldsymbol{Y}_k$ only acts on the angular part of the single-particle wave function, one gets

$$\langle (\tfrac{1}{2}l)\, j || \boldsymbol{Y}_k || (\tfrac{1}{2}l')\, j' \rangle = (-1)^{1/2+l'+j+k} \hat{j} \cdot \hat{j}'$$
$$\times \begin{Bmatrix} l & j & \frac{1}{2} \\ j' & l' & k \end{Bmatrix} \langle l || \boldsymbol{Y}_k || l' \rangle . \tag{F.9}$$

ii) We have now to evaluate $\langle l||\boldsymbol{Y}_k||l'\rangle$. Therefore, we start from the normal matrix element and apply the Wigner-Eckart theorem. So, we obtain

$$\begin{aligned}\langle lm|Y_k^{\kappa}|l'm'\rangle &= \int Y_l^{m^*}(\hat{\boldsymbol{r}})Y_k^{\kappa}(\hat{\boldsymbol{r}})Y_{l'}^{m'}(\hat{\boldsymbol{r}})d\hat{\boldsymbol{r}} \\ &= (-1)^{l-m}\begin{pmatrix} l & k & l' \\ -m & \kappa & m' \end{pmatrix}\langle l||\boldsymbol{Y}_k||l'\rangle\,. \end{aligned} \tag{F.10}$$

We apply some properties of the spherical harmonics. In particular, we use the property (Brink 1962)

$$\begin{aligned} Y_{l_a}^{m_a}(\hat{\boldsymbol{r}})Y_{l_b}^{m_b}(\hat{\boldsymbol{r}}) &= \sum_{l,m}\frac{\hat{l}_a\hat{l}_b\hat{l}}{\sqrt{4\pi}} \\ &\times \begin{pmatrix} l_a & l & l_b \\ 0 & 0 & 0 \end{pmatrix}\begin{pmatrix} l_a & l & l_b \\ m_a & m & m_b \end{pmatrix} Y_l^{m^*}(\hat{\boldsymbol{r}})\,. \end{aligned} \tag{F.11}$$

So, the matrix element (F.10) becomes

$$\begin{aligned} &\int Y_l^{m^*}(\hat{\boldsymbol{r}})Y_k^{\kappa}(\hat{\boldsymbol{r}})Y_{l'}^{m'}(\hat{\boldsymbol{r}})d\hat{\boldsymbol{r}} \\ &= (-1)^m\int Y_l^{-m}(\hat{\boldsymbol{r}})Y_{l'}^{m'}(\hat{\boldsymbol{r}})Y_k^{\kappa}(\hat{\boldsymbol{r}})d\hat{\boldsymbol{r}} \\ &= (-1)^m\int\sum_{k',\kappa'}\frac{\hat{l}\hat{l}'\hat{k}'}{\sqrt{4\pi}}\begin{pmatrix} l & k' & l' \\ 0 & 0 & 0 \end{pmatrix}\begin{pmatrix} l & k' & l' \\ -m & \kappa' & m' \end{pmatrix} \\ &\quad\times Y_{k'}^{\kappa'^*}(\hat{\boldsymbol{r}})Y_k^{\kappa}(\hat{\boldsymbol{r}})d\hat{\boldsymbol{r}}\,. \end{aligned} \tag{F.12}$$

The latter angular integral gives a Kronecker delta $\delta_{kk'}\cdot\delta_{\kappa\kappa'}$ and (F.12) reduces to

$$(-1)^m\frac{\hat{l}\hat{l}'\hat{k}'}{\sqrt{4\pi}}\begin{pmatrix} l & k & l' \\ 0 & 0 & 0 \end{pmatrix}\begin{pmatrix} l & k & l' \\ -m & \kappa & m' \end{pmatrix}\,. \tag{F.13}$$

Identifying (F.13) with (F.10) one immediately has

$$\langle l||\boldsymbol{Y}_k||l'\rangle = (-1)^l\frac{\hat{l}\hat{l}'\hat{k}}{\sqrt{4\pi}}\begin{pmatrix} l & k & l' \\ 0 & 0 & 0 \end{pmatrix}\,. \tag{F.14}$$

ii) Combining the results of (F.9) and (F.14), the total reduced matrix element becomes

$$\begin{aligned} \langle(\tfrac{1}{2}Zl)\,j||\boldsymbol{Y}_k||\,(\tfrac{1}{2}l')\,j'\rangle &= (-1)^{1/2+l'+l+j+k}\frac{\hat{j}\hat{j}'\hat{l}\hat{l}'\hat{k}}{\sqrt{4\pi}} \\ &\times\begin{pmatrix} l & k & l' \\ 0 & 0 & 0 \end{pmatrix}\begin{Bmatrix} l & j & \frac{1}{2} \\ j' & l' & k \end{Bmatrix}\,. \end{aligned} \tag{F.15}$$

We now use the relation (F.5) which leads to

$$\begin{pmatrix} l & k & l' \\ 0 & 0 & 0 \end{pmatrix} \begin{Bmatrix} l & k & l' \\ j' & \frac{1}{2} & j \end{Bmatrix} = (-1)^{j+j'+(1/2)-(3/2)}$$
$$\times \begin{pmatrix} l & \frac{1}{2} & j \\ 0 & -\frac{1}{2} & \frac{1}{2} \end{pmatrix} \begin{pmatrix} j' & k & j \\ \frac{1}{2} & 0 & -\frac{1}{2} \end{pmatrix}$$
$$\times \begin{pmatrix} j' & \frac{1}{2} & l' \\ -\frac{1}{2} & \frac{1}{2} & 0 \end{pmatrix} + (-1)^{j+j'+(1/2)+(3/2)}$$
$$\times \begin{pmatrix} l & \frac{1}{2} & j \\ 0 & \frac{1}{2} & -\frac{1}{2} \end{pmatrix} \begin{pmatrix} j' & k & j \\ -\frac{1}{2} & 0 & \frac{1}{2} \end{pmatrix}$$
$$\times \begin{pmatrix} j' & \frac{1}{2} & l' \\ \frac{1}{2} & -\frac{1}{2} & 0 \end{pmatrix}, \tag{F.16}$$

or, combining the right-hand side in more compact form

$$\begin{pmatrix} j & k & j' \\ -\frac{1}{2} & 0 & \frac{1}{2} \end{pmatrix} \begin{pmatrix} l & \frac{1}{2} & j \\ 0 & -\frac{1}{2} & \frac{1}{2} \end{pmatrix} \begin{pmatrix} j' & \frac{1}{2} & l' \\ -\frac{1}{2} & \frac{1}{2} & 0 \end{pmatrix}$$
$$\times (-1)^{k+1} \left[1 + (-1)^{l+l'+k}\right]. \tag{F.17}$$

Since we know the explicit expression of the Wigner $3j$-symbols

$$\begin{pmatrix} j' & \frac{1}{2} & l' \\ -\frac{1}{2} & \frac{1}{2} & 0 \end{pmatrix} = \frac{(-1)^{l'+1}}{\sqrt{2(2l'+1)}}, \tag{F.18}$$

we can rewrite (F.17) as

$$\begin{pmatrix} j & k & j' \\ -\frac{1}{2} & 0 & \frac{1}{2} \end{pmatrix} \frac{(-1)^{l+l'+k+1}}{\hat{l}\hat{l}'} \frac{1}{2}(1 + (-1)^{l+l'+k}). \tag{F.19}$$

This finally results in

$$\langle (\tfrac{1}{2}l) \, ||\boldsymbol{Y}_k|| \, (\tfrac{1}{2}l') \, j' \rangle = (-1)^{j-1/2} \frac{\hat{j}\hat{j}'\hat{k}}{\sqrt{4\pi}}$$
$$\times \begin{pmatrix} j & k & j' \\ -\frac{1}{2} & 0 & \frac{1}{2} \end{pmatrix} \frac{1}{2}(1 + (-1)^{l+l'+k}). \tag{F.20}$$

In the calculation of this appendix, we have used the conventions:

i) $(1/2l)j$ coupling in that order,
ii) changing the order to $(l1/2)j$ gives an extra phase factor $(-1)^{l+l'+1-j-j'}$. It is that particular convention (that of coupling $\boldsymbol{l} + \boldsymbol{s}$ to form $\boldsymbol{j}$, in that order) that is used throughout the main text except if stated otherwise. The factor $[1 + (-1)^{l+l'+k}]$ induces a selection rule since one should have

$$\text{Parity } \{l + l'\} \equiv \text{Parity operator } \{k\}, \tag{F.21}$$

for non-vanishing matrix elements.

G. The Magnetic Multipole Operator

In the present Appendix G, we derive a more "handlable" expression for the spin part in the magnetic multipole operator of Sect. 4.2, as expressed by equation (4.7). We start from (Brussaard, Glaudemans 1977)

$$\begin{aligned}&\int \nabla(r^L Y_L^M) \cdot \{(\boldsymbol{r} \times \nabla) \times \boldsymbol{\sigma}_i\} d\boldsymbol{r} \\ &\quad = \int \nabla \cdot \boldsymbol{A} - \int \nabla \cdot ((\boldsymbol{r} \times \nabla) \times \boldsymbol{\sigma}_i) r^L Y_L^M d\boldsymbol{r} \,. \qquad \text{(G.1)}\end{aligned}$$

The first part on the right-hand side of (G.1), which is equal to the surface integral

$$\int \boldsymbol{A} \cdot d\boldsymbol{S} \,, \qquad \text{(G.2)}$$

disappears since the spin contribution vanishes through an arbitrary large surface S surrounding the nucleus.

So, (G.1) becomes

$$-\int \nabla \times (\boldsymbol{r} \times \nabla) \cdot \boldsymbol{\sigma}_i r^L Y_L^M d\boldsymbol{r} \,. \qquad \text{(G.3)}$$

We now evaluate $\nabla \times (\boldsymbol{r} \times \nabla)$. We consider the component j, or

$$\{\nabla \times (\boldsymbol{r} \times \nabla\}_j = \sum_{klmn} \varepsilon_{jkl} \varepsilon_{lmn} \partial_k r_m \partial_n \,, \qquad \text{(G.4)}$$

with

$$\varepsilon_{jkl} = +1$$

for an even permutation of (1, 2, 3)

$$\varepsilon_{jkl} = -1$$

for an odd permutation of (1, 2, 3)

$$\varepsilon_{jkl} = 0$$

in *all* other cases.

We also have

$$\sum_a \varepsilon_{abc} \varepsilon_{ade} = \delta_{bd} \cdot \delta_{ce} - \delta_{be} \cdot \delta_{cd} \,. \qquad \text{(G.5)}$$

So, one writes

$$\begin{aligned}\{\nabla \times (\boldsymbol{r} \times \nabla)\}_j &= \sum_{k,m,n} (\delta_{jm}\delta_{kn} - \delta_{jn}\delta_{km}) \partial_k r_m \partial_n \\ &= \sum_k \partial_k r_j \partial_k - \sum_k \partial_k r_k \partial_j\end{aligned}$$

$$= \sum_k (r_j \partial_k \partial_k + \delta_{jk} \partial_k)$$
$$- \sum_k (\partial_j \partial_k r_k - \partial_k \delta_{jk}) \,. \qquad \text{(G.6)}$$

This now becomes in vector notation

$$\{\nabla \times (\boldsymbol{r} \times \nabla)\}_j = r_j \Delta - \nabla_j (\nabla \cdot \boldsymbol{r} - 2) \,, \qquad \text{(G.7)}$$

or

$$\{\nabla \times (\boldsymbol{r} \times \nabla)\} = \boldsymbol{r}\Delta - \nabla(\nabla \cdot \boldsymbol{r} - 2) = \boldsymbol{r}\Delta - \nabla(1 + \boldsymbol{r} \cdot \nabla) \,. \qquad \text{(G.8)}$$

Putting this into the original magnetic multipole operator one gets the result

$$\int (\boldsymbol{r} \cdot \boldsymbol{\sigma}_i \Delta r^L Y_L^M - \nabla(1 + \boldsymbol{r} \cdot \nabla) r^L Y_L^M \cdot \boldsymbol{\sigma}_i) d\boldsymbol{r} \,. \qquad \text{(G.9)}$$

Now, one has

$$\begin{aligned} &\Delta r^L Y_L^M = 0 \,, \\ &\boldsymbol{r} \cdot \nabla r^L Y_L^M = L r^L Y_L^M \,, \end{aligned} \qquad \text{(G.10)}$$

so that finally one obtains the result

$$\int (L+1) \nabla r^L Y_L^M \cdot \boldsymbol{\sigma}_i d\boldsymbol{r} \,. \qquad \text{(G.11)}$$

H. A Two-Group (Degenerate) RPA Model

With a two-gap, unperturbed $1p-1h$ spectrum, we mean that the $1p-1h$ configurations form mainly two groups (both groups containing a number of $1p-1h$ configurations that are close in unperturbed $1p-1h$ excitation energy) with a large interval between the two groups, compared to the average energy separation within the groups themselves.

The unperturbed spectrum can be drawn schematically in Fig. H.1. Here, we call the particle-hole configurations $m, i(1)$ and $m, i(2)$ to differentiate between the groups for a given particle-hole excitation. The analysis goes along the same lines as in Sect. 6.3. The equations (6.39), (6.40) and (6.41) remain valid. We shall, however, separate the sum over $p-h$ states into two parts explicitly, in order to be able to make an easy connection to the situation of Fig. H.1. The dispersion relation corresponding to (6.41) now contains two gaps, characterized by ε_1 and ε_2, respectively, and thus points towards the existence of a second, collective excitation $\hbar\omega_2$ (see Fig. H.2).

We study the solution for $\hbar\omega_1, \hbar\omega_2$ in the limit of two groups at the unperturbed energy ε_1 and ε_2, respectively. So, (6.41) now becomes

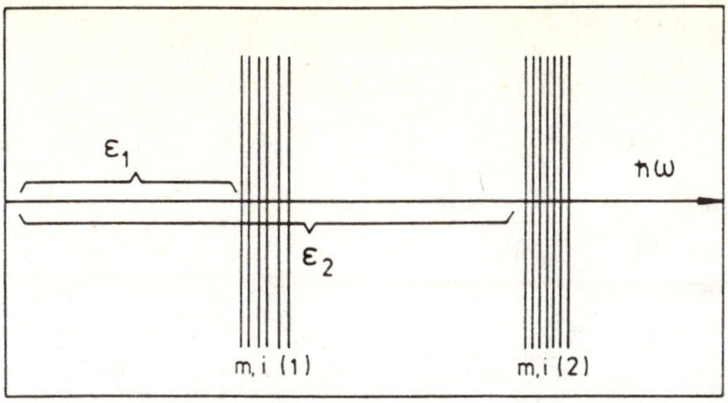

Fig. H.1. The schematic, unperturbed spectrum for $1p - 1h$ states within a two-group system of levels, centered around the energy ε_1 and ε_2, respectively. We have separated the particle-hole notation m, i into two groups, characterized by an extra group index k i.e. $m, i(1)$ and $m, i(2)$

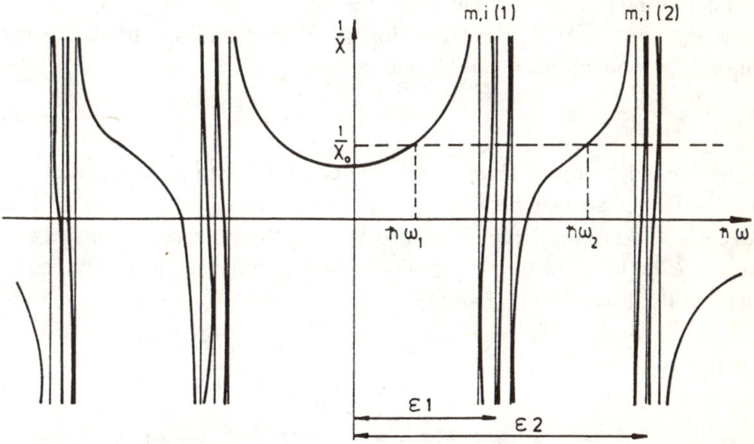

Fig. H.2. The two-group RPA secular equation, corresponding to the unperturbed spectrum of Fig. H.1. The two collective roots are denoted by $\hbar\omega_1$ (lowest one) and $\hbar\omega_2$ (the intermediate root)

$$\frac{1}{\chi} = \frac{2\varepsilon_1 S_1}{(\varepsilon_1^2 - \hbar\omega_\alpha^2)} + \frac{2\varepsilon_2 S_2}{(\varepsilon_2^2 - \hbar\omega_\alpha^2)} \,. \tag{H.1}$$

Solution for $\hbar\omega_1$. To a first approximation, we replace $\hbar\omega_\alpha$ in the second term by ε_1. We then obtain

$$\hbar\omega_1^2 = \varepsilon_1^2 - 2\chi\varepsilon_1 S_1 \frac{1}{1 - \dfrac{2\varepsilon_2 S_2 \chi}{(\varepsilon_2^2 - \varepsilon_1^2)}} \,. \tag{H.2}$$

We remark that the second group of unperturbed states near the energy ε_2 acts coherently with the first group in lowering the excitation energy.

The normalization N_1 becomes

$$N_1 \simeq \left[\frac{S_1}{(\varepsilon_1 - \hbar\omega_1)^2} + \frac{4\varepsilon_1\varepsilon_2 S_2}{(\varepsilon_2^2 - \varepsilon_1^2)^2}\right]^{-1/2}, \tag{H.3}$$

and the wave functions become

$$\begin{aligned} X^1_{m,i}(1) &\simeq \frac{D^{(1)}_{m,i} N_1}{\varepsilon_1 - \hbar\omega_1}\,; \quad X^1_{m,i}(2) = \frac{D^{(2)}_{m,i} N_1}{\varepsilon_2 - \varepsilon_1}\,; \\ Y^1_{m,i}(1) &\simeq \frac{D^{*(1)}_{m,i} N_1}{2\varepsilon_1}\,; \quad Y^1_{m,i}(2) \simeq \frac{D^{*(2)}_{m,i} N_1}{\varepsilon_2 + \varepsilon_1}\,; \end{aligned} \tag{H.4}$$

where the notation means

$$X^1_{m,i}(1)\,, \quad X^1_{m,i}(2)\,, \tag{H.5}$$

the particle-hole X-amplitudes for particle-hole components from group (1) and group (2), respectively. A similar notation for the Y amplitudes is used. For the second root, we then obtain, in a similar way amplitudes

$$X^2_{m,i}(1)\,, \quad X^2_{m,i}(2)\,. \tag{H.6}$$

Now, also the second group of levels (at $\cong \varepsilon_2$) contributes to the collectivity of the root $\hbar\omega_1$ and does so in the ratio $(\varepsilon_2 - \varepsilon_1)/(\varepsilon_1 - \hbar\omega_1)$, relative to the first group of levels. For a not too large difference $\varepsilon_2 - \varepsilon_1$, the influence of the second group is almost as large as for the first group. In second order perturbation theory, we can calculate the transition probability

$$\begin{aligned} |\langle\widetilde{\hbar\omega_1}|\boldsymbol{D}|\tilde{0}\rangle|^2 \simeq \frac{\varepsilon_1}{\hbar\omega_1} S_1 \Bigg[1 &+ \frac{\varepsilon_2 S_2}{\varepsilon_1 S_1}\left[\frac{\varepsilon_1^2 - \hbar\omega_1^2}{\varepsilon_2^2 - \hbar\omega_1^2}\right]\left[\frac{2\varepsilon_2^2 - \hbar\omega_1^2 - \varepsilon_1^2}{\varepsilon_2^2 - \hbar\omega_1^2}\right] \\ &+ \frac{\varepsilon_2^2 S_2^2}{\varepsilon_1^2 S_1^2}\left[\frac{\varepsilon_1^2 - \hbar\omega_1^2}{\varepsilon_2^2 - \hbar\omega_1^2}\right]^2 \left[1 + \left[\frac{\varepsilon_1^2 - \hbar\omega_1^2}{\varepsilon_2^2 - \hbar\omega_1^2}\right]^2 + \ldots\right]\Bigg]\,. \end{aligned} \tag{H.7}$$

Since

$$2\varepsilon_2^2 - \hbar\omega_1^2 - \varepsilon_1^2 > 2(\varepsilon_2^2 - \varepsilon_1^2) > 0\,,$$

all contributions in (H.7) add up coherently.

Solution for $\hbar\omega_2$. We can now, to a first approximation, replace $\hbar\omega_\alpha$ in the first term of (H.1) by ε_2. So, one obtains

$$\hbar\omega_2^2 = \varepsilon_2^2 - 2\chi\varepsilon_2 S_2 \left[\frac{1}{1 + \dfrac{2\varepsilon_1 S_1 \chi}{(\varepsilon_2^2 - \varepsilon_1^2)}}\right]. \tag{H.8}$$

Here, one remarks that the collective excitation at $\hbar\omega_2$ is mainly created by a coherence effect from the states near to ε_2 but that the group of levels near ε_1

counteracts this coherence. This incoherence between the two groups at ε_2 and ε_1 can be made even more explicit.

The normalization factor N_2 is obtained as

$$N_2 \simeq \left[\frac{S_2}{(\varepsilon_2 - \hbar\omega_2)^2} + \frac{4\varepsilon_1\varepsilon_2 S_2}{(\varepsilon_2^2 - \varepsilon_1^2)^2} \right]^{-1/2}, \tag{H.9}$$

with, subsequently

$$\begin{aligned} X_{m,i}^2(1) &\simeq -\frac{D_{m,i}N_2}{\varepsilon_2 - \varepsilon_1}; \quad X_{m,i}^2(2) \simeq \frac{D_{m,i}N_2}{\varepsilon_2 - \hbar\omega_2}; \\ Y_{m,i}^2(1) &\simeq \frac{D_{m,i}^* N_2}{\varepsilon_1 + \varepsilon_2}; \quad Y_{m,i}^2(2) \simeq \frac{D_{m,i}^* N_2}{2\varepsilon_2}, \end{aligned} \tag{H.10}$$

The minus sign shows the orthogonality between the solutions at $\hbar\omega_1$ and $\hbar\omega_2$. Similarly as for $\hbar\omega_1$ the transition probability for the state at $\hbar\omega_2$ becomes

$$\begin{aligned} |\langle\widetilde{\hbar\omega_2}|\boldsymbol{D}|\tilde{0}\rangle|^2 \simeq \frac{\varepsilon_2}{\hbar\omega_2} S_2 \Bigg[1 &- \frac{\varepsilon_1 S_1}{\varepsilon_2 S_2} \left[\frac{\varepsilon_2^2 - \hbar\omega_2^2}{\varepsilon_2^2 - \varepsilon_1^2} \right] \left[2 + \frac{(\varepsilon_2^2 - \hbar\omega_2^2)}{(\varepsilon_2^2 - \varepsilon_1^2)} \right] \\ &+ \frac{\varepsilon_1^2 S_1^2}{\varepsilon_2^2 S_2^2} \left[\frac{\varepsilon_2^2 - \hbar\omega_2^2}{\varepsilon_2^2 - \varepsilon_1^2} \right]^2 \left[1 + \left[\frac{\varepsilon_2^2 - \hbar\omega_2^2}{\varepsilon_2^2 - \varepsilon_1^2} \right]^2 + \ldots \right] \Bigg], \end{aligned} \tag{H.11}$$

where now not all contributions act coherently in the sum. The second solution at $\hbar\omega_2$ therefore is less collective than the first solution. The bigger the gap $\varepsilon_2 - \varepsilon_1$ the stronger the collectivity in the second group becomes.

As an application we study the octupole eigenvalues (3^- levels) in ^{146}Gd (Waroquier 1982). In Fig. H.3 we show both the unperturbed $1p - 1h$ spectrum

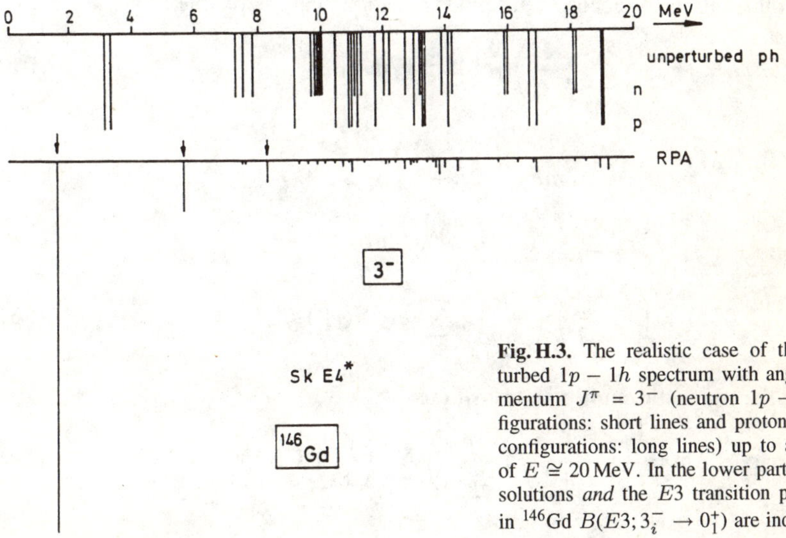

Fig. H.3. The realistic case of the unperturbed $1p - 1h$ spectrum with angular momentum $J^\pi = 3^-$ (neutron $1p - 1h$ configurations: short lines and proton $1p - 1h$ configurations: long lines) up to an energy of $E \cong 20$ MeV. In the lower part, the RPA solutions *and* the $E3$ transition probability in ^{146}Gd $B(E3; 3_i^- \to 0_1^+)$ are indicated (indicating separately the proton and neutron $1p-1h$ states) and the RPA roots. The lowest solution at $\cong$ 1.8 MeV is most collective and contains the higher-lying $1p-1h$ states in a coherent way. Near 5.6 MeV a next, less collective solution is obtained, falling exactly in the gap between the unperturbed states near 3 MeV and near 7.5 MeV. This is a clear, realistic illustration of the discussion given above.

I. The Condon-Shortley and Biedenharn-Rose Phase Conventions: Application to Electromagnetic Operators and BCS Theory

In the present appendix, we shall discuss the two often used phase conventions in defining the single-particle wave functions (Condon, Shortley 1935, Biedenharn, Rose 1953): the Condon-Shortley and Biedenharn-rose phase conventions. Quite some attention has been given in (Rowe 1970), to the possible subtle effects related to the use of these phase conventions. Also, Brussaard and Allaart (Brussaard 1970, Allaart 1971a) have concentrated on the effect of the two phase choices when incorporating pairing correlations (BCS-theory) in the description of nuclear excited states, more in particular concerning the calculation of electromagnetic decay properties. We shall give here a consistent discussion on the phase choices and their use when calculating electromagnetic decay properties using BCS theory.

I.1 Electromagnetic Operators: Long-Wavelength Form and Matrix Elements

Using the discussion of Chap. 4 and the results of (Waroquier et al. 1975), the terms contributing to electric multipole transitions can be denoted by

$$O(\text{el.}, LM) = \frac{1}{\hbar\omega}\sum_i \left(\frac{\tilde{e}_i}{e}\right)\left[r_i^L Y_L^M(\hat{\boldsymbol{r}}_i), H\right] , \tag{I.1}$$

$$\begin{aligned} O'(\text{el.}, LM) = &-\text{i}k\left(\frac{\hbar}{2mc}\right)\left(\frac{1}{2L+3}\right)\left(\frac{2L+1}{L+1}\right)^{1/2} \\ &\times \sum_i 2\left(\frac{\tilde{e}_i}{e}\right) r_i^{L+1}\mathcal{Y}^M_{L(L+1,1)}(\hat{\boldsymbol{r}}_i)\cdot \boldsymbol{p}_i , \end{aligned} \tag{I.2}$$

$$O''(\text{el.}, LM) = \left(\frac{k}{L+1}\right)\left(\frac{\hbar}{2mc}\right)\sum_i g_s(i)\boldsymbol{l}(r_i^L Y_L^M(\hat{\boldsymbol{r}}_i))\cdot \boldsymbol{s}_i , \tag{I.3}$$

and for the magnetic multipole transitions, we have

$$O(\text{mag.}, LM) = -\left(\frac{2}{L+1}\right)\left(\frac{\hbar}{2mc}\right)\sum_i \left(\frac{\tilde{e}_i}{e}\right) \boldsymbol{\nabla}(r_i^L Y_L^M(\hat{\boldsymbol{r}}_i))\cdot \boldsymbol{l}_i . \tag{I.4}$$

$$O'(\text{mag.}, LM) = -\left(\frac{\hbar}{2mc}\right)\sum_i g_s(i)\boldsymbol{\nabla}(r_i^L Y_L^M(\hat{\boldsymbol{r}}_i)) \cdot \boldsymbol{s}_i \ . \tag{I.5}$$

(with $\omega = kc$). Here, we define the vector spherical harmonics (within the Condon-Shortley phase convention) as

$$\mathcal{Y}_{j(l,1)}^{m}(\hat{\boldsymbol{r}}) \equiv \sum_{m_l,\mu} \langle l m_l 1 \mu | j m \rangle Y_l^{m_l}(\hat{\boldsymbol{r}}) \boldsymbol{e}_\mu$$

where $\boldsymbol{e}_\mu$ denote the spherical unit vectors ($\mu = 0, \pm 1$).

An important application is now the evaluation of transition probabilities using the above operators between an initial state $|\mathrm{i}\rangle$ and a final state $|\mathrm{f}\rangle$. Here, difficulties arise when evaluating the so-called single-particle matrix elements of the multipole transition operators, even in the long-wavelength approximation. The electric transition operator $O(\text{el.}, LM)$ still contains many-body effects due to the presence of the nuclear A-body Hamiltonian H. This problem can be solved (Allaart 1971a) by replacing the Hamilton operator, acting on the eigenstates $|\mathrm{i}\rangle, |\mathrm{f}\rangle$ by its eigenvalues E_i and E_f such that

$$\langle \mathrm{f} | O(\text{el.}, LM) | \mathrm{i} \rangle = \frac{E_\mathrm{i} - E_\mathrm{f}}{\hbar\omega} \langle \mathrm{f} | O^{\text{eff}}(\text{el.}, LM) | \mathrm{i} \rangle \ , \tag{I.6}$$

where $O^{\text{eff}}(\text{el.}, LM)$ reduces to the usual, effective electric or Coulomb multipole operator of (4.9). This effective operator now becomes a one-body operator. We draw attention to the difference

$$O(\text{el.}, LM) \neq O^{\text{eff}}(\text{el.}, LM) \ , \tag{I.7}$$

even in the case of photon emission where the factor $(E_\mathrm{i} - E_\mathrm{f})/\hbar\omega$ reduces to +1.

I.2 Properties of the Electromagnetic Multipole Operators Under Parity Operation, Time Reflection and Hermitian Conjugation

Parity. The parity operator can be associated with a unitary transformation operator $\mathcal{P}$, which inverts the spatial coordinates, i.e. one has

$$\begin{aligned} &\mathcal{P}\boldsymbol{r}_i\mathcal{P}^{-1} = -\boldsymbol{r}_i \\ &\mathcal{P}\boldsymbol{p}_i\mathcal{P}^{-1} = -\boldsymbol{p}_i \ , \quad \mathcal{P}\boldsymbol{s}_i\mathcal{P}^{-1} = \boldsymbol{s}_i \ , \\ &\mathcal{P}Y_L^M(\hat{\boldsymbol{r}}_i)\mathcal{P}^{-1} = (-1)^L Y_L^M(\hat{\boldsymbol{r}}_i) \ . \end{aligned} \tag{I.8}$$

Then, we easily derive the parity transformation properties for the multipole operators as for (I.2)–(I.3) having the parity $(-1)^L$ while (I.4), (I.5) have the parity $(-1)^{L+1}$.

Time Reversal. The classical equations of motion for an interacting A-nucleon system have to be invariant with respect to time-reversal. The time-reversal operator $\mathcal{T}$ only affects kinematical quantities such as the momentum $\boldsymbol{p}$ or spin $\boldsymbol{s}$ of the particles. We can associate many properties with the time-reversal operator, so we are free to make a specific choice of phases.

Let $\psi(\boldsymbol{r},t)$ be the eigenfunction of the Schrödinger equation

$$H(\boldsymbol{r},t)\psi(\boldsymbol{r},t) = \mathrm{i}\hbar\frac{\partial\psi(\boldsymbol{r},t)}{\partial t} . \tag{I.9}$$

We define the time-dependent wave functions $\psi_R(\boldsymbol{r},t) \equiv \mathcal{T}\psi(\boldsymbol{r},-t) = \psi^*(\boldsymbol{r},-t)$ (with no spin), as the time-reversed wave function of $\psi(\boldsymbol{r},t)$. It satisfies the original Schrödinger equation, provided

$$\mathcal{T}H(\boldsymbol{r},-t)\mathcal{T}^{-1} = H^*(\boldsymbol{r},-t) = H(\boldsymbol{r},t) . \tag{I.10}$$

The operator $\mathcal{T}$ means in this case simply complex conjugation. So, a time-reversal operator can be defined associated with the properties: (i) the Hamiltonian remains invariant under time reflection, (ii) the function obtained by operating with $\mathcal{T}$ on $\psi(\boldsymbol{r},t)$ remains a solution to the original Schrödinger equation, (iii) the operator $\mathcal{T}$ can be considered as a combination of two operations: complex conjugation (K) followed by a unitary transformation ($\mathcal{T} = U_\mathcal{T}K$), which will be necessary in the case of wave functions with spin.

Let now $|\boldsymbol{r}\rangle$ be a state vector, without spin, invariant under time reflection $\mathcal{T}|\boldsymbol{r}\rangle = |\boldsymbol{r}\rangle$. Since $\mathcal{T} = U_\mathcal{T}K$, we can also determine the effect of time reflection in the momentum representation $|\boldsymbol{p}\rangle$. Here, the momentum $\boldsymbol{p}$ transforms into $-\boldsymbol{p}$ under time reversal. The orbital angular momentum $\boldsymbol{l}$, the spin $\boldsymbol{s}$ and total angular momentum $\boldsymbol{j}$ transform into $-\boldsymbol{l}, -\boldsymbol{s}$ and $-\boldsymbol{j}$, respectively. Since $\mathcal{T}\boldsymbol{j}\mathcal{T}^{-1} = -\boldsymbol{j}$, the operator $\mathcal{T}$ anticommutes with the total angular momentum $\boldsymbol{j}$. Therefore, it is convenient to combine $\mathcal{T}$ with a rotation $\mathcal{R}$ through an angle π about an axis perpendicular to the z-axis, say $\mathcal{R}_y$. State vectors, characterized by the angular momentum j and projection m and being discrete eigenstates of a time-reversal invariant Hamiltonian are therefore also eigenfunctions of the time-reversal operator $\mathcal{T}$. The eigenvalues of this operator are related to the phases of the basic state vectors (Rowe 1970). Two phase conventions are generally used: the Condon-Shortley (CS) (Condon, Shortley 1935) and Biedenharn-Rose (BR) (Biedenharn, Rose 1953) phase conventions. The former assumes the state vectors $|nljm\rangle$ to be invariant under the $\mathcal{PRT}$ operation ($\mathcal{P}$: parity operator); the latter assumes $|nljm\rangle$ invariant under the $\mathcal{RT}$ operation, a combination of $\mathcal{R}_y(\pi)$ with the time-reversal operator. Following the BR convention, we derive $\mathcal{T}|nljm\rangle = \mathcal{R}^{-1}|nljm\rangle$. The rotation around the y-axis can be a positive or negative rotation. We shall always use positive rotation (a rotation that advances a right-handed screw along the axis of rotation). With this choice $\mathcal{R}_y(\pi) = \exp(-i\pi J_y)$, which reduces to $\exp(-\mathrm{i}\pi\sigma_y/2) = -\mathrm{i}\sigma_y$ for intrinsic spin. This operation $\mathrm{i}\sigma_y$ can now be associated with the unitary transformation $U_\mathcal{T}$ and thus

$$\mathcal{T} = \mathrm{i}\sigma_y K . \tag{I.11}$$

So, in a condensed form, we can write

$$\begin{array}{cc} \text{Condon-Shortley} & \text{Biedenharn-Rose} \\ \mathcal{PRT}|nljm\rangle = |nljm\rangle \,, & \mathcal{RT}|\widetilde{nljm}\rangle = |\widetilde{nljm}\rangle \,. \end{array}$$

The state vector $|nlj, m\rangle$ can be decomposed, using $\boldsymbol{l} - \boldsymbol{s}$ coupling

$$\begin{aligned} \text{CS} &: |nljm\rangle = \sum_{m_l, m_s} \langle lm_l, \tfrac{1}{2}m_s|jm\rangle R_{nl}(r) Y_l^{m_l}(\hat{\boldsymbol{r}})\chi_{m_s}^{1/2} \,, \\ \text{BR} &: |nljm\rangle = \sum_{m_l, m_s} \langle lm_l, \tfrac{1}{2}m_s|jm\rangle R_{nl}(r) \tilde{Y}_l^{m_l}(\hat{\boldsymbol{r}})\chi_{m_s}^{1/2} \,, \end{aligned} \tag{I.12}$$

where $\tilde{Y}_l^m \equiv i^l Y_l^m$. Applying the transformation of spherical harmonics under rotation $\mathcal{R}_y(\pi)$ (see Chap. 2), we can easily derive the expressions

$$\begin{aligned} \text{CS} &: \mathcal{T}|nljm\rangle = (-1)^{l+j+m}|nlj-m\rangle \,, \\ \text{BR} &: \mathcal{T}|\widetilde{nljm}\rangle = (-1)^{j+m}|\widetilde{nlj-m}\rangle \,. \end{aligned} \tag{I.13}$$

Let $T_\kappa^{(k)}$ now be a spherical tensor, which transforms according to the irreducible representation $D_{M,M'}^{(L)}(\alpha, \beta, \gamma)$ of the rotation group (Chap. 2). The transformation under time reflection is related to

$$\mathcal{T}T_\kappa^{(k)}\mathcal{T}^{-1} = c_\mathcal{T}(-1)^{k+\kappa}T_{-\kappa}^{(k)} \,. \tag{I.14}$$

The phase factor $c_\mathcal{T}$ is independent of κ but still depends on the choice of phases for $T_\kappa^{(k)}$. It is evident that the choice of $c_\mathcal{T}$ can result into real matrix elements of $T_\kappa^{(k)}$ between wave functions described in either the CS or the BR phase convention. As the time-reversal operator is anti-unitary and since a scalar product remains invariant under a unitary transformation, it follows that

$$\langle \mathcal{T}\varphi|\mathcal{T}T_\kappa^{(k)}\mathcal{T}^{-1}|\mathcal{T}\psi\rangle = \langle\varphi|T_\kappa^{(k)}|\psi\rangle^* \,. \tag{I.15}$$

Applying this property to the matrix elements between single-particle states $|n_1l_1j_1m_1\rangle$ and $|n_2l_2j_2m_2\rangle$ one has

$$\langle j_1m_1|T_\kappa^{(k)}|j_2m_2\rangle = (\hat{j}_1)^{-1}\langle j_2m_2, k\kappa|j_1m_1\rangle\langle j_1||\boldsymbol{T}^{(k)}||j_2\rangle \,, \tag{I.16}$$

and we obtain

$$\langle n_1l_1j_1||\boldsymbol{T}^{(k)}||n_2l_2j_2\rangle^* = c_\mathcal{T}c_{\text{ph}_1}c_{\text{ph}_2}\langle n_1l_1j_1||\boldsymbol{T}^{(k)}||n_2l_2j_2\rangle \,. \tag{I.17}$$

The c_{ph} are phases related to the phase convention in which the spherical harmonics are described and are defined as

$$\mathcal{T}|nljm\rangle = c_{\text{ph}}(-1)^{j+m}|nlj-m\rangle \,, \tag{I.18}$$

where $c_{\text{ph}} = (-1)^l$ in the CS and $c_{\text{ph}} = 1$ in the BR phase convention (see (I.13)). Relation (I.17) means that matrix elements of $T_\kappa^{(k)}$ are real if $c_\mathcal{T} = (-1)^{l_1+l_2}$ in the CS convention and if $c_\mathcal{T} = +1$ in the BR convention.

Applying now the above properties, one can derive the following time-reversal properties

$$
\begin{aligned}
&\mathcal{T} \boldsymbol{r}_n \mathcal{T}^{-1} = \boldsymbol{r}_n \,, \\
&\mathcal{T} Y_L^M(\hat{\boldsymbol{r}}_n) \mathcal{T}^{-1} = (-1)^M Y_L^{-M}(\hat{\boldsymbol{r}}_n) (c_{\mathcal{T}} = (-1)^L) \,, \\
&\mathcal{T} \boldsymbol{p}_n \mathcal{T}^{-1} = -\boldsymbol{p}_n, \;\; \mathcal{T} \boldsymbol{\nabla}_n \mathcal{T}^{-1} = \boldsymbol{\nabla}_n, \;\; \mathcal{T} \boldsymbol{s}_n \mathcal{T}^{-1} = -\boldsymbol{s}_n \,, \\
&\mathcal{T} \boldsymbol{l}_n \mathcal{T}^{-1} = -\boldsymbol{l}_n \,,
\end{aligned}
\tag{I.19}
$$

and thus, the transformation properties for the electromagnetic multipole operators (I.1) to (I.5). They are given in Table I.1.

Table I.1 Intrinsic properties of the electromagnetic multipole operators. The c' factors are, of course, meaningless for the electric multipole operators $O(\text{el.}, LM)$ since this operator contains, in general, many-body effects. With $O(\text{el.}, LM)$, we mean the sum of operators (I.1) to (I.3), with $O^{\text{eff}}(\text{el.}, LM)$ the sum of the operators (I.2), (I.3) and the standard Coulomb operator (Chap. 4, equation (4.9))

$T_\kappa^{(k)}$	Parity	$\mathcal{T} T_\kappa^{(k)} \mathcal{T}^1$	$c_{\mathcal{T}}$	$(T_\kappa^{(k)})^+$	c_H	c' CS	c' BR
$O(\text{el.}, LM)$	$(-1)^L$	$(-1)^M O(\text{el.}, L-M)$	$(-1)^L$	$(-1)^{M+1} O(\text{el.}, L-M)$	-1	–	–
$O(\text{mag.}, LM)$	$(-1)^{L+1}$	$(-1)^{M+1} O(\text{mag.}, L-M)$	$(-1)^{L+1}$	$(-1)^M O(\text{mag.}, L-M)$	$+1$	$+1$	$(-1)^{L+1}$
$O^{\text{eff}}(\text{el.}, LM)$	$(-1)^L$	$(-1)^M O^{\text{eff}}(\text{el.}, L-M)$	$(-1)^L$	$(-1)^M O^{\text{eff}}(\text{el.}, L-M)$	$+1$	$+1$	$(-1)^L$

Hermitian Conjugation. An irreducible tensor of rank k transforms under rotation according to

$$
\mathcal{R} T_\kappa^{(k)} \mathcal{R}^{-1} = \sum_{\kappa'} T_{\kappa'}^{(k)} D_{\kappa',\kappa}^{(k)}(\alpha, \beta, \gamma) \,, \tag{I.20}
$$

where (α, β, γ) correspond to the Euler angles, $\mathcal{R}(\alpha\beta\gamma)$ to a unitary transformation. Taking the Hermitian conjugation of (I.20) and using the unitary property of $\mathcal{R}$ we get

$$
\mathcal{R} (T_\kappa^{(k)})^+ \mathcal{R}^{-1} = \sum_{\kappa'} (T_{\kappa'}^{(k)})^+ D_{\kappa',\kappa}^{(k)^*}(\alpha, \beta, \gamma) \,. \tag{I.21}
$$

Applying properties of the rotation operators and changing the sign of κ and κ', it follows that $(-1)^\kappa (T_{-\kappa}^{(k)})^+$ transforms under rotation in the same way as the $T_\kappa^{(k)}$, or, we can write

$$
\mathcal{R} (-1)^\kappa (T_{-\kappa}^{(k)})^+ \mathcal{R}^{-1} = \sum_{\kappa'} (-1)^{\kappa'} (T_{-\kappa'}^{(k)})^+ D_{\kappa',\kappa}^{(k)}(\alpha, \beta, \gamma) \,. \tag{I.22}
$$

So, the Hermitian adjoint $\boldsymbol{T}^{(k)+}$ of the tensor operator $\boldsymbol{T}^{(k)}$, can be defined as

$$
(T^+)_\kappa^{(k)} = c_{\text{H}} (-1)^\kappa (T_{-\kappa}^{(k)})^+ \,, \tag{I.23}
$$

where c_H is a phase factor independent of κ. It can be fixed by the additional condition for self-adjoint operators

$$(T^+)^{(k)}_\kappa = T^{(k)}_\kappa \, . \tag{I.24}$$

The momentum $\boldsymbol{p}$, the orbital angular momentum $\boldsymbol{l}$, the spin $\boldsymbol{s}$ are self-adjoint operators. The Hermitian conjugation of the vector spherical harmonics simplifies by taking the conjugation of the ordinary spherical harmonics and the spherical unit vectors $\boldsymbol{e}_\mu$ appearing in the definition

$$Y_l^{m_l}(\hat{\boldsymbol{r}}_n)\boldsymbol{e}_\mu = \sum_{j,m} \langle lm_l, 1\mu | jm \rangle \boldsymbol{\mathcal{Y}}^m_{j(l,1)}(\hat{\boldsymbol{r}}_n) \, , \tag{I.25}$$

and results in

$$(\boldsymbol{\mathcal{Y}}^m_{j(l,1)}(\hat{\boldsymbol{r}}_n))^+ = (-1)^{l+1-j+m} \boldsymbol{\mathcal{Y}}^{-m}_{j(l,1)}(\hat{\boldsymbol{r}}_n) \, . \tag{I.26}$$

Since $\boldsymbol{L}Y_L^M(\hat{\boldsymbol{r}}_n)$ and $\nabla \times \boldsymbol{L}Y_L^M(\hat{\boldsymbol{r}}_n)$ can be expressed in terms of the vector spherical harmonics, we can also write that

$$\begin{aligned} (\boldsymbol{L}Y_L^M(\hat{\boldsymbol{r}}_n))^+ &= (-1)^{L+M+1} \boldsymbol{L}Y_L^{-M}(\hat{\boldsymbol{r}}_n) \, , \\ (\nabla \times \boldsymbol{L}Y_L^M(\hat{\boldsymbol{r}}_n))^+ &= (-1)^{L+M+1} \nabla \times \boldsymbol{L}Y_L^{-M}(\hat{\boldsymbol{r}}_n) \, . \end{aligned} \tag{I.27}$$

The Hermitian conjugate of the operators (I.1)–(I.5) can easily be derived (see Table I.1).

The interchange of initial and final states in the reduced single-particle matrix elements is related to the behavior of the operator $\boldsymbol{T}^{(k)}$ under Hermitian conjugation. Using the equations (I.23), (I.24) and applying the Wigner-Eckart theorem, we derive

$$\langle n_1 l_1 j_1 || \boldsymbol{T}^{(k)} || n_2 l_2 j_2 \rangle = c_H (-1)^{j_1 - j_2} \langle n_2 l_2 j_2 || \boldsymbol{T}^{(k)} || n_1 l_1 j_1 \rangle^* \, . \tag{I.28}$$

We can finally obtain the relation

$$\langle n_1 l_1 j_1 || \boldsymbol{T}^{(k)} || n_2 l_2 j_2 \rangle = c' (-1)^{j_1 - j_2} \langle n_2 l_2 j_2 || \boldsymbol{T}^{(k)} || n_1 l_1 j_1 \rangle \, , \tag{I.29}$$

where we have defined

$$c' = c_{\text{ph}_1} c_{\text{ph}_2} c_T c_H \, . \tag{I.30}$$

Hence, the coefficient c' only depends on the considered angular momentum phase convention and on the convention satisfying the tensor operator $\boldsymbol{T}^{(k)}$ under time reflection and Hermitian conjugation. We recall that $c_{\text{ph}} = (-1)^l$ in the CS convention and $c_{\text{ph}} = 1$ in the BR convention. So, in Table I.2, we give the properties of the reduced matrix elements corresponding to the electromagnetic multipole operators (I.1)–(I.5) in both the CS and BR phase conventions. This indicates that in the more standard CS convention matrix elements are real for the operators (I.1)–(I.5) and $O^{\text{eff}}(\text{el.}, LM)$ requiring of course the selection rule $l_1 + l_2 + L$ even (or odd) for electric (or magnetic) operators to be fulfilled. In the BR convention

Table I.2 Properties of the reduced matrix elements of the electromagnetic multipole operators. The state vectors $|nljm\rangle$ are described in the considered phase convention (CS: Condon-Shortley 1935, BR: Biedenharn-Rose 1953)

$T_\kappa^{(k)}$	CS	BR
$O(\text{el.}, LM)$	real	real if L even imag. if L odd
$O^{\text{eff}}(\text{el.}, LM)$	real	real if L even imag. if L odd
$O(\text{mag.}, LM)$	real	real if L odd imag. if L even
$i^L O(\text{el.}, LM)$	real if L even imag. if L odd	real
$i^L O^{\text{eff}}(\text{el.}, LM)$	real if L even imag. if L odd	real
$i^{L-1} O(\text{mag.}, LM)$	real if L odd imag. if L even	real

one has real matrix elements for L even (or odd) for electric (magnetic) multipole operators, respectively. For the operators $i^L O(\text{el.}, LM)$ and $i^{(L-1)} O(\text{mag.}, LM)$, the matrix elements, using the BR convention, all become real! (see Table I.2).

I.3 Phase Conventions in the BCS Formalism

In second quantization, state vectors, single-particle tensor operators etc. are represented in terms of creation- and annihilation operators (see Chap. 5)

$$\begin{array}{cc} \text{CS} & \text{BR} \\ a_\alpha^+|0\rangle \equiv |n_a l_a j_a m_a\rangle\,, & \bar{a}_\alpha^+|0\rangle \equiv |n_a l_a j_a m_a\rangle\,. \end{array}$$

The state vectors are defined in (I.12) in such a way as to transform under time reversal according to their corresponding phase convention. The relation between creation and annihilation operators in the different phase conventions is given by

$$\bar{a}_\alpha^+ = i^{l_a} a_\alpha^+\,; \quad \bar{a}_\alpha = i^{l_a}(-1)^{l_a} a_\alpha\,. \tag{I.31}$$

The transformation under time reflection then becomes

$$\begin{aligned}
\text{CS}\ &: \mathcal{T}a_\alpha^+\mathcal{T}^{-1} = -(-1)^{la}s_\alpha a_{-\alpha}^+ \,, \\
&\quad \mathcal{T}s_\alpha a_{-\alpha}\mathcal{T}^{-1} = (-1)^{la}a_\alpha \,, \\
\text{BR}\ &: \mathcal{T}\bar{a}_\alpha^+\mathcal{T}^{-1} = -s_\alpha\bar{a}_{-\alpha}^+ \,, \\
&\quad \mathcal{T}s_\alpha\bar{a}_{-\alpha}\mathcal{T}^{-1} = \bar{a}_\alpha \,.
\end{aligned} \tag{I.32}$$

Here, s_α denotes the phase factor $(-1)^{j_a-m_a}$.

In a quasi-particle representation, using the BCS transformation (see Chap. 7), one has

$$\begin{aligned}
\text{CS}\ &: c_\alpha^+ = u_a a_\alpha^+ - s_\alpha v_a a_{-\alpha} \,, \\
\text{BR}\ &: \bar{c}_\alpha^+ = \bar{u}_a\bar{a}_\alpha^+ - s_\alpha\bar{v}_a\bar{a}_{-\alpha} \,.
\end{aligned} \tag{I.33}$$

The same transformation (I.32) under time-reversal for the quasi-particle operator exists as well as the relation (I.31) between these operators. The different amplitudes u_a and v_a are related by

$$u_a = (-1)^{la}\bar{u}_a \,, \quad v_a = \bar{v}_a \,. \tag{I.34}$$

Signs can be associated with the amplitudes after studying the expression for the energy gap in the BCS formalism (Baranger 1960):

$$\begin{aligned}
\text{CS}\ &: \Delta_a = \sum_c \frac{\hat{j}_c}{\hat{j}_a} u_c v_c G(aacc;0) \,, \\
\text{BR}\ &: \bar{\Delta}_a = \sum_c \frac{\hat{j}_c}{\hat{j}_a} \bar{u}_c\bar{v}_c\bar{G}(aacc;0) \,,
\end{aligned} \tag{I.35}$$

where $G(abcd;J)$ denotes the antisymmetric two-body matrix element (the precise relation is $G(abcd;J) = -1/2((1+\delta_{ab})(1+\delta_{cd}))^{1/2}\langle ab;JM|V|cd;JM\rangle_{\text{nas}}$). The relation between the pairing matrix elements in both phase conventions reads

$$G(aacc;0) = (-1)^{l_a-l_c}\bar{G}(aacc;0) \,, \tag{I.36}$$

and results in

$$\Delta_a = (-1)^{l_a}\bar{\Delta}_a \,. \tag{I.37}$$

The residual pairing matrix elements $\bar{G}(aacc;0)$ are usually real and positive. The gap equations (I.35) can be transformed into a real non-symmetric eigenvalue problem by introducing the relation $2u_a v_a = \Delta_a/E_a$. In the BR convention, the positive definite matrix $[M]$, appearing in the secular equation

$$\bar{\Delta}_a = \sum_c \bar{M}_{ac}\bar{\Delta}_c \,,$$

with

$$\bar{M}_{ac} = \frac{1}{2}\frac{\hat{j}_c}{\hat{j}_a}\frac{1}{E_c}\cdot\bar{G}(aacc;0) \,, \tag{I.38}$$

Table I.3 Phases of the BCS parameter in the Condon-Shortley (CS) and Biedenharn-Rose (BR) phase conventions, respectively

	CS	BR
u_a	$(-1)^{l_a}$	+1
v_a	+1	+1
Δ_a	$(-1)^{l_a}$	+1
E_a	+1	+1

always involves an eigenvector with positive components and it only has one such eigenvector, applying the Frobenius theorem (Wilkinson 1965). In the BR convention the gaps $\bar{\Delta}_a$ will hence all be positive and so too the amplitudes u_a and v_a. On the other hand, in the CS convention, the energy gap Δ_a and the amplitude u_a can have a negative sign if the parity of the orbital l_a is odd. All results are given in Table I.3. We remark that the sign of Δ_a has no significance since in the BCS expression one always has combinations Δ_a^2 (or $u_a\Delta_a$).

Any single-particle tensor $T_\kappa^{(k)}$ of rank k can then be expressed, using second quantization and the quasi-particle operators, as the sum of three distinct terms:

$$T_\kappa^{(k)}(1) = \delta_{k0}\delta_{\kappa 0}\sum_a \langle n_a l_a j_a||\boldsymbol{T}^{(k)}||n_a l_b j_a\rangle \hat{j}_a v_a^2 \,, \tag{I.39}$$

$$T_\kappa^{(k)}(2) = \frac{1}{\hat{k}}\sum_{ab}\langle n_a l_a j_a||\boldsymbol{T}^{(k)}||n_b l_b j_b\rangle \times (u_a u_b - c'(-1)^k v_a v_b) \times A^0(ab, k\kappa) \,, \tag{I.40}$$

$$T_\kappa^{(k)}(3) = \frac{1}{\hat{k}}\sum_{ab}\langle n_a l_a j_a||\boldsymbol{T}^{(k)}||n_b l_b j_b\rangle \times (u_a v_b A^+(ab, k\kappa) + u_b v_a(-1)^{k+\kappa}A(ab, k-\kappa)) \,, \tag{I.41}$$

with

$$A^0(ab, k\kappa) = \sum_{m_a, m_b}\langle j_a m_a, j_b - m_b|k\kappa\rangle s_\beta c_\alpha^+ c_\beta \,. \tag{I.42}$$

The expressions are the same in the BR convention, on condition that all symbols are described within the BR convention. The operators A^0, A^+ and A denote two-quasi-particle creation operators (see (Waroquier, Heyde 1970) and (Heyde, Waroquier 1971)). The phase relations are

$$\begin{aligned} A^+(ab, k\kappa) &= i^{l_a+l_b}\bar{A}^+(ab, k\kappa) \,, \\ A(ab, k-\kappa) &= (-i)^{l_a+l_b}\bar{A}(ab, k-\kappa) \,. \end{aligned} \tag{I.43}$$

We apply this to the enhanced $E2$ transitions in spherical even-even nuclei as an example. For $E2$ transitions between a $2qp$ state and the BCS $0qp$ vacuum state $|\tilde{0}\rangle$, one has contributions from the term (I.41) only and obtains

$$\begin{aligned}&\langle\tilde{0}|T_{\kappa}^{(k)}A^{+}(ab,J_iM_i)|\tilde{0}\rangle\\&=\delta_{J_ik}\delta_{M_i'-\kappa}(-1)^{k+\kappa}[v_au_b+c'(-1)^k u_av_b]1/\hat{k}\langle j_a||\boldsymbol{T}^{(k)}||j_b\rangle\ .\end{aligned}\tag{I.44}$$

The pairing factor in (I.44) now becomes $(v_au_b+(-1)^k u_av_b)$ and $(|v_au_b|+|u_av_b|)$ in the CS and BR conventions respectively and thus the same final numerical factors result, irrespective of the phase convention used!

J. Basic Concepts of Group Theory

A certain working knowledge of group theory is necessary for an understanding and appreciation of the IBM. Group theory is essential for defining basis states, and for identifying dynamic symmetries and deriving their consequences. Useful, to-the-point references are Lipkin's book (Lipkin, 1965) and Iachello's Gull Lake lectures (Iachello, 1980). Here, we introduce the basic concepts of group theory by working through a simple example: proton-neutron isospin.

Consider the two-dimensional space spanned by orthogonal unit vectors $|\pi\rangle = a_\pi^\dagger|0\rangle$ and $|\nu\rangle = a_\nu^\dagger|0\rangle$, representing the proton and neutron states, respectively, of the nucleon. The general state ψ of the space is a linear combination of $|\pi\rangle$ and $|\nu\rangle$. Unitary transformations on ψ, i.e. essentially, rotations in the $\pi\nu$ space, are said to form the unitary *group* in two dimensions, $U(2)$ for short. The creation and annihilation operators obey fermion anticommutators:

$$\{a_\pi, a_\pi^\dagger\} = 1\ , \quad \{a_\nu, a_\nu^\dagger\} = 1\ ,\tag{J.1}$$

all other commutators being zero.

Forming the bilinear operators

$$a_\pi^\dagger a_\nu\ , \quad a_\nu^\dagger a_\pi\ , \quad a_\pi^\dagger a_\pi\ , \quad a_\nu^\dagger a_\nu\ ,\tag{J.2}$$

we can then make the linear combinations

$$\begin{aligned}&a_\pi^\dagger a_\nu \equiv t_+ \equiv t_1 + it_2\ , \quad a_\nu^\dagger a_\pi \equiv t_- \equiv t_1 - it_2\ ,\\&\tfrac{1}{2}\left(a_\pi^\dagger a_\pi - a_\nu^\dagger a_\nu\right) \equiv t_3\ , \quad a_\pi^+ a_\pi + a_\nu^\dagger a_\nu \equiv B\ .\end{aligned}\tag{J.3}$$

Thus, we have four linearly independent operators t_1, t_2, t_3 and B, which are Hermitian. Next, we calculate all commutators within the latter set. The result is

$$[t_1, t_2] = it_3 \quad \text{and cyclic permutations of 123}\ ,\tag{J.4}$$

$$[B, t_k] = 0\ , \quad k = 1,2,3\ .$$

The four operators are said to *close on commutation* since the right-hand sides of the commutators are linear combinations of the same operators. Likewise, the original operators (J.2), or the set with $t_{\pm}$, close on commutation. As defined by this property, the four operators are said to form a *Lie algebra*, here designated by $U(2)$. The group of transformations on ψ is likewise called a *Lie group*. It is appropriate to remark that the various concepts of group theory introduced here and below are quite general and not restricted to the simple example of $U(2)$.

The four operators are the *generators* of the group $U(2)$. It showed be noted that the operator B commutes with all the generators t_k and, of course, with itself. Such an operator which commutes with all the generators of the group is called a *Casimir operator* of that group. Further, B is called the linear Casimir operator since it is linear in the generators.

In fact, (J.4) shows that the t_k by themselves form a Lie algebra, called $SU(2)$, the S standing for "special". The generators of $SU(2)$ being a subset of those of $U(2)$, the group $SU(2)$ is called a *subgroup* of $U(2)$, denoted as

$$U(2) \supset SU(2)\,. \tag{J.5}$$

Commutation shows that

$$t_1^2 + t_2^2 + t_3^2 \equiv \boldsymbol{t}^2 \tag{J.6}$$

is a Casimir operator of $SU(2)$. It is quadratic in the generators and the only Casimir operator of $SU(2)$. It is also the quadratic Casimir operator of $U(2)$, which thus has two Casimir operators.

The notation $\boldsymbol{t}$ serves to underline the fact that it is *algebraically* similar to the angular-momentum vector $\boldsymbol{J}$: the 123 components of both are Hermitian and obey commutation relations of thc form (J.4). The similarity is purely mathematical; mechanical angular momentum and nucleon isospin are physically completely different. Nevertheless, all results of angular-momentum theory are directly applicable to isospin.

The algebraic similarity between $\boldsymbol{t}$ and $\boldsymbol{J}$ can be expressed in terms of group theory. The vector operator $\boldsymbol{J}$ is the generator of infinitesimal rotations in physical coordinate space. Such transformations are orthogonal transformations in three dimensions and thus form the group $O(3)$. The technical statement of the similarity is that $O(3)$ and $SU(2)$ are homomorphic

$$O(3) \simeq SU(2)\,. \tag{J.7}$$

The generators (J.2), or (J.3), provide a *realization* of $SU(2)$. An example of another realization are the familiar operators of one-particle orbital angular momentum: $L_x = -i\hbar(y\partial/\partial z - z\partial/\partial y)$ etc.

To take the example further, we may now write angular-momentum relations equivalent to those of the isospin $SU(2)$. The basic relations, resulting from the algebra of (J.4), are

$$\left.\begin{aligned} \boldsymbol{J}^2|jm\rangle &= j(j+1)|jm\rangle\,, \\ J_z|jm\rangle &= m|jm\rangle\,, \\ J_{\pm}|jm\rangle &= \sqrt{(j \pm m + 1)(j \mp m)}\,|jm \pm 1\rangle \end{aligned}\right\} \tag{J.8}$$

with $j = 0, 1/2, 1, \ldots$ and $m = j, j-1, \ldots, -j$. The basis $|jm\rangle$, with fixed j and all m, provides a *representation* of the generators, i.e. of the group. The representation is thus labelled by j, and m labels the basis states within the representation.

Generally, a Lie algebra has several commuting generators whose quantum numbers label the basis states. There are then an equal number of commuting Casimir operators, whose eigenvalues give rise to representation labels, and the basis states are similar to $|j_1 j_2 \ldots j_n, m_1 m_2 \ldots m_n\rangle$. The number n is called the *rank* of the algebra or group. Thus, $SU(2)$ or $O(3)$ is of rank 1. Sometimes, additional labels are needed for a complete specification of the basis states.

A representation can be spelled out in terms of matrices. Returning to the simple example, from (J.8), the general matrix elements

$$\left.\begin{aligned} \langle j'm'|\boldsymbol{J}^2|jm\rangle &= j(j+1)\delta_{j'j}\delta_{m'm}\,, \\ \langle j'm'\,|J_z|\,jm\rangle &= m\delta_{j'j}\delta_{m'm}\,, \\ \langle jm'\,|J_\pm|\,jm\rangle &= \sqrt{(j\pm m+1)(j\mp m)}\,\delta_{j'j}\delta_{m',m\pm 1} \end{aligned}\right\} \tag{J.9}$$

result. Clearly the generators $J_\pm$ and J_z do not connect different representations: such representations are said to be *irreducible*. The Casimir operator $\boldsymbol{J}^2$ is diagonal.

To give a detailed example, the $SU(2)$ representation matrices for $j = 1$ are

$$\begin{aligned} (J_+) &= \begin{bmatrix} 0 & \sqrt{2} & 0 \\ 0 & 0 & \sqrt{2} \\ 0 & 0 & 0 \end{bmatrix}, \quad & (J_-) &= \begin{bmatrix} 0 & 0 & 0 \\ \sqrt{2} & 0 & 0 \\ 0 & \sqrt{2} & 0 \end{bmatrix}, \\ (J_z) &= \begin{bmatrix} 1 & 0 & 0 \\ 0 & 0 & 0 \\ 0 & 0 & -1 \end{bmatrix}, \quad & (\boldsymbol{J}^2) &= \begin{bmatrix} 2 & 0 & 0 \\ 0 & 2 & 0 \\ 0 & 0 & 2 \end{bmatrix}. \end{aligned} \tag{J.10}$$

This is the three-dimensional representation of $SU(2)$, or equivalently, of $O(3)$. The same matrices could be written for the abstract vector $\boldsymbol{t}$. However, for $\boldsymbol{t}$, this example would not mean the original proton-neutron states of isospin $T = 1/2$, but an isospin triplet, $T = 1$, such as the charge states of the pion: π^+, π^-, π^0. The two-dimensional representation of $SU(2)$ is called the *fundamental* representation of the group. Similarly, the matrices (J.10) contain the fundamental representation of $O(3)$.

A key problem in quantum mechanics is that of finding the energy eigenvalues, E_i. Group theory can be quite useful in solving this problem in at least two respects. Suppose, in fact, that we write down the Hamiltonian describing the system in terms of the generators G_i of a group G

$$H = \sum_i \alpha_i G_i + \sum_{i,j} \beta_{i,j} G_i G_j + \ldots\,. \tag{J.11}$$

The expansion of H may stop at quadratic, cubic, ... terms. In general, in order to find the eigenvalues of H, one must solve the problem numerically. Group theory helps here in that it provides a basis in which H can be diagonalized. This basis is provided by the irreducible representations of a complete group chain that starts from G

$$G \supset G' \supset \dots \,. \tag{J.12}$$

Since there are, in general, several chains, one can choose any of them and relate them to one another by an appropriate transformation.

The second way in which group theory is useful is in the case in which H can be written simply in terms of the Casimir operators of a complete chain of groups starting from G. This is a special case of (J.11), since now the coefficients in (J.11) must have special values. In this case, the eigenvalue problem for H can be solved analytically, and all properties of the system can be calculated without recourse to numerical studies. These situations are called dynamic symmetries.

As an example, consider the case of $SU(2)$ discussed above. The most general Hamiltonian including up to quadratic terms is

$$H = \sum_i \alpha_i J_i + \sum_{i,j} \beta_{i,j} J_i J_j \,. \tag{J.13}$$

This Hamiltonian can be diagonalized in the basis provided by the group chain

$$SU(2) \supset SO(2) \tag{J.14}$$

and labelled by j and m. In order to calculate the matrix elements of H, we need the matrix elements of the generators J_i. These were obtained in (J.9).

Consider now, as a special case, the Hamiltonian

$$H = -GJ_+J_- \,. \tag{J.15}$$

Then, since

$$J_+J_- = \boldsymbol{J}^2 - J_z(J_z - 1) \tag{J.16}$$

this Hamiltonian can be written in terms of the Casimir invariants of the group chain (J.14). The corresponding problem is analytically soluble and the eigenvalues are obviously

$$E_m = -G\,[j(j+1) - m(m-1)] \;; \quad -j \le m \le j \,. \tag{J.17}$$

The eigenfunctions are the basis states $|j, m\rangle$. Models of the type (J.13), based on the group $SU(2)$, are usually called Lipkin models.

A very well-known example of an application of this type of symmetry consideration to problems in physics (other than nuclear physics) is given by Gell-Mann-Ne'eman's $SU(3)$ (Gell-Mann, 1962; Ne'eman, 1961) where one has

$$SU(3) \supset SU(2) \otimes U(1) \supset SO(2) \otimes U(1) \,, \tag{J.18}$$

and the corresponding mass formula is the Gell-Mann-Okubo-formula (Okubo, 1962)

$$E(I, I_3, Y) = a + bY + c\left[\frac{Y^2}{4} - I(I+1)\right] . \tag{J.19}$$

Further and more precise information can be found in texts on group theory, such as Hamermesh (1962), Parikh (1978) and Gilmore (1974).

References

Abramowitz, M., Stegun, J.A.: *Handbook of Mathematical Functions* (1964) Dover, New York
Alder, K., Winther, A. (1971): Phys. Lett **34B**, 357
Alkhazov, G.D. et al. (1976): Nucl. Phys. **A274**, 443
Allaart, K. (1971): Nucl. Phys. **A174**, 545
Allaart, K., Boeker, E. (1971): Nucl. Phys. **A168**, 630
Allaart, K., Boeker, E., Bonsignori, G., Savoia, M., Gambhir, Y.K. (1988): Phys. Repts. **169**, 209
Anantaraman, N., Schiffer, J.P. (1971): Phys. Lett. **37B**, 229
Arima, A., Kawasada, H. (1964): J. Phys. Soc. Jap. **19**, 1768
Arima, A., Iachello, F. (1975): Phys. Lett. **57B**, 39
Arima, A., Iachello, F. (1975): Phys. Rev. Lett. **35**, 1069
Arima, A., Iachello, F. (1976): Ann. Phys. (N.Y.) **99**, 253
Arima, A., Iachello, F. (1978): Ann. Phys. (N.Y.) **111**, 201
Arima, A., Iachello, F. (1978): Phys. Rev. Lett. **40**, 201
Arima, A., Iachello, F. (1979): Ann. Phys. (N.Y.) **123**, 436
Arima, A., Iachello, F. (1984): in *Advances in Nuclear Physics*, Vol. 13 (139)
Arima, A., Sagawa, H. (1986): Phys. Lett. **173**, 351
Arndt, R.A., Mac Gregor, M.H. (1966): Phys. Rev. **141**, 873

Backenstoss, G. (1993): Phys. Repts. **225**, 97
Baldo, M., Cugnon, J., Lejeune, A., Lombardo, U. (1990): Nucl. Phys. **A515**, 409
Ballhausen, C.J., Gray, H.B. (1964): *Molecular Orbital Theory* (W.A. Benjamin, New York)
Baranger, M. (1960): Phys. Rev. **120**, 957
Baranger, M. (1961): Phys. Rev. **122**, 992
Bardeen, J., Cooper, L.N., Schrieffer, J.R. (1957): Phys. Rev. **108**, 1175
Beiner, M., Flocard, H., Van Giai, N., Quentin, P. (1975): Nucl. Phys. **A238**, 29
Bertin, M.C., Benczer-Koller, N., Seaman, G.E., MacDonald, J.R. (1969): Phys. Rev. **183**, 964
Biedenharn, L.C., Rose, M.E. (1953): Revs. Mod. Phys. **25**, 729
Bjerregaard, J.H., Hansen, O., Nathan, O., Chapman, R., Hinds, S., Middleton, R. (1967): Nucl. Phys. **A103**, 33
Bjørnholm, S., Borggreen, J., Echt, O., Hansen, K., Pedersen, J., Rasmussen, H.D. (1990): Phys. Rev. Lett. **65**, 1627
Bjørnholm, S., Borggreen, J., Hansen, K., Martin, T., Rasmussen, H.D., Redersen, J. (1992): Ecole Internationale Joliot-Curie de Physique Nucléaire (ed. Y. Abgrall, eds. IN2P3) 311
Blin-Stoyle, R.J. (1956): Revs. Mod. Phys. **28**, 75
Bloch, C., Messiah, A. (1962): Nucl. Phys. **39**, 95
Blomqvist, J., Wahlborn, S. (1960): Arkiv. Fys. **16**, 545
Bloom, E.D., Feldman, G.J. (1982): Sci. American **246**, 66
Bohr, A. (1951): Phys. Rev. **81**, 134
Bohr, A. (1952): Mat. Fys. Medd. Dan. Vid. Selsk. **26**, n° 14
Bohr, A., Mottelson, B. (1953): Mat. Fys. Medd. Dan. Vid. Selsk. **27**, n°16
Bohr, A., Mottelson, B. (1955): Dan. Mat. Fys. Medd. Dan. Vid. Selsk. **30**, n°1
Bohr, A., Mottelson, B., Pines, D. (1958): Phys. Rev. **110**, 936
Bohr, A., Mottelson, B. (1969): *Nuclear Structure*, Vol. 1 (W.A. Benjamin, New York)
Bohr, A., Mottelson, B. (1975): *Nuclear Structure*, Vol. 2 (W.A. Benjamin, New York)
Bohr, A. (1976): Revs. Mod. Phys. **48**, 365
Brack, M. (1992): Ecole Internationale Joliot-Curie de Physique Nucléaire (ed. Y.Abgrall, eds. IN2P3) 325

Brack, M., Genzken, O., Hansen, K. (1991a): Z. Phys. **D19**, 51
Brack, M., Genzken, O., Hansen, K. (1991b): Z. Phys. **D21**, 65
Brink, D.M. (1957): Nucl. Phys. **4**, 215
Brink, D.M., Satchler, G.R. (1962): *Angular Momentum* (Oxford Univ. Press, London)
Brody, T.A., Moshinsky, M. (1960): *Tables of Transformation Brackets* (Monografias delo Instituto de Fisica, Mexico)
Broglia, R.A., Federman, P., Hansen, O., Hehl, K., Riedel, C. (1968) Nucl. Phys. **A106**, 421
Brown, B.A., Wildenthal, B.H. (1988): Ann. Rev. Nucl. Sci. **38**, 29
Brown, G.E., Bolsterli, M. (1959): Phys. Rev. Lett. **3**, 472
Brown, G.E. (1964): *Unified Theory of Nuclear Models* (North-Holland, Amsterdam)
Brown, G.E., Kuo, T.T.S. (1967): Nucl. Phys. **A92**, 481
Brown, G.E., Jackson, A.D. (1976): *The Nucleon-Nucleon Interaction* (North-Holland, Amsterdam)
Brown, G.E., Dehesa, J.S., Speth, J. (1979): Nucl. Phys. **A330**, 290
Brueckner, K.A., Eden, R.J., Francis, N.C. (1955): Phys. Rev. **100**, 891
Brueckner, K.A., Gammel, J.L. (1958): Phys. Rev. **109**, 1023
Brueckner, K.A., Thieberger, R. (1960): Phys. Rev. Lett. **9**, 466
Brueckner, K.A., Masterson, K.S. (1962): Phys. Rev. **128**, 2267
Brussaard, P.J., Tolhoek, H.A. (1957): Physica **23**, 955
Brussaard, P.J. (1967): Ned. T. Natuurkunde **33**, 202
Brussaard, P.J. (1970): Nucl. Phys. **A158**, 440
Brussaard, P.J., Glaudemans, P.W.M. (1977): *Shell-Model Applications in Nuclear Spectroscopy* (North-Holland, Amsterdam)

Campi, X., Sprung, D. (1972): Nucl. Phys. **A 194**, 401
Casten, R.F., Warner, D.D., Brenner, D.S., Gill, R.L. (1981): Phys. Rev. Lett. **47**, 1433
Casten, R.F. (1985): Phys. Lett. **152B**, 145
Casten, R.F. (1985): Phys. Rev. Lett. **54**, 1991
Casten, R.F. (1985): Nucl. Phys. **A443**, 1
Casten, R.F. (ed.) (1993): *Algebraic Approaches to Nuclear Structure (Gordon & Breach, New York)*
Cavedon, J.M. et al. (1982): Phys. Rev. Lett. **49**, 978
Clebsch, A. (1872): *Theorie der binären algebraischen Formen* (Teubner, Leipzig)
Condon, E.U., Shortley, G.H. (1935): *The Theory of Atomic Spectra* (Cambridge University Press) Ch. III

Dechargé, J., Gogny, D. (1980): Phys. Rev. **C21**, 1568
De Gelder, P., et al. (1980): Nucl. Phys. **A337**, 285
de-Shalit, A., Talmi, I. (1963): *Nuclear Shell Theory* (Academic, New York)
de-Shalit, A., Feshbach, H. (1974): *Theoretical Nuclear Physics* (Wiley, New York)
De Vries, C., de Jager, C.W., de Witt-Huberts, P.K.A. (1983): FOM Jaarboek, p. 167
Dieperink, A.E.L., Leenhouts, H.P., Brussaard, P.J. (1968): Nucl. Phys. **A116**, 556
Dieperink, A.E.L., de Witt-Huberts, P.K.A. (1990): Ann. Rev. Nucl. Sci. **40**, 239
DOE Nucl. Science Adv. Committee, Dec. 1983: *A Long Range Plan for Nuclear Science*, A report by the NSF
Duval, P.D., Barrett, B.R. (1981): Phys. Rev. Lett. **46**, 1504
Duval, P.D., Barrett, B.R. (1981): Phys. Rev. **C24**, 1272

Eberz, J. et al. (1987): Nucl. Phys. **A464**, 9
Eckart, C. (1930): Revs. Mod. Phys. **2**, 305
Edmonds, A.R. (1957): *Angular Momentum in Quantum Mechanics* (Princeton University Press)
Egmond, A. Van, Allaart, K. (1983): Prog. Part. Nucl. Phys. **9**, 405
Eisenberg, J.M., Greiner, W. (1970): *Nuclear Theory*, Vol. 2 (North-Holland, Amsterdam) p. 93
Elliott, J.P, Flowers, B.H. (1957): Proc. Roy. Soc. **A242**, 57
Elliott, J.P. (1958): Proc. Roy. Soc. **A245**, 128, 562
Elliott, J.P., Harvey, M. (1963): Proc. Roy. Soc. **A272**, 557
Elliott, J.P., Jackson, A.D., Mavromatis, H.A., Sanderson, E.A., Singh, B. (1968): Nucl. Phys. **A121**, 241
Endt, P., Van der Leun, C. (1973): Nucl. Phys. **A214**, 1

Faessler, A., Plastino, A. (1967): Nuovo Cim. **47B**, 297
Fetter, A.L., Walecka, J.D. (1971): *Quantum Theory of Many-Particle Systems* (Mc Graw-Hill, New York)

Flowers, B.H. (1952): Proc. Roy. Soc. **A212**, 248
Flügge, S. (1974): *Practical Quantum Mechanics* (Springer, Berlin-Heidelberg)
Fortune, H.T. (1978): Phys. Lett. **76B**, 267
Frauenfelder, H., Henley, E.M. (1991): Subatomic Physics, 2[d] edition (Prentice Hall)
Frois, B., et al. (1983): Nucl. Phys. **A396**, 409c

Gambhir, Y.K., Rimini, A., Weber, T. (1969): Phys. Rev. **188**, 1573
Gambhir, Y.K., Rimini, A., Weber, T. (1971): Phys. Rev. **C3**, 1965
Gambhir, Y.K., Rimini, A., Weber, T. (1973): Phys. Rev. **C7**, 1454
Gell-Mann, M. (1962): Phys. Rev. **C25**, 1067
Gillet, V., Green, A.M., Sanderson, E.A. (1966): Nucl. Phys. **88**, 321
Gilmore, B. (1974): *Lie Groups, Lie-Algebras and Some of Their Applications* (Wiley-Interscience, New York)
Gloeckner, D.H., Lawson, R.D. (1975): Phys. Lett. **56B**, 301
Goldstein, H. (1980): *Classical Mechanics* (Addison-Wesley, Reading, MA)
Gordan, P. (1875): *Über das Formensystem binärer Formen* (Teubner, Leipzig)
Greiner W., Müller, B. (1989): *Quantum Mechanics, Symmetries* (Springer, Berlin-Heidelber)

Hamada, T., Johnston, I. (1962): Nucl. Phys. **34**, 382
Hamermesh, M. (1962): *Group Theory and its Application to Physical Problems* (Addison-Wesley, Reading, MA)
Haxel, O., Jensen, J.H.D., Suess, H.E. (1949): Phys. Rev. **75**, 1766
de Heer, W.A., Knight, W.D., Chou, M.Y., Cohen, M.L. (1987): Solid State Physics **40**, 93
Heisenberg, W. (1932): Z. Phys. **77**, 1
Helmers, K. (1961): Nucl. Phys. **23**, 594
Herling, G.H., Kuo, T.T.S. (1972): Nucl. Phys. **A181**, 113
Herzberg, G. (1944): *Atomic Spectra and Atomic Structure* (Dover, New York)
Heyde, K., Waroquier, M. (1971): Nucl. Phys. **A167**, 545
Heyde, K., Waroquier, M., Meyer, R.A. (1978): Phys. Rev. **C17**, 1219
Heyde, K., Van Isacker, P., Waroquier, M., Wood, J.L., Meyer, R.A. (1983): Phys. Repts. **102**, 291
Heyde, K., Van Isacker, P., Casten, R.F., Wood, J.L. (1985): Phys. Lett. **155B**, 303
Heyde, K., Sau, J. (1986): Phys. Rev. **C33**, 1050
Heyde, K. (1989): Mod. Phys. A, **A9**, 2063
Holm, G., Borg, S. (1967): Ark. Phys. **36**, 91
Hsieh, S.T., Knöpfle, K.T.K., Mairle, G., Wagner, G.J. (1975): Nucl. Phys. **A243**, 380

Iachello, F. (1980): in *Nuclear Spectroscopy*, Lect. Notes Phys., Vol. 119, edited by G. Bertsch, D. Kurath (Springer, Berlin-Heidelberg)
Iachello, F. (1983): *Lie Groups, Lie Algebras and Some Applications* (Trento Lecture Notes), U.T.F97
Iachello, F., Arima, A. (1988): *The Interacting Boson Model* (Cambridge Univ. Press)
Irvine, J.M. (1972): *Nuclear Structure Theory* (Pergamon, Oxford)
Isgur, N., Karl, G. (1983): Phys. Today **36**, 36

Kerek, A., Holm, G.B., Borg, S., De Geer, L.E. (1972): Nucl. Phys. **A195**, 177
Kerman, A.K. (1961): Ann. Phys. (N.Y.) **12**, 300
Kerman, A.K., Lawson, R.D., MacFarlane, M.H. (1961): Phys. Rev. **124**, 162
Kisslinger, L.S., Sorensen, R.A. (1960): Mat. Fys. Medd. Dan. Vid. Selsk. **32**, n°9
Klingenberg, P.F.A. (1952): Revs. Mod. Phys. **24**, 63
Knight, W.D., Clemenger, K., de Heer, W.A., Saunders, W.A., Chou, M.Y., Cohen, M.L. (1984): Phys. Rev. Lett. **52**, 2141
Kocher, D.C. (1975): Nucl. Data Sheets **16**, 445
Köhler, H.S. (1976): Nucl. Phys. **A258**, 301
Kresin, V.V. (1992): Phys. Repts. **220**, 1
Krewald, S., Klemt, V., Speth, J., Faessler, A. (1977): Nucl. Phys. **A281**, 166
Kuo, T.T.S., Brown, G.E. (1966): Nucl. Phys. **85**, 40
Kuo, T.T.S., Baranger, E., Baranger, M. (1966): Nucl. Phys. **79**, 513
Kuo, T.T.S., Baranger, E., Baranger, M. (1966): Nucl. Phys. **81**, 241
Kuo, T.T.S. (1967): Nucl. Phys. **A90**, 199
Kuo, T.T.S., Brown, G.E. (1968): Nucl. Phys. **A114**, 241
Kuo, T.T.S. (1974): Ann. Rev. Nucl. Sci. **24**, 101

Kuo, T.T.S. (1981): in *Topics in Nuclear Physics I*, (Proceedings, Beijing, China) Lect. Notes. Phys., Vol. 144 (Springer, Berlin-Heidelberg)

Lacombe, M., et al. (1975): Phys. Rev. **D12**, 1495
Lacombe, M., et al. (1980): Phys. Rev. **C21**, 861
Lane, A.M. (1964): *Nuclear Theory*, (Benjamin, New York)
Landau, L.D. (1956): JETP (Sov. Phys.) **3**, 920
Lederer, C.M., Hollander, J.M., Perlman, I. (1968): *Table of Isotopes* (Wiley, New York)
Lipkin, H. (1965): *Lie Groups for Pedestrians* (North-Holland, Amsterdam)
Liu, K.F., Brown, G.E. (1976): Nucl. Phys. **A265**, 385

MacFarlane, M.H. (1966): in *Lectures in Theoretical Physics*, Vol. VIIIc, edited by P.D. Kunz, D.A. Lind, W.E. Brittin (Univ. of Colorado Press, Boulder) 583
MacGregor, M.H., Arndt, R.A., Wright, R.M. (1968): Phys. Rev. **169**, 1128
MacGregor, M.H., Arndt, R.A., Wright, R.M. (1968): Phys. Rev. **169**, 1149
Machleidt, R., Holinde, K., Elster, Ch. (1987): Phys. Repts. **149**, 1
March, N.H., Young, W.H., Sampanthar, S. (1967): *The Many-Body Problem in Quantum Mechanics* (Cambridge Univ. Press, London)
Mayer, M.G. (1949): Phys. Rev. **75**, 1969
Mayer, M.G. (1950): Phys. Rev. **78**, 22
Moszkowski, S.A. (1970): Phys. Rev. **C2**, 402
Mottelson, B.R. (1967): in *Topics in Nuclear Structure Theory* (Nordita, Copenhagen), No. 288
Mottelson, B.R. (1976): Revs. Mod. Phys. **48**, 375
Myers, W.D., Swiatecki, W.J. (1966): Nucl. Phys. **81**, 1

Nadasen, A., et al. (1981): Phys. Rev. **C23**, 1023
Ne'eman, Y. (1961): Nucl. Phys. **26**, 222
Negele, J.W. (1970): Phys. Rev. **C1**, 1260
Negele, J.W., Vautherin, D. (1972): Phys. Rev. **C5**, 1472
Negele, J.W. (1983): Comm. Nucl. Part. Phys. **12**, 8
Neyens, G. et al. (1993): Nucl. Phys., to be published
Nishioka, H., Hansen, K., Mottelson, B. (1990): Phys. Rev. **B42**, 9377

Okubo, K. (1962): Progr. Theor. Phys. **27**, 949
Otsuka, T., Arima, A., Iachello, F., Talmi, I. (1978): Phys. Lett. **76B**, 139
Otsuka, T., Arima, A., Iachello, F. (1978): Nucl. Phys. **A309**, 1

Paar, V. (1979): Nucl. Phys. **A331**, 16
Parikh, J.C. (1978): *Group Symmetries in Nuclear Structure* (Plenum, New York)
Pauli, W. (1925): Z. Phys. **31**, 373
Pedersen, J., Bjørnholm, S., Borggreen, J., Hansen, K., Martin, T.P., Rasmussen, H.D. (1991): Nature **353**, 733
Perkins, D. (1987): *Introduction to High-Energy Physics*, 3^{d} edition (Addison-Wesley, Reading, MA)
Pittel, S., Duval, P.D., Barrett, B.R. (1982): Ann. Phys. (N.Y.) **144**, 168
Puddu, G., Scholten, O., Otsuka, T. (1980): Nucl. Phys. **A348**, 109

Quig, C., Rosner, J.L. (1979): Phys. Repts. **56**, 167

Racah, G. (1942): Phys. Rev. **61**, 186
Racah, G. (1942): Phys. Rev. **62**, 438
Racah, G. (1943): Phys. Rev. **63**, 367
Racah, G. (1949): Phys. Rev. **76**, 1352
Racah, G. (1951): *Lectures on Group Theory* (Inst. for Adv. Study, Princeton)
Racah, G., Talmi, I. (1952): Physica **18**, 1097
Rainwater, J. (1950): Phys. Rev. **79**, 432
Rainwater, J. (1976): Revs. Mod. Phys. **48**, 385
Reid, R.V. (1968): Ann. Phys. (N.Y.) **50**, 411
Richter, B. (1977): Revs. Mod. Phys. **44**, 235
Ring, P, Schuck, P. (1980): *The Nuclear Many-Body Problem* (Springer, Berlin-Heidelberg)
Rose, H.J., Brink, D.M. (1967): Revs. Mod. Phys. **39**, 306
Rotenberg, M., Bivins, R., Metropolis, N., Wooten Jr., J.K. (1959): *The 3-j and 6-j Symbols* (M.I.T. Technology Press, Cambridge, MA)

Roulet, C., et al. (1977): Nucl. Phys. **A285**, 156
Rowe, D.J. (1968): Revs. Mod. Phys. **40**, 153
Rowe, D.J. (1970): *Nuclear Collective Motion* (Methuen, New York)
Rowe, D.J., Ullah, N., Wong, S.S.M., Parikh, J.C., Castel, B. (1971): Phys. Rev. **C3**, 73
Ryckebusch, J. et al. (1988): Nucl. Phys. **A476**, 237

Sagawa, H., Arima, A. (1988): Phys. Lett. **202B**, 15
Sau, J., Heyde, K., Chéry, R. (1980): Phys. Rev. **C21**, 405
Sau, J., Heyde, K., Van Maldeghem, J. (1983): Nucl. Phys. **A410**, 14
Scholten, O. (1980): *The Interacting Boson Model and Some Applications*, Ph. D. Thesis, Univ. of Groningen, The Netherlands
Scholten, O. (1983): Phys. Rev. **C28**, 1783
Segré, E. (1977): *Nuclei and Particles* (Benjamin, Reading, MA)
Shlomo, S., Schaeffer, R. (1979): Phys. Lett. **83B**, 5
Sick, I., McCarthy, J.S. (1970): Nucl. Phys. **A150**, 631
Sick, I. (1973): Nucl. Phys. **A208**, 557
Skyrme, T.H.R. (1956): Philos. Mag. **1**, 1043
Sorensen, R.A., Lin, E.D. (1966): Phys. Rev. **C142**, 729
Speth, J., Van der Woude, A. (1981): in Repts. Prog. Phys. **44**, 719
Start, D.F., Anderson, R., Carlson, L.E., Robertson, A.G., Grace, M.A. (1970): Nucl. Phys. **A162**, 49

Tabakin, F. (1964): Ann. Phys. (N.Y.) **30**, 51
Talmi, I. (1952): Helv. Phys. Act. **25**, 185
Talmi, I. (1971): Nucl. Phys. **A172**, 1
Tape, J.W., Ulrickson, M., Benczer, N., MacDonald, J.R. (1975): Phys. Rev. **C12**, 2125
Ting, S.C.C. (1977): Revs. Mod. Phys. **44**, 251

Van der Werf, S.Y., et al. (1986): Phys. Lett. **116B**, 372
Van Maldeghem, J., Heyde, K. (1985): Phys. Rev. **C32**, 1067
Van Maldeghem, J. (1988): Ph. D. Thesis, University of Gent, Belgium
Van Neck, D., Waroquier, M., Ryckebusch, J. (1991): Nucl. Phys. **A530**, 347
Van Ruyven, J.J., et al. (1986): Nucl. Phys. **A449**, 579
Vautherin, D., Brink, D.M. (1972): Phys. Rev. **C5**, 626
Vold, P.B., et al. (1978): Phys. Lett. **72B**, 311
Vold, P.B., et al. (1978): Nucl. Phys. **A302**, 12

Waroquier, M., Heyde, K. (1970): Nucl. Phys. **A144**, 481
Waroquier, M., Heyde, K. (1971): Nucl. Phys. **A164**, 113
Waroquier, M., Vanden Berghe, G. (1972): Phys. Lett. **41B**, 267
Waroquier, M., Heyde, K. (1974): Z. Phys. **268**, 11
Waroquier, M., Heyde, K., Vincx, H. (1975): Physica **80A**, 465
Waroquier, M., Heyde, K., Vincx, H. (1976): Phys. Rev. **C13**, 1664
Waroquier, M., Sau, J., Heyde, K., Van Isacker, P., Vincx, H. (1979): Phys. Rev. **C19**, 1983
Waroquier, M., Heyde, K. (1981): *6ᵉ Session d'Etudes Biennale de Physique Nucléaire* (Aussois), S17
Waroquier, M. (1982): Hoger Aggregaatsthesis, University of Gent, unpublished
Waroquier, M., Heyde, K., Wenes, G. (1983): Nucl. Phys. **A404**, 269
Waroquier, M., Wenes, G., Heyde, K. (1983): Nucl. Phys. **A404**, 298
Waroquier, M., Bloch, J., Wenes, G., Heyde, K. (1983): Phys. Rev. **C28**, 1791
Waroquier, M., Ryckebusch, J., Moreau, J., Heyde, K., Blasi, N., Van der Werf, S.Y., Wenes, G. (1987): Phys. Repts. **148**, 249
Waroquier, M.: Priv. comm. and to be published
Watanabe, H. (1964): Progr. Theor. Phys. **32**, 106
Weissbluth, N. (1974): *Atoms and Molecules* (Academic, New York)
Werner, S.A.E. (1980): Phys. Today **12**, 24
Wigner, E.P. (1937): Phys. Rev. **51**, 106
Wigner, E.P. (1959): *Group Theory and its Application to the Quantum Mechanics of Atomic Spectra* (Academic, New York)
Wildenthal, B.H. (1969): Phys. Rev. Lett. **22**, 1118
Wildenthal, B.H., Mc Grory, J.B., Halbert, E.C., Graber, H.D. (1971): Phys. Rev. **C4**, 1708
Wildenthal, B.H. (1976): Varenna Lectures **69**, 383

Wildenthal, B.H. (1985): in *Int. Symp. on Nuclear Shell Models*, edited by M. Vallières, B.H. Wildenthal (World-Scientific, Singapore) 360
Wilkinson, J.H. (1965): *The Algebraic Eigenvalue Problem* (Clarendon, Oxford)
Wohlfart, H.D., Shera, E.B., Hoehn, M.V., Yamazaki, Y., Fricke, G., Steffen, R.M. (1978): Phys. Lett. **73B**, 131
Wood, J.L. (1983): Nucl. Phys. **A396**, 245c
Wood, J.L.: Private communication
Wright, R.M., Mac Gregor, M.H., Arndt, R.A. (1967): Phys. Rev. **159**, 1422
Wybourne, B.G. (1974): *Classical Groups for Physicists* (J. Wiley, New York)

Yukawa, H. (1935): Proc. Phys. Math. Soc. Jpn. **17**, 48
Yutsis, A.P., Levinson, I.B., Vanagas, V.V. (1962): *Mathematical Apparatus of the Theory of Angular Momentum* (Israel Program for Scientific Translations, Jerusalem)

Subject Index

Printing: Druckerei Zechner, Speyer
Binding: Buchbinderei Schäffer, Grünstadt